Michail A. Zharkov

Paleozoic Salt Bearing Formations of the World

Editor in Chief A. L. Yanshin

Translated by
R. E. Sorkina, R. V. Fursenko, and T. I. Vasilieva

With 166 Figures

Springer-Verlag
Berlin Heidelberg New York Tokyo 1984

Professor Dr. M. A. ZHARKOV
Institute of Geology and Geophysics
630090, Novosibirsk-90, USSR

The original Russian edition was published by Nedra, Moscow, 1974.
Academy of Sciences of the USSR, Siberian Branch,
Transactions of the Institute of Geology and Geophysics,
Editor in Chief A. L. YANSHIN

ISBN 3-540-13133-7 Springer-Verlag Berlin Heidelberg New York Tokyo
ISBN 0-387-13133-7 Springer-Verlag New York Heidelberg Berlin Tokyo

Library of Congress Cataloging in Publication Data. Zharkov, M.A. (Mikhail Abramovich) Paleozoic salt bearing formations of the world. Translation of: Palezoĭskie solenosnye format͡sii mira. 1. Salt deposits. 2. Geology, Stratigraphic – Paleozoic. I. I͡Anshin, Aleksandr Leonidovich, 1911–. II. Title. QE471.15.S2Z513 1984   553.6′3   84-1291

This work is subject to copyright. All rights are reserved, whether the whole or part of the material is concerned, specifically those of translation, reprinting, re-use of illustrations, broadcasting, reproduction by photocopying machine or similar means, and storage in data banks. Under § 54 of the German Copyright Law, where copies are made for other than private use, a fee is payable to "Verwertungsgesellschaft Wort", Munich.

© by Springer-Verlag Berlin Heidelberg 1984
Printed in Germany

The use of registered names, trademarks, etc. in this publication does not imply, even in the absence of a specific statement, that such names are exempt from the relevant protective laws and regulations and therefore free for general use.

Offsetprinting and bookbinding: Brühlsche Universitätsdruckerei, Giessen.
2132/3130-543210

Preface

The original Russian edition of the monograph *Paleozoic salt formations of the world* was published by Nedra, Moscow, in 1974.

The description of salt basins was given as of 1970–1971 and based on the literature available at that time. Additional evidence was presented in *History of Paleozoic salt accumulation* (Nauka, Novosibirsk 1978), with a brief account of new basins of Paleozoic salt accumulation. These two books complement each other and have one common list of references, the former providing the material which in the latter is the basis for the major trends in the history of evaporite sedimentation in the Paleozoic.

History of Paleozoic salt accumulation was the first of the two books to be published in English (Springer, Berlin, Heidelberg, New York 1981). The present book is the first synthesis dealing with specific Paleozoic salt formations and some salt basins. However, since the first edition was published 10 years ago, new data since then allow more accurate and detailed description of composition, structure, and distribution patterns of salt deposits within Paleozoic evaporite basins. New basins have also been found in some regions of the Earth. The author has attempted to give more complete characterization of Paleozoic salt basins using the data available as of 1981.

As a separate volume, this book provides a description of all known basins of Paleozoic salt accumulation including those discussed in *History of Paleozoic salt accumulation*. In some cases the concepts published earlier have been corrected not, however, to the extent of affecting major conclusions; all significant corrections have been noted specifically. The bibliography of the present book does not contain references listed in *History of Paleozoic salt accumulation;* when these are referred to, the year of publication is underlined.

It is not only difficult but even impossible to systematize and summarize all the data on Paleozoic salt formations available in different countries. The author is aware that some references are probably missing, and that many important problems have not been elucidated because some data have not yet been published. Experts in certain areas of Paleozoic salt accumulation will criticize some of our reconstructions as erroneous and some interpretations of limited data as too daring. The author has tried to be objective and hopes that the synopsis will make a contribution to filling the gap in the study of salt formations.

M.A. ZHARKOV

Contents

Introduction.. 1

CHAPTER I Precambrian Salt Deposits................. 4

Amadeus Basin... 5
Officer Basin.. 9
Precambrian Evaporite Deposits in Other Basins........... 11

CHAPTER II Cambrian Salt Deposits..................... 15

East Siberian Basin... 16
Iran-Pakistan Basin... 45
Mackenzie Basin... 54
Amadeus Basin... 56
Arckaringa Basin.. 58
Cambrian Evaporite Deposits in Other Basins............... 58

CHAPTER III Ordovician Salt Deposits.................. 60

Canadian Arctic Archipelago Basin........................... 61
Ordovician Evaporite Deposits in Other Basins............. 64

CHAPTER IV Silurian Salt Deposits..................... 69

Michigan-Appalachian Basin.................................. 69
Dniester-Prut Basin... 78
Lena-Yenisei Basin.. 81
Canning Basin... 83
Carnarvon Basin... 87
Silurian Evaporite Deposits in Other Basins............... 89

CHAPTER V Devonian Salt Deposits ... 91

North Siberian Basin ... 92
Tuva Basin ... 97
Chu-Sarysu Basin ... 99
Morsovo Basin ... 102
East European Upper Devonian Basin ... 106
West Canadian Basin ... 128
Michigan Basin ... 145
Hudson and Moose River Basins ... 149
Adavale Basin ... 155
Bonaparte Gulf Basin ... 157
Devonian Evaporite Deposits in Other Basins ... 158

CHAPTER VI Carboniferous Salt Deposits ... 161

Chu-Sarysu Basin ... 162
Mid-Tien Shan Basin ... 162
Sverdrup Basin ... 165
Williston Basin ... 170
Maritime Basin ... 173
Saltville Basin ... 181
Paradox and Eagle Basins ... 184
Amazon Basin ... 194
Carboniferous Evaporite Deposits in Other Basins ... 196

CHAPTER VII Permian Salt Deposits ... 203

East European Basin ... 205
Chu-Sarysu Basin ... 241
Moesian Basin ... 242
Central European Basin ... 244
Alpine Basin ... 323
Midcontinent Basin ... 327
Supai Basin ... 357
North Mexican Basin ... 359
Peru-Bolivian Basin ... 361
Permian Evaporite Deposits in Other Basins ... 362

References ... 365

Subject Index ... 397

Introduction

The aim of this study, begun more than 20 years ago, is a comparative analysis, classification, and establishment of types of salt formations as a help in understanding the evolution of salt accumulation throughout geological history. We thus first give as general and systematic a review as possible of all the data available for separate salt formations and basins, based primarily on scattered publications, as specific studies of evaporite sedimentation are few.

We decided to present the material on salt basins of the world as separate volumes and to base our comparative analysis mainly on types of salt formations, their classification and sedimentary environment.

A synthesis of data defining composition, structure and pattern of spatial distribution of salt formations is difficult and laborious. At present it is impossible to characterize all the recognized salt formations in the same way, due to the different state of knowledge of the geological structure in various regions of the Earth, the ambiguity of stratigraphic and tectonic problems, and differing definitions of salt deposits. Original data are, moreover, often not published and are inaccessible to many scientists.

If we admit to the impossibility of an unambiguous and comprehensive description of old basins of salt accumulation in different continents of the Earth, then the difficulties can be overcome by giving a general characterization. It is to be noted that our works on salt formations of the world are the first attempt to synthesize world-wide data, thus making it available in the future for a more fundamental and systematic review.

In characterizing Paleozoic salt basins of the world, the present book focuses on the description of Paleozoic salt series in different regions of the Earth, with evidence of other halogenic formations, in particular the distribution of sulfate rocks (anhydrite, gypsum, etc.); in fact, all data at present available on halogenic deposits of the Paleozoic age.

The stratigraphy of halogenic strata, their thickness, mode of occurrence, particular features of the internal structure of salt series, and the pattern of areal distribution of deposits are discussed for each Paleozoic salt basin. An estimate of the rock salt volume is given for separate salt series of every region.

The nomenclature used should be explained. The terms halogenic, evaporite, and salt-bearing describe sequences, deposits, and rocks. The terms halogenic and evaporite are synonyms, while salt-bearing characterizes only rock, potash and other salts and describes deposits, sequences or series containing salt rocks. The term sulfate refers to anhydrite and/or gypsum and in some cases polyhalitic, glauberite or other rocks, while readily soluble salt rocks of sulfate type are described as sulfate salt rocks.

This work was carried out in the Laboratory of Sedimentary Formations, Institute of Geology and Geophysics, Siberian Branch of the USSR Academy of Sciences, under the supervision of A.L. Yanshin, who organized and provided a scientific basis for studies on salt formations, to elucidate the evolution of salt accumulation throughout geological history; in fact, the present book tends to develop his concepts. The author is grateful to A.L. Yanshin for his generous contribution and constant help.

The author acknowledges the cooperation of the following colleagues: G.M. Drugov, T.M. Zharkova, I.T. Zhuravleva, S.M. Zamaraev, Yu.N. Zanin, A.N. Zolotov, V.S. Isakova, M.L. Kavitsky, V.S. Karpyshev, I.M. Knyazev, I.K. Korolyuk, G.A. Kuznetsov, O.Ya. Kuznetsova, G.G. Lebed', I.A. Malasaev, Ya.G. Mashovich, V.D. Mats, M.G. Minko, V.V. Samsonov, M.A. Semikhatov, R.Ya. Sklyarov, A.I. Skripin, T.N. Spizharsky, P.I. Trofimuk, N.I. Formin, B.A. Fuks, V.V. Khomentovsky, M.A. Tsakhnovsky, E.I. Chechel, and others; on the North Siberian Devonian basin — G.S. Fradkin, Vl.Vl. Menner, R.G. Matuchin and others; on the Devonian basin of the Moscow Syneclise — N.A. Vanin, Ya.G. Lifits, E.V. Mikhailova, B.N. Rozov, V.N. Tikhy; on Devonian deposits of the Pripyat Depression — Yu.I. Lupinovich, V.Z. Kislik, D.M. Eroshina, I.I. Zelentsov, Z.L. Poznyakevich, E.A. Vysotsky, E.V. Sedun and others; on the Devonian and Permian of the Dnieper-Donets Depression — A.D. Britchenko, V.P. Bobrov, S.M. Korenevsky, I.Yu. Lapkin, M.L. Levenshtein, K.S. Supronyuk, A.L. Yudelson and others; on Permian deposits of Solikamsk Cis-Urals area — A.A. Ivanov, M.P. Fiveg, V.I. Kopnin, A.I. Belolikov, B.I. Sapegin, G.A. Dyagilev and others; on Permian formations of the Volga-Ural area and Cis-Uralian Trough — B.I. Blizeev, A.K. Visnyakov, A.I. Otreshko, I.N. Tikhvinsky and others; on the Caspian Syneclise — S.M. Korenevsky, V.A. Ermakov, N.P. Grebennikov, Ya.Sh. Shafiro, N.I. Banera, A.F. Gorbov, V.S. Zhuravlev, R.G. Garetsky, S.A. Svidzinsky, M.D. Diarov, V.S. Derevyagin, F.V. Kovalsky, V.I. Sedletsky, M.M. Muzylevsky, S.A. Tkhorzhevsky, T.N. Dzhumagaliev, Z.E. Bulekbaev; on the Chu-Sarysu Basin — V.I. Ditmar, N.L. Gabai, G.P. Filipiev, R.A. Shakhov, N.B. Golodnova.

The author also thanks foreign geologists who sent the material on the West Canadian Basin — L.L. Price and M.E. Holter; on the Permian Basin of the U.S.A. — H.J. Bissel and G.V. Chilingar; on Permian deposits of West Cumberland — R. Arthurton; on Zechstein deposits of the Central European Basin — R. Meier, R. Kühn, K. Koch, C. Döhner, E. Hoyningen-Huene, Z. Werner, Y. Orska, T.M. Perit, P.I. Suveizdis, and S.M. Korenevsky; on salt deposits of Australia — A.T. Wells.

The book benefited from reviews of the manuscript by M.G. Valyashko, A.A. Ivanov, M.P. Fiveg, Ya.K. Pisarchik, S.M. Korenevsky, I.K. Zherebtsovâ, A.S. Kolosov, V.Z. Kislik, and Yu.I. Lupinovich, who made valuable comments which were taken into consideration. The author is indebted to his colleagues T.M. Zharkova, G.A. Merzlyakov and V.V. Blagovidov. The Cambrian salt deposits of the southern Siberian Platform were studied in collaboration with T.M. Zharkova, who also helped to make a general synthesis on the East Siberian salt basin; G.A. Merzlyakov is a co-author of the section of the Chu-Sarysu Basin of Chapters V and VII and the Central European Basin and East European Basin of Chapter VII.

Introduction

The translation of this book was done by R.E. Sorkina, R.V. Fursenko, and T.I. Vasilieva, to whom the author expresses his heartful thanks.

Certainly it was difficult to collect and summarize all the data available on the Paleozoic salt basins of the world; the book may have errors, so the author will be grateful to all the specialists who send comments and hence help to specify new data on regions of Paleozoic salt accumulation.

Chapter I
Precambrian Salt Deposits

Precambrian halogenic rocks have been recognized in many areas of Eurasia, North America, Africa, and Australia (Fig. 1). As a rule, they are represented by gypsum and anhydrite. Salt beds of more or less confirmed Precambrian age are known only from Australia in the Amadeus Basin. They have been suggested as occurring also in the Officer Basin. New theories that the oldest salt deposits recognized within the East Siberian and Iran-Pakistan Basins are to be considered as Precambrian have been only poorly substantiated.

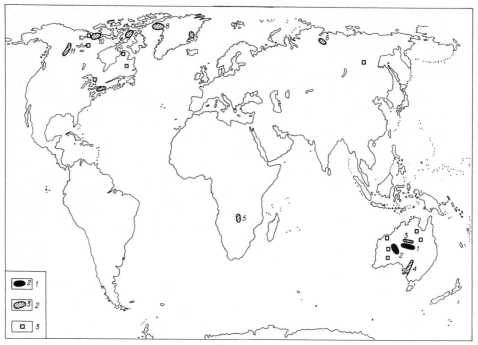

Fig. 1. Distribution of Precambrian evaporite deposits. *1* known occurrence of evaporites (pseudomorphs after rock salt, anhydrite inclusions) of Archean and Proterozoic age; *2* areas and basins of sulfate sedimentation and their numbers (*3* Ngalia Basin, *4* Adelaide Geosyncline, *5* Mufulira Syncline, *6* Kotuikan Basin, *7* Merkebjerg Basin, *8* Thule Basin, *9* Borden Basin, *10* Amundsen Basin, *11* Mackenzie Hills, *12* north-west Adirondack Mountains); *3* Late Precambrian salt basins and their numbers (*1* Amadeus, *2* Officer)

In the East Siberian salt basin most workers now draw the boundary between the Cambrian and Precambrian either at the top or at the base of the Irkutsk horizon, whose upper part in some areas of the Irkutsk Amphitheater contains rock salt units and shows a wide development of sulfate rocks. However, some assign the Irkutsk horizon to Vendian, others, including the present author, to Cambrian. We propose drawing the lower Cambrian boundary not at the base of the first biostratigraphic horizon, which yields fossils with a developed skeleton, but at the base of beds marked by first drastic changes in sedimentary environments resulting in the appearance of organisms with skeletons. In East Siberia this was the beginning of the Irkutsk period, when environments became favorable for the accumulation of evaporite deposits. We assign salt-bearing deposits of the Irkutsk horizon to the Cambrian and they will be discussed in the section devoted to the East Siberian Basin.

In the Iran-Pakistan Basin, salt deposits are known from three localities: (1) in central East Iran, in Kerman; (2) in southern Iran, on some islands in the Persian Gulf and in the north-eastern Arabian Peninsula; (3) in the Salt Range in West Pakistan. In the former two areas, an evaporite sequence is incorporated into the Hormoz Formation; the third represents the Penjab salt-bearing series. The position of both salt sequences seems to be similar due to their occurrence below fossiliferous carbonate deposits where Middle and/or Upper Cambrian trilobites have been found (Gansser 1967, Stöcklin 1966, 1968a, b). *Redlichia,* characteristic of the upper Lower Cambrian, is known from only three localities; the High Zagros, near Kerman, and in the Salt Range. The Lalun Sandstone underlies Cambrian fossiliferous carbonate deposits and overlies a salt sequence in Central and South Iran. In the Salt Range, it may be correlated with purple sandstone also resting on the Penjab salt-bearing series. Red sandstone, 500–1000 m thick, unconformably overlying salt deposits, is assigned to the base of the Cambrian and the Lower Cambrian boundary is drawn at their base. Hence, the Hormoz Formation and Penjab Salt Series are assigned to Precambrian and are placed at the level of the Infracambrian (Stöcklin 1968a, b).

However, trilobites of the genus *Redlichia* were recorded in East Siberia from the Lower Cambrian Olekma horizon (Repina 1966), above the thickest salt-bearing sequences (Irkutsk, Usolye, Belsk) of the Irkutsk Amphitheater. The Lower Cambrian Tolbachan, Elgyan, Usolye, and Irkutsk horizons are stratigraphically important. Their total thickness, which exceeds 2000 m, can be traced below the Olekma horizon yielding *Redlichia.* We thus suggest that salt-bearing sequences of the Iran-Pakistan Basin lie at the same level as those (Irkutsk, Usolye, Belsk) of the Irkutsk Amphitheater and are Cambrian in age. Therefore, they will be discussed in the next chapter.

Amadeus Basin

The Amadeus Basin lies in the center of the Australian continent in southern Northern Territory. It is filled with thick Precambrian and Paleozoic sedimentary rocks. The following formations, in ascending order, are recognized within the Upper Proterozoic composite section: Heavitree, Bitter Springs, Areyonga, and Pertatataka (Wells et al. 1967, 1970, Wells 1980).

The Heavitree Formation is composed mainly of quartzite in places with shale and mudstone beds up to 10 m thick. Sandstone and fine-pebbled conglomerate form the base. The thickness of the formation is 300 m. The Bitter Springs Formation rests conformably on the Heavitree quartzite. It is subdivided into lower and upper members, Gillen and Loves Creek. The Gillen sequence consists mainly of dolomite, especially widespread in the middle and upper parts of the section. They are dark gray, yellowish-gray, and grayish-fallow; mudstone bands occur at the base. White, green, and red siltstone, thin-laminated, platy, and white sandstone lie among dolomites. In some areas, there are grayish-brown and buff dolomites interbedded with green shale and sandstone with crystal molds after rock salt on bedding planes. The sequence contains also gypsum, anhydrite, and rock salt bands and units. The Gillen sequence is 350–400 m thick. The Loves Creek sequence, 700–750 m thick, consists of mudstone with bands of cherts, dolomite, and minor limestone. Mudstone, chert, dolomite, and limestone are respectively reddish-brown, gray and white, yellowish-brown, and dark gray. The total thickness of the Bitter Springs Formation is 100–1150 m.

The overlying Areyonga Formation consists of conglomerate, arkoses, mudstone, and dolomite. The thickness is 350–450 m. The Pertataka Formation, which is 1135–1200 m thick and lies above, contains six members, in ascending order: 1. the Cyclops – well-sorted sandstone 65–70 m thick; 2. Ringwood – limestone, dolomite, and mudstone 165 m thick; 3. Limbla – silicified limestone, mudstone and sandstone 145 m thick; 4. Olympic – dolomite, shale, conglomerate, and sandstone 200–210 m thick; 5. Waldo Pedlar – alternation of sandstone, mudstone, and shale 60 m thick; 6. Julie – dolomite, limestone, sandstone, and mudstone 500–550 m thick.

The above description implies the occurrence of evaporite deposits at the base of the Bitter Springs Formation. A salt-bearing sequence was penetrated by the Ooraminna No. 1 BMR Mount Liebig No. 1 boreholes (Fig. 2).

The Ooraminna No. 1 drilled in Ooraminna Anticline just under Lower (?) Cambrian deposits has penetrated the following formations (Wells et al. 1967):

460.7–1138.9 (678.2 m) – Pertataka Formation
1138.9–1343.8 (204.9 m) – Areyonga Formation
1343.8–1807.1 (501.8 m) – Bitter Springs Formation, Loves Creek Member, dolomite
1807.1–1845.6 (38.5 m) – Bitter Springs Formation, Gillen Member, rock salt

The Erldunda No. 1 borehole in the southern Amadeus Basin penetrated the Bitter Springs Formation at depths of 1228.3–1665.1 m (Wells 1980). The Loves Creek Member lies at depths of 1228.3–1447.8 m (thickness is 219.5 m) and the Gillen at depths of 1447.8–1665.1 m (apparent thickness is 217.3 m). Evaporite rocks in the form of lenses and anhydrite and gypsum inclusions were recorded in the lower Loves Creek Member starting with a depth of 1310.6 m. The upper Gillen Member shows alternate dolomite and anhydrite. A coarse-grained rock salt unit 48.7 m thick lies at its base; boring was abandoned in the salt unit.

The BMR Mount Liebig No. 1 borehole was drilled at the northern margin of the basin on the north-eastern end of the Gardiner Range in the cap rock of a dome, about 200 m in diameter, forming a topographic height of about 40 m. The depth of

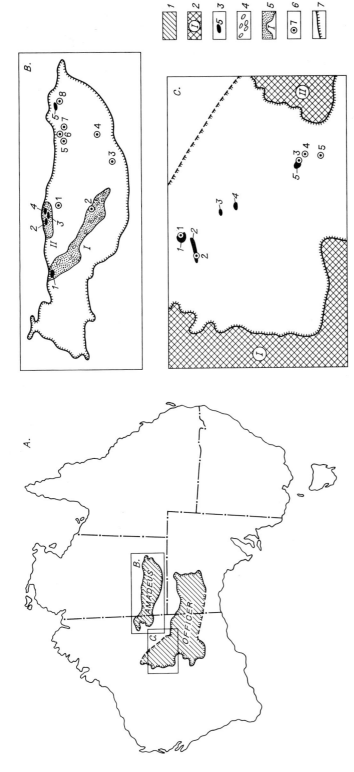

Fig. 2A–C. Amadeus and Officer salt basins. Compiled data from McNaughton et al. (1968), Wells et al. (1970), Wells (1980), Jackson and Van de Graaf (1981). A Sketch map showing location of the Amadeus and Officer Basins. B generalized sketch map of Amadeus Basin, C generalized sketch map of north-west Officer Basin. *1* Amadeus and Officer Basins, *2* outcrops of Proterozoic formations in Yilgarn (*II*) and Musgrave (*III*) blocks bounding Officer Basin to the west and north-east, *3* diapirs and their numbers (in Amadeus Basin, sketch B: *1* Johnstone Hill, *2* Carmichall, *3* Gosses Bluff, *4* Goyder Pass, *5* Ringwood; in Officer Basin, sketch C: *1* Woolnough Hills, *2* Madley, *3* Lake Cohen, *4* Young Range, *5* Brown); *4* anticlines and inferred diapirs, *5* belts of diapiric and anticlinal structures and their numbers (*1* Johnston Hill – Curtin Springs, *II* Domol); *6* wells and their numbers (in Amadeus Basin: *1* BMR Mount Liebig No. 1, *2* BMR Lake Amadeus Nos. 3, 3A, 3B, *3* Erldunda No. 1, *4* Mt. Charlotte No. 1, *5* Orange No. 1, *6* Alice No. 1, *7* Ooraminna No. 1, *8* Alice Springs No. 3; in Officer Basin: *1* BMR Warri No. 20, *2* BMR Madley No. 1, *3* Browne No. 1, *4* Browne No. 2, *5* Yowalga No. 2); *7* basin boundaries

the borehole is 305.9 m. The section was described in detail by Wells (1980). A cap rock, consisting of evaporite breccia, clasts and blocks of gypsum, anhydrite, and dolomite rocks, was penetrated to a depth of 92.0 m. Below is a salt unit composed mainly of brown-orange to dark brown, pink, and variegated coarse-grained salt rock with thin layers, rare clasts, and inclusions of anhydrite and gypsum. The apparent thickness of the salt rock unit is 13.9 m.

Wells and co-authors (1967) believe that the salt unit of the Bitter Springs Formation was penetrated also by the Mount Charlotte No. 1 borehole. This borehole has penetrated Cretaceous, Jurassic, Permian, Carboniferous, Devonian, Silurian, Ordovician, and Cambrian deposits and exposed from a depth of 937.0 m onward the Pertataka, Areyonga, and Bitter Springs Formations (Upper Proterozoic), totaling 1180.6 m in thickness. The Bitter Springs Formation is 511.5 m in thickness. Its lower part contains rock salt bands alternating with anhydrite, dolomite, and siltstone. However, a Precambrian age for these salt-bearing deposits seems questionable. Details are given in the discussion of Cambrian salt-bearing sequences of the Amadeus Basin.

Gypsum and anhydrite interbeds and units were observed within the Bitter Springs Formation in many areas of the Amadeus Basin in cap rocks of diapirs, and are numerous in the Domol District. The following diapirs are known: Carmichall, Gosses Bluff, and Goyder Pass (McNaughton et al. 1968). The cap rock of the Johnstone Hill diapir (Wells et al. 1965) is composed of gypsum with dolomite and limestone clasts and blocks. The diapir is about 2 km in diameter. A mass of gypsum pierces Cambrian, Orodovician, and Permian deposits. There are dolomite blocks with stromatolites and laminated anhydrite-dolomite rocks. Öpik (Wells, Forman, Ranford 1965, p. 40) found fossils in gypsinate dolomite, which allowed him to assign a gypsum-bearing sequence to Cambrian. However, Wells and co-authors (1965) mentioned that material determined by A. Öpik as fossils may be of inorganic origin and resulting from brecciation and crystallization of rocks. The surface of the Gosses Bluff diapir is also composed of gypsiferous rocks. Morphologically, it is similar to salt domes in other regions of salt accumulation, in particular, the Paradox Basin in the U.S.A. (McNaughton et al. 1968).

The second belt of diapirs runs in the central Amadeus Basin from Johnstone Hill in the north-west to Curtin Springs in the south-east (Wells 1980). Gypsiferous deposits crop out in some anticlinal folds, e.g., in Clelend Anticline, where they probably compose a cap rock (McNaughton et al. 1968). Gypsum and anhydrite are known from Mount Murrey, on the north-eastern slope of the Bloods Range and elsewhere (Ranford et al. 1965, Forman 1966). Isolated exposures of the Bitter Springs evaporites occur in the central part of the basin north of Curtin Springs. A number of small boreholes (BMR Lake Amadeus No. 3, No. 3A, and No. 3B) exposed sulfate and carbonate rocks. The reader is referred to Wells' paper (1980) for details.

The large Ringwood Dome with gypsum in the cap rock was found in the north-eastern part of the basin. The BMR Alice Spring No. 3 borehole penetrated dolomite-gypsic breccia with micro- and coarse-grained anhydrite and bituminous dolomite interbeds and inclusions to a depth of 259.7 m (Wells 1980).

According to Wells (1980), the areal extent of evaporite deposits in the Amadeus Basin is 83,000 km^2. Salt sequences may occupy almost the same area. At present, we cannot give even an approximate estimate for salt rock volume of the Bitter Springs Formation since the total thickness of the salt sequence has not yet been penetrated.

Officer Basin

The Officer Basin is located in south-eastern West Australia and north-western South Australia (Fig. 2). To the north, is is separated from the Amadeus Basin by the Musgrave Block where the Middle Proterozoic metamorphic and igneous rocks crop out. Its western boundary runs along exposures of Archean, Early and Middle Proterozoic formations of the Yilgarn block, Nabberu and Bangemall Basins. The Officer Basin is connected with the Eucla, Canning, and Arckaringa Basins, respectively to the south, north-west, and east. Late Precambrian and Cambrian sedimentary strata unconformably overlain by Permian, Cretaceous, Eocene, and Miocene deposits fill the basin. There are diapirs with sulfate and sulfate-carbonate deposits exposed in their cap rocks in some area. The Woolnough Hills, Madley, and Brown diapirs are well-known. The following characterization of evaporite sequences is based mainly on evidence presented by Wells (1980) and Jackson and Van de Graaff (1981).

The base of the Upper Proterozoic formations is represented by Townsend Quartzite, which is composed of cross-bedded, well-sorted, medium-grained, quartz sandstone with rare shale and conglomerate interbeds. The thickness ranges from 110–370 m. On the western margin of the basin, equivalents of Townsend Quartzite are found in the Robert Beds, about 60 m thick.

The overlying deposits are recognized as the Lefroy Beds, which consist of alternate mudstone and sandstone and subordinate claystone. The thickness, which varies greatly along the strike, is about 250 m. The composition of the Lefroy Beds equivalents also varies, hence their different names in various parts of the Officer Basin. Therefore, the equivalents to the north-east in the lower horizons of the Wright Hill Beds are composed of feldspar, sandstone, quartzite, limestone, and siliceous breccia with subordinate siltstone, and have a thickness ranging from 2000–3400 m. In southern parts of the basin, they are known as the Ilma Beds, consisting of sandy, oolitic, carbonate rocks, dolomite, and chert. The Neale Beds are composed of dololutite with small limestone interbeds. The rocks are silicified and contain abundant stromatolites and quartz sandstone in some sections.

In the central Officer Basin, coeval evaporites are assigned to the Brown Beds. They consist of multiple alternate dolomite, limestone, shale, anhydrite, and gypsum. These deposits – the Brown Diapir – were penetrated by two oil wells. The Hunt Oil Brown No. 1 penetrated an evaporite sequence at depths of 132.6–386.8 m. Gray and brown gypsum with shale inclusions and nodules about 101 m thick lie at the base of the exposed section. Above is a limestone and shale unit with lenses and inclusions of gypsum more than 30 m thick. Above these layers is a gypsum unit with limestone interbeds and lenses 70–75 m thick. The upper part of the section, with a thickness exceeding 50 m, consists of limestone with dolomite interbeds and gypsum inclusions. The lower gypsum-bearing unit contains rock salt (Jackson and Van de Graaff 1981).

It is noteworthy that the stratigraphic position of the Brown evaporite sequence is uncertain. It is unconformably overlain by Permian deposits. Seismic surveys suggest that evaporite deposits submerge under the Early Precambrian Table Hill volcanics. The Brown Beds are believed to lie stratigraphically lower than the Babbagoola Beds, thus enabling their correlation with the Madley Beds, which are represented by evaporites.

The Madley Beds are known from the northern Officer Basin where they were penetrated by the BMR Madley No.1 in the cap rock of one of the diapirs. A west-east band of positive folds with exposures of Late Cambrian deposits was recorded there. Six diapirs, three of which had gypsum-bearing deposits, were observed within the Madley Sheet area. A borehole was drilled in the westernmost diapir No. 6. It exposed cap rock deposits of sulfate-carbonate breccia of dolomite, gypsum, and clay, with a thickness of 202.25 m. The rocks are leached and cavernous. A thick salt sequence is suggested to lie below. However, no deep wells have as yet been drilled.

Salt-bearing deposits assigned to the Madley Beds crop out in the isolated Woolnough Hills dome north of the Madley Sheet area. The BMR Warri No. 20 penetrated the following section, in ascending order:

265.48–206.65 m — Rock salt with isolated inlcusions of anhydrite and dolomitic and silicified anhydrite
206.65–203.00 m — Transition zone between a salt unit and overlying deposits of anhydrite cap rock
203.00–160.73 m — Anhydrite cap rock consisting of anhydrite and anhydritic, gray dolomite
160.73–123.35 m — Transition zone between gypsum-rich and anhydrite cap rock
123.35– 0.0 m — Gypsum cap rock

The relationship between evaporite deposits of the Brown and Madley Beds and those of the overlying strata is not quite clear. In the cap rock of the Woolnough Hills dome, the Madley Beds come in contact with the Woolnough Beds along the fault, and therefore they may be older. In this case, the Woolnough Beds may be the equivalent of the Babbagoola Beds. These deposits are assigned to the upper Late Precambrian and Lower Cambrian, although their stratigraphic position has not been identified with certainty.

The Woolnough Beds are composed of gypsum, dolarenite and dololutite, laminated claystone, and sandstone. Their apparent thickness within the Woolnough Hills dome is 250 m.

The Babbagoola Beds were penetrated by the Yowalga No. 2 borehole at 143 m (depth range 846–986 m). Units A, B, and C were recognized, in ascending order: unit A (846–887 m) has alternate sandstone and shale with thin anhydrite and gypsum veinlets and inclusions. Sandstones are buff and dark brown; shales, dark red to buff. Unit B (887–893 m) is composed of crystalline dolomite with anhydrite and gypsum nodules and veinlets. Unit C (893–989 m) has alternate shale and gray, grayish to green, and white siltstone.

In the north-eastern marginal parts of the basin, evaporite deposits have not been recorded. The Lupton Beds, which are composed of massive conglomerate, medium- and thick-platy sandstone with siltstone interbeds, were recognized just above the

Lefroy Beds. The thickness varies from 175–250 m. The Turkey Hill Beds are considered to be the equivalent of the conglomerate occurring in the south-western margin of the basin. A thick, terrigenous sequence, which is subdivided into the Punkerri and Wirrildar Beds, was observed above the Lupton Beds. Their total thickness is approximately 1000 m.

The Early Cambrian formation unconformably overlying Late Precambrian and beds transitional to Early Cambrian are represented, in ascending order, by the Table Hill volcanics, red-buff Lennis Sandstone, and Wanna Beds composed of light sandstone with claystone interbeds.

Thus, evaporite deposits within the Officer Basin are known only from central and northern areas where diapirs have been found. The stratigraphic position of evaporite sequences (Brown, Madley, Babbagoola, and Woolnough Beds) and the relationship to under- und overlying deposits are still uncertain. At present, evaporites can be assigned to two age levels: the lower level is Late Precambrian and incorporates salt sequences of the Madley and Brown Beds; the upper level seems to be either Early or Late Cambrian (Vendian?) with gypsum-carbonate-terrigenous deposits of the Woolnough Beds and probably gypsum-bearing, terrigenous deposits of the Babbagoola Beds may be of different age. The lower stratigraphic level of salt deposits can be correlated with the evaporite Bitter Springs Series of the Amadeus Basin.

Precambrian Evaporite Deposits in Other Basins

The oldest evaporite rocks were reported from the Archean of the Aldan Shield within the Fedorovo Formation of the Iengra Series (Kargatjev 1970, Serdyuchenko 1972, Vinogradov et al. 1976). Purple-pink anhydrite inclusions in calciphyre, diopside-scapolite-feldspar rocks (Kargatjev 1970) have been found in the Bolshoi Yllymakh River Basin. These inclusions account for 15–20% of the rock volume; in some beds, they are 2–10 m thick. Anhydrite rocks are underlain and overlain by hypersthenic biotitic-amphibolitic gneiss; anhydrite inclusions also occur in their contact zones. The age of the Fedorovo Formation is dated 3000–3300 m.y. and 4000 m.y.

In another Archean evaporite terrain in West Australia, pseudomorphs of ankerite after gypsum and anhydrite have been found in the Black Flag and Kalgoorlie Beds (Golding and Walter 1979, Wells 1980). Environments favorable for sulfate sedimentation in the Archean have presumably persisted over a long period of time.

Early Proterozoic (Aphebian) evaporites, confined to deposits older than 1800–2000 m.y., were registered in many regions of North America. They were reported from the Great Lakes area of the upper Huronian Supergroup, where pseudomorphs after gypsum and anhydrite are known from rocks of the Gordon Lake Formation, Cobalt Group (Sims et al. 1981). Evaporites occur as crystal molds after halite or gypsum in the Denault Formation, the Knob Lake Group, within the Labrador Trough (Wardel and Bailey 1981), in the Kasegalik and McLeary Formations, the Belcher Group, Hudson Bay (Bell and Jackson 1974, Ricketts and Donaldson 1981), at the base of the Brown Sound Formation (unit B_1), the Goulburn Group of the Kilohigok Basin, Bathurst Inlet District, Northwest Territories (Campbell and Cecile

1981, Campbell 1978), and in the Stark and Gibraltar Formations in the Great Slave Lake area (Hoffman 1968).

Magnesite, dolomite, or silicious pseudomorphs after gypsum and halite were also registered in Early Proterozoic deposits in Australia. They occur in the Pine Creek Geosyncline among the Celia and Coomalie Dolomites, the Cahill Formation, and in the Koolpin Formation. They are over 1870 m.y. old (Criek and Muir 1979, Wells 1981).

Riphean evaporites have been registered in many regions of the Earth. In North America, they are known from the Grenville Series on the east side of Silvia Lake and the north-western Adirondack Mountains (Brown and Engel 1956). Anhydrite occurs there among schist and marl within unit 10. Its thickness ranges from several meters to 60 m. The thickness of anhydrite interbeds proper is small, and anhydrite occurs mainly as crystals in schist locally gypsinate. Anhydrite is considered deposited concurrently with carbonate. J. Brown suggests that rock salt could have accumulated during the Grenvill epoch, 1200–1500 m.y. ago, in the Adirondack area.

A thick, gypsum-bearing formation was recorded in the Mackenzie Mountains within the Middle Riphean Little Dal Group (Upper Helikian). Its thickness reaches 530 m. It consists mainly of microcrystalline white, off-white, pale-buff-gray, pinkish gypsum, locally clayey and silty, especially in the lower part of the section. The upper part contains red gypsum. There are massive nodular varieties with predominant laminated ones. Clayey, gypsinate dolomite, thin-laminated gypsinate, green and red mudstone, and pink and red sandstone interbeds over 1 m thick occur throughout the entire section. To the north-east, the formation contains a unit of carbonate rocks whose thickness ranges from 7.4–20.9 m (Aitken et al. 1873, Aitken and Cook 1974, Aitken 1981). A higher (Late Riphean) stratigraphic level of evaporites is traced in the Mackenzie Mountains just above the Little Dal Formation within the so-called copper cycle. Pale, buff-gray, clayey, dolomitic gypsum beds are confined to a sand unit of the Redstone River Formation (Aitken 1981). Farther north in the Mackenzie Basin (along the Snake River, a tributary of the Peel River), equivalents of the gypsum formation are found in the Upper Gypsum Valley Formation, which is composed mainly of gypsum and anhydrite (Ziegler 1969). In general, the Middle Riphean gypsum deposits in the Mackenzie Mountains can be traced a distance of more than 500 km.

Riphean gypsum deposits are widespread in the Amundsen Embayment (Northwestern Territories, Canada), on Victoria Lake, and in the Brock Inlier area (Thorsteinsson and Tozer 1961, Young 1981). They occur in the upper Shaler Group and are related to the Minto Inlet, Wynniatt, and Kilian Formations. Gypsum beds of the Minto Inlet Formation are 10–15 m thick and intercalated with dolomite, limestone, gypsinate mudstone, siltstone, and sandstone. The thickness of the formation is approximately 300 m. Thin (up to 1 m) gypsum beds occur in the upper Wynniatt Formation where they are interbedded with shale and limestone. Anhydrite and gypsum of the Kilian Formation are confined to the lower part of the section recognized as the Lower Evaporite Unit. It consists of gray, red, and green mudstone, limestone gypsinate and pure gypsum interbeds not more than 2 m thick.

Evaporites on northern Baffin Island within the Riphean Borden Basin are known from the Arctic Bay Formation where shale had white gypsum efflorescent coatings. They are common in the Society Cliffs Formation and the Uluksan Group. The member SC_1, represented by alternate terrigenous rocks, dolostones, and gypsum rocks was singled out in its lower part (Jackson and Jannelli 1981). This red evaporite member is cyclic in structure and consists of green, red, and black shale, gray-green and pink dolostone, and white gypsum. Gypsum beds are 1.5–3 m thick and form the upper part of the cycles. There are scarce halite crystals.

Late Precambrian gypsum-bearing deposits have also been reported from northwestern Greenland and the Thule Basin (Davies et al. 1963, Berthlsen and Noe-Nygaard 1968, Dawes 1976, Frisch et al. 1978). They are related to the Thule Group, which is subdivided into three formations: Wolstenholme, Dundas, and Narssârssuk. Gypsum beds have been found in the upper Dundas and in two lower members of the Narssârssuk Formation. One of the gypsum beds in a lower member is 8 m thick. In eastern Greenland, the Late Precambrian evaporites, containing gypsum interbeds and halite crystal molds, were registered in the upper Merkbjorn Formation (Katz 1961).

The Precambrian sulfate rocks are also known in the Siberian Platform from the Starorechka Formation on the Kotuikan River where gypsum and anhydrite interbeds up to 2–3 m thick lie among dolomite (Komar 1966, Kulibakina et al. 1972).

Evidence of evaporite sedimentation which occurred during the Middle Proterozoic has been found in many regions of Australia. The data have been recently summarized by Wells (1980). Therefore, pseudomorphs after gypsum and halite in deposits dated 1400–1600 m.y. occur in the McArthur Basin and in Mount Isa. In the former region, the pseudomorphs were found in rocks of the McArthur Group within the Mallapunyah Formation and in the Amelia Dolomite. Transgressive-regressive sedimentary cycles crowned by evaporites could have accumulated over the area more than 5000 km^2 in the Batten Trough. In the Mount Isa area, rocks with pseudomorphs after gypsum were recognized among deposits of the McNamara and Mount Isa Groups. Late Precambrian evaporites, which were registered in Australia in the Ngalia Basin, may also be present in the Adelaide Geosyncline. They are marked by numerous diapirs and siliceous carbonate breccia, which is assigned to the Callanna Beds, crops out in their cap rocks.

In Africa, Precambrian evaporite rocks are known in Zambia within the Mufulira Syncline in the Ore Formation of the lower Roan Group, Katanga Supergroup (Maree 1960, Brandt et al. 1961, Garlick 1981). Small anhydrite inclusions have been found beneath the Ore Formation in underlying argillite. Gypsum and anhydrite become abundant starting from orebody C at the base of the Ore Formation. There is quartzite with a dolomite and anhydrite matrix and thin anhydrite bands among argillite at the eastern margin of the area. In the overlying A/B sediments, anhydrite inclusions occur in the lower dolomite member where they account for 30%. Whereas anhydrite lenses were found deep in argillaceous quartzite overlying the orebody A horizon, gypsum lies near the surface. Evaporites of the Ore Formation may be inferred for the Zambian Copperbelt.

In summary of this brief characterization of the Precambrian evaporite deposits, it is noteworthy to mention scapolite and scapolite-bearing rocks as metamorphic equivalents of once existing salt beds (Kargatiev 1970, Ramsay and Davidson 1970, Serdyuchenko 1972, Dimroth and Kimberley 1976, Badham and Stanworth 1977, Chandler 1978, Lambert et al. 1980, Sidorenko and Rozen 1981).

The above authors suggest that diopside and scapolite-bearing dolomite marble and calciphyre, scapolite-diopside rocks with almost monomineral scapolite interbeds and lenses, and scapolite-diopside-hornblende feldspar gneiss in the Archean and Lower Proterozoic metasedimentary beds have resulted from the metamorphism of old salt-bearing (and saline) sediments. Some feldspar schists similar in composition to anorthosite and containing sulfate-scapolite were also produced by metamorphism of gypsum-bearing arenaceous-argillaceous rocks.

Scapolite-bearing schists, marble, and calciphyre are widespread in the following beds: Archean Iengra Series, Vitim River, Aldan Shield,; in South Australia and Madagascar, in Lower Proterozoic deposits of north-western Queensland, Australia, in the Precambrian beds of Lapland, in granulites of the Kola Peninsula, as well as in the Archean and Early Proterozoic formations of the Canadian Shield. In general, these rocks are metamorphic products of salt-bearing formations.

The above data should be seriously studied for further elucidation of the problem of the Precambrian salt accumulation.

Chapter II
Cambrian Salt Deposits

Cambrian salt-bearing strata are known only from Asia, North America, and Australia (Fig. 3). At present, five Cambrian salt basins are distinguished: East Siberian, Iran-Pakistan, Mackenzie, Amadeus, and Arckaringa. The latter is conventionally assigned to Cambrian and has not been discussed earlier by the author (Zharkov 1981).

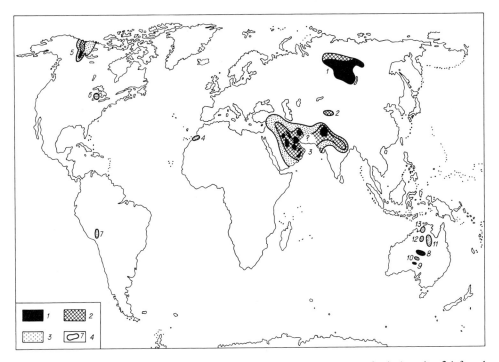

Fig. 3. Distribution of Cambrian evaporite deposits. *1* known areas of salt deposits, *2* inferred areas of salt deposits, *3* areas and basins of sulfate sedimentation, *4* boundaries and number of evaporite basins (*1* East Siberian, *2* Tarim, *3* Iran-Pakistan, *4* Anti-Atlas, *5* Mackenzie, *6* Michigan, *7* Cis-Andean, *8* Amadeus, *9* Arckaringa, *10* Officer, *11* Georgina, *12* Wiso, *13* Daly River)

East Siberian Basin

Salt deposits in East Siberia cover an extensive territory, extending from the Yenisei Range in the west to the Lena Basin in the east (Fig. 4). They have been penetrated by numerous deep holes at different sites within the Angara-Lena Trough, Cis-Sayans-Yenisei, and Tunguska syneclises. Salt sequences are found over the entire area and lie on various stratigraphic levels.

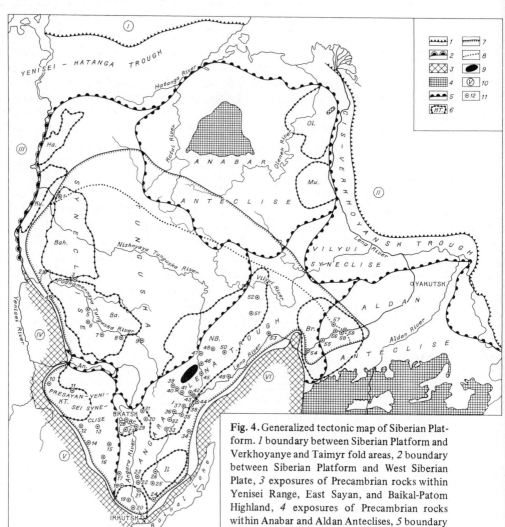

Fig. 4. Generalized tectonic map of Siberian Platform. *1* boundary between Siberian Platform and Verkhoyanye and Taimyr fold areas, *2* boundary between Siberian Platform and West Siberian Plate, *3* exposures of Precambrian rocks within Yenisei Range, East Sayan, and Baikal-Patom Highland, *4* exposures of Precambrian rocks within Anabar and Aldan Anteclises, *5* boundary of major tectonic elements of Siberian Platform, *6* limit of some major structures and their indexes (*Kha* Khantai-Rybinsk Swell, *Ku* Kureika-Baklanin Swell, *Ol* Olenek Arch, *Mu* Muna Arch, *Bakh* Bakhta Arch, *Ba* Baikit Anteclise, *An* Angara fold zone, *KT* Kan-Taseeva Depression, *Br* Bratsk Bulge, *Il* Ilga Depression, *NB* Nepa-Botuoba Anteclise, *Be* Berezovaya Depression); *7* distribution of Cambrian evaporite deposits, *8* inferred northern and north-eastern boundaries of Cambrian rock salt distribution, *9* area of potash salt deposits (Nepa potash basin), *10* tectonic elements flanking Siberian Platform (*I* Taimyr fold area, *II* Verkhoyanye fold area, *III* West Siberian Plate, (cont. p. 17)

Southern and south-western limits of salt strata are quite reliable. They follow mountain structures of the Yenisei Range and the East Sayan and Baikal-Patom Highland. The northern boundary has not yet been outlined. It is well known that thicknesses of salt strata gradually increase towards the central Siberian Platform, and stratigraphically higher, other rock salt horizons appear. In the last few years, salt deposits were penetrated by boreholes in the western Tunguska Syneclise within the Baikit Anteclise and Bakhta Arch (Listvenichnaya, Velmo, Ust-Kamovo, Kuyumba, Taiga, and other fields), in its central zone (Vanavara No. 1 borehole), as well as in the northern and north-western Angara-Lena Trough within the Nepa-Botuoba Anteclise (Danilovo, Preobrazhenka, Verkhnyaya Chona, and other fields). The above evidence suggests that the cental Tunguska Syneclise belongs to the Cambrian salt basin. Northern and north-eastern limits of evaporite distribution are roughly outlined in the Upper Vilyui and Upper Kotui, from where they extend westwards and southwestwards into the Lower Nizhnyaya Tunguska Basin. Carbonate, mostly reefogenic, Cambrian deposits are widespread beyond this area.

Knowledge of Cambrian salt deposits in the East Siberian Basin has much improved during the last decade due to oil exploration. The following characterization results from new evidence presented by the following authors: Zharkov and Skripin (1971), Drugov et al. (1972), Melnikov et al. (1977), Vorobiev and Safronova (1977), Sokolov (1977), Zamaraev and Ryazanov (1972), Zharkov and Chechel (1973), "Geology and potassium content of . . ." (1974), Pisarchik et al. (1975), Zharkova (1976a, b), Chechel and Mashovich (1977), Britan et al. (1977), Chechel et al. (1977, 1980), Melnikov et al. (1978), Savitsky and Astashkin (1978), Chechel et al. (1981), Zamaraev et al. (1981), "Geology of oil and gas . . ." (1981), Kontorovich et al. (1981), Melnikov and Kilina (1981), Sokolov and Khomentovsky (1981), Gavva and Rybiakov (1981), Zharkov et al. (1982), and Tyshchenko and Faizulina (1982).

Cambrian deposits of the East Siberian Basin yield a distinctive stenohaline fauna, hence their assignment to the western (Zharkov and Khomentovsky 1965), or Irkutsk-Tunguska (Geology of oil and gas . . . 1981) facies zone, which is used as a basis for a stratigraphic scheme. So far, the placement of the lower Cambrian boundary has been ambiguous. Recently, the western facies zone was thoroughly discussed by Sokolov and Khomentovsky (1981), who draw the boundary of the Lower Cambrian at the top of the Irkut horizon within the evaporitic series of the Siberian Platform. As noted above, we, as well as other authors, draw the Lower Cambrian boundary at the base, and not at the top of the Irkut horizon (Zharkov and Sovetov 1969, Geology of oil and gas . . . 1981, Savitsky and Ataskin 1978).

◀ Fig. 4 (cont.)

IV Yenisei Range, *V* East Sayan, *VI* North Baikal and Patom Highland); *11* location of wells and deep boreholes throughout the entire Cambrian evaporite succession and their numbers (*1* Tunguska 1-o, *2* Listvenichnaya No. 1, *3* Poligus No. 1, *4* Velmo, *5* Ust-Kamovo 20, *6* Kuyumba, *7* Taiga, *8* Sol' Zavod, *9* Vanavara, *10* Taseeva and Tynys, *11* Pochet, *12* Taishet 1-o, *13* Mironovo 1-o, *14* Nizhneudinsk 1-o, *15* Chebotarikha 1-sp, *16* Tyret, *17* Zima 1-r, *18* Kutulik, *19* Belsk 1-o, *20* Usolye, *21* Bokhan, *22* Osa, *23* Atovo, *24* Bozhekhan 1-sp, *25* Khristoforovo, *26* Zhigalovo, *27* Bratsk 3-r, *28* Bratsk 1-r, *29* Zayarsk 1-o, *30* Ilimsk 1-r, *31* Nizhny Ilimsk 1-o, *32* Ust-Kut, *33* Yuzhny Ust-Kut, *34* Kazachinsk 1-r, *35* Bochakta, *36* Kazarka, *37* Markovo, *38* Krivaya Luka, *39* Tokma 1-r, *40* Nepa 1-r, *41,* Yarakta, *42* Ayana, *43* Karelino, *44* Kirensk 1-o, *45* Sosnino 1, *46* Danilovo 124, *47* Preobrazhenka, *48* Verkhnyaya Chona, *49* Chastin, *50* Peledui, *51* Sredne-Botuoba, *52* Mirny, *53* Murbai, *54* Delgei, *55* Olekma 1-r, *56* Namana 2-r, *57* Dirin-Yuryakh, *58* Kamenka, *59* Russkaya Rechka 1-r)

The upper boundary of the Lower Cambrian in most salt-bearing terrains is well substantiated. It has been drawn at the top of the Namana horizon containing trilobites: *Namanoia namanensis* Lerm., *N. anomalica* Rep., *N. evetasica* Suv., *Bathynotus namanensis* Lerm., *Pseudoalistocare litvinica* Rep., *Antagmella konkinskai* Suv. In many regions, this boundary lies within the mottled limestone marker unit and is based on paleontological evidence. The overlying unit of similar lithology yields trilobites of the following genera: *Proasaphiscus, Schistocephalus, Tankhella, Elrathia, Deltocephalus*, and *Icheriella*, which are assigned to the lower Amga stage (Middle Cambrian) and characterize the Zeledeevo biostratigraphic horizon (Chechel 1969b). The boundary between these different fossil assemblages is often drawn to within 1 m (Grigoriev and Repina 1956, Pisarchik 1963, Zharkov and Khomentovsky 1965, Chechel 1969b).

In different regions, this horizon, starting from the middle of mottled limestone can be correlated with the Litvintsevo Formation in the northern Irkutsk Amphitheater (Pisarchik 1963), the Munok Formation in the area of the Kirenga (Zharkov et al. 1964), Chaya (Zharkov and Chechel 1964), Chuya, Nepa, Nizhnyaya Tunguska, and Bolshoi Patom Rivers, the upper Ichera and Metiger Formations on the Lena River (Bobrov 1964), Zeledeevo Formation on the Lower Angara (Grigoriev and Repina 1956) and their stratigraphic equivalents. Deposits of these formations are correlative in regard to their trilobite assemblages.

Overlying Middle Cambrian deposits, belonging probably to the Mayan stage, comprise the Verkholensk Formation and its equivalents. Such a stratigraphic position of the Verkholensk Formation results from the fact that it is overlain by the Upper Cambrian sediments containing *Kuraspis N. Tschern.* trilobites (Zharkov et al. 1964, Zharkov and Skripin 1971).

Salt-bearing deposits penetrated by deep boreholes were assigned to the Lower Cambrian and the Amga stage (Middle Cambrian). In this part of the section, the Irkutsk, Usolye, Elgyan, Tolbachan, Uritsk, Olekma, Chara, Namana, and Zeledeevo horizons are of regional extent (Zharkov and Khomentovsky 1965, Chechel 1969b). They are discernible over the entire area within the western facies zone. In turn, there are local stratigraphic schemes for different regions of the East Siberian salt basin. In the northern Angara-Lena Trough, the following formations can be distinguished: Zherba, Tinnovka, Nokhtuisk, Yuedeya (Macha), Elgyan, Tolbachan, Olekma, Chara, and Elanka. Within southern areas of the Angara-Lena Trough and the Cis-Sayans-Yenisei Syneclise, the Moty, Usolye, Belsk, Bulai, Angara, and Litvintsevo Formations are distinguished. In the Angara fold zone, the section is divided into the Ostrovnaya, Irkineevo, Klimino, Agaleva, and Zeledeevo Formations. Recently, special stratigraphic schemes have also been compiled for other regions with a distinctive structure of the Lower and Middle Cambrian deposits, e.g., the northern Nepa-Botuoba Anteclise and Baikit Anteclise.

A minute subdivision of Lower and Middle Cambrian salt deposits was proposed by Chechel and co-workers (1977). Fourteen regional marker carbonate units, indexed "R", and 15 salt units, indexed "S", have been recognized above the Irkut horizon. The units are numbered in descending order. Units S_{14}, R_{XIV}, S_{13}, R_{XIII}, S_{12}, R_{XII}, S_{11}, R_{XI}, and S_{10}, are confined to the Usolye biostratigraphic horizon; unit R_X and lower unit S_9, belong to the Elgyan horizon; upper unit S_9 and units R_{IX}, S_8, R_{VIII}, and S_{7a}, to the Tolbachan horizon; unit S_7 to the Uritsk horizon; unit R_{VII} to the Olekma horizon; units S_6, R_{VI}, S_5, R_V, S_4, R_{IV}, S_3, R_{III}, and S_2 to the Chara horizon; the lower part of unit R_{II} to the Namana horizon, and the upper part of unit R_{II}, units S_1, and R_I, to the Zeledeevo horizon.

Marker carbonate or sulfate-carbonate units are very common, and allow a reliable correlation of sections. As a rule, salt-bearing units are not constant in thickness and composition. The main structural features of Cambrian salt deposits become quite obvious from correlation charts (see Figs. 5, 6, 7, 8), as well as from the scheme showing major structural features of the Cambrian evaporitic series (Fig. 9).

Areal extent of Lower and Middle Cambrian evaporitic deposits on the Siberian Platform has become quite evident (Fig. 10). In the south-west, in the piedmonts of the Yenisei Range and East Sayan, they are replaced by mostly terrigenous deposits. In the east, south-east, north-east, north, and north-west, salt strata are replaced along the strike by sulfate-dolomite, dolomite, and finally, calcareous and bituminous-shaly deposits. An extensive zone of archaeocyathean-algal bioherms from 20–30 to 100–150 km in width has been distinguished between bituminous and bituminous-shaly calcareous sequences and dolomite deposits (Khomentovsky and Repina 1965, Zhuravleva 1966, Zharkov 1966, 1970, Pisarchik et al. 1975, Chechel et al. 1977, Savitsky and Astashkin 1978, Geology of oil and gas . . . 1981).

Cambrian evaporitic series in the East Siberian Basin are 2–2.5 km thick (Fig. 11). Maximal values were reported from the south-western Angara-Lena Trough and the central Cis-Sayans-Yenisei and Tunguska syneclises. The following description of salt deposits is more detailed (horizon by horizon) and is provided mainly by Chechel et al. (1977).

Irkut Horizon

At the level of this horizon is a thick, terrigenous-sulfate-carbonate sequence, commonly divisible into two parts: a lower, mostly terrigenous part, and an upper part composed of carbonate, sulfate-carbonate, and sulfate rocks.

The lower part is assigned to the Middle Moty Member. In the central Irkutsk Amphitheater, it is represented by silty, often clayey, sulfate-carbonate rocks and marl. Red sandstone dominates in the Southern Cis-Sayans and Baikal area. In the north-western Cis-Sayans, within the Kan-Taseeva Depression and in the Yenisei Range piedmont, a conglomerate up to 35 m thick lies at the base, overlain by medium- to coarse-grained polymictic sandstone. Along the periphery of the Vitim-Patom Highland, the amount of terrigenous rocks decreases and that of carbonate rocks increases. Equivalents of the Middle Moty Member are known as the Tinnovka Formation, which is mostly of gray and dark gray, sometimes clayey, bituminous, and algal dolomite. Dolomite with clastic and sulfate rock interbeds is widespread along the north-western periphery of the Aldan Anteclise. Rock salt layers appear in the inner areas of the Siberian Platform, adjacent to the Vitim-Patom Highland and covering the the territory of the Berezovaya Depression and the Murbai-Chastin Basin. Their thickness varies from 8–10 m in the Olekma No. 3-r borehole, to 70 m in Murbai No. 1 borehole, and 170 m in Chastin No. 2-ch borehole.

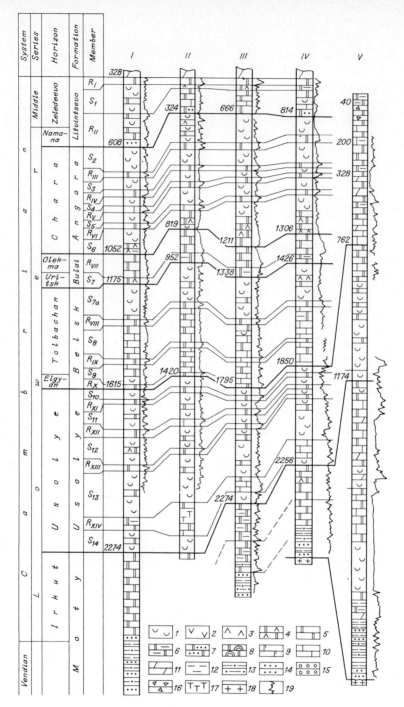

Fig. 5. Correlation chart of Cambrian salt deposits for southern Angara-Lena Trough. After Chechel et al. (1977). Sections: *I* Korkino, well 1-r, *II* Ust-Kut, wells 2-r and 4-r, *III* Markovo, well 23-r, *IV* Karelino, well 3-p, *V* Chastino, well 2-ch. *1* rock salt, *2* gypsum, *3* anhydrite, *4* anhydrite-dolomite and dolomite-anhydrite, *5* dolomite, *6* clayey dolomite, *7* sandy dolomite, *8* algal dolomite, *9* calcareous dolomite, *10* limestone, *11* marl, *12* mudstone, *13* siltstone, *14* sandstone, *15* gritstone, *16* carbonate breccia, *17* trap intrusions, *18* crystalline basement rocks, *19* log curve. *Numerals to the left of each column* indicate the depth of formation occurrence, in meters

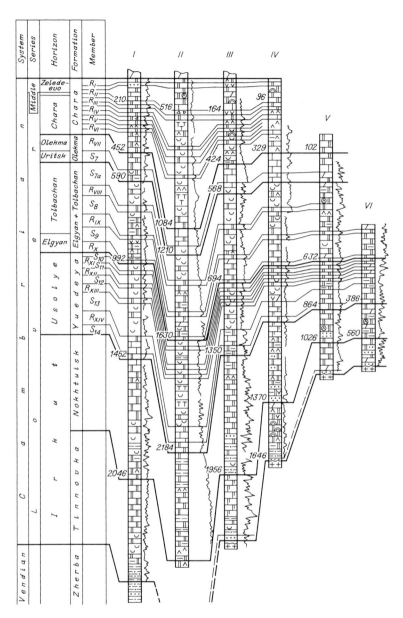

Fig. 6. Correlation chart of Cambrian salt deposits of south-eastern Siberian Platform. After Chechel et al. (1977). Sections: *I* Murbai, well 1-r, *II* Delgei, well 2-r, *III* Olekma composite section, wells 28, 40, 3-r, *IV* composite section of Namana River, wells 8-k, 2-r, *V* composite section of Russkaya Rechka, wells R-3 and 42, *VI* Markha, well 2-k. For legend see Fig. 5

Although the thickness of the lower Irkut horizon in the south-western Siberian Platform does not exceed 80–100 m, it reaches 450–580 m within the Murbai-Chastin Basin and is 300 m on the western flank of the Berezovaya Depression.

The upper Irkut horizon corresponds to the Upper Moty Member of the Irkutsk Amphitheater, which is represented by alternate dolomite, anhydrite, dolomite-

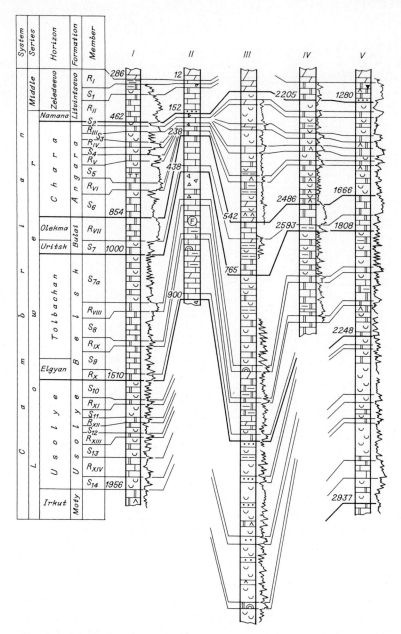

Fig. 7. Correlation chart of Cambrian salt deposits for south-western Siberian Platform. After Chechel et al. (1977). Sections: *I* Taiga area, well 1, *II* Tagara iron deposit, *III* Kan-Taseeva Depression, wells 54 and 57, *IV* Mironovo, well 1-o, *V* Tangui, well 1-sp. For legend see Fig. 5

anhydrite, clay dolomite, and minor marl. Sandstone and siltstone appear in the vicinity of marginal areas. In north-western Cis-Sayans, the upper Ust-Tagul Formation is an equivalent to the Upper Moty Member, whereas in the Kan-Taseeva Depression

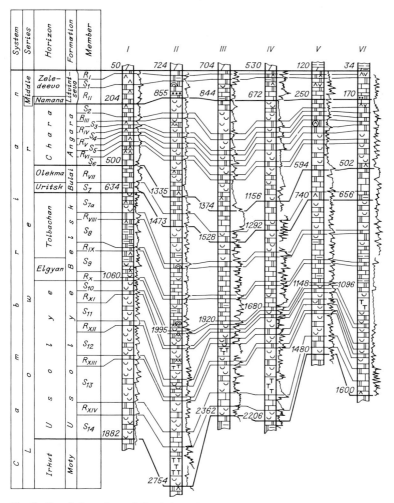

Fig. 8. Correlation chart of Cambrian salt deposits for southern Siberian Platform. After Chechel et al. (1977). Sections: *I* Osa (composite), *II* Zayarsk, well 1-o, *III* Yarakta, well 11-p, *IV* Volokon, well 1-r, *V* Preobrazhenka, well 106-pr, *VI* Sredne-Botuoba, well 2-p. For legend see Fig. 5

and in the Yenisei Range piedmont, the upper Ostrovnaya Formation is an equivalent. Coarse clastic rocks become more abundant and carbonate rocks occur as partings. In the Lower Angara Region and Yenisei Range, the proportion of carbonate rocks increases. Along the periphery of the Vitim-Patom Highland, at the level of the upper Irkut horizon, the Nokhtuisk Formation is distinguished. It consists of dolomite and claystone with sedimentary breccia and siltstone and limestone interbeds. In inner areas of the Siberian Platform, rock salt appears in the section. Within the Murbai-Chastin Basin and Berezovaya Depression, rock salt interbeds up to 2–3 m thick were observed in the Chastin and Olekma boreholes, as well as in Upper Velyuchan and the Sedanovo and Bratsk fields. The thickness of the upper Irkut horizon ranges from 100–600 m.

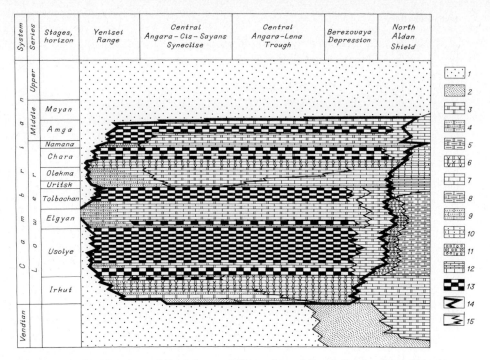

Fig. 9. Generalized lithology of Cambrian evaporitic series and its relation to other sedimentary strata on the southern Siberian Platform. *1* terrigenous red beds, *2* gray quartz sandstone, *3* dolomite, *4* sandy dolomite, *5* clayey dolomite, *6* anhydrite-dolomite and dolomite-anhydrite, *7* limestone, *8* clayey limestone, *9* arenaceous limestone, *10* bituminous limestone and shale, *11* oolitic and oncolitic limestone, *12* calc-bituminous shale, *13* rock salt, *14* boundaries of evaporitic series, *15* boundaries of sedimentary strata

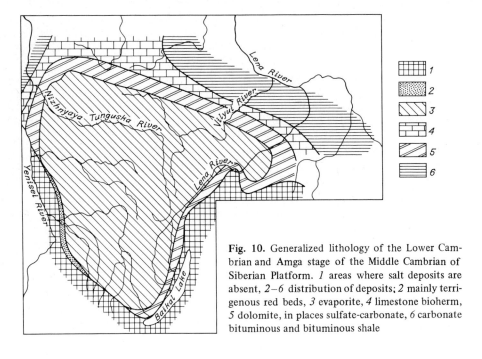

Fig. 10. Generalized lithology of the Lower Cambrian and Amga stage of the Middle Cambrian of Siberian Platform. *1* areas where salt deposits are absent, *2–6* distribution of deposits; *2* mainly terrigenous red beds, *3* evaporite, *4* limestone bioherm, *5* dolomite, in places sulfate-carbonate, *6* carbonate bituminous and bituminous shale

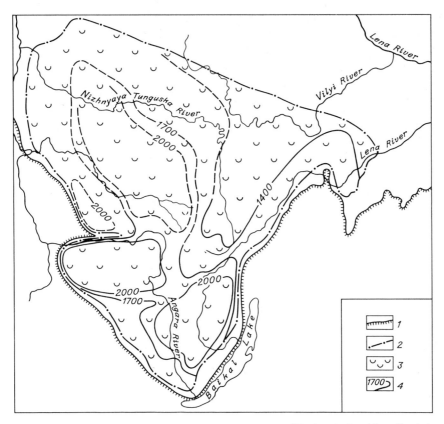

Fig. 11. Thicknesses of Cambrian evaporite series in East Siberian Basin. After Chechel et al. (1977). *1* present continuous distribution of Cambrian deposits, *2* distribution of rock salt, *3* salt-producing zone of East Siberian Basin, *4* isopachs for Lower Cambrian and Amga deposits (Middle Cambrian)

Usolye Horizon

This horizon comprises four salt units (S_{14}, S_{13}, S_{12}, and S_{11}) and four marker carbonate or sulfate-carbonate units (R_{XIV}, R_{XIII}, R_{XII}, and R_{XI}).

Salt unit S_{14} lies at the base of the Usolye Formation and is composed of rock salt, anhydrite, dolomite, and rare, terrigenous rocks. It is known from the inner Cis-Sayans-Yenisey Syneclise, south-western Angara-Lena Trough, and Tunguska Syneclise (Fig. 12a). Anhydrite and dolomite interbeds are confined mainly to the central part of the unit. Thus, the unit can be divided into three parts, namely, lower, salt-bearing; middle, dolomitic-anhydrous-halitic; upper, salt-bearing. The lower part is composed almost entirely of rock salt with rare, seasonal halopelite partings. Halopelite partings, usually dark, are from several millimeters to 5–7 cm in thickness. Rock salt is coarse- to medium-crystalline, colorless, transparent, white, pink, gray, sometimes, off-gray. The maximal thickness is 70–80 m in the southern Irkutsk

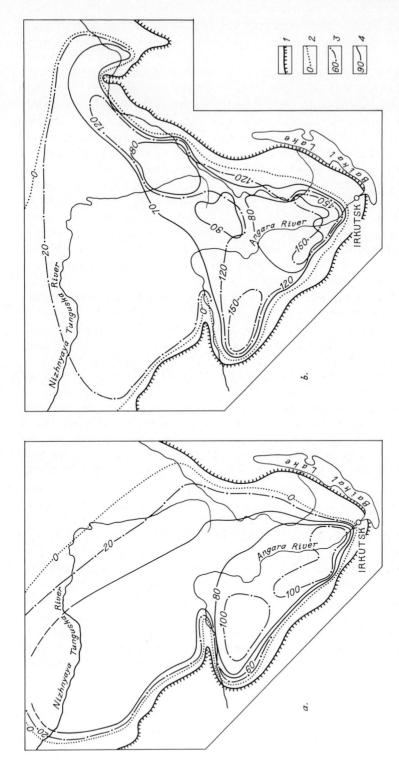

Fig. 12a, b. Maps of total rock salt thickness and salt saturation (%) of units S_{14} (a) and S_{13} (b) in Usolye horizon. After Chechel et al. (1977). *1* continuous distribution of Cambrian deposits, *2* distribution of rock salt, *3* isopachs for rock salt, *4* isolines of rock salt content (%)

Amphitheater. It is 25 and 30 m, respectively, near the town of Nizhneudinsk and in the vicinity of Markovo village. The thickness decreases to 20 m within the Velmo Depression; in the apical part of Nepa-Botuoba Anteclise, it is only several meters. The middle part of the unit is represented by alternate rock salt, anhydrite, and dolomite. Five constant dolomite-anhydrite beds between rock salt layers of the same thickness were observed in the Osa and Parfenovo fields. The total thickness is 40–45 m. In the Tyret, Nukuty, and Kutulik fields, the middle part of the unit consists of four clay anhydrite-dolomite and three rock salt beds. The upper and lower rock salt beds are 2–4 m thick; the middle, 14–15 m. The total thickness of the section may reach 38–42 m. In the Bokhan Field, six dolomite-anhydrite and five rock salt beds are distinguished. The middle rock salt bed (10–12 m) is thickest. There are three of four clay dolomite-anhydrite and two or three rock salt beds in the Atovo, Balykhta, Typta, Zhigalovo, Korkino, and Ust-Kut fields, with a thickness of 30–40 m. In the Kazarka, Bochakta, and Markovo fields, the section consists of two dolomite-anhydrite and one rock salt bed with a thickness of 25–32 m. Northwards, the thickness, due to tapering, decreases to 24 m in the Osharovo borehole and to 20 m in the Preobrazhenka borehole.

The upper salt-bearing part of the unit is rather uniform throughout the southern Siberian Platform, though its thickness varies in different fields. Its thickness is 40–50 m in the Belsk, Usolye, Tyret, and Kutulik fields; in the Atovo, Balykhta, Typta, Zhigalovo, Korkino, and Kasyanka, 30–40 m; in Bokhan, Osa, Parfenovo, and Tulun, 25–35 m; in the Ust-Kut, Kazarka, Bochakta, Preobrazhenka, and Markovo, 8–10 m.

The total thickness of unit S_{14} varies from 30–170 m, being the greatest in the southern Irkutsk Amphitheater (Bokhan, Usolye, Belsk, Tagna), and least within the Nepa-Botuoba Anteclise and Taiga Uplift. Salt content varies widely. Highest values (85–95%) were recorded in the central, south-western, and western Irkutsk Amphitheater. Northwards, salt content decreases to 45–55%, due to tapering rock salt beds and small thickness. Eastwards, low salt content is not accompanied by a decrease in thickness. On the contrary, in some cases, it increases. Therefore, in the Elovka borehole, the thickness is 53 m and salt content amounts to 82%. In the adjacent Bolsherazvodnaya borehole, it is 93 m and salt content equals 22%. The same applies to the Yuzhny Ust-Kut and Kazachinsk fields.

In the south-east, north-east, and north-west, salt unit S_{14} is replaced by sulfate-carbonate deposits. In the south-eastern Irkutsk Amphitheater, their thickness attains 80–180 m, whereas in the north, north-east, and north-west it decreases to 25–50 m. In the latter areas, equivalents of unit S_{14} consist of dolomite (about 50%), sulfate-carbonate rocks (30%), marl (10%), and limestone (10%). Sulfate rocks are rather thin and occur as lenses or inclusions. Equivalents of unit S_{14} along the periphery of the Baikal-Patom Highland, the adjacent Baikal area, and marginal parts of the North Baikal Highland are composed of sedimentary breccia. The thickness does not exceed 30–40 m. Along the western and south-western margins of the Siberian Platform, terrigenous-carbonate equivalents are not more than 15–30 m thick.

Marker unit R_{XIV} is also known as Osa unit. In the central Irkutsk Amphitheater (Osa, Shelonino, South Radui, Bokhan), it is represented by dark gray and brownish-gray dolomite with interbeds of calcareous and clayey dolomite, anhydrite, dolomite-anhydrite, and anhydrite-dolomite. Marl and claystone interbeds appear as the

Cis-Sayans are approached. In northern areas of the Amphitheater and in Baikal area, the section is composed mainly of limestone. The proportion of limestone increases from south to north. In the Berezovaya Depression and in the Aldan Anteclise, the unit is dominated by dolomite, which has a thickness of 40–70 m. Southern and northern areas are marked, respectively, by small and great thicknesses.

Salt-bearing unit S_{13} lies within the inner field of the Irkutsk Amphitheater in the south-eastern Siberian Platform, including the Berezovaya Depression and the central and western Siberian Platform, approximate to the latitude of the town of Turukhansk (Fig. 12b). In these areas, the unit is composed of gray, light gray, white, pinkish, coarse-grained, massive rock salt with "salt-free" rock interbeds: saliniferous clay dolomite, dolomite-anhydrite, anhydrite-dolomite, and light and dark gray anhydrite. Marl and claystone occur occasionally; limestone, rarely. The number of interbeds and thickness of "salt-free" rocks change depending on the area, apparently because of the tapering of rock salt. Therefore, adjacent "salt-free" beds are grouped as one. Regional tapering of rock salt is observed north- and westwards.

In the Osa Field, nine dolomite and ten rock salt beds are distinguished within the unit. In the South Radui Field, their number decreases to eight and nine, respectively. In the Tulun borehole, the uppermost rock salt bed and underlying dolomite, and in Nizhneudinsk borehole, the underlying two lower rock salt beds and two dolomite beds are absent from the section. Thus, in the Nizhneudinsk borehole, the unit is represented by five dolomite and six rock beds. Tapering of the upper and lower parts of the unit is accompanied by a decrease in thickness. Thus, in the Osa Field, the unit is 212 m thick; the Nizhneudinsk, 119 m thick. The structure of the unit in Markovo and in the southern Irkutsk Amphitheater is equivalent, i.e., it includes nine dolomite and ten rock salt beds. In the Krivaya Luka borehole, two lower rock salt layers, as well as overlying dolomite, are absent. The thickness of the third rock salt layer (from below) decreases, and the fourth dolomite layer tapers out. In the Karelino borehole, the third rock salt layer (from the base) tapers out entirely. In the Preobrazhenka section, the fifth rock salt layer thins out. Therefore, overlying and underlying dolomites are grouped together. There are only two dolomite and three rock salt layers in the Sredne-Botuoba borehole.

Salt content of unit S_{13} ranges from 20–85%. Maximal values (75–85%) are recorded in the inner field of the Irkutsk Amphitheater and in the central Berezovaya Depression. In these areas, thickness varies between 100–250 m. In the central, western, and south-eastern Siberian Platform, the thickness is less than 100 m and salt content decreases to 55–65%.

Salt-free equivalents of unit S_{13} are represented either by sulfate-carbonate (southwestern Cis-Sayans, western Baikal area), or by terrigenous-carbonate (south-western marginal areas of the Siberian Platform) deposits. Their thickness ranges from 50 to 150–200 m. Along the periphery of the Baikal-Patom Highland, sedimentary breccia reaches 150 m.

Marker unit R_{XIII} is easily traced within the Irkutsk Amphitheater. It is composed of dolomite and more rarely, limestone. In the south-west, siltstone, sandy, and clayey dolomites appear in the section. The unit is 10–20 m thick.

Salt-bearing unit S_{12} occurs within the Upper Lena, Kan-Taseeva, and Berezovaya depressions, on the Nepa-Botuoba Anteclise, and probably, within the Tunguska

Syneclise, approximate to the latitude of the town of Turukhansk (Fig. 13a). The unit is represented by alternate rock salt, carbonate, and sulfate-carbonate rocks. Rock salt is gray, sometimes pinkish, coarse-crystalline with numerous halopelite interbeds. The thickness ranges from several meters to 30–40 m. "Salt-free" rock beds are 1–2 and 10–20 m thick. In the southern Irkutsk Amphitheater, sulfate-carbonate varieties dominate; in Cis-Sayans, sandy and clay dolomites; in the Kan-Taseeva Depression, sandstone, siltstone, and claystone. In the northern Irkutsk Amphitheater, limestone interbeds have been observed.

Salt and "salt-free" beds differ in number from region to region, due to replacement or tapering out of some parts of the section. The most complete section of this unit, penetrated in the Ust-Kut Field, comprises three successions: lower, composed of four rock salt and three "salt-free" beds; middle, represented by dolomite with rock salt interbeds; and upper, composed of one dolomite and two rock salt beds. North, west, and south of the region, rock salt and "salt-free" beds taper out.

Thickness of the unit varies between 25 and 170 m. Maximal values were reported from the Upper Lena Depression. Salt content of about 90% was recorded in the north-eastern Irkutsk Amphitheater and within the Nepa-Botuoba Anteclise.

Sulfate-carbonate equivalents of unit S_{12} are developed along the south-western, southern, and eastern periphery of the Irkutsk Amphitheater. In the marginal part of the Baikal-Patom Highland, salt deposits are replaced by dolomite and sedimentary breccia, their thickness attaining 30 m. A terriggenous-carbonate complex about 20 m thick runs along the western and south-western margin of the Siberian Platform.

Marker unit R_{XII} is represented by multiple alternation of dolomite and anhydrite and some rock salt lenses. The most complete section was penetrated in the Zayarsk Field, where the unit consists of three beds. The upper is gray and brownish-gray, platy dolomite; in places, saliniferous clay dolomite with anhydrite inclusions. Atop, there is a 2 m anhydrite bed; in the central part, a 1.5 m rock salt interbed was recorded. The middle layer is composed of rock salt, light pink, gray, and medium- to coarse-crystalline, the lower layer is gray, granular, platy, dolomite. In marginal areas of the Siberian Platform, rock salt and anhydrite are absent. In the Cis-Sayans and the Yenisei Range piedmont, sandstone, siltstone, and sandy dolomite appear in the section. Limestone is common in the Baikal area and Berezovaya Depression. The thickness of the unit in central and western areas of the inner field of the Irkutsk Amphitheater is 30–40 m, decreasing to 15–25 m north of Bratsk and to 10 m and less in marginal parts of the Siberian Platform.

Salt unit S_{11} lies in the central Siberian Platform (Fig. 13b). It is composed of alternate rock salt and "salt-free" beds. Rock salt occurs in beds from 1–2 to 30–40 m thick, whereas "salt-free" rocks are 10–15 m thick. In the Cis-Sayans, the latter are represented by dolomite, often sandy, clay, and sometimes algal varieties, with subsidiary gray claystone, siltstone, and sandstone interbeds. In the south-western Kan-Taseeva Depression, they are represented by sandstone, siltstone, and gray, grayish-green, brown, and red claystone. Within the inner field of the Irkutsk Amphitheater, dolomite and anhydrite-dolomite occur, whereas in the Berezovaya Depression, dolomite with thin marl interbeds is common.

The most complete section of the unit, which consists of five rock salt and four sulfate and dolomite beds, has been reported from the central Irkutsk Amphitheater.

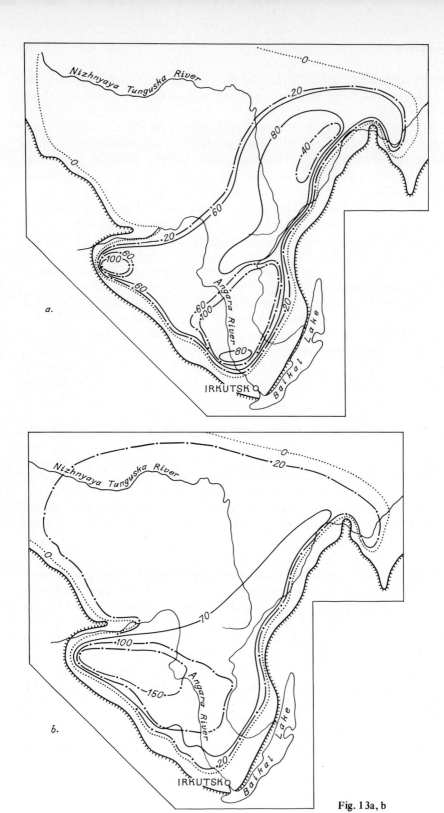

Fig. 13a, b

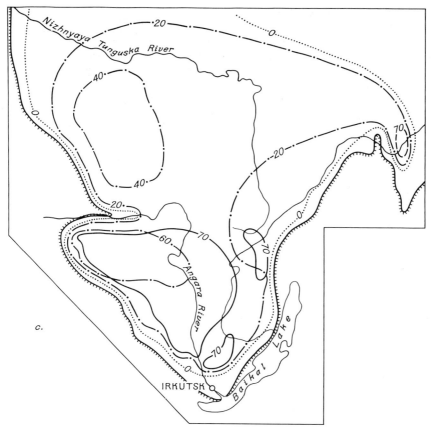

Fig. 13a–c. Maps of total rock salt thickness and salt saturation (%) of units S_{12} (a), S_{11} (b), and S_{10} (c) in Usolye horizon. After Chechel et al. (1977). For legend see Fig. 12

Westwards and northwards, rock salt beds taper out. Thickness and salt content of the unit depend greatly on the structural setting. Maximal (over 100 m) thicknesses are confined to the southern Angara-Lena and Cis-Sayans-Yenisei troughs. Over the rest of the territory, thicknesses decrease to 30–70 m. A high salt content (75–85%) is related to areas with greatest thickness. Within the Berezovaya Depression, Nepa-Botuoba Anteclise, and Tunguska Syneclise, salt content amounts to 50–65%. Along the southern and south-eastern margins of the Siberian Platform, equivalents of this unit are represented by anhydrite, gypsum, dolomite-anhydrite, and dolomite. Sulfate and sulfate-carbonate rocks are dominant. Marl interbeds are subordinate. The thickness is 50–80 m. Along the periphery of the Baikal-Patom Highland, the eastern flank of the Angara-Lena Trough, the Berezovaya Depression, and the south-western margin of the Anabar Anteclise, the equivalents mentioned are composed mostly of dolomite, often calcareous, clayey, and algal, minor sedimentary breccia, limestone, and marl. The thickness does not exceed 40 m. Around the western and south-western periphery of the Siberian Platform, between the Sukhaya Tunguska River mouth and

Irkutsk, clayey and sandy dolomite occurs with interbeds of siltstone, sandstone, sedimentary breccia, and sulfate-carbonate rocks. The thickness is 40–45 m.

Marker unit R_{XI} is composed mainly of gray argillaceous, platy dolomite. It also comprises anhydrite, sulfate-carbonate rock, and algal dolomite partings. In the Baikal area, the unit becomes more calcareous, and in the Cis-Sayans, more arenaceous. In the southern Kan-Taseeva Depression, the section is composed of gray and light gray sandstone. Minor dolomite, siltstone, claystone, thin rock salt, and anhydrite interbeds were recorded. In the northern part of the depression, the amount of dolomite and rock salt increases. Within the Berezovaya Depression and Nepa-Botuoba Anteclise, the unit is composed of dolomite with rare, red marl interbeds with anhydrite inclusions. In the central and western Siberian Platform, dolomite and calcareous dolomite dominate. The thickness ranges from 10–20 m, increasing to 30 m only within the Kan-Taseeva Depression.

The areal extent of salt-bearing unit S_{10} is approximately the same as that of the above described units within the East Siberian Basin (Fig. 13a). It is composed of alternate rock salt, anhydrite, anhydrite-dolomite, dolomite, and scarce terrigenous rocks. In the northern Irkutsk Amphitheater, the section can be divided into three parts. The lower one comprises two rock salt beds, 10.7 and 8.0 m thick with an anhydrite-dolomite and dolomite bed which separates them. The central part begins with dolomite-anhydrite with anhydrite interbeds, followed by 6.0 m of rock salt, and atop the alternation of dolomite and dolomite-anhydrite. The upper part is composed chiefly of rock salt about 20 m thick. In the Kan-Taseeva Depression, the thickness of the lower and upper rock salt beds increases greatly, namely, to 60 m and 69 m, respectively. Widespread among "salt-free" rocks are sandstone, siltstone, and claystone. The greatest thickness of the unit (about 150 m) was recorded in the Kan-Taseeva Depression. In the north-western Irkutsk Amphitheater and in the western Siberian Platform, it decreases to 80–100 m; the remaining area, to 30–60 m. Salt content as high as 70% is characteristic of central, northern, and western areas of the inner field of the Irkutsk Amphitheater. Over the remaining territory, it ranges from 50–60%, decreasing to 20–30% only along the periphery of the Irkutsk Amphitheater. Similar to facies, equivalents of the unit in the southern and eastern Irkutsk Amphitheater are represented by dolomite, dolomite-anhydrite and anhydrite 40–60 m thick; along the periphery of the Patom Highland, by sedimentary breccia, dolomite, and limestone, with a thickness not exceeding 30 m; and in the south-western Siberian Platform, by clay dolomite, siltstone, and sandstone 10–20 m thick.

Elgyan and Tolbachan Horizons

Three marker (R_X, R_{IX}, and R_{VIII}) and three salt-bearing (S_9, S_8, and S_{7a}) units can be distinguished in this part of the section.

Marker unit R_X is dominated by gray and brownish-gray dolomite, massive, and thick-plated, sometimes algal, clayey, and saliniferous. There are several interbeds of limestone, sulfate-carbonate rocks, siltstone, and sandstone. Some rock salt beds also occur in the Kan-Taseeva Depression and in sections along the Lena River near the village of Markovo. The thickness of the unit varies between 10–15 and 20 m.

Salt-bearing unit S_9 has a limited extent and composes northern and central parts of the Irkutsk Amphitheater inner field (Fig. 14a). It is represented by alternate rock salt, limestone, dolomite, sulfate-carbonate, and terrigenous rocks, whereby limestone and dolomite dominate. Terrigenous rocks, rock salt, and anhydrite are subordinate and account for about 20–30%. Rock salt beds commonly occur in the lower part of the unit; sometimes in the upper part, but of a limited extent. The thickness is 120–140 m and the salt content ranges from 10–20 to 33–38%.

Marker unit R_{IX} is composed of dolomite and limestone 10–20 m thick.

Salt-bearing unit S_8 is composed chiefly of dolomite and limestone with minor rock salt, anhydrite, anhydrite-dolomite, and terrigenous rock interbeds. The unit is located within two separate areas. One area encloses the Kan-Taseeva Depression and the northern Irkutsk Amphitheater; the other lies within the Nepa-Botuoba Uplift and Berezovaya Depression (Fig. 14b). The unit consists of two parts. The lower one is represented by a uniform alternation of dolomite and limestone; the upper one contains rock salt and sulfate rocks. The unit is composed of dolomite (40–60%), limestone (20–50%), rock salt (10–20%), anhydrite, and anhydrite-dolomite (not more than 15%). The maximal salt content (35%) was recorded in the Berezovaya Depression area. The thickness varies between 110 and 150 m.

Facies equivalents of the unit are represented by dolomite, marl, and siltstone (80–120 m thick) in the foothills of the Yenisei Range and in the Cis-Sayans; and by dolomite and limestone (140–160 m thick) in the south-eastern, eastern, and north-eastern Siberian Platform.

Marker unit R_{VIII} is composed mostly of dolomite and limestone. In the Cis-Sayans and in the southern Irkutsk Amphitheater, dolomite prevails; however, limestone is predominant over the rest of the area. In the central Amphitheater, several anhydrite, anhydrite-dolomite, and rock salt interbeds have been registered. In the Cis-Sayans and the piedmont of the Yenisei Range, some terrigenous rock interbeds were also found. The thickness ranges from 15–25 m.

Salt-bearing unit S_{7a} occupies a rather large area within the inner Siberian Platform (Fig. 14c). It is marked by frequent alternation of rock salt and "salt-free" beds, ranging in thickness from several dozen centimeters to 20–30 m. The thickest rock salt beds are confined to the lower part of the unit; "salt-free" beds, to the upper one. "Salt-free" rocks are dolomite, and rare dolomite-anhydrite, anhydrite-dolomite, and anhydrite. Terrigenous rocks are common in the south-western Irkutsk Amphitheater. The most complete section of the unit comprises seven "salt-free" and eight rock salt beds. Such a structure of the unit is characteristic of the Zayarsk, Kasyanka, Mironovo, Karelino, Krivoluka, Markovo, Bochakta, Yuzhny Ust-Kut, and Korkino fields. In other regions of the southern Siberian Platform, the number of beds in the unit decreases. Maximal thicknesses (170–240 m) were reported from the Lena River and northern Irkutsk Amphitheater; and minimal values (80–120 m), from the Cis-Sayans, central and north-eastern regions. The maximal salt content (65–75%) was recorded in the northern and north-eastern Irkutsk Amphitheater and in the Berezovaya Depression.

"Salt-free" equivalents of unit S_{7a} occur along the periphery of the salt accumulation area. They are composed of dolomite, dolomite-anhydrite, anhydrite, and limestone in the south-eastern and eastern Irkutsk Amphitheater and have a thickness

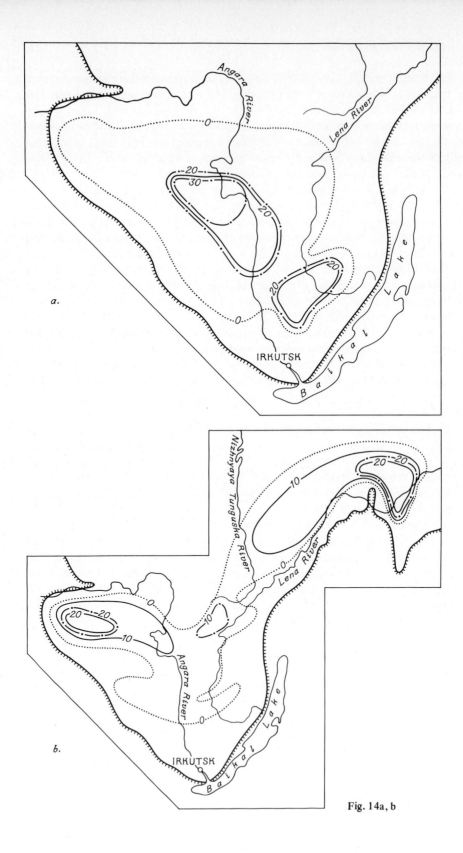

Fig. 14a, b

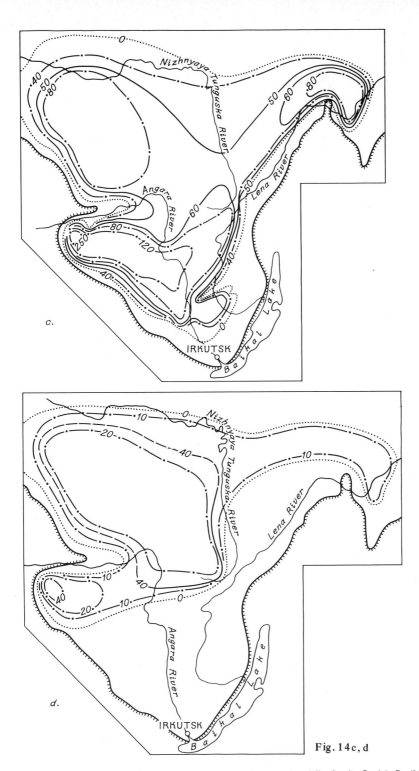

Fig. 14a–d. Maps of total rock salt thickness and salt saturation (%) of units S_9 (a), S_8 (b), S_{7a} (c), and S_7 (d) in Elgyan, Tolbachan, and Uritsk horizons. After Chechel et al. (1977). For legend see Fig. 12

ranging from 50–60 to 120–130 m. In the Baikal area, along the periphery of the Baikal-Patom Highland and along the periphery of the Aldan Anteclise, they are composed of dolomite, limestone, and sedimentary breccia 60–100 m thick. In the piedmont of the Yenisei Range and in the Cis-Sayans, dolomite, siltstone, and sandstone 20–150 m thick are registered.

Uritsk Horizon

Salt-bearing unit S_7 lies at the level of this horizon. It occurs in the northern Irkutsk Amphitheater in the Kan-Taseeva, Berezovaya depressions, and probably, in the Tunguska Syneclise. Its northern limit follows the Nizhnyaya Tunguska Valley (Fig. 14a). The lower part of the unit is composed mostly of dolomite; the upper part is composed of alternate dolomite, rock salt, and sulfate-carbonate rocks. Rock salt accounts for 20–40% and the thickness ranges from 40–50 to 170 m. The Kan-Taseeva Depression is marked by the greatest thicknesses.

Facies equivalents of the salt-bearing unit remain the same; terrigenous-carbonate strata 30–120 m thick in the south-western and western Siberian Platform; sulfate-carbonate and carbonate strata 40–60 m thick along the periphery of the North Baikal and Patom Highlands.

Olekma Horizon

The Olekma horizon comprises deposits placed into marker unit R_{VII} in the East Siberian basin. In the southern Irkutsk Amphitheater, it is represented by dolomite. In northern, north-eastern, and western regions, it consists of limestone, which dominates in the Berezovaya Depression, on the western flank of the Aldan Anteclise and the Lower Angara River. In the Cis-Sayans and in the piedmont of the Yenisei Range, clay dolomite, marl, and claystone are common. The unit is 70–90 m thick.

Chara Horizon

The Chara horizon can be divided into five salt-bearing (S_6, S_5, S_4, S_3, and S_2) and four marker carbonate or sulfate-carbonate (R_{VI}, R_V, R_{IV}, and R_{III}) units.

Salt-bearing unit S_6 is widespread within the Cis-Sayans-Yenisei Syneclise and the Angara-Lena Trough (Fig. 15a). The unit consists of two parts. The lower part is represented by multiple alternation of dolomite, anhydrite, and dolomite-anhydrite with rock salt interbeds, whereas the upper part is dominated by rock salt. In the northern Irkutsk Amphitheater, the unit becomes almost entirely salt-bearing. The greatest thickness of the unit (up to 300 m) was reported from the Surinda-Gazhenka Depression, where is comprises thick potash salt beds (Chechel et al. 1980, 1981, Zharkov et al. 1982). There, in the last years, a large area of potassium accumulation — the Nepa Basin — was outlined.

The Nepa Basin occupies the Nizhnyaya Tunguska-Nepa Interfluve. It extends from the south-west to the north-east for about 240 km and has a width of about

120 km. Salt-bearing unit S_6, to which potash salts are confined, is distinguished in the Nepa Basin under the name of the Gazhenka Member. On the basis of the rock distribution pattern, it can be divided into five horizons, in ascending order: lower anhydrite, lower rock salt, potassium-bearing, upper rock salt, and upper anhydrite. Salt deposits have accumulated in a large basin, unfilled by sediments. Their emplacement coincides with the beginning of the unit deposition. The above is confirmed by an abrupt change in thickness and structure of the lower anhydrite horizon. In the center of the Nepa Basin, it is not less than 1 m and consists of deep-sea laminite anhydrite, whereas shallow massive anhydrite 30–60 m thick is developed along the margin. The depression on the sea floor favored flowing down of heavy brines; in turn, their evaporation led to the deposition of the lower rock salt and potassium-bearing horizon.

The potassium-bearing horizon proper has a rather complex structure. The available evidence allowed the distinction of three types of sections: carnallite, sylvite-carnallite, and sylvinite. The section of carnallite is observed in marginal areas. Its lower part includes a halite-carnallite zone, about 10 m thick, with thin sylvite-carnallite interbeds. A carnallite zone is 20–30 m thick when it occurs above. The upper part of the horizon is again halite-carnallite in composition; its thickness rarely exceeding 50 m. The total thickness of the potassium-bearing horizon in the section of carnallite attains 90–100 m.

Sylvite-carnallite section is characteristic of internal parts of the Nepa Basin, bounded probably on all sides by carnallite deposits. This type of a section comprises rocks of carnallite and halite-carnallite composition along with sylvite-carnallite rocks and rather thick sylvinite beds. Sections of sylvite-carnallite are subdivided into two subtypes. Occurrences of the first subtype may be inferred for the north-east. Sylvinite beds are confined in this area mainly to the upper potassium-bearing horizon and total 8–16 m. In the second subtype, sylvinite beds occur generally in the lower potassium-bearing horizon. Quite distinct in these sections is a massive sylvinite bed reaching 18.6 m in thickness.

The sylvinite section of the potassium-bearing horizon is observed within the internal Nepa Basin. It is composed exclusively of sylvite-bearing rocks. As a rule, several (2–7) sylvinite beds 3–28 m thick are recorded.

There is a massive sylvinite bed 3.2–13 m thick in the middle potassium-bearing horizon.

Sylvinite rich in KCl has been found in the central Nepa Basin. Potassium chloride content amounts to 30–50%. Sylvinite ores have extremely low $MgCl_2$ content (hundredths of a percent, very rarely 0.5%) and insoluble residue (fractions of a percent, sometimes more than 1%). The potassium-bearing horizon lies at depths of 600–900 m.

Outside the Nepa potassium basin, the thickness of salt-bearing unit S_6 decreases rapidly, and as a rule, varies between 60 and 100 m. At the same time, the thickness of its lower sulfate-carbonate part increases. Salt content of the section rarely attains 20–30%.

Facies equivalents of salt-bearing unit S_6 are represented by the following deposits: sulfate-carbonate in the southern, eastern, and northern Irkutsk Amphitheater; carbonate in the Baikal area, along the periphery of the North Baikal and Patom Highlands

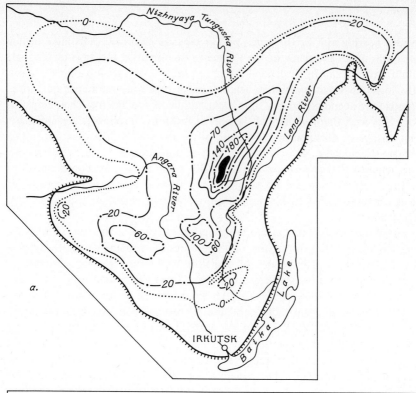
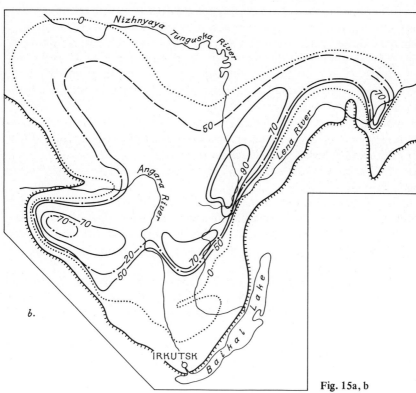

Fig. 15a, b

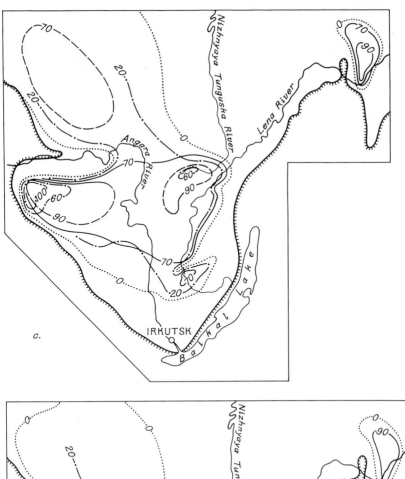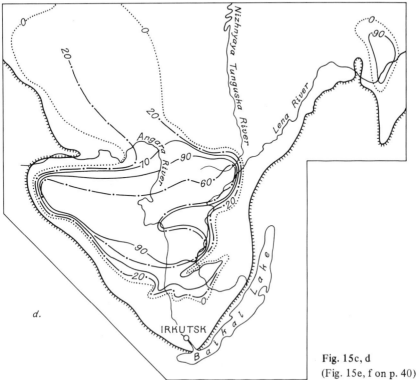

Fig. 15c, d
(Fig. 15e, f on p. 40)

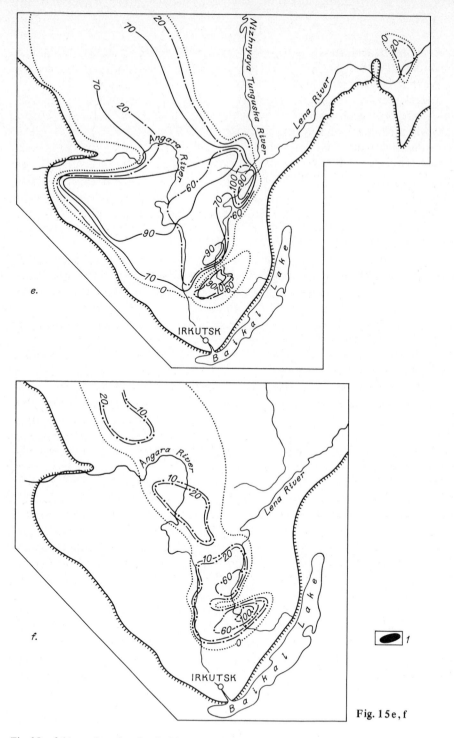

Fig. 15a–f. Maps of total rock salt thickness and salt saturation (%) of units S_6 (a), S_5 (b), S_4 (c), S_3 (d), S_2 (e), and S_1 (f) in Chara and Zeledeevo horizons. After Chechel et al. (1977). *1* area of potash salt distribution. For other symbols see Fig. 12

and the Aldan and Anabar Anteclises; and terrigenous-carbonate and terrigenous deposits in the western and south-western parts of the platform. Their thickness ranges from 80–150 m.

Marker unit R_{IV}, over most of the Siberian Platform, is composed of brownish-gray, sometimes gray, massive dolomite. Some regions are underlain by limestone, anhydrite, and rarely, rock salt. The thickness is 16–22 m.

Salt-bearing unit S_5 is easily discernible in the Kan-Taseeva Depression, Nepa-Botuoba Anteclise, and Berezovaya Depression (Fig. 15b). The section can be divided into three parts: lower – rock salt, anhydrite, anhydrite-dolomite; middle – dolomite; and upper – mainly rock salt alternating with anhydrite. Salt content ranges from 25 to 75–80%, with maximal ranges in the Kan-Taseeva and Berezovaya Depressions, as well as in the central Nepa-Botuoba Anteclise. The thickness of the unit varies between 30 and 100 m.

In the west and south-west, salt deposits are replaced by red, terrigenous rocks about 100 m thick. In the Cis-Sayans and Baikal area, sulfate-carbonate deposits are common, whereas limestone, dolomite, and carbonate breccia 30–50 m thick are found along the periphery of the Patom Highland.

Marker unit R_V is composed of dolomite and limestone with rare anhydrite interbeds and rock salt lenses. The thickness is 20–25 m.

Salt-bearing unit S_4 composes two isolated areas (Fig. 15c). The first area covers the Cis-Sayans-Yenisei Trough and the Upper Lena Depression; and the second the Berezovaya Depression. The unit is composed of alternate dolomite, dolomite-anhydrite, anhydrite, and rock salt. Dolomite-anhydrite and claystone often form thin (3–4 mm) interbeds. Dolomite is extremely rare. Its occurrence increases slightly towards the southern boundary of the Irkutsk Amphitheater. There, the unit also contains limestone interbeds up to 3 m thick. In the Ilga Depression, borate minerals are abundant, while inclusions of native sulfur are scarce. Salt content shows a wide range. In the Tulun Cis-Sayans and on the western Siberian Platform, it reaches 60–75%. It decreases to 10–30% towards the margins of the salt basin. The highest salt content of the unit (80–98%) was recorded in the central Kan-Taseeva and Berezovaya Depressions. There, the unit is composed of gray, medium-crystalline rock salt with thin (not more than 1 m) anhydrite, anhydrite-dolomite, siltstone, and sandstone interbeds. Terrigenous rocks are known only from the Kan-Taseeva Depression. The thickness of the unit is 45 m; however, it reaches 120–160 m in the Kan-Taseeva Depression.

"Salt-free" equivalents of unit S_4 have been registered on the Siberian Platform margins, where they are represented by carbonate, terrigenous-carbonate, and other deposits. Sulfate rocks occur near the flanks of the Berezovaya Depression. They consist of gray and dark gray gypsum, sometimes clayey, with thin dolomite interbeds; anhydrite and dolomite account for 80%. The thickness is 14 m. Sulfate-carbonate deposits are common in the central and south-eastern Siberian Platform, forming a narrow band along the Lena-Kirenga watershed. They are represented by clayey, algal dolomite, anhydrite (not more than 10%), anhydrite-dolomite and dolomite-anhydrite (40%). The thickness is 20–25 m. Sedimentary breccia and carbonate rocks make up the southern and eastern margins of the Irkutsk Amphitheater, the Baikal-Patom Highland, the north-western Aldan Anteclise, and probably, run as a narrow band through the central Siberian Platform between the Aldan Anteclise and the Turukhansk Uplift. In addition to sedimentary breccia, they include dolomite and limestone. The thickness attains 20 m. Terrigenous-carbonate deposits in the south-eastern Cis-Sayans are mainly dolomite with siltstone, marl and limestone interbeds, and sandy dolomite with claystone interbeds in the north-western Cis-Sayans. Their thickness is about 20 m. Reddish-brown marl with anhydrite and gypsum inlcusions and lenses was observed along the periphery of the Yenisei Range, north of the Angara River. Terrigenous rocks, represented by alternate sandstone and siltstone, are widespread at the western margin of the Kan-Taseeva Depression. Their thickness is 70 m.

Marker unit R_{IV} is traced throughout the entire salt basin and no marked changes in its composition have been recorded. It consists of dolomite and limestone with minor anhydrite, marl, siltstone, and sandstone. High proportions of anhydrite and anhydrite-dolomite have been registered in the Cis-Sayans and Baikal area. Red and red-brown marl, siltstone, and claystone are common in the Kan-Taseeva Depression. The unit is 10–20 m thick.

Salt-bearing unit S_3 is known from the Cis-Sayans-Yenisei Syneclise, the south-western Angara-Lena Trough, and in the Berezovaya Depression (Fig. 15d). Its relatively high salt content is in the range of 45–95%. It is only in the marginal parts of the basin that it decreases to 20–30%. The highest (85–95%) salt content was estimated for the Kan-Taseeva and Berezovaya Depressions, in the northern Lena area of the Irkutsk Amphitheater inner field. In all these areas, the unit is composed of two rock salt beds, which are separated by a 2–3 m thick dolomite bed. Claystone, dolomite-anhydrite, and anhydrite interbeds appear at various levels.

The thickness of the unit usually does not exceed 50–70 m, except for the Kan-Taseeva Depression, where it increases to 100 m.

Sulfate-carbonate equivalents (anhydrite-dolomite, anhydrite, and dolomite) occur along the southern periphery of the Irkutsk Amphitheater, in the Kirenga-Lena watershed, and in the central and south-eastern Siberian Platform. Anhydrite and gypsum (93%) are widespread along the western and eastern flanks of the Berezovaya Depression. Gypsum and anhydrite are, as a rule, gray, dark gray, greenish-gray, fine- and medium-grained, often clayey. The thickness of some beds attains 5–7 m. The total thickness of the sulfate-carbonate and sulfate equivalents may reach 20–30 m. In the Baikal area, along the periphery of the Baikal-Patom Highland and north-western Aldan Anteclise, equivalents of salt-bearing unit S_3 are represented by dolomite, limestone, and breccia; their thickness is not more than 10 m. In the south-western Siberian Platform, salt deposits are replaced by red marl and then by sandstone and siltstone with rare anhydrite inclusions up to 30 m thick.

Marker unit R_{III} is dominated by dolomite, limestone, and dolomite-anhydrite with anhydrite interbeds. In the Cis-Sayans and in the Baikal area, the proportion of anhydrite increases. The thickness attains 1–2 m. There, rock salt beds up to 3 m thick also occur. In the south-western Siberian Platform, equivalents of the unit are represented by red siltstone and marl. The unit is about 25 m thick.

Salt-bearing unit S_2 is observed mainly in the western Siberian Platform within the Cis-Sayans-Yenisei Syneclise and in the western Tunguska Syneclise. The Berezovaya Depression is referred to as another isolated area (Fig. 15e). The unit is composed of rock salt with anhydrite, dolomite-anhydrite, and dolomite interbeds. Its salt content amounts to 90–95%. The thickness varies between 50 and 60 and 90 and 100 m. Facies changes display the same trend. Westwards and south-westwards, the unit is replaced by terrigenous red beds; and south-eastwards, eastwards, and north-eastwards by sulfate-carbonate and carbonate deposits.

Namana and Zeledeevo Horizons

Marker units R_{II} and R_I and salt-bearing unit S_1 lie at this stratigraphic level. Marker unit R_{II} is composed of sulfate-carbonate over most of the basin. Four anhydrite beds (their thicknesses in ascending order: 24.7 m, 22.1 m, 36 m, and 8.1 m) and four dolomite beds (their thicknesses: 24.7 m, 35.4 m, 12.1 m, and 12.4 m) can be distinguished in the unit. The total thickness of the unit is 162.2 m. Anhydrite beds taper out towards the margin of the basin. To the east, limestone appears in the lower part of the unit, its proportion gradually increasing. In the south-west and west, equivalents of the unit, red marl, sandstone, and siltstone, are assigned to the Verkholensk Formation.

Salt-bearing unit S_1 is located in the south-western Angara-Lena Trough and in some parts of the Tunguska Syneclise (Fig. 15f). The unit consists of three parts: lower, salt-bearing; middle, sulfate-carbonate with rock salt interbeds; and upper, also salt-bearing. The highest salt content (75–85%) was recorded in sections of the Ilga Depression. It is also there that the unit has the greatest thickness (160 m). To the east and north-east, salt deposits are replaced by sulfate-carbonate rocks. Two anhydrite beds, separated by dolomite, are easily traceable within the Angara-Lena Trough. Farther east and north-east, dolomite and limestone occur with scarce anhydrite interbeds and inclusions. In the Cis-Sayans and the piedmont of the Yenisei Range, facies equivalents of the unit are tentatively assigned to the Verkholensk Formation. They are represented by red marl and siltstone.

Marker unit R_I consists of limestone, dolomite, dolomite-anhydrite, and anhydrite. It crowns the section of the Cambrian evaporitic series in the East Siberian basin. Its thickness varies between 10 and 40 m.

Lower and Middle Cambrian deposits are overlain by the thick, red Verkholensk Formation, especially widespread in the south and south-west of the Siberian Platform. The larger, upper part of this formation belongs to the Mayan stage, Middle Cambrian, and to the Upper Cambrian. Sulfate, sulfate-carbonate, and possibly, salt-bearing strata are assumed to occur at this stratigraphic level in the central inner areas of the platform (Zharkov 1966, Zharkov and Skripin 1971).

The available data on the Cambrian salt deposits of the Siberian Platform allow only approximate estimates of accumulated rock salt (Table 1). Table 1 shows that only the minimal possible total rock salt thicknesses were used. Thus, the average rock salt thickness of the Usolye horizon for the entire area was taken as 200 m; however, it often exceeds 300–400 m for borehole areas. The mean thickness for the Chara horizon was 100 m and where data are available from deep boreholes in the Irkutsk Amphitheater, it reaches 150–300 m. Such low figures were estimated thicknesses, since no deep holes have yet been drilled over a vast salt-accumulation area;

Table 1. Estimates of rock salt volume of the Lower Cambrian and Amga stage of the Middle Cambrian of the East Siberian Basin

Series	Horizon	Unit	Salt accumulation area (km^2)		Average total thickness of rock salt (km)	Volume (km^3)			Volume of rock salt in salt-bearing sequences	
			Determined	Inferred		Established	Assumed	Total	Sequence	Volume (km^3)
Middle	Zeledeevo	S_1	5×10^5	—	0.05	2.5×10^4	—	2.5×10^4	Litvintsevo	2.5×10^4
Lower	Chara	S_2	4×10^5	2×10^5	0.02	8×10^3	4×10^4	1.2×10^4		
		S_3	4×10^5	2×10^5	0.02	8×10^3	4×10^3	1.2×10^4		
		S_4	4.5×10^5	2×10^5	0.02	9×10^3	4×10^3	1.3×10^4	Angara	1.2×10^5
		S_5	8×10^5	6×10^5	0.02	1.6×10^4	1.2×10^4	3.8×10^4		
		S_6	1×10^6	5×10^5	0.03	3×10^4	1.5×10^4	4.5×10^4		
	Uritsk	S_7	2×10^5	6×10^5	0.03	6×10^3	1.8×10^4	2.4×10^4		
	Tolbachan	S_{7a}	1×10^6	7×10^5	0.05	5×10^4	3.5×10^4	8.5×10^4	Belsk	1.3×10^5
		S_8	8×10^5	3×10^5	0.01	8×10^3	3×10^3	1.1×10^4		
	Elgyan	S_9	4×10^5	6×10^5	0.01	4×10^3	6×10^3	1×10^4		
	Usolye	S_{10}	1×10^6	1×10^6	0.02	2×10^4	2×10^4	2×10^4		
		S_{11}	1×10^6	1×10^6	0.05	5×10^4	5×10^4	1×10^5	Usolye	3×10^5
		S_{12}	1×10^6	1×10^6	0.02	2×10^4	2×10^4	4×10^4		
		S_{13}	1×10^6	1×10^6	0.05	5×10^4	5×10^4	1×10^5		
		S_{14}	1×10^6	1×10^6	0.01	1×10^4	1×10^4	2×10^4		
	Irkut		0.5×10^{11}	0.5×10^4	0.02	1×10^2	1×10^2	2×10^2	Irkutsk	2×10^2

thus, mean thicknesses of rock salt at the above mentioned stratigraphic horizons may prove smaller than those in areas where boreholes have been drilled. Consequently, 575.2 km² of the Lower Cambrian and the Amga (Middle Cambrian) rock salt may be considered as a minimal estimate. Besides, rock salt volume could have been as large as 1×10^4 km³ during the Mayan stage (Middle Cambrian) and Upper Cambrian. Altogether, the East Siberian Basin may contain over 585,000 km³ of rock salt.

Iran-Pakistan Basin

The Iran-Pakistan Salt Basin situated on the south-western Asiatic continent can be outlined tentatively using isolated exposures of halogenic deposits in Iran, Oman, and Pakistan (Fig. 16).

The basin is not well-known due to a complex geological structure and deep-lying Late Cambrian and Cambrian beds. They were reported in the following studies: Lees (1927, 1931), Gee (1934), Bailey (1948), O'Brien (1957), Humphrey (1958), Kent (1958, 1970), Morton (1959), Mostafi and Frey (1961), Stöcklin (1961, 1962, 1968a, b, 1974a, b), Asrarullah (1963), Naquib (1963), Stöcklin et al. (1964, 1965), Beydoun (1966), Krishnan (1966, 1968), Greenwood and Bleackley (1967), Hajash (1967), Mina et al. (1967), Sokolov and Movshovich (1968), Ruttner et al. (1968),

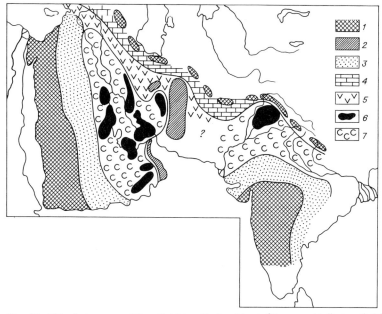

Fig. 16. Lithofacies map of Iran-Pakistan Basin. *1* land, *2* basement inliers probably formed in the Mesozoic. Areas of distribution of: *3* mainly terrigenous red beds, *4* mainly dolomite and limestone, *5* sulfate-carbonate deposits (dolomite, gypsum), *6* and *7* salt deposits (*6* known rock salt, *7* inferred)

Stepanov (1969), Kamen-Kaye (1970), Kolchanov et al. (1971), Voskresensky et al. (1971), Ala (1974), Aliev and Zabanbarg (1974), Auden (1974), Falcon (1974), Glennie et al. (1974), Haynes and McQuillan (1974), Zamel (1975), Kent and Hedberg (1976), Chmyrev et al. (1977), Colman-Sadd (1978), Aliev et al. (1979), Geology and mineral resources of Afghanistan (1980), Konishchev (1980), Murris (1980).

Rather thick salt beds in southern Iran, in the Persian Gulf, in the north-eastern Arabian Peninsula, Oman, central East Iran, Kerman, and Shirhesth are marked by evaporite exposures in more then 200 salt domes. In southern Iran, in the Persian Gulf, and in the north-eastern Arabian Peninsula, they are concentrated in four areas: Shiraz, Hormoz, South Persian Gulf, and Oman-Zufar (Fig. 17). The areas inferred by Stöcklin (1968a) form, in general, a wide south-north zone bounded by the Qatar-Mud Thrust in the west and by Zagros Fault in the east. In central East Iran, the North Kerman salt-bearing area is bounded by the Lut Swell to the east and the Neyband Fault along the western boundary of the rise. Kent (1970) has specified the areal extent of diapir structures. Thus, the salt domes of the Shiraz area were placed into the group of the High Zagros domes. The following groups were established in the Hormoz area: (1) Dashti; (2) Central Fars; and (3) South-Eastern Fars. The latter embraces plugs situated in the Bastak area, southern Iran, and near Hormoz and Qeshm Islands in the Persian Gulf. Kent named the salt structures of the South Persian area, the Abu-Dhaby domes; those in the Oman-Zufar area he called the Oman domes.

Murris (1980) recognized two salt-bearing areas, South Oman-Dhofar and Fahud, among the Oman-Zufar salt domes. The remaining areas in the Persian Gulf and in southern Iran are grouped into two areas: the Southern and Northern Persian Gulfs. Murris regards the above four areas as separate salt basins with boundaries following the areas of salt dome distribution. Salt domes owe their characteristic patterns to peculiar lithology and varying thickness of salt strata.

Stöcklin (1974b) placed all these areas, including North Kerman, into a single sedimentary basin shown in Fig. 17. This major basin had apparently occupied an extensive stable region embracing not only the Arabian Platform, but the Persian Gulf and entire Iran. According to Stöcklin, it was bounded by the Lut Swell and Oman High to the east. However, it is quite possible that the Oman-Lut zone, as an area of the basement emergence, resulted in Mesozoic time from tectonic movements along the Dibba and Masirah transform faults (Murris 1980). In this case, a salt basin could run farther east and join an area of coeval salt deposits in Pakistan. The concept of the existence of two separate salt basins, namely, the Iran-Arabian and Pakistan, separated by the Lut-Oman High, now appears quite feasible.

Salt deposits in South Iran, the Persian Gulf, and Oman are known only from salt domes and anticlines, i.e., they are strongly deformed. A general record of the section remains uncertain.

Evaporite sequences in southern Iran and on islands in the Persian Gulf are grouped into the Hormoz Formation called after the island of the same name. Outcrops of salt deposits composing domes up to 8 km in diameter occur in central Hormoz Island. In addition to rock salt, there are also dolomite blocks, gypsum and anhydrite inclusions, metamorphic and igneous rocks, including diabase, clasts, and rather large blocks of Tertiary, Jurassic, and older rocks. Therefore, deposits consisting mainly of rock salt, gypsum, anhydrite, and blocks of other rocks are called the Hormoz Formation.

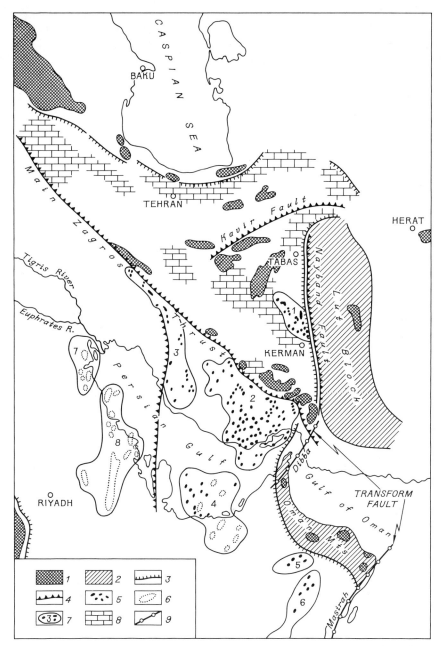

Fig. 17. Lithofacies map of Iran-Arabian zone in Iran-Pakistan Basin. Compiled from the data of Stöcklin (1968a, 1974b) and Murris (1980). *1* areas of no sedimentation (land) or where deposits eroded, *2* Lut and Oman Highs, probably formed in the Mesozoic, *3* basin boundary, *4* major faults and thrusts, *5* salt domes composed of Hormoz Formation, *6* interpretive buried salt anticlines, *7* salt-bearing areas and their numbers (*1* North Kerman, *2* Hormoz, *3* Shiraz, *4* South Persian, *5* Fahud, *6* South Oman, Dhofar, *7* Arabian, *8* Kuwait); *8* outcrops of dolomite, *9* transform faults

One of the best sections of the Hormoz Formation was described by Richardson and Long on the basis of a block in the Hamayran Dome in front of Qeshm Island. The section was reported for the first time by Bokkh in 1929, and later by Kent (1970). In descending order, the section consists of brown shale and marl (60 m) underlain by green micaceous sandstone, intercalated with limestone and marl (180 m), and even lower purple and green, saline marl with gypsum (75 m); rock salt lies at the base.

Lees (1927) described the section of the Hormoz Formation from a block on the Al Buza Salt Plug. The succession in ascending order is as follows: pink and purple sandstone with green claystone interbeds and gypsum bands up to several tens of meters thick at the base; and dark, sandy, dolomitic limestone, and claystone yielding Middle Cambrian trilobites. A clear-cut succession was established from a large block on the Chan Benu Salt Plug, south-eastern Fars (Kent 1970). The beds in the upper part of the section, in descending order, are: (1) dark-brown and yellowish dolomite (130 m); (2) black, odorous dolomite (5–15 m); (3) white, laminated gypsum (10–15 m); and (4) purple, red, and gray claystone (60 m). The succession differs in various parts of the block at lower levels. In one section, red claystones rest on pale-buff quartz sandstone (150 m) with conglomerates of sedimentary pebble in a sandstone matrix (300 m) beneath them. In the other portion, the same red claystone overlies laminated, pink, and yellow claystone (60 m); below this, purple-red clay and marl with hematite inclusions are found.

The understanding of the general formation structure is difficult, since sections of the Hormoz Formation differ greatly both in these and other blocks on various salt domes.

The stratigraphic position of the Hormoz Formation is quite obvious in the Bazun Pir area, North Shiraz, and High Zagros (Stöcklin 1974b). There are no outcrops of salt-bearing deposits, but their presence is inferred due to salt springs. The overlying deposits are represented by the following sequences:

1. Dolomite and dolomitic limestone intercalated with breccia and variegated (dominantly purple) mica schist;
2. Shales, purple micaceous with dolomite interbeds;
3. Sandstone, arkose, purple and pink;
4. Quartz sandstone and pink to white quartzite;
5. Dark, siliceous dolomite; black and blue-black limestone with *Redlichia* and stromatolites in alternate yellow, green, and pink shale with pseudomorphs after rock salt crystals;
6. Dolomite and limestone with shale interbeds. The lower part contains *Iranoleesia, Anomocare, Lioparella*, while *Chuangia, Briscoia, Indahoia, Billingsella* were found in the upper part.

Unit 3, correlative to the Lalun Formation, and Units 5 and 6, yielding the Lower and Middle Cambrian fossils, are recognized in the section. The lower boundary of the Cambrian is drawn at the base of Unit 3 (Lalun Formation). It becomes evident that salt-bearing deposits occur much lower, and hence, their recent assignment to the Infracambrian or Vendian (Stöcklin 1968a, b, 1974a, b, Stepanov 1969, Kent 1970, Murris 1980).

A similar stratigraphic position of salt deposits has been reported from central East Iran, North Kerman area (Stöcklin 1961, 1968a, Ruttner et al. 1968). They were assigned to the Ravar Formation, which is composed of complexly dislocated rocks, represented by rock salt and gypsum with limestone, dolomite, shale, amphibolite, and diabasic blocks and clasts. The rocks crop out in apexes of diapir structures; they are more than 40 in the area between Kerman and Tabas. The Ravar Formation was established within a diapir of the same name north of Kerman. The overlying section contains:

1. Sandstone, light pink and red, 500 m;
2. Sandstone, quartzite-like, 20 m;
3. Shale, purple-red, 150 m;
4. Black limestone with hyolithid and brachiopods, 20 m;
5. Shale, purple-red, 15 m;
6. Sandy limestone, 5 m;
7. Shale, purple-red, 120 m;
8. Limestone, dolomitic, reddish-brown with interbeds of red and green shale, 100 m;
9. Quartzite-like sandstone intercalated with shale, often sandy and dolomitic, and gypsinate marl, 170 m.

Bed 1 of the section is correlatable with the Lalun Formation, while overlying deposits are assigned to Lower-Middle Cambrian. According to Stöcklin (1968a), the occurrence of the salt-bearing Ravar Formation (an equivalent of the Hormoz Formation) beneath the deposits mentioned above, implies their Precambrian age.

In the Oman area, salt-bearing deposits composing plugs of diapir structures and containing rock salt, gypsum, anhydrite, terrigenous and carbonate rocks belong to the Huqf Group. The underlying Precambrian carbonate and terrigenous formations belong also to this group (Tschopp 1967, Glennie et al. 1974, Murris 1980).

As mentioned above, the Salt Range in Pakistan is another area with exposures of a salt sequence in the Iran-Pakistan Basin (Gee 1934, Bailey 1948, Schindewolf 1954, Schindewolf and Seilacher 1955, Asrarullah 1963, Krishnan 1966, 1968, Gansser 1967, Sokolov and Movshovich 1968, Voskresensky et al. 1971, Auden 1974, Stoneley 1974). Evaporite deposits are grouped into the Penjab Series and belong to the oldest deposits exposed in the Salt Range. The following description of the Cambrian formations of the Salt Range was given by Gansser (1967).

Gypsum and anhydrite lie at the base with intercalated marl and salt rock beds above them. The salt sequence is overlain by a thin unit of gypsum and anhydrite in alternation with dolomite. The same unit contains traps up to 1 m thick, as well as black partings of paper bituminous shale called "dysodile". Such shales are typical of evaporite cap rocks of many oil fields in Iran. Gypsum is directly overlain by a section of purple sandstone starting with thin-laminated clayey, sandstone, and red-buff shale with thick interbeds of purple, fine-grained, quartz sandstone. Cross-bedding is very common; ripple marks, mud cracks, and pseudomorphs after rock salt crystals are visible on bedding planes. The thickness of sandstones is 150 m. Higher, there are gray-green Neobolus shales with a thin, quartz, conglomerate bed at the base. Above them is a zone of shale consisting of thin interbeds of dolomite, oolitic dolomite, and gray, pyritized shale. The Lower Cambrian trilobites *Redlichia* have been found in the Neobolus shale (Schindewolf and Seilacher 1955). The shale is overlain by a dolomite sequence 80 m thick, where shale interbeds similar to the Neobolus shale occur; they also yield fossils. The section is crowned by red-purple, thin-laminated, clayey sandstone and shale with abundant crystal molds after rock salt and thin dolomite interbeds below.

Krishnan (1966) reported that the Cambrian begins in a similar section of the Salt Range with a unit of purple sandstone overlain by the Neobolus shale with Middle Cambrian trilobites and brachiopods in the upper part. Above is a magnesian "sandstone" (apparently dolomite) followed by a unit of shale with crystal molds after rock salt. He correlates the purple sandstone with the upper Vindhyan sequence from Agra and environs of Delhi, Hindustan Peninsula; hence, the Precambrian age of an underlying salt series. It is subdivided into three sequences: the lower one, gypsum-dolomite, composed of massive gypsum and red, gypsinate varve up to 300 m thick; the middle one, salt marl, buff, red, and purple, with rock salt bands up to 240 m thick; and the upper one, gypsum-dolomite, massive, white and gray gypsum with bituminous shale interbeds and minor diabase. The thickness of the upper sequence does not exceed 30 m.

It is obvious that the Cambrian section of the Salt Range is correlatable with those of the Iran-Pakistan Basin farther west, in particular, with the North Kerman, Iran. The Penjab Series is compared with the Hormoz Formation. There is also a good correlation between the unit of purple sandstone and the Lalum Formation; the lower Neobolus beds and the carbonate sequence yielding trilobites *Redlichia*. This implies that the stratigraphic position of the Hormoz Formation and that of the Penjab Series are similar.

A similar section of salt deposits was penetrated by the Karampur well drilled on the north-western slope of the Hindustan Platform, 90 km east-south-east of Multan (Sokolov and Movshovich 1968, Voskresensky et al. 1971, Meyerhoff and Meyerhoff 1972, Auden 1974). An evaporite sequence, 880 m thick, composed of rock salt with dolomite and anhydrite bands and clay interbeds rests on metamorphic basement rocks. Voskresensky et al. (1971) reported purple sandstone (160 m), dolomite sandstone (220 m), and clay with pseudomorphs after salt (150 m) above the section.

The Karampur well shows that salt deposits extend far southward from the Salt Range and may underlie the Cis-Sulaiman Foredeep and the north-western slope of the Hindustan Peninsula. Salt deposits southwards and south-eastwards are replaced by red Vindhyan deposits with thick gypsum beds, which have been registered in the Nagaur area 320 km south-east of Karampur (Auden 1974). Gansser (1964) suggested a widespread distribution of halogenic deposits that might extend both into the Himalayas Foreland zone and under the Himalayas proper, since the Vindhyan deposits, slightly resembling purple sandstone of the Salt Range and known from the northern Hindustan Peninsula, could grade northwards into deposits of the evaporite facies. This evaporite belt served as a horizon which promoted thrusting of the Lower Himalayas onto the northern extension of the Hindustan Shield.

The Penjab Series proper shows the alternation of gypsum, anhydrite, dolomite, red marl, and rock salt. Salt rocks are intercalated with peculiar rock salt interbeds with inclusions of red halite known as khallar and marl interbeds with inclusions of potash salt.

Salt deposits from three areas were studied in detail, Khewra, Warcha, and Kalabagh, where rock salt is mined.

In the Khewra area, a section of the salt sequence is as follows (Asrarullah 1963):

1. Upper gypsinate dolomite   0–15.2 m
2. Upper salt-bearing marl   122.0 m
    a) bright red marl with thin salt interbeds   76.2 m
    b) dull red marl with a gypsum unit (4.6 m) at the top   45.8 m
3. Middle gypsum unit   45.7 m
4. Lower or main sequence of salt marl   610.0 m
5. Lower gypsinate dolomite   above 100.0 m

The base is not exposed.

The lower or main sequence of salt marl is subdivided into two lithologic units: the Baggy and Pharwala complexes of seams, which are composed of closely alternating interbeds of marl, rock salt, and khallar. Lithologically, each complex can be divided into members.

Thus, the North Baggy Member 7.6–15.2 m thick with a rock salt seam having inclusions of red, earthly halite (khallar) 3 m thick in its upper part is at the top of the Baggy complex. Below is the Baggy Member proper (10.5–45.5 m), which is separated by a 4.5 m khallar seam into two parts. Beneath the Baggy Member proper lies the Sujjowal Member (4.6–21.3 m), which is composed of red and white rock salt. The latter is mined mainly for salt in the Khewra area. The underlying deposits of the Baggy Complex are represented by thin interbeds of rock salt, marl, and khallar totaling 30 m in thickness.

The Pharwala Complex is subdivided into the Upper, Middle, and South Pharwala and the so-called bed of a new lower horizontal adit.

A thick section of the Penjab Series is exposed in the Warcha area. Similar to the Khewra area, it is composed of alternate seams of rather pure rock salt, marl, and khallar. The thickness of the rock salt seams often reaches 15 m and those of khallar and marl 5–6 m. The section of salt sequence in the Kalabagh area is the same, though rock salt and marl beds are thinner and more deformed.

Besides the Hindustan Peninsula areas, equivalents of evaporite deposits in marginal near-shore zones of the Iran-Pakistan Basin have also been found in the Arabian Peninsula. In southern Oman deposits known as the Huqf Group are subdivided into four formations (Morton 1959, Beydoun 1966), namely, two clastic and two dolomite. The First Clastic Formation consists of red siltstone and sandstone with gypsum interbeds, inclusions, and lenses in the lower and upper parts; the thickness is 315 m. The First Dolomite Formation is composed of gray, sandy, and siliceous dolomite with a gypsum inclusion in the upper horizons; the thickness is 299 m. The Second Clastic Formation consists of lower and upper parts, respectively with siltstone and marl, separated by a unit of oolitic and clayey limestone; the thickness is 299 m. The Second Dolomite Formation is composed of clayey and sandy dolomite with breccia and oolitic limestone interbeds; the thickness is 224 m.

The equivalents mentioned belong to the Ghabar Group in south-western Hadramaut. Here, the proportion of sand increases, volcanic rocks (tuffs, tuffaceous conglomerates) appear, and carbonate units are thinning. Thus, the equivalents of the

First Dolomite Formation assigned to the Shabb Formation are 13–42 m thick. The thickness of the Second Dolomite Formation (Harut Formation) reaches 31–48 m (Beydoun 1966). Clastic equivalents known as the Minhamir and Khabla Formations are rich in tuff. The Khabla Formation also contains dolomite, silty sandstone, siltstone, and gypsum interbeds. The thickness of the two formations varies from 95–143 to 142–201 m.

In Aden, the Ghabar Group has conglomerates at the base, followed by thin-laminated limestone with gypsum lenses, and cherts above. Even higher, there are silty sandstone and siltstone intercalating with dolomite and rare gypsum lenses. Sandy dolomite, quartzite, and thin-laminated shale compose the upper part of the section (Greenwood and Bleackley 1967). In Saudi Arabia, the Saq Formation, which is composed of red sandstone with gypsum inclusions and lenses, may be considered as an equivalent of the salt-bearing deposits.

Carbonate rocks – dolomite and limestone – become predominant in the northern Iran-Pakistan Basin. One of the most complete sections is known from the Soltanieh and Elborz Mountains (Stöcklin et al. 1965, Stöcklin 1974a, b). The following formations are recognized, in ascending order:

Bayandor Formation — Pink sandstone and sandy mica schist with siliceous dolomite interbeds. The thickness ranges up to 500 m.

Soltanieh Formation — The lower part consists of siliceous dolomite with several units of calcareous shale, shale, and chert. The upper part is composed of massive dolomite with cherts and stromatolites. The thickness is 100–1200 m.

Barut Formation — Dolomite and dark, odorous limestone with cherts and stromatolites intercalating with purple and variegated mica schist. The thickness is 400–700 m.

Zaigun Formation — Silty and sandy shale, mica schist with thin dolomite interbeds. The thickness is 300 m.

Lalun Formation — Purple and pink arkoses, fine- to medium-grained, platy in the lower part and cross-bedded and massive in the upper part; a shale unit at the top. The thickness is 400–600 m.

Mila Formation — The thickness is 450–500 m. It is subdivided into:
1. white, coarsely cross-bedded quartzite (50 m);
2. dark silicified dolomite with interbeds of yellow, green and pink marl and shale with pseudomorphs after rock salt (200 m);
3. platy, white, and gray dolomite and limestone with sandstone, marl, and shale interbeds mainly in the upper part. Fauna: *Iranoleesia, Anomacarella, Lioparella, Dorypyge, Prochuangia, Briscoia, Idohoia, Kaolishania, Billingsella, Drepanura, Chuangia, Archeoorthis, Saukia, Quadraticephalus* (200–250 m).

The salt-bearing deposits of the Hormoz and Ravar Formations are supposed to occur at the level of the Soltanieh Dolomite or lower. The Bayandor, Soltanieh, Barut, and Zaigun Formations are considered by most authors as Infracambrian or

Vendian. The lower Cambrian boundary is drawn at the base of the Lalun Formation, which is also placed at the level of the Lower Cambrian. A Middle-Late Cambrian age of the Mila Formation is based on the fossil assemblage. The above section of the Soltanieh and Elborz Mountains can be correlated with the earlier discussed section of the Bazun-Pir area of the High Zagros where trilobites *Redlichia* have been found.

Carbonate equivalents of evaporite deposits are widespread in northern and central Iran (see Fig. 17). They apparently extend eastwards into Afghanistan where coeval beds were reported from the Helmand-Archandab High (Geology and mineral resources of Afghanistan 1980). In the Kohe-Kaftarhan Range, the equivalents of the Lalun Formation are easily recognizable. They consist of red sandstone and siltstone with gravelstone and small-pebbled conglomerates and dolomite interbeds up to 1500 m thick. This red series is underlain by a sequence of dolomite and limestone with cherts and volcanics; its thickness is 900–1000 m. It is overlain by a dominantly carbonate sequence with a quartz sandstone unit at the base and is approximately 500 m thick. The following fossils have been found there: *Agnostus, Crepicephalus, Schoriella, Billingsella,* and others, thus suggesting the Middle and Late Cambrian age of the rocks.

Limestone and dolomite with units of variegated sandstone and siltstone 400–900 m thick occur in the Helmand-Archandab High.

The peculiar features of areal extent and facies change of deposits within the Iran-Pakistan Basin may be summarized as follows (see Fig. 16). Two major salt zones can be recognized, namely, the Iran-Arabian to the west and Pakistani to the east. South of the salt zones, in the Arabian and Hindustan Peninsulas, salt-bearing deposits are replaced by clastic formations, mainly redstones marking near-shore areas of an evaporite basin near large land masses. North of the salt zones, salt-bearing deposits grade into sulfate-carbonate and carbonate formations. The belt of carbonate sedimentation runs across northern Iran into Afghanistan and outlines an extensive carbonate plateau which separates an evaporite basin from the open sea of normal salinity, located apparently even farther north.

A group of islands seems to be situated between the open sea and the belt of carbonate accumulation.

Facies and structure of sections in the Iran-Arabian and Pakistani parts of the basins are much the same, despite the present separation by the Lut-Oman Block. The southern near-shore zones and salt and carbonate areas also have a common or similar stratigraphic succession. These similarities imply the formation of a single major sedimentary basin once situated within the Arabian-Iran-Hindustan Platform, which was part of Gondwanaland during Late Precambrian and Cambrian, in the area discussed.

The Hormoz, Ravar, and Penjab evaporites are dated mainly as Late Precambrian in age. This conclusion is further supported by the much lower occurrence of salt deposits than that of carbonate beds with *Redlichia*. Furthermore, the horizon yielding *Redlichia* within the East Siberian salt basin is located in the middle Lower Cambrian with the thickest Cambrian salt-bearing deposits beneath it. These deposits are the Belsk, Usolye, and Irkutsk sequences. The redstone of the Lalun Formation and the underlying salt deposits of the Hormoz Formation and its equivalents may lie at the same level. Hence, the assignment of salt deposits to the Cambrian in the Iran-Pakistan Basin is further substantiated.

The areal extent of the Iran-Pakistan Basin was probably the same as that of the East Siberian Basin. The volumes of rock salt accumulated in both basins were apparently also comparable.

Mackenzie Basin

A large Cambrian salt basin which has not yet been outlined is located in the Mackenzie River Basin, Northwest Territories, Canada (Fig. 18). It was described by the following authors: Stewart (1945), Heywood (1955), Lefond (1969), Tassonyi (1969), Douglas et al. (1970), Aitken and Cook (1974), Meijer-Drees (1975), Norford and Maqueen (1975), and Aitken et al. (1973).

A rock salt sequence of Cambrian age was penetrated by a number of deep boreholes in the Norman Wells area. One of the wells, the Imperial Vermilion Ridge No. 1, 40 km east of the Norman Wells, penetrated a salt sequence at a depth of 1027–1701 m (674 m); and another drilled 16 km east of Norman Wells reached

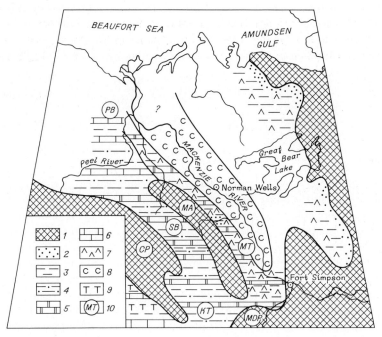

Fig. 18. Lithofacies map of Salina River Formation and its equivalents in Mackenzie Basin. Compiled from the data of Douglas et al. (1970), Aitken et al. (1973), Aitken and Cook (1974). *1* areas where deposits are absent (probable land or areas of post-sedimentary erosion), *2* mainly sandstone, *3* mudstone, *4* shale, *5* dolomite, *6* limestone, *7* gypsum and anhydrite, *8* rock salt, *9* volcanic rock, *10* indexes referring to tectonic elements (*MT* Mackenzie Trough, *MA* Mackenzie Arch, *SB* Selwyn Trough, *MDP* MacDonald Platform, *PB* Peel Basin, *CP* Cassiar Platform, *KT* Kechika Trough)

a depth of 1455–1807 m (352 m). Evaporite deposits were found in the Salina River Formation. In the Norman Wells area, the latter is subdivided into two lithologic units, namely, the lower salt sequence where rock salt dominates and the upper, containing claystone with rare evaporite deposits.

The salt sequence consists mainly of white halite with rare patches of light orange to red rock salt. The upper part of the section contains horizons with a regular alternation of clayey and saline laminated claystone. At lower levels, rock salt is more contaminated by clay material. In general, the salt sequence contains the following rocks: anhydrite, pale buff-gray gypsinate mudstone, light brown to cream anhydrite dolomite, brownish-gray, intensely dolomitized, clayey siltstone, and red and green shale. Dark green-gray and green, evenly bedded shale dominate the lower 60 m of the section. The contact with the underlying Mount Cap Formation, which was penetrated by the Imperial Vermilion Ridge No. 1 well, is easily discernible, despite the absence of a drastic change in rock content.

A mudstone sequence composing the upper Salina River Formation consists mainly of red, yellow, and green partly dolomitic mudstone with gray gypsinate and evenly laminated mudstone interbeds. There is an alternation of the rocks with light anhydrite, locally with cream and dark brown partial anhydrite or gypsinate dolomite in places with heavy contamination from clayey or silty material, especially in the upper part.

In the boreholes, the contact between the salt-bearing Salina River Formation and overlying deposits is rather sharp and apparently unconformable. In the Imperial Vermilion Ridge No. 1 well, the section is as follows (Steward 1945, Tassonyi 1969):

853.2–1027.0 m	(173.8 m) – mudstone sequence. Green and cream shale intercalated with dolomite and gypsum. The number of interbeds increases downwards.
1027.0–1341.7 m	(314.7 m) – rock salt with rare anhydrite, mudstone, and dolomite partings.
1341.7–1362.4 m	(20.7 m) – siltstone with rock salt and gypsum interbeds.
1362.4–1419.2 m	(56.8 m) – rock salt with small dolomite and siltstone interbeds.
1419.2–1592.6 m	(173.4 m) – siltstone intercalated with gypsum and rarely with rock salt.
1592.6–1607.9 m	(15.3 m) – rock salt.
1607.9–1685.5 m	(77.6 m) – alternation of siltstone, shale, and rock salt.
1685.5–1701.0 m	(15.5 m) – rock salt.

Salt-bearing deposits are related to the Mackenzie Trough running south-north along the Mackenzie River. Their maximal thickness is 850 m. Rock salt thins out towards the trough margins and the Salina River Formation becomes rich in clastics. South of the Norman Wells area in the Keele River Basin, the formation is composed of red and green gypsiferous shale, marl, and siltstone with pink, white, and gray gypsum interbeds and lenses, as well as thin dolomite and breccia partings. Gypsiferous quartz sandstone appears farther east in the vicinity of the Mackenzie Arch (Aitken and Cook 1974). The thickness of the formation in the Keele Basin varies from 100–200 m. The Salina River Formation rests unconformably on the older erosion surface, e.g., Lower Cambrian. Proterozoic deposits fill topographic and structural depressions.

The maximal thickness of the formation reaches 162 m in the Franklin Mountains, Norman Range. It consists of red mudstone, red and green shale, dolomite, and gypsum. Red and green gypsiferous shale and mudstone and pale, buff-pink and gray gypsum dominate the area along the Mackenzie Mountain front (Aitken et al. 1973).

The Salina River Formation tapers out in the apical part of the Mackenzie Arch. The overlying Franklin Mountain Formation rests directly on Proterozoic deposits.

Equivalents of the Salina River Formation 30—60 m thick, which are composed of red and green shale and marl with pseudomorphs after rock salt crystals on bedding planes and few dolomite and gypsum interbeds, occur east of the Mackenzie Trough on the Interior Platform near Great Bear and Great Slave Lakes.

The stratigraphic position of the Salina River Formation is based on its resting unconformably on the Lower-Middle Cambrian clastic-carbonate deposits of the Mount Cap Formation. The latter is in turn unconformably overlain by the basal red beds of the Mount Cap Formation, which are believed to be Upper Cambrian-Lower Ordovician in age. Therefore, the Salina River evaporites are believed to be Late Cambrian or Late-Middle Cambrian in age.

The Cambrian evaporite basin was possibly fairly large and could have been 800—1000 km long and more than 500 km wide. A salt-bearing zone might have occupied a considerable area of the Mackenzie Trough and extended even into the Beaufort Sea shelf.

Amadeus Basin

A salt sequence occurring among the Lower Cambrian deposits has been recognized in addition to the above discussed deposits of the Bitter Springs Formation of the Amadeus Basin. It was penetrated by three wells: Mount Charlotte No. 1, the Alice No. 1, drilled in the north-eastern part of the basin, and the Orange No. 1, drilled south-east of the previous well (see Fig. 2).

The Cambrian formations of a composite section of the eastern and north-eastern Amadeus Basin, where salt-bearing deposits have been registered, are assigned to the Pertaoorrta Group and lower Larapinta Group. The former is subdivided into the following sequences: (1) Arumbura Sandstone; (2) Todd River Dolomite; (3) Chandler Limestone; (4) Giles Creek Dolomite; (5) Shannon Formation; (6) Hugh River Shale; (7) Jay Creek Limestone; and (8) Goyder Formation. Pacoota Sandstone is recognized in the lower Larapinta Group (Wells et al. 1967, 1970, Brown et al. 1968). Relationships between formations and sequences within the Pertaoorrta Group are complex and uncertain. The Cambrian section changes greatly across the area; it is mostly carbonate to the east, while to the west shale dominates.

Salt-bearing deposits are related to the Chandler Limestone and belong to the upper Lower Cambrian. The Chandler Limestone was registered only in some areas of the eastern Amadeus Basin; in particular, the western Rodinga Sheet area, which is on the western flank of the Ooraminna Anticline. These areas were penetrated by a number of deep wells. In the western Rodinga Sheet area, the Chandler Limestone rests directly on the Arumbura Sandstone or on Precambrian deposits, where they are overlain by the Jay Creek Limestone. On the Ooraminna Anticline flanks and in other synclinal zones, the Chandler Limestone is underlain by the Todd River Dolomite and overlain by the Giles Creek Dolomite. The thickness of the Chandler Limestone varies from 60–140 m, and reaches 165–225 m where evaporite sediments are present. It crops out as dark gray, laminated limestone and dolomite with lens-like interbeds and inclusions of gray cherts. These rocks are usually strongly folded and brecciated. Anhydrite and rock salt with red-brown shale interbeds, which are not exposed, are common in submerged areas within the Chandler Limestone.

Wells and co-workers (1967) gave a general description of salt-bearing deposits of the region.

The Cambrian deposits of the Pertaoorrta Group were penetrated by the Alice No. 1 well within the interval of 916.2–2283.0 m beneath the Pacoota Sandstone, whose base lies at a depth of 9 16.2 m. According to Wells and co-workers (1967), the well section of the group consists of three sequences, in descending order:

916.2–1174.3 m (258.1 m) – Goyder Formation, sandstone and siltstone.
1174.3–2209.7 m (1035.4 m) – Jay Creek Limestone, limestone, dolomite, and shale with evaporite interbeds at the base, namely, rock salt, anhydrite intercalated with claystone. Their thickness is 158.6 m.
2209.7–2283.0 m (73.3 m) – Arumbura Sandstone. The complete section is not exposed.

However, others propose a different subdivision for the Pertaoorrta Group in the Alice No. 1 well. The following formations were recognized within a depth interval of 916.2–2283.0 m, in descending order:

1. Goyder Formation – 259.3 m
2. Shannon Formation – 497.2 m
3. Giles Creek Dolomite – 356.9 m
4. Chandler Evaporite – 158.6 m
5. Todd River Dolomite – 21.4 m
6. Arumbura Sandstone – 73.4 m

The Chandler Evaporite is composed entirely of rock salt with rare anhydrite and shale interbeds.

A Cambrian section in the Mount Charlotte No. 1 well differs in thickness and in rock composition from that of the Alice No. 1 well. The base of the Pacoota Sandstone is marked at a depth of 471.2 m. Another section beneath it was reported by Wells and co-workers (1967):

471.2– 710.7 m (239.5 m) – Jay Creek Limestone.
710.7– 937.0 m (226.3 m) – Chandler Limestone, predominantly rock salt with anhydrite and claystone interbeds.
937.0–1425.3 m (488.3 m) – Pertatataka Formation.
1425.3–1606.1 m (180.8 m) – Areyonga Formation.
1606.1–2117.6 m (511.5 m) – Bitter Springs Formation with a rock salt unit at the base.

The comparison of the two boreholes suggests that the Cambrian section in the Mount Charlotte No. 1 well is considerably condensed, despite the great thickness of a salt sequence assigned to the Chandler Limestone. So, the Goyder Formation, Todd River Dolomite, and Arumbura Sandstone are absent from the section, while the Jay Creek Limestone, an equivalent of the Shannon Formation and Jay Creek Dolomite, is much thinner there. In general, the Cambrian salt deposits of the Amadeus Basin are rather widespread. They could be found both in the eastern and northern parts of the basin (diapirs of the Domol District). The area of the Cambrian evaporites reaches 41,000 km^2 and their maximal thickness is 225 m (Wells 1980). If an average thickness is taken of 100 m for the entire area in question, then the volume of the Cambrian salt in the Amadeus Basin will equal 4.1×10^3 km^3 [1].

Arckaringa Basin

Salt-bearing deposits of probable Cambrian age are recognized in the southern Arckaringa Basin located in western South Australia (Wells 1980). They were penetrated by the Wilkinson No. 1 well in the Tallaringa Trough. A rock salt sequence was drilled at depths of 573–690 m. It has been tentatively assigned to the Cambrian Observatory Hill Beds, which can be considered as equivalents of the Early Cambrian Chandler Limestone of the Amadeus Basin.

The extent and mode of occurrence of salt deposits of probable Cambrian age in the Arckaringa Basin have not as yet been studied.

Cambrian Evaporite Deposits in Other Basins

In addition to the salt basins of Cambrian age, evaporite deposits are known from Cambrian sedimentary sequences in other regions of the Earth.

Gypsum interbeds were recognized among Lower Cambrian deposits in the western Anti-Atlas, Ida or Gnidif, north-western Tafraout (Destombes 1952).

A thick sequence of gypsum and anhydrite was reported from Cambrian deposits in the Andes foothills, South America (Benavides 1968), where occurrence of rock salt seems quite probable. This sequence is part of the Limbo Formation exposed along the Cochabamba-Villa Tunari Road, which marks the eastern margin of the Bolivian Andes. A lower basal sequence is composed of light-colored anhydrite 150 m thick continuing up the section into clayey anhydrite and claystone. The sequence may have a thickness of 200–300 m. Above is an argillaceous sequence consisting of chert and mica schist with calcareous mudstone and dolomite interbeds with a thickness of 100–300 m. An upper sequence 900 m thick, is composed of mudstone, sandstone, and conglomerates. Cambrian age of the Limbo Formation (and hence, of anhydrite sequence) is based on the occurrence below deposits yielding *Lingula* and *Orthis,* which are assigned to the Ordovician.

---

[1] This figure is much higher than our earlier estimate, which was 1×10^2 km^3 (Zharkov 1981)

Many localities of Cambrian gypsum and anhydrite are known in Australia. They were reported from the Daly River, Georgina, Wiso, and Officer Basins (Wells 1980). In the Georgina Basin, gypsum interbeds occur among the following formations: Arrinthrunga, Ninmaroo of Late Cambrian age; in the Mail Change Limestone; and probably in the Selwyn Range Limestone and the Mungerebar Limestone of Middle Cambrian age. Small gypsum interbeds were penetrated by the Frewena No. 1 borehole at depths of 110–149 m in red and purple mudstone, gray limestone, and dolomite of the Middle Cambrian Wonarah Beds. Anhydrite lenses and interbeds are also exposed in the Alcoota area in the Central Mount Stuart Beds. In the Wiso Basin, gypsum interbeds were observed in Middle Cambrian deposits of the Merrina Beds containing siltstone and dolomite. In the Officer Basin, pseudomorphs after evaporite minerals are common among clayey and calcareous dolomite in units II and III, Observatory Hill Beds.

Rodgers (1970) suggested a distribution of Cambrian salt deposits in the Central and South Appalachians, USA. This was based on the occurrence of peculiar breccia along the Pulaski Fault in the Range and Valley Province in Virginia. According to many authors, this breccia is similar to that of the Haselgebirge, Austrian Alps, which resulted from intrusion and leaching of salt deposits. Furthermore, Rodgers infers the presence of Cambrian rock salt on the basis of paleogeographical reconstructions and occurrence of halogenic rocks among Cambrian sedimentary strata in the Appalachians adjacent to the Palaski area.

Cambrian sulfate rocks were also observed in the Michigan Basin, North America. To the north-west of the central part, the Brazos State Foster No. 1 well penetrated an anhydrite sequence which is composed of alternate anhydrite, dolomite, sandstone, and claystone, 425 m thick. It was placed into the Munising Group and assigned to the Upper Cambrian by Vary and co-workers (1968) and Catacosinas (1973). The anhydrite sequence apparently occupies a small area of the northern Michigan Basin.

Cambrian salt deposits were also registered in the Tarim Basin, China (Khu Bin et al. 1965, Meyerhoff and Meyerhoff 1972). The data of Khu Bin and co-workers imply that rock salt and anhydrite lie in the Lower Cambrian deposits beneath the beds with *Redlichia;* they occupy the same stratigraphic position as the lower salt strata of the East Siberian and Iran-Pakistan Basins.

Chapter III
Ordovician Salt Deposits

The available information allows the recognition of a single Ordovician salt basin — the Canadian Arctic Archipelago Basin. The Canning Basin in Australia may also be Ordovician in age, but the stratigraphic position of salt deposits occurring therein is still unknown. They are assigned to Late Ordovician/Early Devonian, but may be older. Recently, Wells (1980) assigned these salt deposits to the Early Devonian and Late Silurian. This opinion seems to be more substantiated, although previously the Canning Basin was tentatively placed into the Ordovician (Zharkov 1974a, 1981). This is discussed in Chapter IV, which deals with Silurian salt deposits.

Sulfate deposits of Ordovician age are known from the Williston, Lena-Yenisei, Severnaya Zemlya, Baltic, Moose River, South Illinois, Anadarko, Canning, and Georgina Basins (Fig. 19).

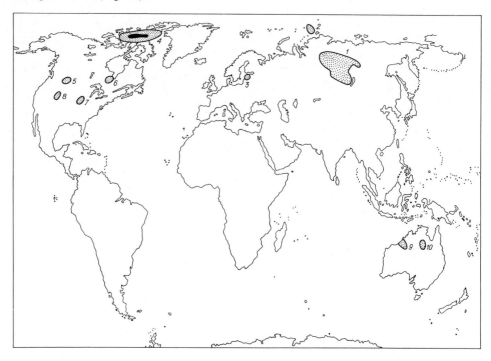

Fig. 19. Distribution of Ordovician evaporite deposits. Evaporite basins: *1* Lena-Yenisei, *2* Severnaya Zemlya, *3* Baltic, *4* Canadian Arctic Archipelago, *5* Williston, *6* Moose River, *7* South Illinois, *8* Anadarko, *9* Canning, *10* Georgina. For legend see Fig. 3

Canadian Arctic Archipelago Basin

Within the Canadian Arctic Archipelago Basin, Thorsteinsson and Tozer (1970) recognized the following geological provinces: Canadian Shield, Arctic Platform, Franklinian Eugeosyncline, Franklinian Miogeosyncline, Boothia Uplift, Sverdrup Basin, and Arctic Coastal Plain.

Ordovician evaporite deposits are known from the Arctic Platform and the Franklinian Miogeosyncline. Only the southern and south-western boundaries of the evaporite basin are known along outcrops of Precambrian formations. The north-western boundary of the basin is fairly precise. It extends between the miogeosynclinal and eugeosynclinal terranes of the Paleozoic in the northern part of Ellesmere Island. As for the northern and north-eastern boundaries, they are not yet defined and are provisionally drawn along the southern boundary of the Sverdrup Basin. However, Lower Paleozoic and Ordovician deposits are buried under sedimentary strata underlying the Sverdrup Basin. The evaporite basin may extend farther north. The eastern boundary of the basin is unknown, but appears to continue into the Arctic Coastal Plain (Fig. 20).

Ordovician deposits of the Canadian Arctic Archipelago are better known from Ellesmere, Devon, Bathurst, Cornwallis, and Somerset Islands. Furthermore, they are exposed in four deep boreholes, one of which (Bathurst Caledonian River J-34) penetrates the saliniferous bed. Data from non-Russian literature were used to characterize the Ordovician deposits (Fortier et al. 1963, Greiner 1963, Kerr 1967a, 1968, 1974, 1975, Trettin 1969, 1971, Mossop 1972a, b, 1973, 1979, Kerr et al. 1973, Miall 1974a, b, 1975, Mayr 1975, 1976, 1978, 1980, Reinson et al. 1976, Miall and Kerr 1977, 1980).

At the base of the Ordovician on Ellesmere and Cornwallis Islands, the Copes Bay Formation, whose lower part may be placed into the Upper Cambrian, was identified. It consists chiefly of silty and argillaceous, locally gypsinate, limestone; conglomerates can be observed in the lower formation. The thickness varies within a range of 100–1200 m. The formation is overlain by a bed of evaporite deposits identified as the Baumann Fiord Formation (Kerr 1967a). The formation mainly crops out on Ellesmere Island along the coasts of Trold Fiord, Flat Pebble, and Stearfish Bays, as well as in the central part of the island and in its western regions on the Bache Peninsula. To the west, the Baumann Fiord evaporites subside under younger deposits and are exposed only in the central part of Cornwallis Island in the Panarctic Deminex Cornwallis Central Dome K-40 borehole. The Baumann Fiord Formation has been discussed by Kerr (1967a) and Mossop (1972a, b, 1973, 1979).

The formation was divided into three members: A, B, and C. Lower member A consists mainly of beds of coarsely crystalline anhydrite interlayered with nodular anhydrite and thin-bedded anhydrite-carbonate rocks. Lenses of micrograined limestone 0.6–3 m thick occur locally. Silicification associated with anhydrite is rare. Member B contains green to gray, thin- to thick-platy micrograined limestone interlayered with isolated red quartz sandstone and red argillaceous limestone. Anhydrite interbedded with anhydritized limestone dominates member C section; these rocks are thin-bedded. The thickness of the Baumann Fiord Formation varies from 200–500 m.

In northern Ellesmere Island, the Baumann Fiord Formation continues northward to the margin of the Franklinian Miogeosyncline, wedges out, and is replaced by the

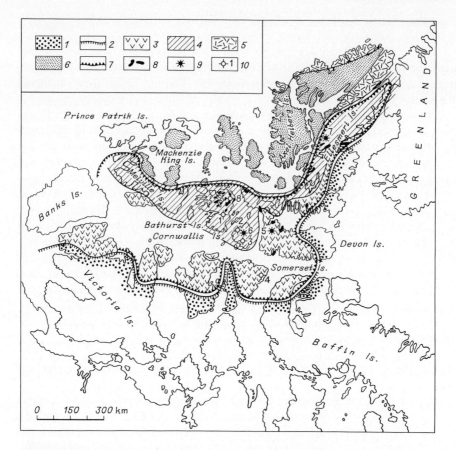

Fig. 20. Index map of Canadian Arctic Archipelago. Compiled from the data of Fortier et al. (1963), Kerr (1967a, b, 1968, 1974, 1975), Trettin (1969, 1971), Thorsteinsson and Tozer (1970), Mossop (1972a, b, 1973, 1979), Miall (1974a, b, 1975), Mayr (1975, 1980), and others. *1* Precambrian outcrops, *2* areal distribution, *3* Arctic Platform, *4* Franklinian Miogeosyncline, *5* Franklinian Eugeosyncline, *6* Sverdrup Basin, *7* Ordovician evaporite basin, *8* outcrops of Ordovician sulfate rocks, *9* diapir domes (*5* Haughton, *6* Central Cornwallis, *7* Vesla), *10* boreholes through Ordovician deposits (*1* Bathurst Caledonian River J-34, *2* Panarctic Deminex Cornwallis Central Dome K-40, *3* Lovitos Cornwallis Resolute Bay L-41, *4* Panarctic Deminex Garner O-21, *5* Young Inlet D-21)

lower Eleanor River Formation (Thorsteinsson 1974). The Eleanor River Formation overlies the Baumann Fiord evaporites elsewhere in the interiors of the Franklinian Miogeosyncline and the Arctic Platform. The Eleanor River Formation consists of dark to light gray limestone varying from thin-platy to massive, silicified, locally dolomitized types. The formation reaches 200 m in thickness.

Overlying Ordovician deposits are united in the Cornwallis Group, including the Bay Fiord, Thumb Mountain, and Irene Bay Formations (Kerr 1967a, 1974, Thorsteinsson 1974, Mayr 1975, 1978, 1980). The formations can be traced along the strike of the Franklinian Miogeosyncline from Bathurst Island to North Greenland, but in the western basin they are buried under Silurian/Devonian deposits.

The lowermost Bay Fiord Formation of the Cornwallis Group contains evaporite deposits composing the second halogenic sequence in the Ordovician section of the Canadian Arctic Archipelago. Rock salt beds are also exposed in this sequence. The Bay Fiord Formation consists chiefly of thin-bedded argillaceous limestone, siltstone, anhydrite, and dolomite. Its thickness varies from 300–600 m.

The Bay Fiord Formation in the Haughton Dome on Devon Island is described as a gypsum-bearing sequence of the Cornwallis Formation (Greiner 1963), consisting of three units. Dark to light gray, thin-bedded to thin-platy gypsinate mudstone and dolomite mudstone interbedded with thin (up to 30 cm) light gray dolomite occur in the lower unit (about 120 m thick). The upper 15 m of the unit contains abundant dolomite and mudstone. The second unit includes dolomite which is thin-platy at the base and massive at the top; this unit is 30 m thick. The third unit, 65 m thick, consists of gray dolomite, mudstone, and gypsum. The gypsum-bearing sequence in the Haughton Dome is tightly folded and extremely variable in composition. Locally, it contains abundant sulfate rocks.

In the Prince Alfred Bay region on Devon Island, the gypsum-bearing sequence of the Lower Cornwallis Group (probably, Bay Fiord Formation) includes 80 m thick thin-platy gypsum in the lower part and is interbedded with green, blue, and dark gray gypsum, calcareous siltstone, gray mudstone, and gypsinate mudstone in the upper part (about 100 m thick); gypsum dominates the upper part of the section.

On Bathurst Island, the Bay Fiord Formation contains thick beds of rock salt. They are penetrated by the Bathurst Caledonian River J-34 and Young Inlet D-21 boreholes. The visible thickness of the Bay Fiord Formation is 1446.9 m and 570.0 m in the first and second boreholes, respectively. Its upper part is represented by shale, anhydrite, and dolomite. Rock salt interbedded with anhydrite, dolomite, and shale occurs below. The saliniferous part of the section is penetrated to 1001.9 m (Kerr 1974, Mayr 1980). The formation is divided into five units.

Unit 1 consists of interbeds of dolomite, shale, anhydrite, and rock salt. Dolomite is chiefly light to dark gray, finely crystalline, more or less anhydritized or argillaceous. Shale is dark gray to brown and dolomitic. Anhydrite is white and finely crystalline. Rock salt is usually pure, but sylvine grains are locally present (up to 2%). The apparent maximal thickness of the unit is 1001.9 m. Unit 2 includes dolomite, dolomitic shale, and anhydrite; dolomite dominates. It is dark brown to gray, aphanocrystalline or finely crystalline. The thickness varies from 156.1–168.5 m. Unit 3 is represented by gray dolomitic and calcareous mudstone. The thickness ranges from 67.3–119.5 m. Unit 4 consists of dark brown aphanocrystalline and finely crystalline dolomites interbedded with dolomitic shale. The thickness of unit 4 ranges from 96.6–136.6 m. Unit 5 includes limestone and shale. The thickness varies from 20.4–24.4 m.

The Thumb Mountain Formation, which forms the central part of the Cornwallis Group, reaches 450–500 m in thickness. In the central parts of the Franklinian Miogeosyncline, the formation includes thick-platy, dark gray to brown, argillaceous limestone, locally dolomitized at the top. In a number of sections on Bathurst Island, the formation contains light gray dolomite and green mudstone. Locally, dolomite composes individual units up to 125 m thick, which are usually brown to chocolate-brown. The Irene Bay Formation is represented by thin-platy, gray to green-gray limestone with lenses, bands, and beds of green mudstone. Its thickness varies from 35–55 m.

The section of Ordovician deposits is crowned with the Allen Bay Formation, which developed on Ellesmere, Cornwallis, and Devon Islands, or with its equivalent Cape Phillips Formation on Bathurst Island and in the northern regions of the Franklinian Miogeosyncline on Ellesmere Island. The upper members of the formations are assigned to the Silurian. The Ordovician deposits of the Cape Phillips Formation include chiefly dolomite, argillaceous limestone, mudstone, and silicified, argillaceous limestone. Gray to yellow, brown, and yellow to gray dolomite dominates the Allen Bay Formation. Its thickness varies from 300–1200 m.

The salt sequence appears to have a considerable areal distribution within the Ordovician evaporite basin of the Canadian Arctic Archipelago. The sequence apparently extends farther north, into the Sverdrup Basin. The areal extent of rock salt is still unknown, but it appears to include Bathurst and Cornwallis Islands, exceeding 100,000 km^2. If average thickness of rock salt in the area is at least 100 m, by taking the visible thickness of the salt sequence into account, the content of rock salt in the basin then seems to exceed 1×10^4 km^3. This estimate is certainly lower than the actual content of salt, but it should be viewed as a first approximation.

Ordovician Evaporite Deposits in Other Basins

Besides the Canadian Arctic Archipelago Basin, evaporites have also been found in Ordovician sequences in basins of Asia, Europe, North America, and Australia (see Fig. 19). They are represented by gypsum and anhydrite, which occur as interbeds, and locally as rather thick units among carbonate sequences of red, variegated, terrigenous formations. As yet, no rock salt beds have been found in these basins. In those having rather thick sulfate beds, salt deposits are supposed to be present in deeper, unfamiliar regions.

Ordovician evaporite deposits are better known in the Williston Basin (Andrichuk 1959, Porter and Fuller 1959, Fuller 1961, Patterson 1961, Carlson and Anderson 1965, Lefond 1969).

From known Ordovician deposits, the Williston Basin appears to be a vast subisometric depression with an areal extent of about 1200 km^2, including North Dakota, Montana, and South Dakota (U.S.A); and Manitoba, Saskatchewan, and Alberta Provinces (Canada). The thickest Orodovician deposits (500–550 m) occur in northwestern North Dakota near Williston. The Winnipegosis, Red River, Stony Mountain, and Stonewall Formations are recognized (in ascending order) in the composite section (Fig. 21).

The section including the Red River, Stony Mountain, and Stonewall Formations consists mainly of carbonate deposits with evaporites, as a rule anhydrites. These formations form the Bighorn Group.

At the base of the Red River Formation, a thin basal bed composed of sandstone, siltstone, and mudstone interbedded with dolomite can be traced; it is recognized as Hecla or "transition" beds in different regions. The beds are underlain by Whitewood

Ordovician Evaporite Deposits in Other Basins 65

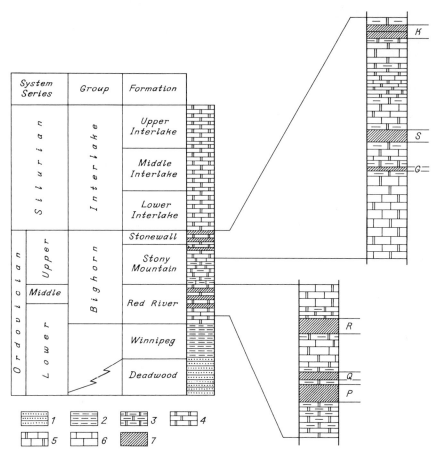

Fig. 21. Subdivision of Ordovician deposits of Williston Basin. After Porter and Fuller (1959), Fuller (1961), and Patterson (1961). *1* sandstone, *2* mudstone, shale, *3* clayey dolomite, *4* dolomite, *5* marly dolomite, *6* limestone, *7* anhydrite. Indexes of anhydrite beds: *P, Q, R, G, S*, and *K*

carbonate beds consisting of dolomite and limestone, about 30 m thick. The upper Red River Formation has a rhythmic structure. Three rhythms, each ranging from limestone and dolomite to anhydrite are easily discernible. The lower rhythm, which is the thickest (up to 30–35 m) and most extensive, ends with anhydrite bed P. The central rhythm reaching 15 m in thickness contains anhydrite bed Q at the top, while the upper rhythm about 10 m thick ends with anhydrite bed R. The location of the beds is shown on Fig. 22.

The lower Stony Mountain Formation in the central Williston Basin consists chiefly of dark gray to brown mudstone interbedded with crinoid and coral, clayey limestone. Nodular dolomite interbeds occur at the base of the formation to the west. The amount of clay material increases southward and northward. The upper part of the Stony Mountain Formation recognized as the Gunton Formation is about 30 m thick and consists of dolomitic limestone. Anhydrite interbeds with thickness

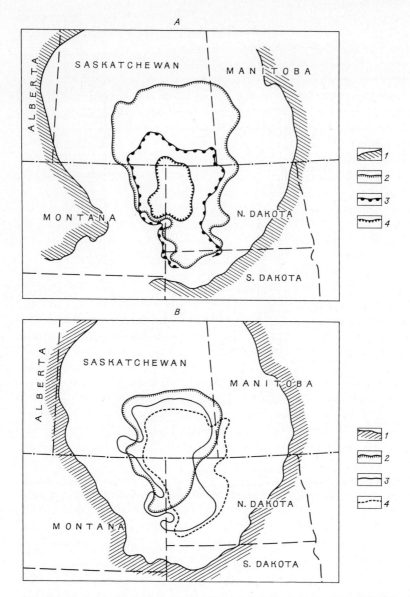

Fig. 22A, B. Areal distribution of anhydrite beds in Ordovician deposits of Williston Basin. After Fuller (1961). **A** Areal distribution of Upper Red River evaporite beds: *1* eroded tract of Upper Red River beds. Depositional limit: *2* P anhydrite, *3* Q anhydrite, *4* R anhydrite. **B** Areal distribution of Upper Stony Mountain and Stonewall evaporite beds: *1* eroded tract of Upper Stony Mountain and Stonewall Formations. Depositional limit: *2* G anhydrite, *3* S anhydrite, *4* K anhydrite

rarely reaching 1–1.5 m occur at the top of the formation; the anhydrite interbeds are marked Q and their location is shown on Fig. 22. The Gunton Formation forms a separate sedimentary rhythm ending with anhydrite.

The overlying Stonewall Formation includes two carbonate-evaporite rhythms, each starting with dolomitic limestone overlain by dolomite and ending with laminated anhydrite. Two upper anhydrite beds are marked S and K. Their occurrence is shown on Fig. 22. It should be noted that some investigators assign anhydrite K to the Silurian and draw the Ordovician/Silurian boundary at the base of the uppermost carbonate-evaporite rhythm.

The Ordovician deposits of the Williston Basin are overlain by the Interlake Carbonate Formation dated as Silurian in age and represented essentially by dolomite with anhydrite and anhydrite-dolomite interbeds.

The available information indicates that Ordovician deposits of the Williston Basin contain only anhydrite. The Stonewall Formation is believed to contain rock salt bands exposed in boreholes at depths over 2900 m along the Cedar Creek Anticline in south-eastern Montana, U.S.A. (Lefond 1969). However, the evidence must be substantiated.

Within the Lena-Yenisei Sulfate Basin, sulfate-bearing rocks have been found in the north and north-western Siberian Platform, as well as in its central and south-eastern regions. They are developed mainly in the Norilsk region (north-western Siberian Platform) where sulfate rocks are present throughout the Ordovician section of the Iltyk, Guragir, Angir, Amarkan, and Zagornino Formations (Ordovician Stratigraphy . . . 1975). Gypsum and anhydrite usually occur among carbonate, variegated terrigenous carbonate, or marl-clayey, sediments. They vary mainly from 0.2–0.5 m; individual beds reach 1 m in thickness. In the Nizhnyaya Tunguska River Basin, sulfate rocks within the Orodvician sequences are exposed in the Turin and Tutoncha boreholes. In the former borehole, they occur as bands and lenticular inclusions in a Lower Ordovician limestone-dolomite sequence, while in the Tutoncha borehole, anhydrite was found in the lower part of the dolomite sequence and in the Lower Ordovician limestone sequence. In the northern Siberian Platform, gypsum beds in the Moiero River Basin occur in the central member of the Irbukla Formation and in the Kochakan Formation. Gypsum bands, up to 0.2 m thick, have also been found in the Maimecha River Basin in deposits of the Tompok and Bysyuryakh Formations. In the Vilyui River Basin, Middle/Upper Ordovician deposits contain sulfate rocks. The Stan Formation has gypsum intercalations, several centimeters thick. Pink to white gypsum beds compose the entire Kharyalakh Formation; the thickness of the upper horizons varies from 1–15 m. In the Markha-Morkoka district, a gypsum sequence marked by the presence of gypsum beds interlayered with mudstone, marl, and dolomite is considered to be an equivalent to the Kharyalakh Formation (Ordovician Stratigraphy . . . 1975). In the Berezovaya Depression, gypsum is reported from the Lower Ordovician Tochilnin Formation and from the Upper Ordovician Ilyun Series. In the Irkutsk Amphitheater, gypsum bands and lenses lie among red beds of the Upper Ordovician Bratsk Formation. Hence, in the Lena-Yenisei Basin, sulfate rocks form a wide belt running from the north-western and northern Siberian Platform (Norilsk, Moiero, and Maimecha Rivers) south-eastwards to the Berezovaya Depression.

In the Severnaya Zemlya Basin, sulfate rocks dominate the dolomite-marl and gypsum-limestone members of the Komsomol Formation (Egiazarov 1970).

In the Baltic Basin, gypsum lenses, inclusions, and thin layers, about 10 cm thick, occur among dolomite and terrigenous-carbonate sediments in the Ievsk superhorizon. Moreover, small gypsum inclusions and lenses, as well as gypsification or rocks can be observed in deposits of the Itfer, Tallin, and Porkun horizons (Selivanova 1971).

The Moose River Basin is situated on the south-western coast of the Hudson Bay, Canada. Gypsum bands and beds occur in the Ordovician dolomite sequence (Norris and Sanford 1969). In the South Illinois Basin, bands and thin beds of anhydrite (up to 0.5–1 m thick) occur in the Middle Ordovician Ioachim Formation (Bond et al. 1968). In the Anadarko Basin, a considerable amount of anhydrite occurs in the West Spring Creek Formation where it forms individual interbeds and beds intercalated with dolomite and dolomitic marl, mudstone, and sandstone, or occurs in the rocks as lenses and inclusions (Reedy 1968, Latham 1973).

In the Georgina Basin, Australia, a thin gypsum band of Upper Cambrian/Lower Ordovician age occurs in terrigenous-carbonate rocks of the Ninmaroo Formation (Wells 1980).

In the Canning Basin, two deep boreholes, Parda No. 1 and Wilson Cliffs No. 1, drilled on the north-west and on the south-east of the basin, respectively, expose anhydrite inclusions and lenses among shale, limestone, and dolomite of the Goldwyer Formation, dated as Middle Ordovician in age (Wells 1980).

Chapter IV
Silurian Salt Deposits

The Late Silurian age is widely accepted as the period of considerable salt accumulation (Lotze 1957a, 1965, Ivanov and Levitsky 1960, Pierce and Rich 1962, Strakhov 1962, Borchert and Muir 1964, Kozary et al. 1968, Lefond 1969). This is based chiefly on information concerning the Michigan-Appalachian Basin where thick salt sequences are widespread in North America. The Michigan-Appalachian Basin was proved to be a vast unique area of Silurian salt accumulation in the world. However, it is not as large as many other salt basins of the Cambrian, Devonian, and Permian age.

Three salt-producing Silurian evaporite basins, namely, the Dniester-Prut, Lena-Yenisei, and Carnarvon Basins can now be outlined.

The Canning Basin, where Silurian salt deposits are probable but unconfirmed, is also discussed. Wells (1980) advocates an Early Devonian age of these deposits. Silurian evaporites are also known from Severnaya Zemlya, Pechora, Baltic, Canadian Arctic Archipelago, Moose River, and Williston Basins (Fig. 23).

Michigan-Appalachian Basin

The Michigan-Appalachian Basin occupies a vast territory including New York, Pennsylvania, West Virginia, Ohio, and Michigan States, U.S.A, as well as the south-western extremity of Ontario Province, Canada.

Salt deposits fill in the Michigan and the Appalachian Depressions on the northwest and on the south-east, respectively. The depressions are separated by the Findlay and Algonquin Arches. The forthcoming description of the salt deposits of the Michigan-Appalachian Basin is based on the detailed studies of many authors (Alling 1928, Martens 1943, Landes 1945, Pepper 1947, Dellwig 1955, Fettke 1955, Kreidler 1957, Alling and Briggs 1961, Pierce and Rich 1962, Sanford 1965, 1968, Fergusson and Prather 1968, Vary et al. 1968, Dellwig and Evans 1969, Lefond 1969, Rickard 1969, Colton 1970, Matthews 1970, Poole et al. 1970, Kahle and Floyd 1972, Treesh 1973, Briggs and Briggs 1974, Mesolella et al. 1974, Dennison and Head 1975, Patchen and Smosna 1975, Miller 1976, Gill 1977, 1979, Gill et al. 1978, Mesolella 1978, Smosna and Patchen 1978, Wold et al. 1981).

Distribution of Silurian salt deposits in the Michigan-Appalachian Basin is shown in Fig. 24. In the Michigan Depression all evaporite rocks belong to the Salina Group,

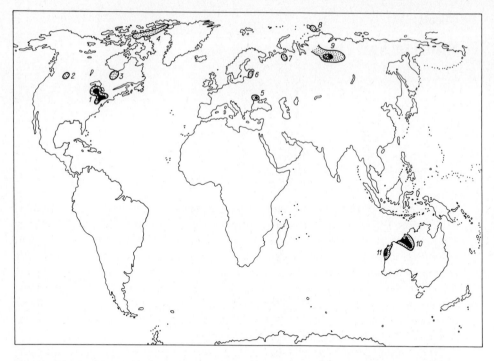

Fig. 23. Distribution of Silurian evaporite deposits. Evaporite basins: *1* Michigan-Appalachian, *2* Williston, *3* Moose River, *4* Canadian Arctic Archipelago, *5* Dniester-Prut, *6* Baltic, *7* Pechora, *8* Severnaya Zemlya, *9* Lena-Yenisei, *10* Canning, *11* Carnarvon. For legend see Fig. 3

which is assigned chiefly to Late Silurian, and only its lower part may be assigned to Middle Silurian. The Salina Group is divided into seven units labeled A through G, in ascending order. Units C, E, and G are essentially dolomitic or dolomite-argillaceous in composition, while units B, D, and F are salt-bearing. In the central Michigan Depression, unit A is subdivided into A-1 and A-2, each containing evaporite and carbonate members. The Salina Group is overlain by the Bass Island Formation represented mainly by dolomite, locally known as Unit H. The evaporite deposits are underlain by the Niagara Group including carbonate rocks of complex facies composition. The Niagara Group consists of Clinton, Lockport, and Guelph Formations.

In the Appalachian Basin, salt deposits contain the Vernon and Syracuse Formations, correlative respectively, with units A-2, B, C, and units D, E, F, belonging to the Salina Group of the Michigan Depression. Unit G is an equivalent of the Camillus Formation. The Bertie and the lower Rondout Formations correspond to the level of the Bass Island Formation. In the central and north-western Appalachian Basin, thick carbonate deposits underlying evaporites are assigned to the Lockport Formation whose stratigraphic range is much wider than that of the formation of the same name in the Michigan Depression; it may be correlated with the Niagara Group as a whole and with Unit A-1 of the Salina Group.

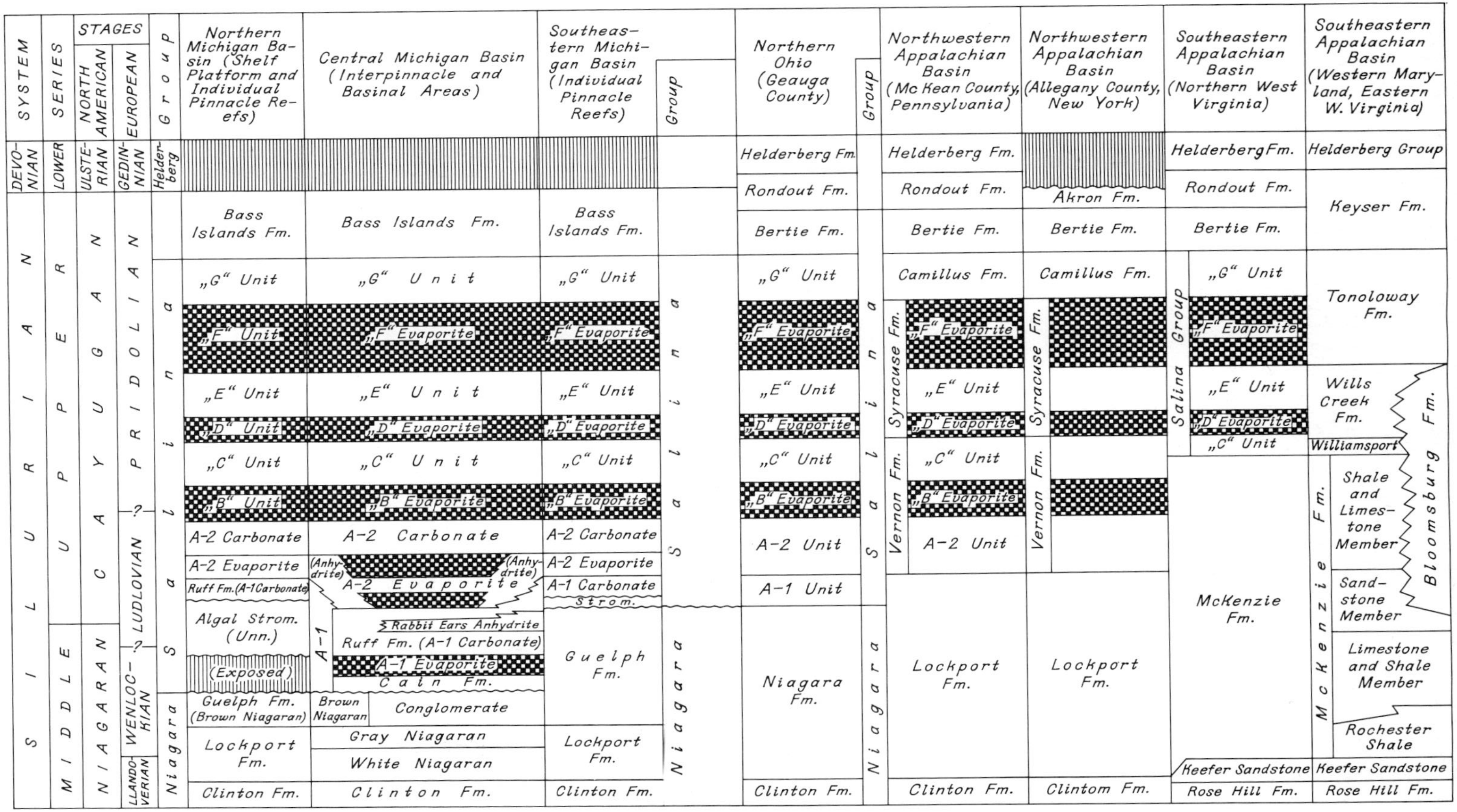

Fig. 24. Subdivision and correlation of Silurian evaporite deposits of Michigan-Appalachian Basin. *Squares* show stratigraphic levels at which rock salt was stated

In the south-eastern and eastern Appalachian Basin, Middle and Late Silurian deposits show considerable facies change. Dark shale interbedded with argillaceous limestone and siltstone become widespread in the section. Sandstone bands and beds can also be seen. The Rose Hill Formation, the Keefer Sandstone, and the McKenzie Formation are recognized at the level of the Niagara Group. The upper McKenzie Formation can be considered as an equivalent of units A-1, A-2, and B of the Salina Group or of the lower Vernon Formation. South-eastward, evaporites wedge out and are replaced by terrigenous, clayey, and carbonate sequences. The Williamsport, Wills Creek, Tonoloway, and Keyser Formations are equivalents of the upper Vernon, Syracuse, Camillus, Bertie, and Rondout Formations. Red, terrigenous rocks assigned to the Bloomsbury Formation dominate the Appalachian Basin margins. The stratigraphic position of the Bloomsbury Formation is similar to that of the upper McKenzie, Williamsport, and Wills Creek Formations, i.e., it is an equivalent of most evaporites of the lower Salina Group.

In the Michigan Depression, the relationship between reefs and associated carbonates of the Middle Silurian Niagara Group, on the one hand, and the overlying Salina Group evaporites, on the other hand, are well known (Huh 1973, Briggs and Briggs 1974, Mesolella et al. 1974, Gill 1977, 1979, Gill et al. 1978, Mesolella 1978). The studies show the existence of four facies zones peculiar in composition and structure of their carbonate deposits: (1) platform shelf carbonates; (2) a barrier reef belt or carbonate bank along the platform margin; (3) a pinnacle-reef belt coinciding with the platform slope; and (4) deeper water basinal carbonates in the Michigan Depression occurring during Niagaran time (Fig. 25). Such environmental conditions took place at the close of the Wenlockian and resulted in the formation of a continuous barrier which surrounded the sediment-starved basin. Then, mostly in Late Silurian time, evaporites of 900 m accumulated in the basin.

Halogenic deposition began in the Michigan Depression mainly in the interpinnacle and basinal areas (Fig. 26). Unit A-1, a basal part of the Salina Group, is subdivided into the Cain Formation, A-1 Evaporite, the Ruff Formation or A-1 Carbonate, and A-2 Evaporite.

The Cain Formation incorporates thin (not more than 7–8 m) carbonate, called A-0 Carbonate (Huh 1973, Gill 1977, 1979). It consists of dark gray to black, argillaceous mudstone, rich in bituminous material. A unit of anhydrite, rock salt, limestone, and dolomite interbeds overlies A-1 Evaporits. Limestone and anhydrite are massive or laminated; they usually contain large crystals and inclusions of halite. In general, rock salt is light and coarsely ·crystalline with thin, gray anhydrite and dolomite interbeds. Orange rock salt occurs as well. Sylvine was observed in a thin zone of orange salts in Grand Traverse County, in the north-western Michigan Depression and in the Pan American State Union 1–14 well (Anderson and Egleson 1970, Matthews 1970). In some specimens, K_2O reaches 40%. A zone of potassium accumulation is 28 m thick; sylvinite beds occupy an area of about 36,000 km^2. The upper A-1 Evaporite consists mainly of finely laminated, gray anhydrite. The greatest thickness (150–160 m) of A-1 Evaporite was found in the central Michigan Depression; it is a single salt occurrence.

At the Michigan Depression margins, in the pinnacle reef belt, the thickness of A-1 Evaporite decreases to 5–7 m. It is represented mainly by nodular mosaic and

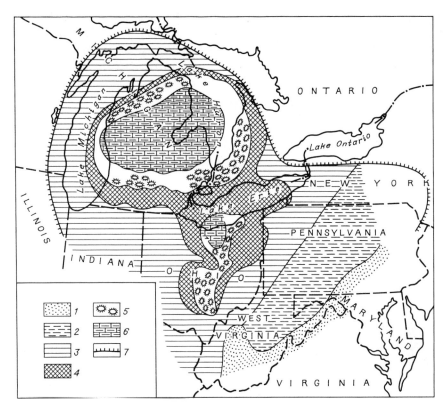

Fig. 25. Lithofacies map of Niagaran deposits of Michigan-Appalachian Basin. Compiled from the data of Sanford (1965, 1968), Huh (1973), Briggs and Briggs (1974), Mesolella et al. (1974), Patchen and Smosna (1975), Gill (1977, 1979), Briggs et al. (1978), Gill et al. (1978), Mesolella (1978), Smosna and Patchen (1978), and others. *1* areas of terrigenous rock distribution, *2* depositional limit of shale interbedded with limestone, *3* depositional limit of platform shelf carbonates, *4* barrier reef belt, *5* pinnacle-reef belt, *6* terrain of deep water basinal carbonates, *7* present boundary of Silurian deposits

laminated (massive) anhydrites (Gill 1977). Another locality of A-1 Evaporite is in Ohio, where dolomitized anhydrite about 6 m thick fills in the interpinnacle areas and the inner parts of the former deep-sea basin (Mesolella 1978).

Dense, brown dolomite, which was replaced stratigraphically higher by dark gray to black, shaly limestone or by limy shale, composes the base of A-1 Carbonate or the Ruff Formation in the Michigan Depression. The upper part of A-1 Carbonate consists of dense, fine-grained, argillaceous, and thin-bedded carbonate, locally brecciform, and somewhat anhydritized or salinized. The thickness of A-1 Carbonate does not exceed 25 m in the depression's interior, and reaches 30 m or even 50 m in the pinnacle-reef belt. Within the barrier reef belt, the thickness of A-1 Carbonate decreases to 25 m; it becomes difficult to differentiate between the latter and the Niagara Formation. Micro-laminated mudstone, planar stromatolite, various pelletal wackestone carbonate, breccia, and flat pebble conglomerate compose A-1 Carbonate (Gill 1977).

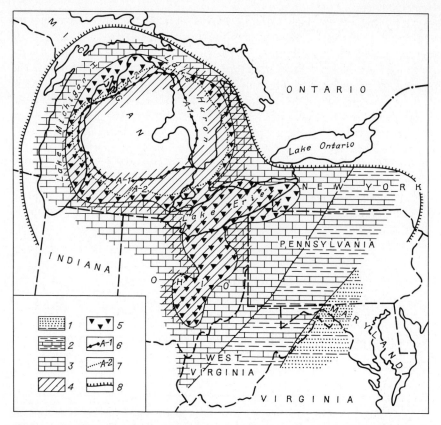

Fig. 26. Lithofacies map of basal Salina Group (Unit A) of Michigan-Appalachian Basin. Compiled from the data of Landes (1945), Alling and Briggs (1961), Sanford (1965), Lefond (1969), Huh (1973), Mesolella et al. (1974), Patchen and Smosna (1975), Gill (1977, 1979), Briggs et al. (1978), Mesolella (1978), Smosna and Patchen (1978), and others. Areal distribution of: *1* essential terrigenous rocks, *2* shale and argillaceous limestone, *3* essential dolomite, *4* barrier reefs and pinnacle reefs of Niagaran time, *5* anhydrite. Depositional limit of: *6* A-1 Evaporite, *7* A-2 Evaporite, *8* present boundary of Silurian deposits

Nodular anhydrite, called Rabbit Ears Anhydrite is known from the middle A-1 Carbonate of the interpinnacle areas. Two anhydrite members, the Lower and the Upper Rabbit Ears, each 1.5–2 m thick, can usually be traced as they taper on the reef slopes (Gill 1977).

In the Appalachian Basin, A-1 Unit is similar in composition to carbonate of the Niagara Formation. The A-1 Unit is composed chiefly of thin-grained, somewhat argillaceous dolomite. Shaly and anhydritic rocks, as well as anhydrite bands and inclusions occur only in a belt running from western New York State through northwestern Pennsylvania to eastern Ohio. The remaining north-western Appalachian Depression contains equivalents of the A-1 Unit, which are assigned to the Lockport Formation. South-eastward, dolomite is replaced by dark shale with partings of dark, argillaceous limestone assigned to McKenzie Formation. Farther south-east, age equivalents of the A-1 Unit are terrigenous, red beds (lower Bloomsbury Formation).

Salt deposits confined to the A-2 Unit are known only from the Michigan Depression (see Fig. 27). The A-2 Evaporite and A-2 Carbonate have been recognized. The A-2 Evaporite consists mostly of coarsely crystalline rock salt with anhydrite and dolomite interlaminations. Dense anhydrite bands up to 1 m thick are common in the upper part. The thickness varies from 30 to 100–130 m. At the depression margins, A-2 Evaporite thins out and is dominated by anhydrite. Rock salt occurs only in the interpinnacle areas. Reefs are usually overlain by a thin anhydrite unit. The A-2 Evaporite appears to fill in all the rough parts in the bottom of the Michigan Depression, both in the inner zone and in the pinnacle reef belt, as well as in the barrier reef belt and outside, in the adjacent carbonate platform. The A-2 Evaporite covers all the reef carbonate structures formed during Late Niagaran and Early Cayugan time like a mantle.

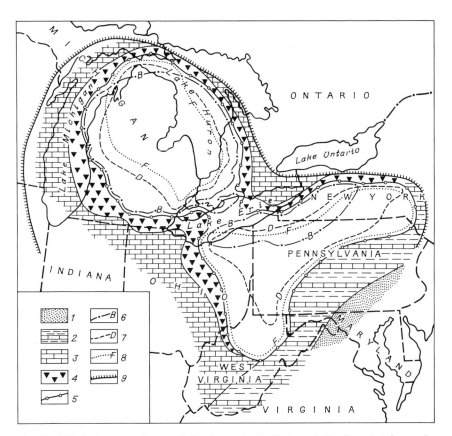

Fig. 27. Lithofacies map of upper Salina Group (units B through G). Compiled from references listed in Fig. 26. Areal distribution of: *1* predominantly terrigenous rocks, *2* shale and argillaceous limestone, *3* carbonate rocks (mainly dolomite), *4* anhydrite. Depositional limits: *5* salt deposits of upper Salina Group, *6* salt, B Unit, *7* salt, D Unit, *8* salt, F Unit, *9* present limits of Silurian deposits

Overlying A-2 Carbonate is much more persistent in composition. In the Michigan Depression, it includes dark gray and brown dolomite with limestone bands. The thickness ranges from 30—50 m. In the Appalachian Basin, carbonate is replaced by red and green shale which composes the lower Vernon Formation in New York State, or by shale and limestone belonging to the upper McKenzie Formation in the south-western depression.

Salt-bearing deposits associated with units B, D, and F lie both in the Michigan and in the Appalachian Basin (see Fig. 27). In the Michigan Depression, unit B consists of rock salt with minor anhydrite and dolomite. The thickness of the unit varies from 25—30 m at the margin to 100—130 m in the internal zones. Anhydrite, anhydritic dolomite and shale become more widespread towards the depression periphery. In the Appalachian Basin, salt B Unit was found only in the north and north-west. It consists of rock salt interbedded with shale and the thickness reaches 60—70 m. Two regions, in northern Ohio and in western New York State, containing the greatest thickness of salt deposits, can be recognized.

In the Michigan Depression, the C Unit consists of green to gray shale and shaly dolomite interbedded with anhydrite. The thickness ranges from 20—50 m. In the Appalachian Basin, the deposits of the upper Vernon Formation may be considered as equivalents of the C Unit. In western New York State, they include red and green shale 30 m thick. Shale and argillaceous dolomite are known from Ohio. Their thickness does not exceed 15—20 m. In southern and south-western West Virginia, the C Unit is replaced by the Williamsport Sandstone composed of green and brown sandstone with siltstone and shale interbeds, and in places with limestone beds. The thickness varies from 10—15 to 25—30 m (Smosna and Patchen 1978).

In the Michigan Depression, the thickness of D Evaporite ranges from 20—50 m. In the central deeper parts, it is represented mainly by rock salt with thin, buff dolomite interbeds. In the Appalachian Basin, the D Unit has a much greater areal extent than the underlying B Unit. The D Unit was registered in western New York State, in north-western Pennsylvania, in eastern Ohio, and in northern West Virginia. It consists of gray, orange, and brown, coarsely crystalline rock salt interbedded with gray, brown, shaly dolomite. The thickness varies from 5 to 25—30 m.

The E Unit is known both from the Michigan and the Appalachian Basin. It contains gray and red shale with dolomite, shaly dolomite, and anhydrite interbeds. The thickness ranges from 10—40 m. In the south-eastern marginal zones of the Appalachian Depression, salt-brearing deposits of the D Unit thin out, whereas the E Unit is replaced by silty, calcareous shale assigned to the Wills Creek Formation in West Virginia. Farther east, red terrigenous beds (upper Bloomsburg Formation) occur at the same stratigraphic level.

In the Michigan-Appalachian Basin, F Evaporite embraces the thickest upper unit of salt-bearing deposits. In the Michigan Depression, it includes salt beds intercalated with shale, dolomite, and anhydrite. Here the thickness of F Evaporite exceeds 400 m. In the Appalachian Basin, F Evaporite has the greatest areal extent. In the western areas, in Ohio, it can be divided into four zones. The lower zone includes salt deposits known as F-1 or Second Salt. The thickness of this member ranges from 20—25 to 40—50 m. The second zone about 7 m thick contains gray and brown-gray anhydrite interbedded with massive dolomite. The third zone is composed of white and gray, coarsely crystalline, massive rock salt. It is marked F-2 or named First Salt. Its thickness reaches 20–30 m. The fourth upper zone 5–6 m thick consists of dolomite interbedded with anhydrite.

The F Evaporite has the greatest thickness – about 200 m – in eastern New York State and in Pennsylvania, where it composes the upper Syracuse Formation. In northern West Virginia, the thickness of salt deposits reaches 50–60 m. On the east, south, and west, the salt zone is rimmed by anhydrite; on the south and south-east, anhydrite is replaced by shaly, silty, and sandy limestone and dolomite, known as the Tonolway Formation.

The G Unit crowning the Salina Group section is dominated by gray, argillaceous dolomite interbedded with green and red shale and gray and brownish-gray anhydrite in most parts of the Michigan-Appalachian Basin. In the north-western Appalachian Basin, their equivalents form the Camillus Formation composed of gray dolomitic and calcareous shale, shaly dolomite, dolomitic limestone with gypsum lenses and inclusions. The thickness of the deposits ranges from 10–15 to 200 m.

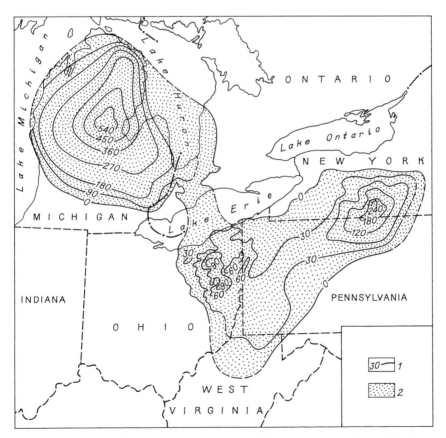

Fig. 28. Isopach map of Salina Group rock salt in Michigan-Appalachian Basin. After Pierce and Rich (1962). *1* isopach contours of rock salt in meters, *2* areal extent of rock salt

In the Michigan Depression, the Silurian section is capped by the Bass Island Formation including chiefly light brown dolomite and anhydrite. Its thickness varies from 60–200 m. In the Appalachian Basin, its equivalents are the Bertie Formation and the lower Rondout Formation. The Bertie Formation consists of dolomite and shale with anhydrite bands and lenses. The thickness does not exceed 30 m. In eastern and central New York State, the Rondout Formation is composed of argillaceous, shaly dolomite; the thickness ranges from 8–18 m. The remaining Appalachian Basin is dominated by dolomite.

The Late Silurian evaporite assemblage of the Michigan-Appalachian Basin contains a total of five salt-bearing sequences separated by sulfate-carbonate or shale-carbonate units. They reach considerable thickness — about 1000 m and 750 m — in the central Michigan and central Appalachian Basin, respectively. The maximal total thickness of rock salt in these depressions is 540 m and 250 m, respectively (Fig. 28). The upper salt beds lie at depths of 100–2400 m; the maximal depth was recorded in the south-eastern piedmont area of the Appalachian Basin.

According to Pierce and Rich (1962), the area under salt amounts to approximately 260,000 km^2; however, Lefond (1969) reports the figure of 214,000 km^2. The minimal possible volume of salt can be as high as 2.6×10^4 km^3.

Dniester-Prut Basin

The Dniester-Prut Basin occupies the south-western margin of the East European Platform within the Moldavian Plate. The basin is bounded in the north-east by the Ukrainian Shield, and in the east and south probably by the uplift zone postulated for the Black Sea and extending into Rumania. In the south-west, the basin was separated from the open sea of normal salinity by a shallow carbonate zone, developed within the Podolian Saddle and probably extending into northern Dobruja. The basin was first described as evaporite by Zavidonova (1956) and later by Shulga and Krandievsky (Atlas of Paleogeographical . . . 1960, Atlas of Lithologo-Paleogeographical . . . 1960).

The following description of the basin is based on their data as well as those of Bobrinskaya et al. (Stratigraphy of Sedimentary . . . 1964), Vyalov (1966), Edelstein (1969), Tesakov (1971), Nikiforova et al. (Reference Key . . . 1972), and Grachevsky and Kalik (1976), Tsegelnyuk (1976).

Silurian deposits within the Dniester-Prut Basin and in Podolia are divided into Kitaigorod, Muksha, Ustie, Malinovtsy, and Skala horizons. The Kitaigorod horizon in Moldavia consists of dark gray, argillaceous limestone, locally dolomitized, interbedded with mudstone and marl, and in some cases with siltstone and sandstone at the base. The horizon varies from 25–30 m in thickness. In East Podolia, deposits of the Kitaigorod horizon are identified as a formation of the same name. In occurrences near the villages of Verkhnyakovtsy, Shidlovtsy, Yurkovtsy, Ivanovka, and Darakhov, as well as in exposures along the Dniester River, the formation is represented by dark, gray, nodular limestone with marl interlayers. The formation ranges from 39–68 m in thickness. Farther north-west in Volyno-Podolia, the Kitaigorod Formation includes, along with limestone and marl, dark and green-gray mudstone and dolomite bands. Westward (in Lutsk, and in the villages of Pelcha, Romashkovka, and Godomichy) dolomitized limestone, dolomite, and dolomitic marl appear in the section. Dolomites and dolomitized carbonate rocks then become subordinate. Toward the Lvov Depression, mudstone, siltstone, and black, pelitomorphic limestone dominate the Kitaigorod Formation. In the deeper part of the Lvov Depression, equivalents of the Kitaigorod Formation are known to form the Dublyany Formation, including essentially black, argillaceous siltstone with layers of black to dark gray limestone and locally sandstone; the formation is 44–80 m thick.

Within the Dniester-Prut Basin, the overlying deposits include the Muksha, Ustie, Malinovtsy, and lower Skala horizons and form a single sequence of halogenic rocks, predominantly anhydrite. In different parts of Moldavia, the sequence is represented by dolomite, dolomitized limestone, limestone, marl, and mudstone interbedded with anhydrite and bentonite clay. The thickness of the sequence is 250–300 m. Anhydrite occurs chiefly at the level of the Ustie, Malinovtsy, and lower Skala horizons, while dolomite and dolomitic limestone with rare lenses and inclusions of anhydrite occur chiefly in the Muksha horizon. Zavidonova (1956), using Eremenko's data, reported the presence of rock salt bands in the Malinovtsy horizon near the village of Sarateny-Vek, Nisporen District. The thickness and areal extent are unknown. The same evidence reported by Zavidonova was cited by Shulga (Atlas of Paleogeographical ... 1960). However, in later works rock salt was not reported in the Silurian deposits of the Dniester-Prut Basin. Even in deep boreholes in Nisporen and adjacent areas, only thin bands and inclusions of anhydrite were reported (Stratigraphy of Sedimentary ... 1964, Edelstein 1969). Hence, Zavidonova's information is not completely reliable. The possible presence of rock salt can be noted only in the Dniester-Prut Basin.

A sulfate-bearing carbonate sequence of the Muksha, Ustie, Malinovtsy, and lower Skala horizons in north-western Moldavia (the villages of Brinzeny, Morosheshty) is known as the Pigai Formation, consisting chiefly of dolomite and dolomitic marl with bands and inclusions of anhydrite. Farther north-west in the Dniester River Basin, equivalents of the sulfate-bearing carbonate sequence are divided into the Bagovitsy, Malinovtsy, and Prigorodok Formations. The Bagovitsy Formation consists of two members: the lower Muksha Member, comprising dolomitized limestone, dolomite marl, and dolomite, 13–25 m thick, and the upper Ustie Member 20–26 m thick, represented by dolomite and dolomitic marl with rare limestone layers. The Malinovtsy Formation includes nodular and massive limestone, platy dolomite, and marl, Its thickness in the Dniester region ranges from 112–142 m. The Malinovtsy Formation is subdivided into several dolomite and dolomitic marl members. One of them, the Perepelitsy Member along the Dniester River and in the Zbruch and Zhvantsy Rivers interfluve, is assigned to the lower Malinovtsy Member. Within the basins of the Dniester, Zbruch, Zhvantsy, Stryp, and Safet Rivers, the upper Malinovtsy Member includes deposits of Isakovtsy beds represented by dolomite and dolomitized limestone, interbedded with gypsum and anhydrite in the vicinity of Darakhov. The Prigorodok Formation also contains dolomite and dolomitic marl. Its thickness varies from 22–60 m. Near the village of Verkhnyakovtsy and the town of Darakhov, the formation also contains gypsum and anhydrite layers. Hence, evaporite rocks are known to stretch from the Dniester-Prut interfluve beyond the Dniester Valley to the town of Darakhov and possibly farther north-west.

West of the above region, the Muksha, Ustie, Malinovtsy, and lower Skala sections are mainly limestone. Equivalents of the Bagovitsy Formation form the Stryp Formation, containing dark gray to black limestone; marl and argillaceous limestone begin to dominate the Malinovtsy Formation, which is equivalent to the Prigorodok Fomation. Black, argillaceous limestone, black mudstone and siltstone, forming the Kulichkov, Peremyshl, and Zavadov Formations, occur in a deeper zone of the Lvov Depression and in the Cis-Carpathian Depression.

The upper Skala deposits in Moldavia are represented by limestone and marl interlayered with mudstone and dolomite, up to 200 m thick. In Podolia, these deposits

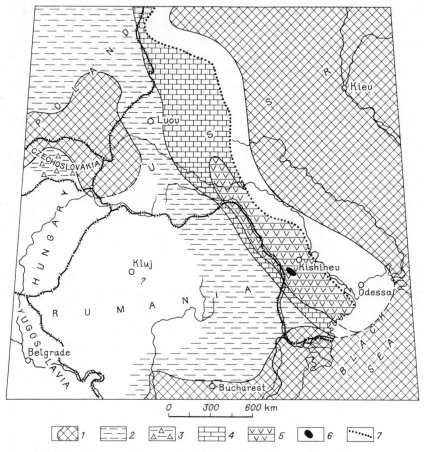

Fig. 29. Litho-paleogeographic map for the Ludlovian of the south-western margin of the East European Platform and adjacent area. Compiled from data taken from publications by Zavidonova (1956), Atlas of Litho-Paleogeographic Maps ... (1960), Atlas of Paleogeographic Maps ... (1960), and publications by Vyalov (1966), Edelstein (1969), Nikiforova et al. (Reference ... 1972), and Tsegelnyuk (1976). *1* land, *2–5* areas dominated by shale and graptolitic shale *(2)*, cherty shale *(3)*, limestone and dolomite *(4)*, dolomite and sulfate rocks *(5)*, *6* inferred areas of rock salt development, *7* recent boundaries of deposits

are subdivided into the Rashkov and Zvenigorod Formations. The former includes limestone, dolomitic marl, dolomite, and clay; the latter consists chiefly of nodular and platy limestone. In both formations, the thickness of argillaceous rocks increases westward toward the Lvov Depression; and finally, dark gray to black mudstone interbedded with dark limestone becomes dominant. This argillaceous sequence is divided into the Glinyan and Poltvin Formations.

Thus, within the Dniester-Prut Basin, evaporite deposits occur in the center of the Silurian sedimentary sequence at the level of the Muksha, Ustie, Malinovtsy, and lower Skala horizons. They are mainly Wenlockian/Ludlovian in age.

The Dniester-Prut evaporite basin was engulfed and connected with the sea in the south-west (Fig. 29) as suggested by facies changes that occur, when a sulfate-dolomite sequence in the central basin grades into dolomite, limestone, and limestone-argillaceous sequences. Deep-water black mudstone and limestone occur in the central Lvov Depression. Recently, they were penetrated by deep boreholes in southern Moldavia, in Dobruja (in the Yargorin, Baurchin, and Mantov areas), in the Odessa region (Orekhovo area), and in Rumania near the towns of Constanta, Călărasi, Borsul-Verde, Popeşti, Hirlău, and Rădăuti (Grachevsky and Kalik 1976). These regions were once covered by a deep open sea situated west and south-west of the evaporite basin. The latter was separated from the sea by a rather narrow strip of shallow water, in which carbonate reef structures are believed to have developed (Grachevsky and Kalik 1976). Conditions might have been favorable at one time for salt accumulation in the evaporite basin. However, there is as yet no reliable evidence to support accumulation of salt sequences in the Dniester-Prut Basin.

Lena-Yenisei Basin

In many areas of the Siberian Platform, gypsum and anhydrite are present in Lower and Upper Silurian deposits. According to data reported by Predtechensky and Tesakov (1979), Predtechensky (1980), and Tesakov (1981), the oldest sulfate-bearing sequence lies in the Nyuya and Berezovaya Depressions, where it forms the Utokan Formation, consisting of gray and variegated, silty limestone interlayered with dolomite and gypsum. The Utokan Formation belongs to the Khaastyr horizon of the Upper Llandoverian. The same sulfate-bearing sequence appears to continue into the Vilyui River Basin where gypsum and gypsinate dolomite were found at the same stratigraphic level in the limestone-dolomite unit of the Khaastyr horizon.

Two sulfate-bearing sequences can be recognized in the Upper Silurian. One of them, of Ludlovian age, is known in the Norilsk region and probably stretches farther east into the Moiero River Basin and south into the Turukhan region. In the northwestern Siberian Platform, the sequence is known as the Kongda Formation, in which thin gypsum and gypsinate dolomite bands occur in limestone, dolomite, and marl; in the Moiero River Basin, the sequence is represented by the Yangada Formation, the central part of which is characterized by gypsification of rocks, and contains lenses, inclusions, and bands of gypsum. Both formations occur at the level of the Tukal horizon. The second uppermost sulfate-bearing sequence is rather well

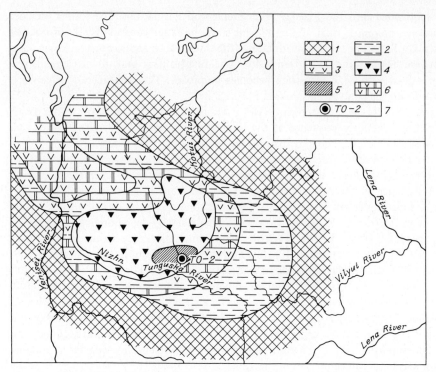

Fig. 30. Lithofacies map of Kholyukhan horizon (Přidolian) in Lena-Yenisei Basin. After Tesakov (1981) and supplemented with new data. *1* land. Depositional limit of: *2* marl and argillaceous dolomite, *3* dolomite and marl with interbeds and inclusions of gypsum, *4* dolomite, marl, and gypsum, *5* probable rock salt, *6* mainly dolomite with interbeds and lenses of gypsum, *7* location of the Turin T0-2 borehole

developed in the north-western and central Siberian Platform (Fig. 30). The sequence is found in the Norilsk region south-eastward to the Moiero-Morkoka Rivers Basin and probably occurs in the central Tunguska Syneclise. The sequence lies at the level of the Upper Silurian Kholyukhan horizon. Gypsum beds locally reach 4–5 m in thickness; in the Moiero River Basin and in the Norilsk region, their total thickness is no less than 50 m.

Salt deposits appear to occur in the lowest parts of the Tunguska Syneclise in the upper sulfate-bearing (Kholyukhan) sequence (Sokolov et al. 1977). The Turin T0-2 borehole is believed to penetrate rock salt beds up to 4 m thick, which were established from geophysical data and from the presence of rock salt crystals in dolomite-anhydrite and anhydrite. If this information is confirmed by future investigations, then the Silurian Lena-Yenisei Basin can be regarded as salt-producing.

Canning Basin

The Canning Basin is situated in north-western Australia (Fig. 31). It stretches from south-east to north-west, embracing a vast area of the Great Sandy Desert. The basin exceeds 900 km in length and ranges from 300–600 km in width. On the north-east and on the south-west, the basin is bounded by outcrops of Precambrian formations in the Kimberley Block and in the Pilbara Block, respectively. The deep Fitzroy Trough and the Kidson Sub-Basin are characteristic structural elements on the north-east and on the south, respectively. In the deepest parts, the depth to the basement ranges from 7–9 km. The basin is filled in with thick Ordovician, Silurian, Devonian, Carboniferous, and Permian strata overlain by thin Jurassic and Cretaceous beds. Salt-bearing formations lie at the base of the section studied; however, the age of the sediments is still unkown. As mentioned above, their characteristics are quite tentative and their Silurian age could be questioned.

Evidence on salt-bearing deposits of the Canning Basin is given in publications by Veevers and Wells (1961), Singleton (1966), Johnstone et al. (1967), Veevers (1967), Glover (1973), and Wells (1980). The forthcoming description is based on these works.

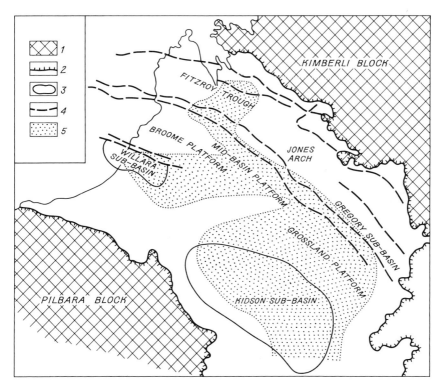

Fig. 31. Index tectonic map of Canning Basin. Compiled from the data of Veevers (1967) and Wells (1980). *1* Precambrian outcrops, *2* limit of Canning Basin, *3* limits of sub-basins, *4* major fault systems, *5* maximal extent of halite

Evaporite deposits are confined to the Carribuddy Formation. The area under evaporites is about 2.1×10^5 km^2; they were penetrated by eight petroleum boreholes (Fig. 32).

In the Kidson Sub-Basin, the pre-Permian section is divided in the following sections, in ascending order:

Nambeet Formation	– Tremadocian (?), Arenigian;
Willara Formation	– Arenigian;
Goldwyer Formation	– Llanvirnian and, possibly, Llandeilian;
Nita Formation	– Llanvirnian and Llandeilian;
Carribuddy Formation	– Late Silurian or Early Devonian;
Tandalgoo Red Beds	– Gedinnian;
Mellinjerie Limestone	– Frasnian and, possibly, Givetian.

Evidently, the stratigraphic position of the Carribuddy Formation is inferred from its occurrence between Middle Ordovician and Early Devonian fossiliferous deposits. Data

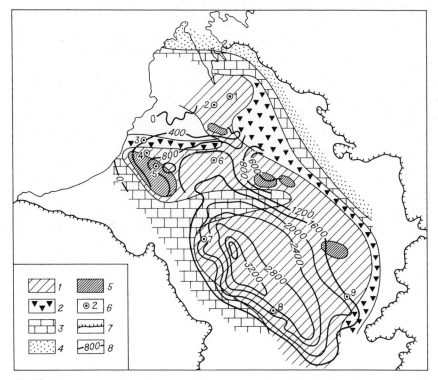

Fig. 32. Isopach map and maximal extent of evaporite minerals in Carribuddy Formation, Canning Basin. After Wells (1980). Maximal extent of evaporite minerals: *1* halite, *2* gypsum, anhydrite, *3* dolomite, *4* nonevaporite, *5* salt structures, *6* evaporite boreholes and their numbers (*1* Grant Range No. 1, *2* Frome Rocks No. 1, *3* Parda No. 1, *4* Willara No. 1, *5* Munda No. 1, *6* McLarty No. 1, *7* Sahara No. 1, *8* Kidson No. 1, *9* Wilson Cliffs No. 1), *7* limits of Canning Basin, *8* isopach contours, in meters

presented by Wells (1980) suggest a probable Early Devonian age for the upper Carribuddy Formation. However, most of the lower part of the formation, incorporating almost all salt deposits, may be much older, at least, Late Silurian [1].

A complete section of the Carribuddy Formation was penetrated by the Kidson No. 1, McLarty No. 1, Willara No. 1, Wilson Cliffs No. 1, and Parda No. 1 boreholes. These sections are discussed briefly.

In the Kidson No. 1 borehole, the top of the Carribuddy Formation is exposed at a depth of 2570.1 m. It is overlain by 733.1 m thick Tandalgoo Red Beds. The Carribuddy Formation is divided into five units designated A, B, C, D, and E.

Unit A, 371.2 m thick, is penetrated over a depth interval of 2570.1–2941.3 m. It consists of dolomite, claystone, siltstone, and sandstone with anhydritic dolomite bands and anhydrite veinlets and lenses.

Unit B, 526.4 m thick, is penetrated over a depth interval of 2941.3–3467.1 m. It is dominated by rock salt with thin claystone bands and anhydrite inclusions.

Unit C, 438.6 m thick, is penetrated over a depth interval of 3467.7–3905.5 m. It contains anhydritic, calcareous, saline, sandy claystone with limestone interbeds and anhydrite inclusions and lenses.

Unit D, 166.1 m thick, is penetrated at depths of 3905.5–4071.2 m. It is composed of rock salt interbedded with claystone and dolomite.

Unit E, 208.1 m thick, is penetrated at depths of 4071.2–4279.3 m. It is composed of claystone and dolomite with rare, thin bands and lenses of rock salt and anhydrite.

The total thickness of the Carribuddy Formation is 1709.2 m. It is underlain by Middle Ordovician deposits. As noted above, the Carribuddy Formation contains two salt-bearing units, D and B, separated by a thick, essentially clayey Unit C. Carbonate-clayey sequences can be recognized both at the base and at the top of the formation.

The McLarty No. 1 borehole penetrates the Carribuddy Formation over a depth interval of 452.0–1687.3 m. The thickness is 1235.3 m. All five units are discernible in the section.

Unit A, 223.4 m thick (depth interval 452.0–675.4 m), is composed of dolomite with shale interbeds at the top, while the base is dominated by shale and siltstone. Halite and anhydrite intercalations occur near the base.

Unit B, 739.4 m thick (675.4–1414.9 m), is composed of rock salt with minor mudstone and traces of anhydrite.

Unit C, 96.6 m thick (1414.9–1511.5 m), contains mudstone with halite inclusions.

Unit D, 66.4 m thick (1511.5–1577.9 m), is represented by alternate rock salt and mudstone; the lower part consists of mudstone with halite and anhydrite inclusions.

Unit E, 109.4 m thick (1577.9–1687.3 m), contains mudstone, anhydrite, and anhydritic dolomite at the top and limestone with minor shale and anhydrite at the base.

---

1 In earlier publications (Zharkov 1974, 1981), salt deposits of the Canning Basin were tentatively assigned to the Ordovician

This section shows that the thickness of salt-bearing Unit B greatly increases, whereas that of the lower salt-bearing Unit D decreases, as compared to the Kidson No. 1 borehole.

Composition and structure of the Carribuddy Formation changes essentially in the Willara Sub-Basin, where it is penetrated by the Willara No. 1 borehole over a depth interval of 1255.2–1873.9 m. The apparent thickness of the formation is 618.7 m. It is unconformably overlain by Permian, terrigenous deposits and underlain by a Middle Ordovician, carbonate-argillaceous sequence. Salt content of the section is very high despite a condensed thickness. According to Wells (1980), the upper Unit A becomes saliniferous in this section, while lower Unit D wedges out. The following subdivision is proposed for the section:

Unit A, 203.2 m thick (1255.2–1458.4 m), incorporates rock salt and anhydrite interbeds with a 24.4 m thick claystone bed.

Unit B, 278.0 m thick (1458.4–1736.4 m), consists of rock salt and anhydrite with intercalations of claystone, siltstone, and sandstone.

Unit C is 137.5 m thick (1736.4–1873.9 m); its upper part (to a depth of 1786.1 m) is composed of dolomite and anhydrite interbeds with limestone beneath.

The Sahara No. 1 borehole has penetrated only the upper Carribuddy Formation, represented by units A and B. Singleton (1966) described the following section:

Unit B is 87.2 m thick (depth interval 2121.6–2034.4 m).

1. White, red to brown, massive, and nodular anhydrite interbedded with rust-red, red to brown, locally light grayish-green, anhydritized mudstone with inclusions of salt crystals. . . . . 32.3 m
2. Light and reddish crystalline salt with gray, red, massive, and nodular anhydrite and brownish-buff to greenish-gray mudstone laminations. . . . . 54.9 m

Unit A, known as Stottid Shale, is 306.6 m thick (depth interval 2034.4–1727.8 m).

1. Red to brown, grayish to green, silty, calcareous mudstone; gray, rust-brown to beige, fine-grained, clayey, and sandy limestone; white and rust, locally red and purple, well and moderately rounded, anhydritized sandstone. Rocks are salinized; anhydrite lenticules and partings can be encountered . . . . 149.5 m
2. Red to brown, locally greenish-gray, slightly calcareous, and sandy mudstone with anhydrite inclusions and lenses. . . . . 85.4 m
3. Interbedded red to brown, orange to red, grayish to purple, locally green, calcareous mudstone; red to buff, locally gray to green siltstone with anhydrite inclusions and sandstone lenses; buff, light gray, rust, clayey limestone . . . . . 71.7 m

The Tandalgoo Red Beds, dominated by red to brown sandstone, siltstone, and locally mudstone, occur above.

A peculiar section of the Carribuddy Formation was penetrated by the Wilson Cliffs No. 1 borehole. Unit A consists mainly of dolomite 250.9 m thick. Unit B is free of rock salt. Unit C is 338.9 m thick and incorporates gray, calcareous shale with anhydrite nodules and lenses. Only Unit D contains salt bands. It is composed chiefly of halite with anhydrite traces, shale and sandstone interbeds. The thickness is 128.0 m. Unit E is 36.6 m thick and dolomitic in composition. The total thickness of the Carribuddy Formation is 754.4 m.

The Parda No. 1 borehole on the northern side of the Willara Sub-Basin shows that salt deposits are absent from that section of the Carribuddy Formation. The thickness of the formation decreases to 203.6 m. Its upper part (171.9 m) is composed of red to brown claystone with anhydrite interbeds and inclusions, while the lower part (31.7 m) is moderate to very anhydritic and contains green and grayish to green claystone and limestone with anhydrite crystals.

Generally, salt deposits occur in the central Canning Basin, namely in the Kidson and Willara Sub-Basins and on the Grossland, Broome, and Mid-Basin Platforms. They are also known from the Fitzroy Trough. Salt-bearing Unit D probably occupies the deepest parts of the basin and thins out towards their margins. Unit B is more extensive; it also runs into the Willara Sub-Basin, where Unit A also becomes salt-bearing. The maximal thickness, 3600 m, of the Carribuddy Formation was recorded in the Kidson Sub-Basin.

Salt structures were reported from some areas of the Canning Basin (Fig. 32). One of the salt domes — Frome Rocks — is located in the Fitzroy Trough. The Frome Rocks No. 1 borehole penetrated a 532.5 m salt unit beneath a 464 m cap rock consisting of dolomitic breccia with gypsum. The age of the salt deposits has not yet been established; they could be tentatively assigned to the Carribuddy Formation. Veevers and Wells (1961) suggested that rock salt penetrated by the Frome Rocks No. 1 borehole may be as old as Precambrian and hence, equivalent to the salt-bearing Bitter Springs sequence of the Amadeus Basin.

It should be emphasized that not only the age of salt deposits in the Canning Basin, but their composition, structure, and facies change in evaporites pose serious problems. Nevertheless, at present, one can say that the Canning Basin was once a large salt-producing reservoir. Its area exceeded 4.53×10^5 km^2, and that under evaporites was over 2.1×10^5 km^2 (Wells 1980). The total thickness of rock salt in the basin might have exceeded 500–600 m. The volume of rock salt in the basin could be rather high, approximately 1.5×10^4 km^3.

Carnarvon Basin

The forthcoming description of the Carnarvon Basin is based on data published by Wells (1980), which suggested the presence of salt deposits incorporated in the Dirk Hartog Formation [2] in the southern part of the basin, in the Gascoyne Sub-Basin.

The Carnarvon Basin is situated in western Australia, where it embraces coastal regions filled in by Ordovician, Silurian, Devonian, Permian, and Cenozoic sediments; their total thickness probably exceeds 6000 m. To the east, the basin is bounded by outcrops of Precambrian rocks in the Pilbara Block, and to the west, it extends into adjacent shelf zones of the Indian Ocean (Fig. 33).

Salt deposits were penetrated by the Yaringa No. 1, Hamelin Pool No. 1, and Hamelin Pool No. 2 boreholes. The area under the salt deposits is about 1300 km^2,

---

[2] In earlier publications (Zharkov 1974a, 1981), the Carnarvon Basin was determined as sulfate-bearing

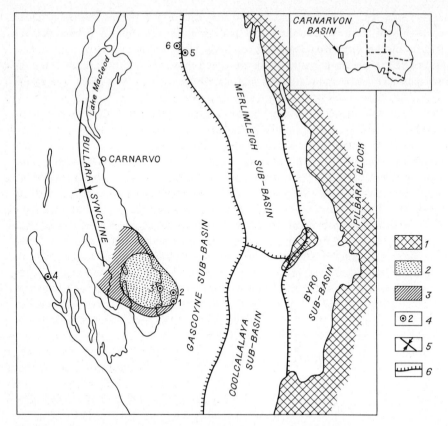

Fig. 33. Location of wells and known and possible extent of salt in Carnarvon Basin. After Wells (1980). *1* Precambrian outcrops, *2* known extent of salt, *3* possible extent of salt, *4* boreholes and their numbers (*1* Hamelin Pool No. 2, *2* Yaringa No. 1, *3* Hamelin Pool No. 1, *4* Dirk Hartog No. 17B, *5* Quail No. 1, *6* Wandagee No. 1), *5* syncline axis, *6* limits of sub-basins

including both near-shore and partly offshore territory of the Shark Bay. Structurally, the salt terrane is confined to the Bullara Syncline and probably extends farther north from the drilling sites. A number of boreholes in other parts of the Carnarvon Basin penetrated Late Silurian anhydrite. Anhydrite was reported from the Dirk Hartog No. 17B, Quail No. 1, and Wandagee No. 1 boreholes in the southern Carnarvon Basin, Gascoyne, and Merlinleigh Sub-Basins, and from the Pendock I.D. No. 1 borehole in the northern part of the basin.

The most complete section of the Dirk Hartog Formation was penetrated by the Yaringa No. 1 borehole. Henderson and Shannon assign a carbonate-sulfate-salt series exposed over a depth interval of 855.3–1526.4 m (571 m thick) into this formation. Playford and co-workers assign the other two overlying units, namely the psammitic red beds 529.1 m thick (depth interval 326.1–855.2 m) and an argillaceous-carbonate unit with anhydrite interbeds 179.8 m thick (depth interval 146.3–326.1 m) to the Dirk Hartog Formation as well. In this case, the thickness of the formation is 1380.0 m.

According to Playford and co-workers, the Dirk Hartog Formation consists of three units: the upper argillaceous-carbonate sequence with evaporites; the middle and the lower sequences with psammitic and carbonate-sulfate-salt, respectively. The above units were penetrated by the Hamelin Pool No. 1 and 2 boreholes. Wells recognized the upper and the middle units as Unknown A and Unknown B, respectively, thus excluding them from the Dirk Hartog Formation.

The identified rock salt beds are confined to the lower unit or to the Dirk Hartog Formation, with regard to reports of Henderson and Shannon. In turn, they can be divided into four units, namely A, B, C, and D, in descending order. These units are represented by alternate dolomite, shale, anhydrite, and rock salt. Salt beds are present in all the units, but differ in number. Unit B has the highest salt content and is named the Yaringa Evaporite Member. The Yaringa No. 1 borehole revealed seven salt beds in this member, totaling approximately 33.2 m in thickness. In the Hamelin Pool No. 1 borehole, the number of salt beds in Unit B decreases to four, and their total thickness is 17 m. However, there are only three salt beds, totaling about 5 m in thickness, in the Hamelin Pool No. 2. Rock salt in Units A and D was reported only from the Hamelin Pool No. 1 borehole; in Unit C, only from the Yaringa No. 1 borehole. Minor sylvite was observed in some halite beds. The upper salt beds occur at depths of more than 1120 m.

The Dirk Hartog Evaporite Formation is underlain by the Tumblagooda Sandstone of Ordovician age and overlain by Cretaceous deposits. The evaporites are Late Silurian (Middle and Late Ludlovian).

As a whole, the Carnarvon Basin is not well-known. The limited number of boreholes enables the outlining of only the south-eastern part of the salt-producing zone. Pandery suggested its extension further north-west and north from the Yaringa location. The Dirk Hartog No. 17B, Wandagee No. 1, and Quail No. 1 boreholes show that the salt-producing zone is confined by zones of argillaceous-sulfate-carbonate sedimentation to the south-west and north-east.

At present, it is impossible to give even an approximate estimate of the volume of rock salt in the Carnarvon Basin.

Silurian Evaporite Deposits in Other Basins

At present, Silurian anhydrite and gypsum are also known from the Severnaya Zemlya, Pechora, Baltic, Canadian Arctic Archipelago, Moose River, and Williston Basins.

In the Severnaya Zemlya Basin, gypsum bands and lenses are reported from the upper part of the Silurian carbonate sequence on Oktyabrskaya Revolyutsiya Island (Egiazarov 1970). The thickness of the gypsum-bearing unit is 75–80 m, but interbeds and lenses of gypsum and gypsinate rocks reach 15 m and can be traced along the strike for a distance of 300–400 m. The gypsum-bearing unit appears to be Přidolian in age.

In the Baltic Basin, sulfate rocks were recorded throughout the entire Wenlockian (Paprenyai horizon) and in the lower part of Ludlovian deposits (the lower Pagegyai horizon) (Pashkevichus 1961, Gailite et al. 1967). Sulfate rocks occur along the south and east margins of the basin. They are developed in south Estonia, east Latvia, and along the eastern and southern margins of Lithuania. This belt includes dolomite, dolomitized and dolomitic marls and rare interbeds of limestone, as well as inclusions, lenticular bands of gypsum, and gypsinate rocks. Gypsum was reported from the vicinity of the towns of Prenyai and Zhezhmuryai in Lithuania; Mezhtsiems, Aluksne, Akniste, and Druvas in Latvia. Sulfate rocks are not very thick and are apparently subordinate. The entire Paprenyai-Pagegyai dolomite sequence can be regarded as sulfate-bearing.

The Pechora sulfate-bearing basin is situated along the north-western margin of the East European Platform and includes the interfluve of the upper Pechora, Lemiu, and Ilyich Rivers (Pershina 1972, Filippova 1973, Atlas . . . 1972). Sulfate rocks are found in two sequences, one of which is situated at the level of the Kosyu and lower Adak horizons, the second belonging to the Lower Silurian upper Filippielsk horizon. Both sequences consist chiefly of dolomite and dolomitic marl with interbeds, lenses, and inclusions of gypsum.

In the Canadian Arctic Archipelago Basin, gypsum and anhydrite compose a rather thick unit of the lower part of the Read Bay Formation. Separate bands and lenses of gypsum reach 10 m; the total thickness of the unit exceeding 20–25 m. A sulfate-bearing unit was reported from south-eastern Ellesmere Island and Somerset Island (Fortier et al. 1963, Christie 1964). The extent of sulfate rocks is confined to the Moose River Basin. Gypsum and anhydrite occur as lenses and thin bands in the Kenogami River Dolomite of Late Silurian to Early Devonian age (Norris and Sanford 1969, Sanford and Norris 1975). In the Williston Basin, anhydritized dolomites are known from the Interlake Formation (Porter et al. 1964).

Chapter V
Devonian Salt Deposits

Devonian salt-bearing deposits are known at present from the following basins: North Siberian, Tuva, Chu-Sarysu, Morsovo, East European Upper Devonian, West Canadian, Michigan, Hudson, Bonaparte Gulf, and Adavale Basins. Besides, Devonian evaporites represented by gypsum and anhydrite were observed also in the Chulym-Yenisei, Minusinsk, Kuznetsk, Teniz, Turgai, Ili, Moesian-Wallachian, Canadian Arctic Archipelago, Moose River, Illinois, Central Iowa, Tindouf, Canning, Arckaringa, Carnarvon, and Bancannia Basins (Fig. 34).

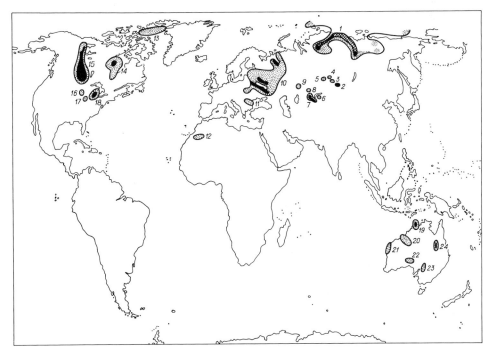

Fig. 34. Distribution of Devonian evaporites. Evaporite basins: *1* North Siberian, *2* Tuva, *3* Minusinsk, *4* Chulym-Yenisei, *5* Kuznetsk, *6* Ili, *7* Chu-Sarysu, *8* Teniz, *9* Turgai, *10* Morsovo and East European Upper Devonian, *11* Moesian-Wallachian, *12* Tindouf, *13* Canadian Arctic Archipelago, *14* Hudson, *15* West Canadian, *16* Central Iowa, *17* Illinois, *18* Michigan, *19* Bonaparte Gulf, *20* Canning, *21* Carnarvon, *22* Arckaringa, *23* Bancannia, *24* Adavale. For legend see Fig. 3

North Siberian Basin

Many Devonian evaporite strata have been recently found in the north of East Siberia, in the Verkhoyansk-Chukotka District, in Severnaya Zemlya, and on Vrangel Island; the strata may have formed within a single evaporite basin (Kalinko 1951, 1953, 1959, Egiazarov 1959, 1970, 1973, Bogdanov and Chugaeva 1960, Ivanov and Levitsky 1960, Menner 1962, 1965, 1967a, b, Fradkin 1964, 1967, Bgatov et al. 1967, Glushnitsky 1967, 1977, Cherkesova 1968, 1973, Salt ... 1968, Menner and Fradkin 1969, 1973, Glushnitsky and Menner 1970, Merzlyakov 1971, Menner et al. 1973a, b, Kolosov et al. 1974, Matukhin and Menner 1974, 1975, Divina and Matukhin 1975, Kameneva 1975, Sokolov and Matukhin 1975, Fradkin et al. 1977, Kislik et al. 1977, Sokolov et al. 1977).

Salt formations have been recognized in four regions of the North Siberian Basin: in the Norilsk Region, north-western Siberian Platform; in the central Tunguska Syneclise; and in the Khatanga and Kempendyai Depressions (Fig. 35).

In the Norilsk Region rock salt was found in the Zubovo, Manturovo, and Fokin Formations. Rock salt beds in the Zubovo Formation were exposed in the GOS-4 borehole on the right bank of the Neralakh River (Sokolov et al. 1977). The beds occur in the central unit of the Zubovo Formation. The unit is composed of alternating red and green mudstone, gray argillaceous dolomite, light gray anhydrite, and rock salt. Two salt beds, 4.3 m and 5.8 m thick, can be observed. Single grains of

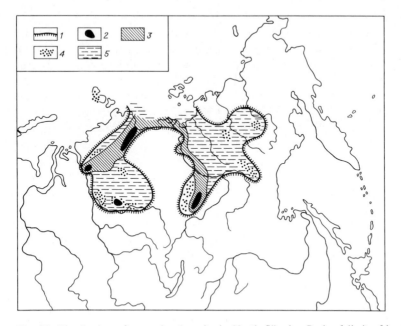

Fig. 35. Distribution of evaporite deposits in North Siberian Basin. *1* limit of basin, *2* known extent of salt, *3* possible extent of salt, *4* known areas of sulfate deposition, *5* area of possible evaporite deposition

sylvite occur in rock salt. Coeval salt deposits are thought to occur to the south in the Nizhnaya Tunguska River Basin where they are associated with the Nima Formation (Sokolov et al. 1977). Drilling data of the Turin TO-2 borehole show that the lower and central parts of the formation contain rock salt beds up to 17 m thick, in variegated mudstone and siltstone. This evidence suggests that the Zubovo salt formation is distributed over much of the north-western Siberian Platform and western Tunguska Syneclise. Salt deposits of the Zubovo Formation are usually referred to as the Early Devonian (Downtonian/Gedinnian), but some geologists assign them to the Late Silurian (Glushnitsky 1977).

Glushnitsky and Menner (1970) gave the following description of the Manturovo Formation section (in ascending order):

1. Argillite, red, poorly carbonate, silty with lenses of sandstone, fine-grained, feldspar-quartz and nodules of silty phosphorite. . . . .   3–8 m
2. Argillite, gray, green-gray and red-buff, silty of dolomitic, thinly and obscurely bedded with rare bands of gray clayed limestone and dolomite, black argillite with phosphate nodules, gypsum and anhydrite inclusions . . . .   15–25 m
3. Argillite, dolomitic, and clayey dolomite, variegated, with gypsum and anhydrite patches and interbeds, locally with a dolomite and anhydrite unit. . . .   50–70 m
4. Argillite, dolomitic; intercalated clayey dolomite and anhydrite, with a lenticular rock salt bed 5–8 m thick stretching for several hundred of meters, found in the Kharyalakh map area. The thickness of sulfate interbeds reaches 5 m, in places 15 m. . . . .   50–80 m
5. Dolomite, gray, green-gray, red-brown clayey dolomite argillite intercalated with thick gypsum or anhydrite . . . .   20–50 m

In the Mikchanda River Basin the Manturovo Formation is composed of a multiple alternation of rock salt beds up to 2–47 m thick, anhydrite, marl, and argillite. Its maximum thickness is 190 m. The formation is assigned to the Eifelian-Lower Givetian.

Menner and Fradkin (1973) have subdivided the Fokin Formation into two members. The lower member consists of thinly intercalated anhydrite, dolomite marl, and dolomite. The thickness of sulfate interbeds varies from 0.01–2 m. Sulfate and clayey-carbonate breccias are common in its lower part. A 26-m-thick salt unit was found on the west side of Pyasino Lake at the level of the breccias. Rock salt is intercalated with gray anhydrite and dolomite there. The thickness of the lower member of the Fokin Formation reaches 220–260 m. The upper member, 200–230 m in thickness, is separated by a 3–5 m thick fossiliferous bed from the lower one. The higher horizons are dominated at first by gray clayey-dolomite and sulfate rocks, then becoming rich in sand-silty admixture, a number of red-buff and gray bands increases while dolomite marl gives way to marl; breccia is also common (Menner and Fradkin 1973).

Rock salt beds are supposed to stretch from Norilsk Region towards the Khatanga Trough and Taimyr, where thick salt sequences may occur at a great depth (Salt. . . 1968, Menner and Fradkin 1973).

Salt-bearing deposits of the Kempendyai Depression are grouped into the Kygyltuus Formation assigned to the Upper Devonian (Fradkin 1964, 1967). A section of the formation in a well drilled on the south-western flank of the Kempendyai Brachyanticline contains many rock salt beds whose thickness varies from 3–10 to 100 and even 450 m. The formation was penetrated over the depth interval 2010–2840 m; however, its normal thickness is estimated at 630 m.

Fradkin (1967) presented the following description of the formation. The lower part (280 m thick) is dominated by rock salt (up to 70%); gypsum, marl, mudstone, siltstone, and tuffite occur in the form of thin interbeds (5–10 cm) and lenses. The central part (100 m thick) is composed of saline mudstone (75%) and siltstone (20%) with anhydrite, marl, and ash tuff interbeds. The upper part of the section (250 m thick) is marked by a relatively uniform alternation of rock salt beds with siltstone and sandstone; dolomitized marl, dolomite, anhydrite, tuff, and tuffite are subordinate. The overlying Namdyr Formation is composed of variegated sandy-marly and tuffaceous rocks with rare anhydrite and dolomite interbeds at the top.

The Upper Devonian salt deposits crop out also in cap rocks of salt structures in the Kempendyai and Kundyai Rivers basins. They are believed to occupy the central Kempendyai Trough, their thickness exceeding 4000 m (Menner and Fradkin 1973). However, salt sequences have not been found in adjacent areas where the Upper Devonian deposits crop out (Ygyattan and Berezovaya Depressions); the equivalents of the Kygyltuus and Namdyr Formations (Vilyuchan and Andylakh Formations) contain only gypsum interbeds.

The Nordvik-Khatanga District, northern East Siberia, has been known for a long time as an area of Devonian salt sequences. There are several exposures of rock salt in salt domes in that area; besides, geological and geophysical studies have shown the occurrence of salt-bearing deposits over a large area of the Khatanga River basin (Baranov 1946, 1947, Lappo 1946, Lappo and Kusov 1947, Kalinko 1951, 1953, 1959, Ivanov and Levitsky 1960, Menner 1965, Menner and Fradkin 1973). The total thickness of the sequence is unknown, and the apparent thickness reaches 500 m, although it has been penetrated by deep boreholes to a depth of 1250 m. The salt-bearing part of the section is overlain by a sulfate sequence of a variable thickness (0–140 m on the Nordvik Dome and 250–300 m on the Kozhevnikov Dome) which is composed of gypsum and anhydrite along with interbeds and inclusions of dolomite, dolomitized limestone, breccia, and gabbro-diabase. The above deposits are limestone- and dolomite-yielding fossils allowing the assignment of enclosing rocks to the Frasnian (Upper Devonian). Mirabilite beds up to 2 m thick and of epigenic origin have been found above the salt sequence in the Nordvik Dome. The majority of researchers consider the age of a saliniferous part of the section as Late Eifelian – Early Givetian.

Geophysical and geological data permit the assumption that salt-bearing deposits exist both south-east and south-west of the Nordvik District; they probably reach the South Yenisei Basin and could also join the salt sequences of the Norilsk District (Menner and Fradkin 1973).

Sulfate deposits are extremely widespread in the North Siberian Basin (Fig. 36). They are known almost throughout the Devonian section. The oldest sulfate-bearing sequence is associated with the Zubovo Formation in the north-western Siberian Platform, at Norilsk, in the coeval Koldin Formation, developed in the lower parts of Devonian deposits along the north-eastern slope of the Tunguska Syneclise in the Kotui River Basin and in the Olenek-Vilyui Rivers interfluve. Beds of anhydrite and gypsum in the Zubovo Formation commonly reach 5–10 m, locally even 20 m;

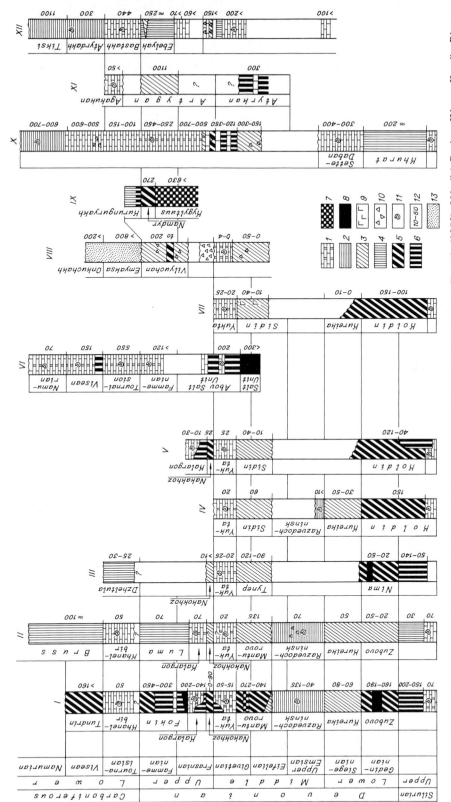

Fig. 36. Correlation of Devonian evaporite sections in Siberian Platform and adjacent areas. After Menner and Fradkin (1973). *I* Norilsk Region, *II* lower Kureika River, *III* middle Nizhnaya Tinguska River, *IV* middle Kotui River, *V* Changada and Kuntanakha Rivers, *VI* Khatanga Trough, *VII* Olenek-Vilyui Interfluve, *VIII* Ygyattan Trough, *IX* Kempendyai Trough, *X* Sette-Daban Range, *XI* Dzhardzhan Uplift, *XII* Kharyalakh Range. Litofacies complexes: *1* calc-dolomitic, *2* terrigenous and terrigenous-carbonate, *3* red terrigenous-carbonate, *4* gray dolomitic and terrigenous-carbonate, *5* red sulfate, *6* gray sulfate, *7* red salt, *8* gray salt, *9* inclusion of gypsum, *10* admixture of tuffaceous material and lava sheets, *11* marine fossils, *12* thickness, in meters, *13* gray terrigenous

and lenses, inclusions, and bands of sulfate rocks occur throughout the section. In the Koldin Formation gypsum is less common. In the western and central Tunguska Syneclise, and in the Nizhnaya and Podkamennaya Tunguska Rivers Basins, bands and lenses of gypsum and anhydrite occur in deposits of the Nima Formation at the same stratigraphic level. Hence, in the west, north-west, and central Siberian Platform rather thick (up to 200 m) sulfate-bearing deposits occur ubiquitously at the base of the Devonian. They seem to form a single sulfate-bearing sequence known as the Zubovo Formation. The same sequence is likely to include the lower Kureika Formation at Norilsk, i.e., its lower sulfate-bearing unit.

Additionally, the Lower Devonian gypsum-bearing deposits are known in two regions of the North Siberian Basin: the Taimyr Peninsula, in the lower Taryan Basin, and in Severnaya Zemlya. In the former region a gypsum-bearing unit of banded gypsum, argillaceous dolomite, and dolomitic clay, over 60 m thick, is recognized at the base of the Devonian section. In Severnya Zemlya, on Oktyabrskaya Revolyutsiya Island, interbeds and lenses of gypsum, up to 35–40 m thick, occur in dolomite of the carbonate sequence assigned to the upper Lower Devonian, corresponding probably to the Dezhnev Formation.

Some sulfate-bearing sequences can be observed in Middle Devonian deposits of the North Siberian Basin. One of them includes the Sidin Formation containing variegated mudstone, marl, dolomite, and siltstone with gypsum in the Kotui River Basin and in the Olenek-Vilyui interfluve. The Sidin sulfate-bearing sequence is regarded as an equivalent to the Manturovo sequence at Norilsk, where anhydrite and gypsum are very common as well. The Sidin Formation also appears to be developed in the central Tunguska Syneclise, where inclusions of anhydrite occur in the Tynepa Formation at the same stratigraphic level. The second sulfate-bearing sequence can be recognized as the upper part of the Yukta (or Makusov) Formation in the Norilsk Region where beds of gypsum and/or anhydrite occur in limestone and dolomite. The third sequence is known in Severnaya Zemlya. It includes the upper Rusanov Formation consisting of dolomite, dolomitized limestone, marl, and gypsum with a total thickness of 135 m; gypsum locally reaches 8–10 m. The fourth sulfate-bearing sequence including deposits of the Atyrkan Formation, where beds and units of gypsum and anhydrite occur, is known in the northern Verkhoyansk Range, in the Uel-Siktyakh River Basin. The fifth sequence is widespread in the Sette-Daban Range, southern Verkhoyansk Range. Sulfate rocks are associated there with the upper sequence of the Burkhala Formation. The sequence has been recently identified as the separate Zagadochnino Formation (Khaiznikova 1970, Sedimentary and Volcanogenic-Sedimentary Formations . . . 1976). The sixth Middle Devonian sulfate-bearing sequence can be recognized as the upper Sebechan Formation known in the basin of the Sebechan, Khobochalo, and Danynie Rivers in the Tas-Khayakhtakh Range. Finally, the seventh Middle Devonian sulfate-bearing sequence includes rocks of the Vayakh Formation on the eastern slope of the Omulev Mountains where gypsum and anhydrite are very common.

Upper Devonian sulfate-bearing sequences are known in the north-western and central Siberian Platform, in Severnaya Zemlya, and on Vrangel Island. In the Siberian Platform sulfate rocks are most widespread in the Norilsk Region, where they make up the essential part of the Nakokhoz, Kalargon, and Fokin Formations.

The Nakokhoz sulfate formation is most extensive. It extends from the Norilsk Region farther east into the Kotui River Basin. A sulfate-bearing type of the Kalargon section is recognized north and west of Norilsk. Rock salt beds are present in the Fokin Formation. However, they occur in the lower part; only there can a salt-bearing sequence be identified. As for the upper part of the Fokin Formation, consisting chiefly of gray and red to brown marl with gypsum and anhydrite, it can be regarded as a separate upper Fokin sulfate-bearing sequence.

In the Kempendyai Depression, the upper part of the Devonian section refers to the Namdyr Formation composed of red siltstone and sandstone interbedded with dolomite, limestone, tuff, and tuffite with anhydrite bands. In the Ygyattan Basin the Vilyuchan Formation contains sulfate rocks of the same age. It is probable that the Namdyr sulfate-bearing sequence is widespread in the Vilyui Syneclise. In Severnaya Zemlya the sulfate-bearing sequence includes the Matusevich Formation where lenses of gray to dirty gray gypsum are present in dolomite and marl units among variegated sandstone and siltstone. A separate Upper Devonian sulfate-bearing sequence can be recognized on Vrangel Island. It is developed in the upper reaches of the Khishchnikov River and made up of sandstone, limestone, and shale interbedded with gypsum. Until recently these deposits were referred to as Early Carboniferous (Tilman et al. 1970), but later the Late Devonian age was confirmed (Kameneva 1975).

Thus, Devonian evaporite deposits are recognized over a vast territory above 2,500,000 km^2 of northern East Siberia and the north-eastern USSR. Approximate estimates give a figure of 1.35×10^4 km^3 as a minimal possible volume of rock salt.

Tuva Basin

In the Tuva Basin the Devonian salt-bearing sequence has long been known in foothills of the western Tannu-Ola Range in southern Tuva (Matrosov 1974, Levenko 1955, 1956, 1960, Lepeshkov et al. 1958, Ivanov and Levitsky 1960, Pastukhova 1960, 1965, Zaikov et al. 1967, Minko 1972, Meleshchenko et al. 1973, Kolosov et al. 1977) (Fig. 37).

The Tuz-Tag deposit has been mined there for rock salt for several centuries. Salt deposits of the Tuva Basin belong to the Ikhedushiingol Formation of the Upper Eifelian–Lower Givetian age (Zaikov et al. 1967). They crop out and were drilled at the Tuz-Tag deposit and 8 km farther east at the Torgalyg deposit. The thickness of the salt sequence is more than 600 m. Zaikov and co-workers have distinguished three salt units within the salt sequence at the Torgalyg deposit, viz., lower (36 m), middle (48 m), and upper (above 120 m). These salt units contain clayey and silty interbeds several meters thick; they are separated by a unit of gypsinate saline clay 130–170 m thick. Clay is commonly folded and brecciated. At the Tuz-Tag deposit rock salt is intercalated with salt- and anhydrite-bearing clays and anhydrites up to 15–20 m thick; the salt content reaches about 68% (Ivanov and Levitsky 1960).

Minko (1972) suggests that the salt unit of the Tuz-Tag deposit is an equivalent of the uppermost, third, unit at the Torgalyg deposit. In this case, the total thickness

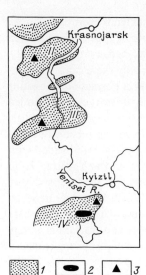

Fig. 37. Distribution of Middle Devonian deposits in Tuva and adjacent areas. After Minko (1972). *1* Middle Devonian basinal deposits, *2* areal extent of salt in Tuz-Tag and Torgalyg deposits, *3* gypsum deposits. Deposits: *I* Nazarovo, *II* Chulym-Yenisei, *III* Syda-Erba, *IV* Tuva

of the Middle Devonian salt sequence in the Tuva Basin can be no less than 900 m. A composite section of the salt-bearing part of the Ikhedushiingol Formation prepared by Minko (1972) using borehole data is shown in Fig. 38.

Salt deposits of the Tuva Basin are overlain and underlain by red, mainly terrigenous formations. The presence of complex iron chloride — rinneite — is a peculiar feature of the salt sequence (Kolosov and Pustyl'nikov 1967). Upper horizons contain small sylvite interbeds several millimeters thick and potassium mineral inclusions scattered in rock salt (Ivanov and Levitsky 1960, Kolosov and Pustyl'nikov 1967, Zaikov et al. 1967).

We have good reason to assume the presence of rather thick salt deposits between the Tuz-Tag and Torgalyg deposits (Zaikov et al. 1967). In this case the area covered by rock salt accounts for 18–20 km^2.

However, many workers suggest a much wider distribution for the Devonian salt sequence. Ivanov (in: Ivanov and Levitsky 1960) emphasizes that the sequence goes farther west into Mongolia. Zaikov and Onufrieva (Zaikov et al. 1967) point to several sites for deep occurrence of rock salt. They are: the upper Kurbun-Shivi Creek area, 13 km east of the Tuz-Tag deposit, the area between the Elegest River to the north-east and Uryk-Nur Lake to the south-west. Pinneker (1968) believes that the Samagaltai Lake area has high potential in the search for rock salt, where it is exposed on the left side of the Tee-Khem River at the foot of Mount Khaiyrkan slope. This site is 160 km from Tuz-Tag and hence the areal extent of rock salt may exceed several ten thousands of kilometers.

The Tuva Basin seems to extend far south and south-west into Mongolia and its now accepted boundaries are valid only for the north-eastern marginal part of this salt basin.

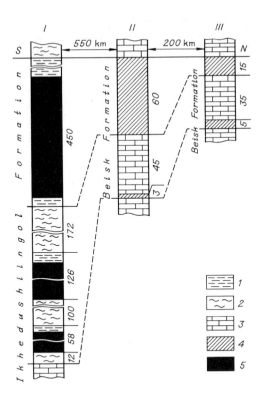

Fig. 38. Correlation of sections of Middle Devonian halogenic deposits in *I* Tuva, *II* Minusinsk (Khamazas gypsum deposit). *III* Chulym-Yenisei (Dadonkov gypsum deposit) Depressions. After Minko (1972). *1* clay, *2* siltstone, *3* limestone and marl, *4* gypsiferous deposits, *5* rock salt. Thickness is given in meters

At present it is impossible to estimate the volume of rock salt. In areas where salt-bearing deposits have been found, the average area is 10,000 km² and the average thickness is 100 m; the volume of rock salt equals 1×10^3 km³. The actual volume of rock salt with allowance made for an area of salt accumulation in Mongolia is probably much greater.

Chu-Sarysu Basin

The Chu-Sarysu Depression is situated in western Central Kazakhstan. To the south-west and north-west it is bounded by the Karatau and Ulutau Ranges, while by the Chu-Ili Mountains and Kirghiz and Kendyktas Ranges to the north-east and south and south-east, respectively (Fig. 39). The depression is filled in by thick Middle and Upper Paleozoic deposits with rock salt units found at different stratigraphic levels. The majority of workers (Petrushevsky 1938a, b, Zaitsev and Pokrovskaya 1948, Aleksandrova and Borsuk 1955, Ditmar 1961, 1963, 1965, 1966, Varentsov et al. 1963, Shakhov 1965a, b, 1968, Gulyaeva et al. 1968, Kunin 1968, Visloguzova et al. 1968, Orlov et al. 1969, Blagovidov 1970, Li and Mailibaev 1971, Mikhailov et al. 1973, Sinitsyn et al. 1977) assign them to the Upper Devonian or Lower Carboniferous. Salt deposits were observed in the following three sub-basins: Dzhezkazgan, Tesbulak,

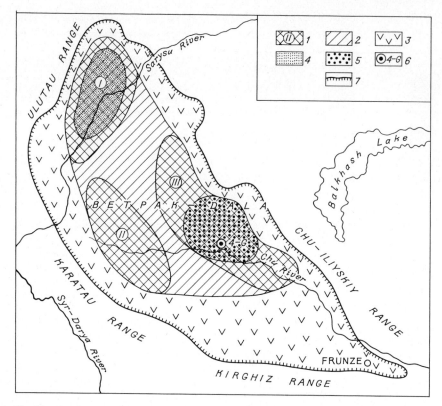

Fig. 39. Distribution of Devonian-Carboniferous salt deposits in Chu-Sarysu Basin. *1* known areal extent of salt in sub-basins and their numbers (*I* Dzhezkazgan, *II* Tesbulak, *III* Kokpansor), *2* possible area extent of salt, *3* areas where gypsum and anhydrite occur among terrigenous-carbonate deposits, *4* Sarysu salt dome region, *5* Betpak-Dala salt dome region, *6* location of the 4-G well, *7* limit of Chu-Sarysu Basin

and Kokpansor, all penetrated by deep boreholes. There are a number of salt domes grouped under the name Betpak-Dala (Ditmar 1961, 1966), Bestyube (Kunin 1968, Visloguzova et al. 1968), or Lower Chu in the central Kokpansor sub-basin, in eastern Betpak-Dala. All in all, 25 salt structures have been recognized; the most well known among them are the Bestyube and Kentaral Domes.

Well 4-G on the northern flank of the Bestyube Dome has penetrated a salt sequence and exposed sub-salt deposits. However, geophysical data show that the well penetrated only a small part of salt-bearing deposits (810 m thick) building up an overhang of a salt plug (Ditmar 1966, Gulyaeva et al. 1968). The total thickness of the salt sequence could exceed 1000 m. According to Shakhov (1965) and Filipiev, in the upper part of the sequence (depth interval 1070–1260 m), rock salt is intercalated with mudstone, anhydrite, dolomite, dark gray marl, and variegated sandstone, while its lower part consists of thick rock salt beds alternated with rare mudstone and siltstone interbeds and in places with anhydrite. Salt-bearing deposits are underlain by red siltstone, sandstone, and mudstone. Their apparent thickness is 622 m, and they are overlain by dark gray and gray bituminous limestone and marl with mudstone interbeds of the same color.

Salt deposits in Bestyube District occur at different depths. On salt domes, rock salt beds come to the surface locally or lie at depths not more than 100–300 m. Between domes, a depth of 1500–2000 m for cap rock was inferred from geophysical data obtained by Povzner and Kunin (1968), and Khabibulin et al. The areal extent of salt deposits has not been estimated. Kunin (1968) suggests the presence of a rock salt sequence in a salt dome trough, irregular in shape and measuring 40 × 50 km. Ditmar (1966) confines the extent of salt sequences in Betpak-Dala to the Tesbulak Creeks area to the north and Tastin and Talas Uplifts to the south. Salt deposits are supposed to run north-westwards where they join salt sequences of the Sarysu District. A stratigraphic position of the salt sequence is determined by its occurrence beneath the Tournaisian (Carboniferous) carbonate formations. Underlying red beds as a rule are nonfossiliferous and tentatively placed into the Upper Devonian according to their position in the section.

The second area where numerous salt domes have been established is the Dzhezkazgan sub-basin. It is the Sarysu District (Zaitsev 1940, Ditmar 1961, 1965, 1966, Kotlyarov et al. 1965, Gulyaeva et al. 1968, Orlov et al. 1969). The Sarysu deformations are typical domes with piercement cores and, possibly, salt plugs in the cap rock. It has been confirmed in the case of the Rakhmetnura Dome where deformed gypsiferous rocks composing the cap rock are exposed in the center, and where, at depth, there is a deformed rock salt unit .The salt plug of the Rakhmetnura Dome was traced by a seismic survey to a depth of 2000 m. The Bureinak, Kok Tyube and, possibly, Kadzham Nura, Shukuntai Nura, and other domes have similar structures. The gypsiferous deposits of many domes show a tectonic contact with a Visean carbonate unit which accounts for the pre-Visean age of salt deposits in the area. The salt sequence is assigned either to the Lower Carboniferous (Zaitsev 1940, Orlov et al. 1969) or Upper Devonian, or Upper Devonian-Lower Carboniferous (Varentsov et al. 1963, 1964, Ditmar 1966).

Wells drilled on the Rakhmetnura Dome have revealed the peculiar structural features of the salt sequence. According to Orlov and co-workers (1969) the sequence consists of an upper part composed of anhydrite intercalated with mudstone, dolomite, and limestone over 100 m thick, and a lower rock salt section. The apparent thickness of the salt sequence is 778 m. Rock salt accounts for 77%–99% of the section. Salt deposits cover an area of about $10-15 \times 10^3$ km^2.

Two salt sequences have been penetrated by deep wells in the Tesbulak sub-basin (Sinitsyn et al. 1977). The lower sequence is composed of rock salt with dolomite and anhydrite interbeds. Its thickness reaches 500 m. It belongs to the Upper Devonian. The upper sequence consists of rock salt alternating with mudstone and anhydrite and is about 350 m thick. It is tentatively identified as Early Carboniferous. The sequences are separated by carbonate-terrigenous deposits that are up to 100 m thick.

The net areal extent of the Upper Devonian-Lower Carboniferous rock salt in the Chu-Sarysu Basin probably exceeds 50×10^3 km^2. An average thickness based on seismic data for the area discussed totals of 100 m. In this case, the volume of rock salt in the Chu-Sarysu Basin must equal 5×10^3 km^3.

Morsovo Basin

The Morsovo salt basin encompasses the central, eastern, and south-western parts of the East European Platform (Tolstikhina 1952, Sarkisyan and Teodorovich 1955, Devonian... 1958, Ivanov and Levitsky 1960, Strakhov 1962, Pomyanovskaya 1964, Tikhy 1964, 1967, 1973, Khizhnyakov and Pomyanovskaya 1967, Lyashenko and Lyashenko 1967, Aronova et al. 1967, Tikhomirov 1967, On the Potassium... 1970, Utekhin 1971, Blagovidov et al. 1972, Filippova and Krylova 1973, Golubtsov 1973, Gurevich et al. 1973, Kirikov 1973, Summet 1973, Kislik et al. 1976a, b, 1977). Because of the distribution of evaporite deposits (Fig. 40) confined to the Morsovo horizon and its equivalents, viz., the lower and middle Narova horizon of the Pripyat Depression and Baltic area as well as part of the Lopushany Formation, Lvov Depression, the majority of workers assign them to the Eifelian age (Middle Devonian).

The basin occupies the central Moscow Syneclise and stretches as a narrow embayment westward into the Baltic area, south-eastward into the Pripyat and Lvov Depressions, and south-eastward along the Ryazan-Saratov Trough towards the Caspian Depression. The embayment-like extension of the basin is seen also in the Dnieper-Donets Depression, where coeval halogenic deposits do occur at a greater depth.

Salt deposits in the Morsovo horizon have been found in a large number of boreholes in two areas, i.e.: (1) the central Moscow Syneclise and (2) the Pripyat Depression.

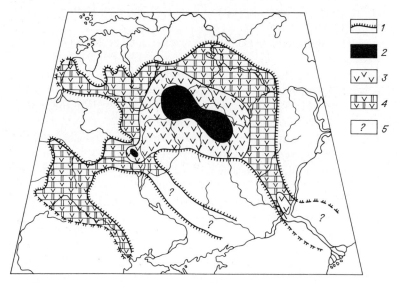

Fig. 40. Distribution of evaporite deposits in Morsovo Basin, East European Platform. *1* limits of evaporites, *2* known areal extent of salt, *3* predominantly sulfate deposits interbedded with anhydrite-dolomite and dolomite, *4* areas where laminae of sulfate rocks lie among carbonate and terrigenous-carbonate strata, *5* areas of possible evaporite occurrence at great depths

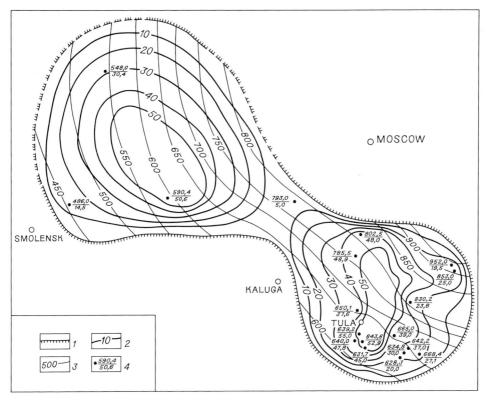

Fig. 41. Isopach map of depths of occurrence of Morsovo horizon salt in Moscow Syneclise. *1* limits of salt, *2* isopach contours of salt, in meters, *3* isolines showing depths to top of rock salt, in meters, *4* wells: *numerator* indicates depth to top of occurrence from sea level, in meters; *denominator* indicates thickness of rock salt, in meters

In the Moscow Syneclise an area occupied by a rock salt sequence of the Morsovo horizon stretches from south-east to north-west in a band about 500 km long and 170–200 km wide (Fig. 41). The southern boundary of the rock salt distribution goes south of the towns of Tula and Vyazma while the northern boundary follows the line connecting such towns as Zaraisk, Serpukhov, Zubtsov. To the north-east a salt field is situated between the towns of Nelidovo and Toropets and to the south-east in the vicinity of the towns of Mosolovo and Gorlovo, Ryazan District. The salt sequence has been studied in great detail.

Rock salt exhibits the greatest thicknesses in two areas, viz., (1) Vyazma to the north-west and (2) Tula-Serpukhov to the south-east. In both areas the thickness slightly exceeds 50 m; a maximal thickness (55 m) was recorded near the village of Novobasovo south of Tula. The depth to the rock salt top is 450 –950 m, the top of the rock salt unit plunging monoclinally from south-west to north-east.

In all the sections, light gray, buff-gray or dark gray rock salt builds up irregularly with alternating interbeds one to several centimeters, and rarely reaching 10–15 cm;

hence the distinct laminated pattern of the unit is observed. Rock salt is often represented by zonal-granular milkwhite halite. The presence of thin (a fraction of a millimeter to several centimeters) anhydrite partings emphasizes the laminated structure of the salt unit. Annual sets of rocks are easily discernible. Thicker anhydrite or anhydrite dolomite interbeds are not very common; as a rule they are confined to the middle and upper parts of the rock salt unit.

Rock salt interbeds rich in potash minerals (On the Potassium . . . 1970) were observed in a salt unit near the town of Yartsevo, Smolensk District.

The spatial position of the Morsovo salt member in the Moscow Syneclise has some peculiar features (Tolstikhina 1952, Devonian . . . 1958, Ivanov and Levitsky 1960, Strakhov 1962). First, the rock salt member is not underlain by a thick anhydrite unit. In areas with high salt content and considerable thicknesses of rock salt (Novobasovo, Petelino, Tula, Yasnaya Polyana, Serpukhov and others) the rock salt member rests directly on a clay member with thin dolomite and rare clayey anhydrite interbeds in the uppermost horizons. The thickness of the member varies from several meters to 25 m. Mudstone and clayey dolomite often contain abundant fossil fish, lingulids, and estheriids.

A clay unit underlying salt is confined to the central part of the salt accumulation terrane, and towards the southern and northern margins it is enriched in coarser terrigenous material. To the north-west the clay unit becomes more carbonate and terrigenous and in the Vyazma, Yartsevo, and Nelidovo areas it is built up of marl and dolomite with sandstone interbeds. At the same time, rather thick anhydrite interbeds are common in the uppermost strata underlying salt-bearing deposits. In Smolensk District a thickness of anhydrite reaches 1–2 m, despite a considerable proportion of clay material.

The Morsovo salt member in the Moscow Syneclise is overlain everywhere by an anhydrite unit of the same thickness. The thickness of the latter is maximal in areas of highest salt content, i.e., 23–35 m. In marginal parts the thickness of anhydrite decreases to 10–15 km and the thickness of underlying rock salt decreases as well.

In the central Moscow Syneclise a sulfate unit is overlain by a marl-dolomite unit assigned in some regions to the Mosolovo (Devonian) horizon and in others to the Morsovo horizon. Even higher there is a richly fossiliferous limestone member of the Mosolovo horizon.

The decrease in the thickness of the salt member towards the margins of the basin is accompanied by a facies change. So, from south to north, namely, from the town of Yartsevo to Krasnyi Kholm, the salt member is replaced along the strike by anhydrite which, in turn farther north, grades into marl and dolomite very similar in composition to rocks of the Mosolovo horizon. The same trend is observed for the Morsovo horizon from Tula towards Gorlovo town. The area of the Morsovo horizon sulfate member is much more extensive than that of the salt member. Sulfate rocks of the Morsovo horizon were drilled both in the western and north-western (Toropets, Kuvshinovo, Maksatikha, Krasnyi Kholm, and other boreholes), and northern (Shatura, Pereslavl-Zalessky), south-eastern (Ryazan, Mosolovo, Gorlovo, Ryazhsk, and other towns) and southern (Plavsk, Baryatino, and other towns) parts of the syneclise. Anhydrite interbeds in all the areas mentioned are several meters thick. As a rule they compose the lower Morsovo horizon, whose thickness is within 20–40 m (Tikhomirov 1967, Utekhin 1971).

Sulfate rocks have been distinguished among the Morsovo deposits in a zone of the Ryazan-Saratov Trough south-east of the Moscow Syneclise. They were penetrated by boreholes in Morsovo, Zubovaya Polyana, Kaverino, and other regions. The sulfate part of the section does not exceed 40 m. Anhydrite interbeds up to several meters thick are common in the lower Morsovo horizon where they are intercalated with dolomite and dolomite marl, whereas the upper part of the horizon consists of dolomite and dolomite marl with rare limestone interbeds. Farther south-east anhydrite among the Morsovo deposits is known from the Volga Monocline, Don-Medveditsa Swell, and eastern Khoper Monocline (Karpov 1970). The thickness of the sulfate part increases there to 100 m, while the thickness and proportion of anhydrite increases as well.

Sulfate rocks are widespread in equivalents of the Morsovo horizon in the Baltic area in places where coeval formations are placed into the Narova horizon (Kayak 1962, Tikhyi 1964, Polivko 1967, Tikhomirov 1967). In Lithuania gypsum and anhydrite interbeds among marl and dolomite of the lower Narova horizon have been penetrated by the Stonshkyai, Kauno-Vone, Aregala, Krekepava, and Shvenchenas boreholes. The thickness of a gypsum member of the Narova horizon varies from 40 to 70—80 m. In some districts of Lithuania (Adze, Sturi, Stachunyai, Laanemetsa, Blidene, and others) the lower Narova horizon contains interbeds and nests of gypsum and anhydrite among a lower carbonate-clayey member with a variable thickness of 20—45 m (Berzin and Ozolin 1967, Polivko 1967).

The second salt region of the Morsovo Basin is the Pripyat Depression. Thin rock salt interbeds have been found there among the Narova horizon deposit in the Vishan Field (Lithology ... 1966, Shevchenko 1971). Like the Morsovo horizon of the Moscow Syneclise rock salt in the Vishan Field occurs in the lower horizon of a salt-bearing section at the base of a sulfate-dolomite-marl member of the Narova horizon; this member can be traced for a great distance in the Pripyat Depression outside the salt accumulation area. The thickness of rock salt beds does not exceed 1 m. One can suggest the occurrence of salt beds only in the western Rechitsa monoclinal terrace.

In other parts of the Pripyat Depression (Lithology ... 1966) the Narova horizon is composed exclusively of sulfate rocks, namely, anhydrite and anhydrite-dolomite which build up a sulfate-dolomite unit. Its thickness on the Rechitsa monoclinal terrace (villages of Davydovka and Tishkovka, towns of Vishi and Rechitsa) varies from 16—35 m (Shevchenko 1971). The anhydrite content is not higher than 3—5%. In other areas of the Pripyat Depression a sulfate section of the Narova horizon is almost the same, i.e., near the village of Gavrilchitsa a sulfate-dolomite-marl unit is 23 m thick with a sulfate content of 3%; in the town of Starobino it is 24 m and 2%; in Chervonaya Sloboda it is 50 m and 8%; in Kopatkevichi field it is 31 m and 7%; in Anisimovka it is 25 m and 5%. The sulfate member of the Narova horizon was also observed near Polotsk, Mogilev, and some other areas of Byelorussia; however, its thickness decreases there to 10 m (Makhnach 1958, Golubtsov and Makhnach 1961, Makhnach and Kurochka 1964, Lithology ... 1966).

The Lvov Depression is another area of Eifelian (Middle Devonian) halogene deposit distribution which can be tentatively placed in the Morsovo Basin. Anhydrite, gypsum, and anhydrite-dolomite interbeds and units have been found among the middle Lopushany member which is considered equivalent in age to the Narova horizon, the Pripyat Depression, and to the Morsovo horizon, the Moscow Syneclise (Pomyanovskaya 1964, Khizhnyakov and Pomyanovskaya 1967, Tikhomirov 1967). The thickness of the member varies from 34–42 m. In the Lopushany area it consists of gypsinate buff-gray dolomite with interbeds of greenish-gray mudstone, siltstone, and sandstone. The member becomes more gypsiferous to the east and to the north-east, and gypsum begins to dominate the section. To the west the member is primarily carbonate whereas gypsum and anhydrite are minor.

Sulfate rocks in the Eifelian deposits were reported from south-western Poland where they have been penetrated by deep holes near Lyublin (Pajchlowa 1967). At present we have good reason to suggest that the area of Middle Devonian halogenic rock accumulation extended far south of the Lvov Depression and may have embraced regions south-west of the Ukrainian Shield, including the Prut Basin. Middle Devonian sulfate deposits (Safarov and Kaptsan 1967, Kaptsan and Safarov 1969) found in the Dobruja Trough contribute to the assumption. They were exposed at a depth of more than 2500 m by wells on the Baurchin Anticline and are composed of alternating dark gray, rare white-greenish anhydrites, sugar-like crystalline, massive and laminated, fractured with partings of dolomitized limestone and gray, locally brownish and creamy dolomite, dense, crypto- and microcrystalline, fractured, with cracks "healed" by anhydrite veinlets, and dark gray, greenish mudstone, dense, hard, and uncalcareous, totalling 360 m.

Recent data show that the area occupied by rock salt in the Moscow Syneclise is no less than 80,000 km^2. If the mean thickness of salt for the entire area is taken as 10 m which is quite permissible being the minimal possible value, then the volume of rock salt accumulated during Eifelian time in the Moscow Syneclise could be no less than 800–900 km^3. With allowance made for rock salt of the Pripyat Depression and other regions the total volume of salt in the Morsovo Basin must be about 1000 km^3.

East European Upper Devonian Basin

Upper Devonian time was marked by several periods favorable for salt accumulation in different areas of the East European Platform (Fig. 42). One of these areas is situated between the Voronezh Massif and the Byelorussian Shield on the north and the Ukrainian Shield on the south. The Dnieper-Donets and Pripyat graben-like Depressions on the south-east and north-east, respectively, separated by the Bragin (Bragin-Chernigov) Uplift, can be recognized there. Salt-bearing sequences occupy a narrow zone 950–970 km long and 80–130 km wide; the areal dimension of rock salt is about 65,000–70,000 km^2.

Another region of salt accumulation was the Timan (Pritimansky) Trough. The areal extent of the salt series has not been outlined there.

Sulfate rocks such as anhydrite and gypsum occur elsewhere among the upper Devonian deposits. They were recorded in the Baltic and Moscow Syneclises, western Volga-Ural Region, western Pechora Syneclise. Evaporite deposits are discussed below for each individual major tectonic element.

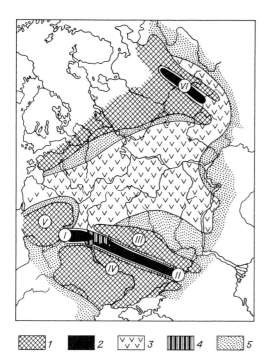

Fig. 42. Distribution of Upper Devonian evaporites on East European Platform. *1* Upper Devonian absent, *2* salt basins, *3* areas of wide distribution of sulfate deposits, *4* Bragin (Bragin-Chernigov) Uplift, *5* areas of sporadic distribution of sulfate rocks. Tectonic structures: *I* Pripyat Depression, *II* Dnieper-Donets Depression, *III* Voronezh Massif, *IV* Ukrainian Shield, *V* Byelorussian Massif, *VI* Timan Trough

Pripyat Depression

Upper Devonian salt-bearing deposits in the Pripyat Depression are rather well known (Bruns 1956, Fursenko 1957, Kirikov 1959, 1963a,b, 1973, Shcherbina 1959, 1960a, b, 1961, 1962, Golubtsov and Makhnach 1961, Bayazitov 1963, Gorkun 1964, Kislik and Lupinovich 1964, 1968, Makhnach and Kurochka 1964, Kislik 1966, Kurochka 1966a, b, Lithology ... 1966, Eroshina 1968, 1981, Lupinovich et al. 1969a, b, 1970, Eroshina and Vysotsky 1972, 1975, Shchedrovitskaya 1973, Eroshina et al. 1976, Kislik et al. 1976a, b, 1977, Protasevich 1976a, b, Shablovskaya et al. 1976, Bordon et al. 1977, Garetsky and Konishchev 1977, Korzun 1977, Zelentsov and Zingerman 1977, Eroshina and Kislik 1980, Kityk et al. 1980, Kityk and Galabuda 1981, Petrova and Sedun 1981, Vysotsky et al. 1981). Salt deposits have been recognized in an area of about 26,000 km^2 extending for 120–130 km and 150–200 km from north to south and from west to east, respectively. For the distribution of Upper Devonian deposits in the Pripyat Depression see Fig. 43.

Several evaporite horizons were distinguished in the Upper Devonian section. The oldest of them is the middle Shchigrov horizon composed of anhydrite and anhydrite-dolomite interbedded with argillaceous dolomite, dolomitic marl, and argillaceous limestone. This part of the section is recognized as a separate sulfate-bearing unit, the thickness of which ranges from 7–20 m in the eastern and southern parts of the Pripyat Depression, and reaches 20–22 m in the central part of the western Pripyat Depression (Golubtsov and Poznyakovich 1963, Lithology ... 1966).

Fig. 43. Composite section of Devonian salt deposits of Pripyat Depression. *1* sandstone, *2* clay, *3* limestone, *4* dolomite, *5* marl, *6* dolomitic marl, *7* clayey marl, *8* anhydrite, *9* gypsum, *10* rock salt, *11* potash salt, *12* tuff and tuffite. Horizons: *p + nr* Pyarnu and Narva, *lz* Luzhsk, *sch* Shchigrov, *sem* Semiluki, *pet* Petin, *vor* Voronezh, *evl* Evlanovo, *lv* Liven, *zd-el* Zadon-Elets, *lb-d* Lebedyan-Dankov

Another level of evaporite distribution starts at the base of the Voronezh horizon built up of anhydrite and anhydrite-dolomite rocks; their abundance gradually increases towards the top of the Podsolevaya [3] sequence. In the eastern and southern regions of the Pripyat Depression (towns of Rechitsa, Strelichevo, Narovlya, and village of Anisimovka) anhydrite and anhydrite-dolomite beds are known only from the upper Voronezh horizon. A sulfate-bearing unit 12–37 m thick including intercalations of marl, argillaceous dolomite, limestone, and anhydrite can be traced there. In the western Pripyat Depression sulfate rocks occur throughout the entire Voronezh horizon 32–48 m thick. The proportion of sulfate rocks increases at the level of the Evlanovo horizon.

On the whole, the Voronezh-Evlanovo level of sulfate rock distribution marks the initial stage of a major cycle of evaporite sedimentation which terminated in the accumulation of the so-called lower salt sequence of the Pripyat Depression. Most investigators assign an Evlanovo-Liven age to it. The forthcoming description of the sequence is based on the publications by Eroshina and Vysotsky (1975), Eroshina and Kislik (1980), and Eroshina (1981).

The lower salt-bearing sequence is represented by saliniferous and terrigenous-sulfate-carbonate types of a section. Salt deposits compose most of the depression (21,000 km^2). To the north- and south-west they are replaced by salt-free deposits occurring also in the Gorodok-Hatets tectonic step. To the south-west, on the southern flank, and to the east salt deposits are replaced by sandy silt and volcanogenic sediments, respectively. The depth to the top of the lower salt sequence ranges from 1040–4252 m. Small depths are recorded in the south-eastern and north-western parts of the depression and the depth to the top increases along the flanks on the south and north.

Over most of the area the thickness of the sequence is 400–600 m varying from 96–200 m in the north-western to 800–910 m in the southern and south-eastern parts of the depression. In some localities a great thickness of 1200–1670 m is due to salt tectonics.

The sequence consists of rock salt (locally with potash horizons) interbedded with sulfate-carbonate-clayey rocks. In general, the salt content of the section amounts to 45%–52%, and in peripheral areas it decreases to 10%–20%. The thickness of salt-free units varies from 1–20 m, and that of rock salt units reaches 40 m; 10–25 m thick units are common.

Eight rhythm-units with a salt-free unit at the base of each have been distinguished in the section. Many of them are persistent and can serve as litho-geophysical markers (Fig. 44).

Lower rhythm-unit I is underlain by a basal anhydrite-clay-carbonate unit (9–31 m thick) at the base of which a lower boundary of the salt sequence has been drawn. Breccia-conglomerate, sandstone, and carbonate oolite interbeds up to 10–15 cm thick are common along the contact with underlying deposits.

Over most of the basin the basal unit is overlain by a rock salt unit with a maximal thickness (68–86 m) on the south-east. Rock salt is absent in the western part of the depression and on consedimentary uplifts.

---

3 Podsolevaya sequence means the one lying under a salt unit. Prefix pod is the Russian for under

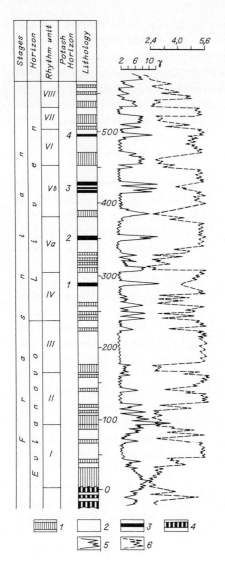

Fig. 44. Composite section of lower (Upper Frasnian) salt sequence. After Eroshina (1981). *1* salt-free units; *2* salt units, *3* potash horizons, *4* Podsolevaya sequence, *5* gamma log curve, *6* neutron log curve

At the base of rhythm-unit II lies a 10–15 m thick unit consisting of dark gray dolomitic and calcareous-dolomitic marl and argillaceous dolomite with rare thin anhydrite interbeds. The rocks contain substantial tuffaceous material; in the east the section is made up almost exclusively of tuff and tuffite. Higher horizons are marked by the alternating rock salt units up to 20–25 m thick and salt-free rocks.

Rhythm-unit III starts with a clay-carbonate-sulfate unit, the thickness of which varies from 8–10 m in the north-western and western parts of the depression to 25–32 m in the south and east. The unit consists chiefly of anhydrite with rare clayey dolomite and limestone intercalations. Effusives are present on the east, and in the Vyshemirovo-Nadvin area the entire unit is built up of tuff and tuffite with marl and anhydrite interbeds; the thickness increases there to 50—57 m. The upper part of rhythm-unit III includes rock salt with rare anhydrite bands; the thickness of rock salt in places reaches 60 m.

The lower part of rhythm-unit IV consists of three salt-free units 2—10 m thick, separated by rock salt units up to 15 m thick which wedge out locally and give way to anhydrite. The salt-free units are essentially clayey-carbonate in composition; anhydrite interbeds are scarce. The section ends with a rather thick (up to 40—65 m) salt unit with a potash (sylvinite) horizon in its center on the west and north-west of the depression.

The thickness of rhythm-unit V ranges from 10—150 m. Two cycles can be distinguished in the section; a salt-free unit and a rock salt unit dominate, respectively, the lower and the upper part of each cycle. Potash horizons 1.5—3.5 m thick mark the central parts of the cycles. The lower and upper horizons are confined to the southern and western Pripyat Depression, respectively. Salt units have maximal thicknesses in areas of potash salt occurrence. A 6—10 m thick unit made up of carbonate-clayey and volcanogenic rocks lies at the base of the upper cycle.

The thickness of rhythm-unit VI varies from 50—80 m. Its upper part contains a salt member about 50 m thick, where a potash horizon 2.5 m thick was found in the Zarechinsk field.

Over much of the Pripyat Depression rhythm-unit VII (30—60 m thick; 120 m thick in the south) is composed of rock salt units 10—15 m thick interbedded with salt-free units up to 4—5 m, locally 14 m thick. Salt-free rocks consist of dolomitic and calcareous-dolomitic marls, as well as of sulfate-carbonate-argillaceous rocks. Anhydrite interbeds can be encountered. In the central, southern, and eastern areas salt content amounts to 55%—60%, but decreases substantially to the west and north.

Rhythm-unit VIII is composed of irregularly alternating salt-free and rock salt strata. Salt-free rocks are variable in composition: dolomitic limestone, dolomite, marl, anhydritic rocks, anhydrite, sulfate-carbonate-terrigenous rocks, siltstone, and sandstone are present. Many rock salt units (2—5 m, locally 10 m thick) are known from the south-eastern and southern areas. Anhydritic rocks intercalated with halite-anhydritic rocks locally replace salt beds in the western and south-western, and, rarely, in the central parts of the depression. There is no rock salt to the north and north-west.

Four potash horizons associated with rhythm-units VI, V, and IV have been found in the lower salt sequence. The areal extent of the potash horizons, and thicknesses of the related salt units are shown in Fig. 45.

The first lowermost potash horizon forms three sublatitudinal terranes such as South Vishan, Komarovichi, and Kopatkevichi, with an areal dimension totalling about 1450 km^2 (see Fig. 45a). The depth of occurrence ranges from 2250—3453 m. The horizon consists chiefly of rock salt beds (10—20 cm thick) alternated with halopelite interbeds (up to 4 cm thick) and three sylvinite interbeds 4—6 cm thick spaced 50—90 cm apart.

The second potash horizon is confined to the lower cycle of rhythm-unit V. The horizon is traced in a salt unit developed in the central and southern Pripyat Depression where it reaches a maximal thickness (see Fig. 45b). The horizon was penetrated by wells in the Elsk, West Valava, and Nikolaevka fields. The depth of occurrence

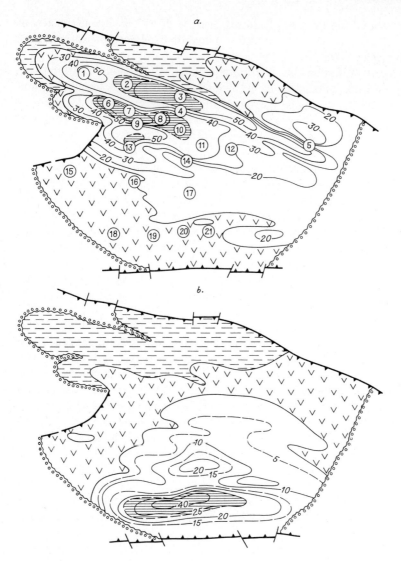

Fig. 45a–d. Thicknesses of salt units and distribution of related potash horizons (**a** 1st horizon, **b** 2nd horizon, **c** 3rd horizon, **d** 4th horizon) of lower salt sequence in Pripyat Depression. After Eroshina and Kislik (1980). *1* isopachs of salt unit including potash horizon in meters, *2* potash horizon present, *3* salt unit absent, *4* rhythm-unit absent, *5* effusive-sedimentary rocks present, *6* depositional limit of lower salt sequence, *7* faults flanking Pripyat Depression. *Encircled numerals* show prospecting and exploration areas: *1* Lyuban, *2* North Kalinovo, *3* South Vishan, *4* Oktyabrsk, *5* Rechitsa, *6* Kalinovo, *7* Chervonaya Sloboda, *8* Komarovichi, *9* Zarechinsk, *10* Savich, *11* North Bobrovichi, *12* Nikulino, *13* Kopatkevichi, *14* Gorokhov, *15* Turovo, *16* Shestovichi, *17* Mozyr, *18* West Valava, *19* Nikolaevka, *20* Elsk, *21* East Elsk

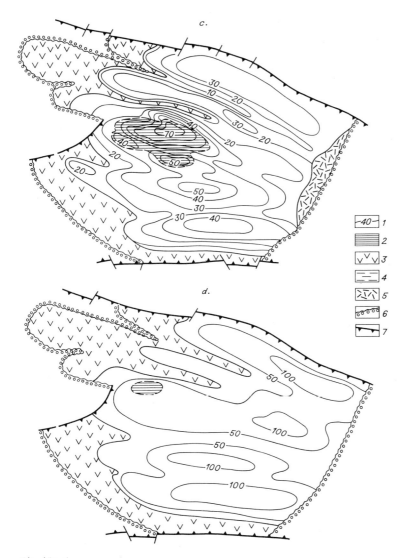

Fig. 45c, d

ranges from 2708–4126 m. The proposed areal distribution of the horizon measures about 630 km². A 1.5–3.0 m thickness is most common. The horizon consists of two potash (probably sylvinite) beds separated by rock salt.

The third horizon is confined to the upper cycle of rhythm-unit V and enclosed in a salt unit developed almost throughout the entire depression, except for its north- and south-western areas and probably for a zone following the southern flank of the depression. The potash horizon is traced in the central and western parts of the depression embracing Zarechinsk, Savich, North Bobrovichi, Kopatkevichi, Gorokhov, and other areas (see Fig. 45c). Its areal extent measures about 1500 km^2. The variable structure is marked by a repeated wedging out of potash beds. The most complete section comprises six potash salt beds 0.5 m thick interbedded with rock salt. The upper beds show successive thinning out down to three or two and, hence, a decrease in thickness from 19 to 3.5—5 m.

The fourth potash horizon, 1.5 m thick, was drilled by a single 1-P Zarechinsk well at a depth of 3094 m. It is confined to rhythm-unit VI and occupies an area of about 130 km^2 (see Fig. 45d).

Over much of the Pripyat Depression the lower salt sequence is overlain by the Mezhsolevaya[4] sequence assigned to the Zadon and Elets horizons of the Famennian.

The thickness of the Mezhsolevaya sequence ranges from 30—100 m in the northwest to 300—500 m in the center and exceeds 1000 m in the north. It is built up of intercalated units of argillaceous rocks, marl, limestone, dolomite, in places anhydrite, mudstone, sandstone, and tuffite. Lithological composition changes regularly across the area. In the southern and south-eastern areas these deposits contain carbonate, argillaceous, and terrigenous rocks. The amount of terrigenous rocks decreases in the north; deposits become essentially carbonate-argillaceous in composition. The proportion of volcanics increases in the east. A peculiar type of the section has been recognized in the north-east; it is a thick (about 1500 m) volcanogenic-sedimentary sequence. At some localities the section contains rock salt units and beds. Halogenic rocks are minor in the Mezhsolevaya sequence. Sulfate rocks as anhydrite intercalations are present in the lower sequence. Their proportion increases slightly in the central and northern parts, where rock salt interbeds occur too, i.e., in the Davydovka and Korenev fields.

Overlying Upper Devonian deposits are known in the Pripyat Depression under the name of the upper salt sequence. The depth of occurrence varies from 200 m on local uplifts to 3000 m in submerged parts due to strongly deformed deposits. The thickness of the sequence ranges from several tens of meters to 3260 m and above (Fig. 46). A maximal thickness was recorded near the northern side of the Pripyat Depression. In the center it ranges from 600 m to 2—2.5 km. In the south it amounts to 700—2000 m and decreases around the periphery. In some areas of the Pripyat Depression initial thicknesses of sediments composing the upper salt sequence are greatly obscured due to salt tectonics.

The sequence consists of rock salt units including potash salt horizons interbedded with salt-free rocks of different composition (clay, marl, limestone, dolomite, locally anhydrite, siltstone, and sandstone). In the north-eastern part of the depression volcanogenic-sedimentary rocks are present among deposits of the upper salt sequence. The number and thickness of salt-free rock units increases in the upper part of the section, whereas the area is greater near the flanks of the Pripyat Depression. The upper salt sequence is divided into the lower halite or potash-free and the upper clayey-halite or potash-bearing members.

The lower member contains mainly white and light gray rock salt with anhydrite intercalations. The thickness of individual rock salt units amounts to 7—30 m, and locally exceeds 300 m. No potash salt has been reported as yet. Salt-free units are

---

4 Mezhsolevaya sequence means in Russian the one lying between two salt units

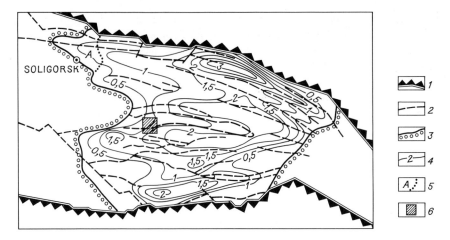

Fig. 46. Isopach map of upper salt sequence of Pripyat Depression. After Lupinovich, Kislik, Zelentsov, and Eroshina (1970). *1* deep faults flanking the depression, *2* major faults in the depression, *3* depositional limits of salt, *4* isopachs of upper salt sequence, in kilometers, *5* Starobino field, *6* Petrikovo field

a few meters thick and locally over 30 m. They comprise anhydrite, dolomite, dolomite-anhydritic, and halite-anhydritic rocks, locally limestone, clay, marl, siltstone. The salt content of the lower member accounts for 80%–96%. The member extends throughout the depression, but on the north-west its thickness substantially decreases (down to 30–40 m).

The upper clayey-halite (potash-bearing) member consists of gray, yellow, and fulvous rock salt with distinct seasonal layering. The thickness of rock salt units ranges from 3–7 to 35–45 m. Some units contain potash salts such as sylvite and carnallite. Salt-free units are built up of carbonate clay, marl, clayey dolomite. Their thickness ranges from several meters to 30–40 m and in places reaches 170–280 m. The salt content of the upper member accounts for 50%–70%. In the eastern and south-eastern areas of the depression the member is absent or represented only by its lower part.

The structure of the clayey-halite (potash-bearing) member in the north-western part of the depression is illustrated by Figs. 47 and 48. The composite section of the member contains about 60 potash horizons or potash occurrences. They are grouped into four potash-bearing stages (Kislik et al. 1976a, b, Protasevich 1976a, b).

The first stage is formed by three lower potash horizons (IV, V, and VI, as well as VIII-P; in the north of the depression they are marked 1-c, 2-c, and 3-c) and occupies mainly the north-western part of the depression. The horizons were reported from the Starobino and Petrikovo fields, and from Malyn, Vishan, Ostashkovichi, Pervomay, and other areas. Two upper horizons were recognized in the Northern and Central zones. The Northern zone is controlled by the Starobino and Shatilkovo Depressions, and the Central zone is confined chiefly to the Petrikovo Depression. The areal dimension of the Northern and the Central zones measures 4100 km^2 and

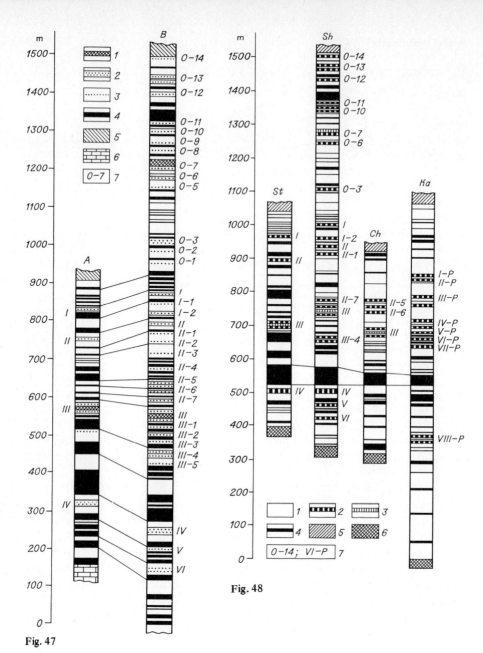

Fig. 47

Fig. 47. Structure of upper salt sequence in north-western Pripyat Depression. After Lupinovich et al. (1969b). *A* composite section of salt sequence of Starobino field explored in detail, *B* composite section of salt sequence east of Starobino field. *1* carnallite rocks, *2* sylvinite, *3* rock salt (*dots* show sylvite inclusions), *4* halopelite, *5* Nadsolevaya sequence made up of carbonate-clayey deposits, *6* Podsolevaya sequence made up of anhydrite-dolomite deposits, *7* indexes of potash horizons

Fig. 48. Correlation of composite sections of potash member in different parts of western Pripyat Depression. After Lupinovich et al. (1970). *St* Starobino field, *Sh* Shatilkovo Depression (east of Starobino field), *Ch* Chervonaya Sloboda Nose, *Ka* Kalinkovichi Depression (in Petrikovo area). *1* rock salt, *2* sylvinite, *3* carnallite rocks, *4* halopelite, *5* Nadsolevaya sequence, *6* Podsolevaya sequence, *7* indexes of potash horizons

900 km², respectively. Areal dimensions of horizons 1-c, 2-c, and 3-c amount to 4000 km², 1400 km², and 2200 km², respectively. The net thickness of the potash horizons of the first stage in the Northern zone averages 24 m (up to 60 m on the north). The Central sublatitudinal zone is approximately 60 km long and 25 km wide. There is only a single potash horizon VIII-P there. On the whole, the potash horizons of the first stage are rather thick (up to 10–40 m) and rhythmic in structure. They consist of halopelite, rock salt, and sylvinite beds. In the Petrikovo field horizon VIII-P contains carnallite units, 2–7 m thick, replaced by sylvinite along the strike.

The second stage 150–400 m thick lies 200–400 m above the first. 28 potash horizons and shows have been recognized there (horizons I, II, and III of the Starobino field; 0-1 to III-5 of the Shatilkovo Depression; 0-1-P to VII-P of the Petrikovo field). The horizons of this stage have a rather large areal extent and correspond to a phase of maximal potash accumulation when more than half of the potash salt volume was formed in the Pripyat Depression. These horizons occupy the Northern Pribortovaya, Shatilkovo, Starobino, Kopatkevichi, Malodushin, Petrikovo, Shestovichi, Oktyabrskaya, Kalinkovichi, Zhitkovich, Buinovich, Mozyr, and Elsk Depressions. The second stage of potash accumulation totals 16,000 km². However, there are numerous "windows" where potash salt is absent (as a rule, on local salt domes).

The potash horizons are, respectively, 70 m and 60 m in the western Elsk (in the southern Valava area) and in the Shatilkovo Depresssions. Their total thickness in the eastern Starobino and Shestovichi Depressions, and in the Pritok Anticline is relatively great.

The potash horizons of the second stage show a maximal extent and considerable thickness, the latter ranging from 0.5–1.0 to 30 m. The thickest horizons are III, II-7, IV-P, and VI-P; they exhibit a complex structure and different composition. Sylvinite, carnallite-sylvinite, sylvite-carnallite rocks build up these horizons. Potash horizons III, II-7, II-6, I-P, II-P, and III-P are carnallite-bearing.

The third stage 300–500 m thick lies 150–200 m above the second. All in all 17 potash horizons (0-2 to 0-10 in the Shatilkovo Depression, and 0-2-P to 0-6-P in the Petrikovo field) have been distinguished. Their areal extent is about 10,000 km² including Starobino and Shatilkovo Depressions, and some parts of the Kopatkevichi, Mozyr, and Buinovich Depressions. A maximal total thickness (over 40 m) was recorded in the north-eastern part of the Pripyat Depression. Some horizons are 2.5–3.0 m, in places 10 m thick. They are usually simple and single-bedded in structure with one or several sylvinite layers. Sylvinite dominates potash rocks. Carnallite rocks have not been reported.

The fourth stage of potash accumulation is confined to the uppermost part of the member and situated 250–300 m above the third one, its thickness exceeding 500 m. Eighteen horizons (0-2 to 0-14) occupy primarily the axial and most subsided parts. Main terranes of the potash horizons are the Shatilkovo and Starobino Depressions. Some horizons have been reported from the Oktyabrskaya and Buinovich Depressions.

A total areal extent of the fourth stage is about 5000 km². The horizons are 1–2 to 5–12 m thick. The thickest of them are 0-7, 0-13, and some others; a single-bedded structure predominates. The horizons are an irregular alternation of halopelite, rock salt, and sylvinite. Locally the number of sylvinite beds vary from three to eight.

Their constituent potash minerals are sylvite and rare carnallite. The thickness of the carnallite unit in horizon 0-7 amounts to 7–8 m.

Sylvinites of the Pripyat Depression are of two types, namely, the Starobino type (red rocks with distinct structural-textural features and bedding), and the Petrikovo type (variegated rocks with obscure bedding, irregular texture, and carnallite inclusions). Sylvinites of the Starobino type are more abundant. Variegated sylvinite was observed in the Petrikovo field, central Pripyat Depression (Yurovich and other areas) and it is typical only of the second stage of potash accumulation.

The Upper Devonian section in the Pripyat Depression is crowned by the Nadsolevaya [5] sequence subdivided into three units: the lower 60–130 m gypsum-bearing unit; the middle unit consisting of 90–110 m massive clay and marl; and the upper 50–150 m shale unit (Shablovskaya et al. 1976).

In the Pripyat Depression, rock salt constitutes mainly the lower and the upper salt sequences. The areal extent of the lower sequence is about 20,000 km^2; the average thickness of rock salt in the area is taken to be 500 m (200–1000 m), and salt content accounts for 50%. Using above values we obtain a volume of 5000 km^3 for rock salt present in the lower salt sequence. The areal distribution of the upper salt sequence totals about 26,000 km^2. The sequence is 1.5 km thick on the average and has a salt content of 60%, a total thickness of rock salt in the area being 0.9 km, and hence the volume of rock salt in the upper salt sequence must amount to 23,400 km^3. Thus, a net volume of rock salt in the Pripyat Depression may account for 28,400 km^3.

Dnieper-Donets Depression

Upper Devonian salt deposits in the Dnieper-Donets Depression have not been studied in such detail as those of the Pripyat Depression. This was mainly due to their strong deformation, as well as to the fact that in the inner areas of the Dnieper-Donets Depression the Upper Devonian salt sequences lie at a great depth and crop out only in cap rocks of salt domes.

Three or four salt sequences are believed to occur among the Devonian deposits of the area (Fig. 49); two of them were penetrated by deep holes in the north-western Dnieper-Donets Depression in normal position, and the third sequence makes up salt plugs of some diapir folds in the inner part of the depression. As regards the lowermost, fourth, sequence, there are no outcrops, and its presence is inferred only from geophysical evidence (Chirvinskaya and Zabello 1969). Some investigators (Kityk 1970, Kityk and Galabuda 1975, 1981, Kityk et al. 1980) believe that the depression contains only two Devonian salt sequences.

The third salt sequence, lowermost among those so far recognized, was drilled by deep wells in the Romny, Chernukha, Isachkov, Novo-Senzhara, and other anticlines, where it builds up salt plugs piercing the Upper Devonian, and locally all overlying deposits. The sequence is tentatively assigned to the lower Frasnian as it pierces the upper Frasnian deposits based on fossils found in the limestone blocks of cap rock

---

5 Nadsolevaya sequence means the one lying above the salt unit. Prefix nad is the Russian for above

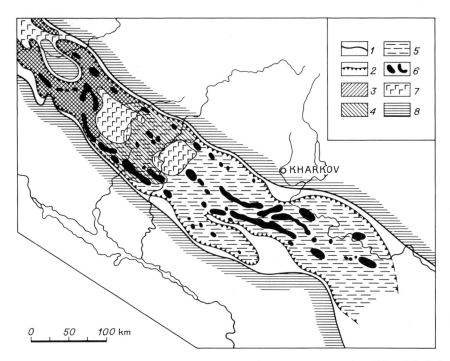

Fig. 49. Distribution of Devonian salt strata in Dnieper-Donets Depression. After Chirvinskaya and Zabello (1969). *1* recent outline of Devonian deposits, *2* depositional limit of salt, *3* areal extent of Elets (upper) salt sequence, *4* areal extent of Evlanovo-Liven (lower) salt sequence, *5* area where two upper sequences, an underlying third, and probably, fourth salt sequence occur, *6* junction of salt strata of different age, *7* areal distribution of Devonian effusive-terrigenous deposits, *8* Devonian deposits absent

breccia (Novik 1952, 1954, Kutsyba 1954, 1959, Kirikov 1962, 1963b, Chuprin et al. 1968). The stratigraphic position of the sequence is inferred from a comparison with the Devonian deposits of the Pripyat Depression. Kirikov (1962, 1963b) places it at the level of the upper Shchigrov horizon, i.e., at the level where a sulfate rock unit has been recognized in the Pripyat Depression. Taking into account the presence in the Pripyat Depression of another salt accumulation level confined to the Narova horizon of the Eifelian, where rock salt interbeds have been discovered, the third salt sequence of the Dnieper-Donets Depression could occur stratigraphically much lower, viz., beneath the Shchigrov horizon. It is also feasible to suggest the presence of two salt sequences near the base of the Devonian section in the Dnieper-Donets Depression; according to Kirikov (1963b) they lie at the level of the Shchigrov and Narova horizons, respectively.

The third salt sequence, presumably upper Shchigrov, occurs at a depth of 4–5 km in the central Dnieper-Donets Depression. However, in cap rocks of salt stocks its top almost comes to the surface (40–50 m), whereas towards the limbs the depth increases to 3000–3500 m. The third sequence consists mainly of rock salt interbedded with anhydrite, clay-carbonate, and effusive rocks, and dark gray bituminous shale (Kutsyba 1959). It is the sequence that fills in the deepest zones of the depression and forms numerous salt domes and stocks. Rock salt in salt stocks can exceed 3–4 km. The original thickness of the sequence remains uncertain; some investigators believe that it could reach 1000 m (Halogenic ... 1968).

The second (from the base) salt sequence was penetrated in the normal position by a number of deep holes in the north-western Dnieper-Donets Depression. It shows a good correlation with the lower salt sequence of the Pripyat Depression and is widely known as the lower or Frasnian sequence (Britchenko et al. 1968, Chuprin et al. 1968, Kityk 1970, Makhnach et al. 1970, Kityk and Galabuda 1975, 1981, Khomenko 1977, Galabuda 1978). The sequence is dated as Evlanovo/Liven or Voronezh/Liven in age. Probably it is the sequence that Pashkevich and Pistrak (1964) and later Pistrak et al. (1972) have recognized as the Isachkov Formation and placed into the Voronezh and Evlanovo horizons.

The lower salt sequence consists, as a rule, of rock salt with numerous interbeds and units of terrigenous, carbonate, and sulfate rocks, such as dolomite, limestone, marl, mudstone, siltstone, anhydrite, and anhydrite-dolomite. Britchenko and co-workers (1968) have recognized two types of sections for the lower salt sequence in the north-western Dnieper-Donets Depression. A section of the Adamovo type crops out in the Adamovo, Velikobubnovsk, Shapovalov, and Kalaida fields. The sequence reaches 350 m. It can be divided into the lower, marl-saliniferous, and the upper, dolomite-anhydritic parts. The thickness of the lower salt sequence increases to 1700–1800 m in the section of the second, saliniferous or Priluksk-Kholmsk type, which was drilled in the Kholmsk, Velikozagorovsk, Antonovo, Priluksk, Krasnopartisansk, West Nezhinsk, and other areas in the low inner parts of the depression. The sequence is composed mainly of rock salt with interbeds of sulfate, sulfate-carbonate, and sulfate-terrigenous rocks.

Some sections in the lower salt sequence contain interbeds of volcanogenic and volcanogeno-terrigenous rocks. For example, in the Bereznyansk field, the lower 340 m thick salt sequence in well 305 is subdivided into three parts in ascending order: (1) limestone and tuff limestone with tuffstone, tuff siltstone, and tuffite interbeds, and thin (1–2 m) rock salt interbeds totalling 89 m; (2) rock salt with subordinate anhydrite, dolomite, and limestone interbeds amounting to 125 m; (3) anhydrite, anhydrite-dolomite, and dolomite with single rock salt units totalling 126 m (Kityk 1970).

The thickness of the lower Evlanovo-Liven salt sequence decreases to 150–200 m toward the depression margins. Its composition changes as well: rock salt is not to be found in the section, and the sequence is composed of anhydrite, dolomite, and limestone with terrigenous rock interbeds.

The upper salt sequence of the Dnieper-Donets Depression appears to lie at the level of the Elets horizon of the Famennian (Makhnach et al. 1970, Kityk and Galabuda 1975, 1981, Khomenko 1977). The above fact implies a lower stratigraphic position for the upper salt sequence as compared to that of the same name in the Pripyat Depression.

The upper salt sequence was distinguished in normal position only in marginal parts of the Adamovo, Videltsev, Maksakovo, and other fields. Its thickness decreases due to the omission of rock salt units in cap rocks of salt structures, and increases on flanks where salt content becomes much higher as well (Makhnach et al. 1970). The upper salt sequence ranges from 0–320 m in thickness. It consists of rock salt interbedded with marl, anhydrite, siltstone, sandstone, and locally with tuffaceous rocks. The thickness is believed to increase south-eastward. Within the section of the so-called Adamovo type Britchenko and co-workers (1968) subdivide the upper salt sequence into three parts: (1) lower dolomite and anhydrite; (2) middle, essential salt; and (3) upper anhydrite and dolomite interbeds.

The preceding discussion implies only a tentative outline of the salt sequences of the Dnieper-Donets Depression, whereas their original thicknesses, salt content, and the internal structure of deposits remain unknown for most of the depression. The Devonian deposits of the depression are 1–5 km thick, and the height of salt plugs locally reaches 7–11 km. The original thickness of rock salt is still uncertain. The available data suggest that the third salt sequence, presumably upper Shchigrov, can attain 1000 m; the thickness of the lower, Evlanovo/Liven salt sequence increases in troughs to 1700–1800 m; the upper, Elets salt sequence can be 350–400 m thick. This suggests that the total thickness of the Devonian salt series in the Dnieper-Donets Depression initially could have exceeded 3000 m, and that no less than 2000 m accounted for rock salt. Some workers believe that the original thickness of the Devonian halogenic rocks in the areas of wide development of salt plugs was about 1000 m (Kityk 1970).

The areal distribution of salt deposits within the depression has not yet been determined. It concerns mainly the extent of rock salt in the south-east. There, Devonian salt strata were reported from the Slavyansk area, and some investigators (Pistrak 1963) suggest the continuation of Devonian salts farther south-east beneath coal measures of the Donets Basin. Taking into account the preceding information, an areal distribution of Devonian salt deposits in the Dnieper-Donets Depression of 40,000 km^2 is a minimal estimate which can be undoubtedly used to assess the volume of the Devonian rock salt in the region.

Thus, with an area of 40,000 km^2 and a mean thickness of rock salt of 1000 m, a minimal possible volume of the Devonian rock salt in the Dnieper-Donets Depression may run as high as 40,000 km^3.

Timan Trough

The evidence on salt deposits distributed in the south-western Timan Trough concerns only the Seregov Salt Dome (Pakhtusova 1957, 1963a, b, Ivanov and Levitsky 1960, Zoricheva 1966, 1973, Privalova et al. 1968). Some data on an areal distribution of salt sequences in the Timan Trough were obtained from geophysical and tectonic studies (Lyutkevich 1955, 1963, Tszu 1964, Dedeev and Raznitsyn 1969).

Salt deposits occur in the Vychegda Basin which is a part of the Timan Trough stretching from south-east to north-west along the entire Timan Range (Fig. 50).

In the north-western Vychegda Basin two salt domes, namely the Seregov and Chusovaya, have been found; they are marked by intense gravity lows. The presence of rock salt in the cap rock of the Seregov Salt Dome was determined from drilling data; many wells drilled there showed that the thickness of rock salt exceeds 1200 m (Pakhtusova 1963a; Privalova et al. 1968).

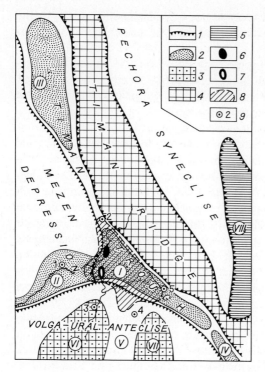

Fig. 50. Distribution of salt strata in south-eastern Timan area. *1* contours of major structures, *2* troughs, *3* arches, *4* folded structures in Timan area, *5* Cis-Ural Trough, *6* known areal extent of rock salt (Seregov Dome), *7* inferred areal extent of salt strata within Chasov gravity low, *8* inferred areal extent of salt deposits, *9* wells (*1* Yarensk, *2* Veslyansk, *3* Sysolsk, *4* Lopydin, *5* Anyba). *Encircled numerals* refer to major tectonic elements: *I* Vychegda Trough, *II* Yarensk Trough, *III* Safonovo Trough, *IV* Solikamsk Trough, *V* Kirov-Kazhim Trough, *VI* Sysolsk Arch, *VII* Komi-Permyak Arch, *VIII* Upper Pechora Depression

According to Leontiev (in: Privalova et al. 1968), the Seregov Dome is round in shape. The dimension of its ice cap is 1.5 × 1.9 km; it is brecciform, 216–516 m thick, and contains dark violet and greenish-gray clay, mudstone, siltstone, pink and gray sandstone, as well as dolomite underlain by a rock salt unit. The latter is composed of pink, red to dark red, locally light gray rock salt irregularly interlayered with terrigenous and carbonate rocks. Salt-free rocks commonly consist of inclusions and clasts as large as a few centimeters. Individual interbeds of terrigenous rocks are 2–7 m thick at a maximal thickness of 51.4 m. The greatest net thickness of these interbeds reaches 67 m. The proportion of salt-free rocks in individual wells accounts for 0.7%–18% of the total salt sequence. Rock salt contains inclusions and small patches of sylvite; its content usually does not exceed a few percent (Ivanov and Levitsky 1960, Privalova et al. 1968).

The Seregov Salt Dome flanks are built up of Upper Carboniferous (?), Permian, and Triassic rocks. Salt pierces the Upper Carboniferous (?) and Lower Permian rocks; their contacts run along the faults. The oldest sequence penetrated by wells on the domal flanks consists of gray dolomite 70–80 m thick. Light blue to gray micrograined limestone containing cherts which grade into siliceous dolomite up the section lies above; the sequences is 70–75 m thick. Pakhtusova (1963b) suggests that tha lower dolomitic sequence and the upper limestone-dolomitic sequence ought to be placed into the upper Carboniferous and into the Sakmarian (Lower Permian), respectively. She assigns an Upper Permian age to the overlying deposits. They chiefly contain red beds with gray terrigenous-carbonate rocks (dolomite and siltstone) as well as lenses and interbeds of anhydrite and gypsum at the base. This lower part of the section, which includes carbonate and sulfate interbeds and has a thickness of 70 m, could actually be Artinskian and Kungurian (Lower Permian).

Thus, the data available suggest that the age of the salt sequence in the Seregov Salt Dome is older than the Lower Permian and even older than the Upper Carboniferous. These data support the probable Devonian age of the salt sequence in the south Timan area (Kalberg 1948, Pakhtusova 1957, 1963b, Ivanov and Levitsky 1960). It is safe to assign these salt deposits to the Frasnian (Upper Devonian).

As noted in the preceding material, the distribution of salt sequences of probable Upper Devonian age in the south-western Timan area remains uncertain. Geophysical data suggest that salt fills in most of the Vychegda Basin. The presence of salt is inferred from apparent gravity lows. An approximate area covered by rock salt measures 60–100 km long and 30–40 km wide, i.e., about 2000 km^2. The salt sequence is 2.5–3 km thick (Dedeev and Raznitsyn 1969). However, salt deposits may go beyond this area. It may be bounded by the Timan Range; the Volga-Ural Anteclise limbs, and some uplifts in the marginal eastern part of the Mezen Depression to the east, south, and west, respectively. Isolated deep wells drilled in this area show an increase in thickness of the Upper Devonian deposits. Salt deposits are thought to occur in the deepest parts of the Vychegda Basin, in the eastern Yarensk Trough, and possibly in the northern Kirov-Kazhim Trough, i.e., over an area of about 20,000 km^2. Salt deposits are also likely to occupy the Safonovo Trough where geophysical data suggest a thickness of 3500–4000 m (Dedeev and Raznitsyn 1969).

A rather approximate estimate for rock salt volume in the south-western Timan area is as follows: with allowance for extent based on geophysical evidence of 2000 km^2 and an average rock salt thickness of 500 m (a maximal thickness is 2.5–3 km), the volume of rock salt will amount to 1000 km^3.

Baltic Syneclise

Halogenic rocks are rather common in the Upper Devonian deposits of the Baltic Syneclise. They were reported from the Plyavin, Salaspils, Daugava, Ogr, Baus, and Amul Formations of the Frasnian in the northern Baltic Syneclise of Latvia (Gailite 1963, Liepinsh 1963a, b, 1973, Ulst 1963, Berzin and Ozolin 1967, Tikhomirov 1967), and from such Frasnian formations as the Plyavin, Tatul, Istra, Pamush, and Pakryoi in the southern Baltic Syneclise of Lithuania (Narbutas 1959, 1961, 1964, Zhaiba et al. 1961, Tikhomirov 1967, Vodzinskas and Kadunas 1969, Summet 1971, 1973). All the above formations contain gypsum and anhydrite interbeds, lenses, and inclusions.

The oldest deposits among the Frasnian deposits of Latvia are gypsum and anhydrite confined to the Plyavin Formation in the form of thin bands in dolomites. Gypsum is particularly widespread in the Salaspils Formation. Over much of the Baltic Syneclise the Salaspils Formation can be divided into three members, the central which is mostly gypsiferous. For example, near the village of Saulkalne, the member consists of fibrous and layered gypsum; gypsum beds are about 0.7 m thick. Gypsum is mainly interbedded with dolomitic marl. Gypsum of the Salaspils Formation in the western Daugava River Valley occurs chiefly as two varieties. Dark gray layered gypsum composing the main gypsiferous sequence dominates the section. It builds up 2 m thick interbeds and lenses. Layered gypsum consists of sparry gypsum interbedded with fibrous. The sparry gypsum is light gray to brown owing to the organic matter content. White or yellowish-pink fibrous gypsum occurs as bands or lenses or fills in joints and cavities. The bands are 2–15 cm, but usually 2–6 cm thick (Liepinsh 1963a).

The Salaspils gypsum is observed in some areas of the Polish-Lithuanian Depression too, near Riga, Slok, and Kemeri. Gypsum is exposed along the valleys of the Slotsene, Abava, Venta, and Tebra Rivers. The richly gypsiferous lower and central Salaspils Formation has been penetrated by wells near the village of Zalenieki. Some gypsum beds reach 1.3 m in thickness. The wells drilled in the Rundale, Alanda, Akmene, and Kapseda areas have exposed, layered and fibrous gypsum interbedded with clay, dolomitic marl, and dolomite. In the latter area gypsum lenses are up to 1.2 m thick.

The Akmene, Alanda, Liepaya, Remte, Ezere, Dzhukste, and Zalenieki wells have penetrated gypsinate rocks among the Daugava Formation deposits in the Polish-Lithuanian Depression. Gypsum occurs there in the form of layered and fibrous interbeds or fills in cavities and joints; it is encountered as individual crystals, patches, nodules, and lenses. A gypsinate part of the Ogr Formation occurs in the central Polish-Lithuanian Depression. Gypsum intercalations and lenses are common in the section. In the same areas fibrous gypsum as lenses and veinlets was reported from the Baus and Amul Formations.

Gypsum is also widespread among the Frasnian deposits in Lithuania, within the southern and south-eastern Polish-Lithuanian Depression. Gypsum-bearing deposits of the Plyavin Formation occupy the lowermost stratigraphic position where gypsinate dolomite and minor gypsum intercalations fill in cavities and joints (Vodzinskas and Kadunas 1969). Sulfate rocks occur throughout the Tatul Formation at many localities in Lithuania (Zhaiba et al. 1961, Narbutas 1964). Two gypsum units 13–16 m thick each separated by a 4–6 m thick gypsinate clay and marl unit were observed in the Tatul Formation near the towns of Birzhai and Pasvalis. Each gypsum unit shows a high gypsum content (up to 80%). Coarsely crystalline so-called sparry and layered gypsum interbeds 2–5 m thick, locally with admixture of dolomitic material alternate with rare 0.7–1.2 m thick dolomitic interbeds associated chiefly with the central and lower parts of the section. The upper gypsinate unit has a lower gypsum content (70%–75%) and thinner gypsum interbeds (up to 2 m). Gypsum is interbedded with dolomitic marl 0.4–1.0 m thick (Zhaiba et al. 1961). Thin 0.1–0.15 m gypsum intercalations can be observed in the central member of the Tatul Formation as well. South of the Birzhai-Pasvalis area the proportion of gypsum decreases in the Tatul Formation, whereas, to the west, its amount and thickness increases and reaches 46 m (Stanyuchai well). In the Kedainyai and Aregala areas a section of the Tatul Formation consists entirely of marl and dolomite with clay and gypsum interbeds. Anhydrite appears in the section near Palanga. Gypsiferous deposits of the Tatul Formation crop out in a 12–20 km wide and 40-km-long zone stretching south-westward between the towns of Birzhai and Pumpenai. It is this area where gypsum pits are being presently mined in Lithuania; the largest of them are Kirdonis and Karaimishkyai (Narbutas 1961).

In the Istra Formation gypsum generally fills in cavities, caverns, and joints, and forms patches and veinlets in dolomitic rocks. Such gypsinate dolomite was penetrated by wells in the Aregala, Raseinyai, and Palanga fields (Zhaiba et al. 1961, Vodzinskas and Kadunas 1969). Sulfate rock, essentially gypsum, is rather widespread in the Pamush Formation in south-western and western Lithuania. The boreholes drilled on the Raseinyai, Aregala, and Palanga fields have exposed numerous gypsum lenses and inclusions in marl, clay, and dolomite among the deposits of the Pamush Formation. In the Palanga borehole gypsum bands reach a thickness of 0.2–0.5 m, and gypsum in the section totals 12%–15% (Zhaiba et al. 1961). Gypsum plays an important part in the Pakryoi Formation in the Ionishkis and Palanga fields.

Moscow Syneclise and the Volga-Ural Area

Within the Moscow Syneclise and the Volga-Ural area, halogenic rocks among the Upper Devonian deposits differ with respect to stratigraphic levels and regions. Sulfate formations are most widespread among deposits of the Evlanovo and Liven horizons of the Frasnian and the Lebedyan and Dankov horizons of the Famennian.

Halogenic deposits, found in the Sargaev horizon of the Lower Frasnian, occupy the lowest stratigraphic position in an Upper Devonian section. They have been reported from Pskov and Novgorod regions, between Moscow and Soligalich, within the Upper Kama Basin and in the Cis-Urals, Perm area. In the first region, gypsum dominates the Chudovo beds, in the Kudeb Basin, where it occurs among dark greenish-gray clay and dolomitic marl. There are several pits, in which gypsum is mined.

Minor gypsum shows among calcareous-dolomitic rocks and dolomite in the Sargaev horizon were found in the Nekrasov borehole in the northern Moscow Syneclise (Utekhin and Sorskaya 1971). In the Upper Kama Basin, rare anhydrite inclusions in limestone and dolomite of the Sargaev horizon have been penetrated by the Glazov borehole, whereas in the Cis-Urals, Perm area, similar inclusions were observed in boreholes drilled in Severokamsk and Krasnokamsk (Lyashenko et al. 1970).

In two regions, viz., in the western Moscow Syneclise and in the Upper Kama Basin, sulfate rocks have been found in the Semiluki horizon of the Frasnian. To the west, in the Koloshka Basin, small gypsum crystals in limestone were observed in the uppermost Svinord beds, whereas the middle part of the same beds of the Kudeb Basin contains one or two gypsum interbeds up to 1.3 m thick (Summet 1971). In the Upper Kama Basin, near the towns of Glazov, Borodulino, and in adjacent areas (Kudymkar, Maikor, Polom, and other areas), thin gypsum and anhydrite inclusions and lenses occur among the Semiluki and Mendym dolomites (Lyashenko et al. 1970). Dolomite with anhydrite patches in the Frasnian Mendym beds have been also found on the eastern slope of the Nema-Loina Arch.

Much more common are sulfate rocks in deposits of the Frasnian Voronezh horizon. They occupy most of the western and northern Moscow Syneclise. Gypsinate rocks, mainly dolomite and dolomitized limestone, have been found near the towns of Vyazma, Toropets, Shar'ya, and Velikie Luki; in the last two regions, gypsum lenses and inclusions are several centimeters thick. Anhydrite and gypsum patches in marl, clay, and conglomerate have also been reported from the Upper Voronezh beds near the town of Gorkiy from the areas of the Kotelnich Arch and the town of Sovetsk. In the Upper Kama Basin, the entire section of the Voronezh horizon is sulfate in nature. Anhydrite and gypsum patches and inclusions were found there in Akarshura, Burakovo, Yakushur-Bod'ya, Glazov areas. Gypsum and anhydrite impregnations in dolomite of the Voronezh beds were also reported from the Tokmovo Arch area (e.g., Prudy Village) and from the eastern Samarskaya Luka near Morkvasha.

Highly sulfate are deposits of the Evlanovo and Liven horizons (Frasnian) over much of the Moscow Syneclise and the Volga-Ural area. In the western Moscow Syneclise, sulfate rocks at this stratigraphic level were reported from the south-eastern Pskov Region near the towns of Kun'ya and Bilovo, where nodules, patches, and interbeds of pinkish-white gypsum up to 10 cm thick occur among silty clays; similar occurrences are known from the areas near the towns of Vyazma, Yartsevo, Toropets, Redkino, Shar'ya. In the eastern Moscow Syneclise and on the Kotelnich Arch, gypsinate rocks, as well as anhydrite lenses and patches among dolomites of the Evlanovo and Liven horizons have been found near Balakhna, Kotelnich, and in Sovetsk, Chichino, Akul areas as well. In the Volga-Ural area, gypsum and anhydrite inclusions and sulfate are typical of rocks from the Kazan District, the Kazan-Kazhim Trough zone (Yanga-Aul), in the Upper Kama Basin, Tokmovo Arch (Tokmovo, Issa), and in the western Zhiguli-Pugachev Arch near Krasnaya Polyana Village. Sulfate rocks are very common in almost all the Famennian (Upper Devonian) deposits of the Moscow Syneclise and Volga-Ural area. Gypsinate rocks of the Zadon horizon, as well as a presence of anhydrite and/or gypsum inclusions were reported from south-eastern Pskov Region, western and north-western Moscow Syneclise, viz., Redkino, Krasnyi Kholm, Rybinsk; northern Moscow Syneclise, viz., Soligalich and Shar'ya; south-eastern and eastern part of the Syneclise, near Uren', Vetluga, Shar'ya, and Kotelnich, along the western slope of the Tokmovo Arch (Poretskoe, Balakhonikha, Lyskovo, Sundyr', Marpasad), on the southern summit of the Tatar Arch and Volga area near Saratov. Sulfate rocks among deposits of the Elets horizon are known almost from the same localities. In south-eastern Pskov Region, where it borders on the Smolensk Region (Usvyati, Usmyni and other villages), the Elets horizon contains numerous lenses, stringers, and interbeds of white and smoky, finely crystalline gypsum up to 0.15 m thick. Gypsum and anhydrite inclusions were observed in areas near Moscow, Nepeitsino, the Kotelnich Arch, over much of the Tokmovo Arch (Tokmovo, Issa, Kikino, Komarovka, Lyskovo, Poretskoe), the southern Tatar Arch-Leninogorsk, Sulin, and other areas.

Many papers (Ivanov and Levitsky 1960, Makhlaev 1964, Tolstikhina 1952, Strakhov 1962 etc.) emphasize the sulfate nature of rocks from the Lebedyan and Dankov horizons (Famennian). The abundance of sulfate rocks there allowed Ivanov and Strakhov to distinguish a halogenic (sulfate-bearing) formation of the Dankov-Lebedyan age. Sulfate rocks − gypsum and anhydrite − are confined mainly to the axial part of the Moscow Syneclise, to a belt, running from south-west of Tula and Kaluga, to north-east, through Moscow and Ivanovo, to Chukhloma, Manturovo, Shar'ya. Sulfate rocks in the Dankov-Lebedyan horizon often account for 40%−60%, locally even for 70%−80% (Ivanov and Levitsky 1960). South-west and north-east of the belt, the proportion of sulfate rocks decreases; however, lenses and inclusions, as well as thin anhydrite and gypsum interbeds among dolomite, marl, and clayey dolomite, can be found throughout the section. They occur both in Smolensk Region, and in the Kalinin, Yaroslavl, and Vologda regions, in the Ryazan-Saratov Trough, in the Tokmovo and Kotelnich arches areas, and over most of the Volga-Ural area, including the Kazan-Kazhim Trough, Nema-Loina Arch, Upper Kama Basin, in the northern and southern Tatar Arch (in Krasnovka, Yanchikovo, Golyushurma, Baitugan areas), in the Sergiev-Abdulin Depression (Balykla, Buzbash, Borovka, Rydaevka), in the western Melekes Depression (Bugrovka, Filippovka, Pichkassy), within the Zhiguli-Pugachev and Orenburg arches (Orekhovka, Nikolskoe, Chubovka, Zolnyi Ovrag, Zhigulevskoe, Berezovka, Syzran, Krasnaya Polyana, Pokrovskoe, Pugachev, Balakovo, and other areas).

Sulfate deposits of the Dankov-Lebedyan sequence outline a large epicontinental basin on the Russian Platform, which, in fact, can be considered as an intermediate one among salt basins which existed in the Dnieper-Donets and Pripyat, as well as Timan troughs, on the one hand, and the open sea of normal salinity, situated in place of the eastern Russian Platform, on the other hand. In the east, an area of sulfate sedimentation was consistently an extensive zone of reefogenic carbonate deposits.

Pechora Syneclise

In the Pechora Syneclise, sulfate rocks in the Upper Devonian have been found in the Ukhta and Upper Izhma regions, as well as in the Pechora Range. In the first two regions, halogenic formations are confined to the Evlanovo and Liven horizons (Frasnian) and, partly, to the Upper Famennian, whereas in the latter they occur among the Famennian deposits.

In Ukhta Region, gypsiferous strata are associated with the Ust-Ukhta Formation, placed by Tikhomirov (1967), at the level of the Liven horizon (Frasnian). Along the Izhma and the lower Sed'-Yu Rivers, the Ust-Ukhta Formation, in addition to clays, contains thick gypsum layers of great extent, especially numerous in its middle section. Ivanov and Levitsky (1960) state, using Zamyatin's and Vollosovich's evidence, that a belt of gypsiferous deposits outcrops along the upper Vezha-Vozha and the lower Sy-Vozha Rivers, stretching for 8–10 km and having a width of about 2–3 km. There are several deposits (Ust-Ukhta, Veselyi Kut, etc.) mined for gypsum in the Ukhta Region. The beds are 3 m thick and upward. The proportion of sulfate rocks in the Ust-Ukhta Formation greatly increases south-east of Ukhta. So, in Chemkis-Iol' Region, anhydrite interbeds appear in the section, which, along with gypsum, alternate with clay, dolomite, and limestone.

In Upper Izhma Region, sulfate rocks are confined to deposits of the same age and extend there directly from the Ukhta Region. In the north-western Upper Izhma Region, the Upper Frasnian formations, assigned to the Evlanovo-Liven sequence, are composed of gypsum and anhydrite alternating with clay, locally dolomite and limestone. The thickness of gypsum and anhydrite interbeds decreases to the south-east, and dolomite and limestone become predominant; however, sulfate rocks disappear when traced farther in the same direction. Besides, a sulfate-bearing sequence is widespread in Upper Izhma Region, occupying a higher stratigraphic position. It rests on the Izhma limestones and is made up of dolomite with dolomitized limestone, gypsum, and anhydrite interbeds. According to Kushnarev, Raznitsyn, and others, this sequence corresponds to the Dankov-Lebedyan beds of the central Russian Platform (Nalivkin 1963).

In the Pechora Range area, in the Kamensk, South Lyzhsk, North Lyzhsk, and Kyrtyiol anticlines, slightly west of the middle Pechora River, sulfate rocks lie among the Upper Famennian deposits within the Kamensk beds. According to the evidence provided by Nalivkin (1963), gypsum interbeds build up the lower part of the section, and alternate with marl, clay, limestone, and dolomite. It is most likely that Upper Famennian sulfate-bearing rocks extend farther north, into the lower Pechora Basin (Atlas . . . 1969).

To finalize the discussion of the East European Upper Devonian evaporite basin, it should be emphasized that conditions favoring halogenic sedimentation recommenced repeatedly over a rather extensive area. Evaporite deposits were formed both in the western Russian Platform, within the Baltic Syneclise, and in its central, eastern, north-eastern, northern and south-western parts, viz., Moscow Syneclise, Volga-Ural area, Timan Trough, Pechora Syneclise, Pripyat, and Dnieper-Donets troughs. However, it is only in three basins, viz., Timan, Pripyat, and Dnieper-Donets, that the Upper Devonian salt-bearing series were accumulated. Data available allow only an approximate estimate of the volume of salts accumulated on the East European Platform during Late Devonian time. It may exceed 6.9×10^4 km^3.

West Canadian Basin

The West Canadian evaporite basin is one of the largest basins in North America. It occupies an area lying between the Rocky Mountains to the west and the Canadian Shield to the east and runs from northern North Dakota and Montana, U.S.A., through Saskatchewan and Alberta provinces to the southern Northwest Territories, Canada. There are many papers dealing with halogenic deposits of the basin (Law 1955, Andrichuk 1958, 1960, Pearson 1960, 1963, Belyea 1960, 1964, Sandberg 1962, Grayston et al. 1964, Lane 1964, Schwerdtner 1964, Norris 1965, 1967, Carlson and Anderson 1966, Keyes and Wright 1966, Klingspor 1966, 1969, Basset and Stout 1967, Danner 1967, Griffin 1967, Harding and Gorrel 1967, Jordan 1967, 1968, Kent 1967, Sandberg and Mapel 1967, Langton and Chin 1968, McCamis and Griffith 1968, Prather and McCourt 1968, Wardlaw 1968, Fuller and Porter 1969, Holter 1969, Douglas et al. 1970, Price and Ball 1971, Anderson and Swinehart 1979, Worsley and Fuzesy 1979, etc.).

Based on changes in the thickness and structure of Devonian sedimentary strata in the West Canadian Basin, the following tectonic elements can be distinguished (from south to north): (1) Saskatchewan sub-basin, which is often subdivided into the Saskatchewan Depression proper and the southern Williston Depression, to the southwest with the Swift Current Platform inbetween; (2) Meadow Lake Escarpment; (3) Central Alberta sub-basin; (4) Peace River-Athabasca Arch; (5) Northern Albèrta sub-basin or Peace River sub-basin; (6) Tathlina High (Fig. 51). These mainly sedimentary tectonic elements are easily discernible in the Middle Devonian deposits,

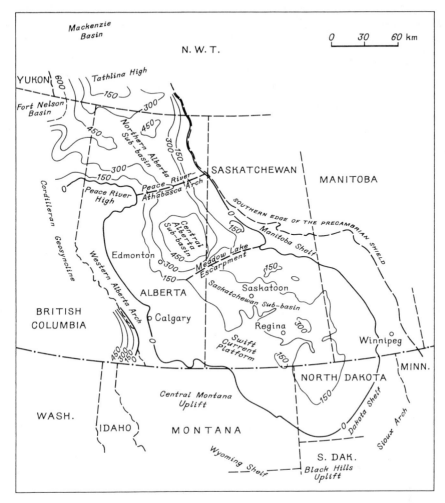

Fig. 51. Isopach map of Elk Point Group and related tectonic elements. After Holter (1969). Thickness is given in meters

containing the thickest salt strata. Each tectonic element in the basin is marked by peculiar lithological features of Devonian deposits, which gave grounds for the compilation of special stratigraphic schemes (Fig. 52).

The lower Devonian deposits for the entire West Canadian Basin are distinguished as the Elk Point Group, divided into the Lower and Upper Elk Point. This group has been assigned to a different age. Some workers place it into the Lower-Middle Devonian, others into the Middle Devonian.

The overlying deposits in the Saskatchewan sub-basin can be divided into two groups, namely, Manitoba and Saskatchewan. The Manitoba Group comprises the Dawson Bay and Souris River Formations. In central Alberta, the first formation is placed into the Elk Point Group. The lower Manitoba Group is assigned to the Givetian

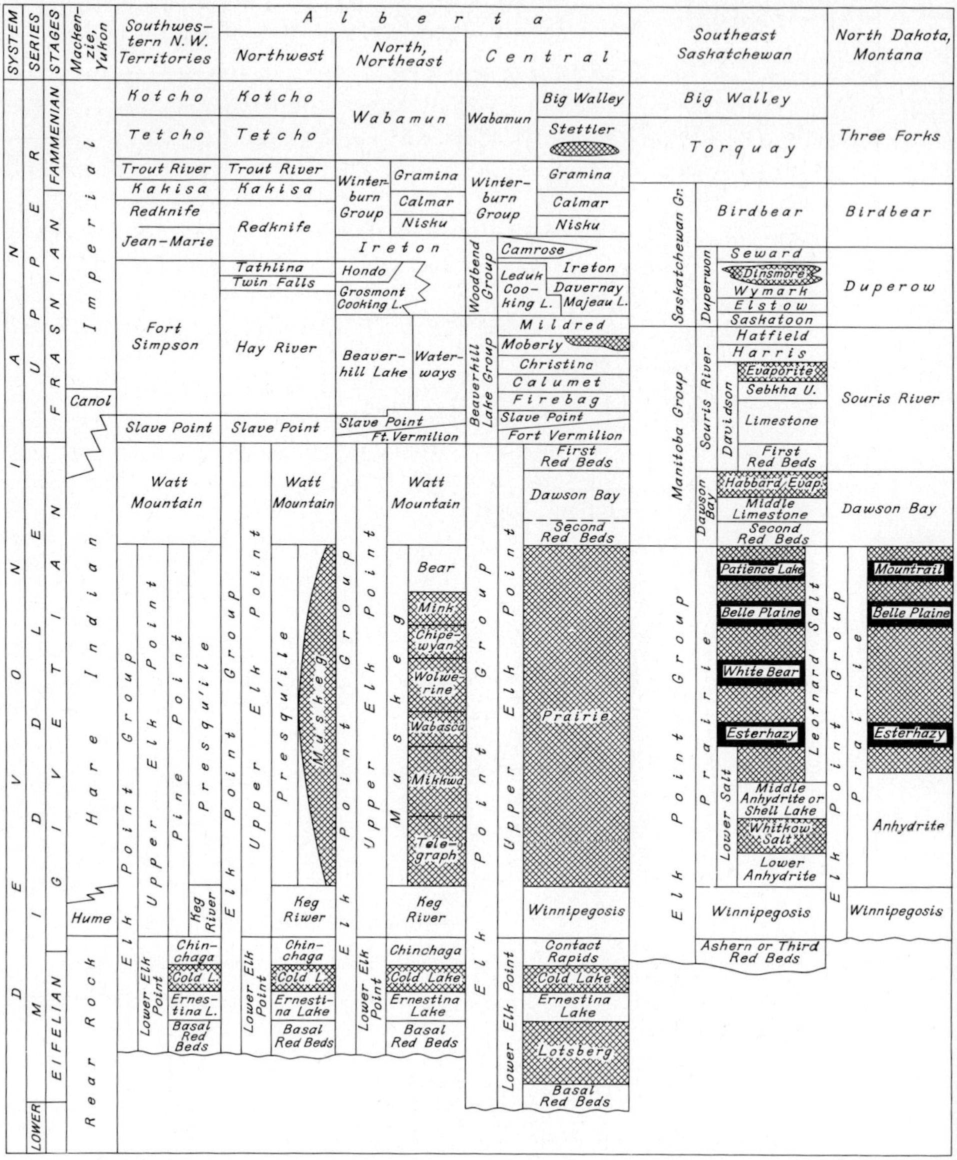

Fig. 52. Correlation chart of Devonian deposits in West Canadian evaporite basin also showing stratigraphic position of salt strata (*black* for potash salts, *oblique squares* for rock salt). Compiled from the data of Belyea (1964), Grayston et al. (1964), Basset and Stout (1967), Jordan (1967), Norris (1967), Sandberg and Mapel (1967), Holter (1969), Klingspor (1969), Prive and Ball (1971), Anderson and Swinehart (1979), and Worsley and Fuzesy (1979)

(Middle Devonian), and its upper part, the Souris River Formation, to the lower Frasnian (Belyea 1964, Belyea et al. 1964, Basset and Stout 1967, Kent 1967). The Saskatchewan Group is placed at the level of the upper Frasnian. Its equivalents in Alberta are the Woodbend and the lower Winterburn Groups. The Famennian is represented as follows: by the Three Forks Formation in North Dakota and Montana, by the Torquay and Big Valley Formations in the Saskatchewan sub-basin, and by the Wabamun Group in the Central Alberta sub-basin. Eight salt-bearing units have been distinguished in the Devonian section of the West Canadian Basin; in ascending order they are: Lotsberg, Cold Lake, Prairie (Muskeg), Black Creek, Hubbard, Davidson, Dinsmore, and Stattler Formations. The thickest and most widespread are the Lotsberg, Cold Lake, and Prairie Formations. In the Saskatchewal Depression the latter contains potassium salt beds.

The Lotsberg Formation occurs only in the Central Alberta sub-basin (Fig. 53). It is underlain by the Basal Red Beds, composed chiefly of red dolomitic mudstone, dolomite, siltstone, and sandstone about 75—80 m thick. Rock salt dominates this formation. Its lower section contains 30—60 m. This is a red mustone and marl unit dividing

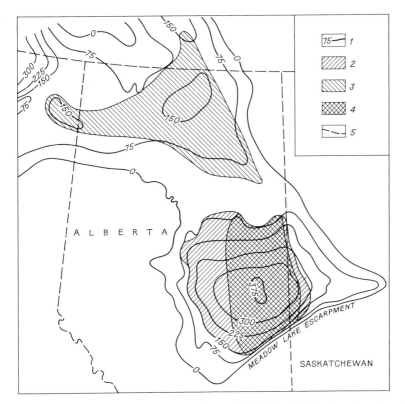

Fig. 53. Isopach map and areal extent of Lower Elk Point salt. After Grayston, Sherwin, and Allen (1964). *1* isopach contours showing thickness, in meters, *2* areal extent of Lotsberg salt, *3* areal extent of Gold Lake salt; *4* areal extent of Lotsberg and Gold Lake salts, *5* boundaries of states

the Lotsberg Formation into two unequal members: the lower one, about 20–30 m thick; and the upper one, ranging in thickness from 150–250 m. In the deepest parts the thickness is as great as 300 m. The area under rock salt does not exceed 140,000 km^2, and the volume may account for about 14,000 km^3.

In the Central Alberta sub-basin, the Lotsberg Formation is overlain by the Ernestina Lake Formation. It is divided into three members, namely, the lower one, composed of red marl and mudstone; the middle one, mainly carbonate; and the upper one, anhydritic. The carbonate unit in the Central Alberta sub-basin contains algal limestone; in northern Alberta, crystalline dolomite with crystal molds after rock salt is found. The thickness of the Ernestina Lake Formation reaches 20–25 m.

The salt-bearing Cold Lake Formation occurs in the Central Alberta and Northern Alberta sub-basins, and north of the Tathlina High. It is dominated by rock salt, in places alternating with red dolomitic clay. Its thickness ranges from 45 m in the south to 80 m in the north. Rock salt pinches out towards the West Alberta Arch. Red dolomitic marl and mudstone have primarily been found there. The areal extent of rock salt in the Cold Lake Formation equals 180,000 km^2; the volume of salt may reach 5400 km^3.

In central Alberta, the Contact Rapids Formation crowns the Lower Elk Point section and shows a greater extent than all the underlying deposits. It is composed of gray and green mudstone, clayey dolomite, and dolomitic marl. In the north and north-west, near the Peace River–Athabasca High and in the north-western Northern Alberta sub-basin, equivalents of the Contact Rapids Formation are distinguished as the Chinchaga Formation, with anhydrite interbeds and units; however, up the section it becomes mainly sulfate-bearing.

As a whole, the Lower Elk Point deposits are composed of three sedimentary cycles: the lower one, confined to the base of the Lotsberg Formation; the thickest, middle one, making up the upper part of the same formation; and the upper one, corresponding to the Cold Lake Formation. A peculiar feature of salt accumulation, characteristic of the two lower cycles, is the absence of rather thick sulfate beds both inside the formation and in under- and overlying deposits. The mode of occurrence of the Lower Elk Point salt deposits and the facies changes pattern are shown in Fig. 54.

The Lower Elk Point salt deposits are separated from the thick Upper Elk Point evaporites not only by the Contact Rapids and Chinchaga Formations, but by the overlying Keg River and Winnipegosis Formations, as well. They are composed of different carbonate rocks, often biogenic, forming reefal and often carbonate buildups. Their thickness ranges from 15 m in interreef zones to 180–200 m on carbonate banks and reefs. As a rule, interreef zones contain microgranular dolomite, and black shale. Reefs are built up of dolomite, oolitic, or oncolithic, containing corals, stromatoporoids, and algae, and consisting of microgranular pseudo-oolithic dolomite yielding scarce fossils. In reefogenic structures dolomite is usually secondary. Original limestone is minor, and its thickness does not exceed 15 m.

The Prairie and Muskeg Formations are age equivalents and closely related owing to facies change. Their thickness varies from 300 m in depression zones between reefal buildups, composed of the underlying Keg River or Winnipegosis carbonates, to 75 m above the buildups mentioned. Two zones of greater thickness can be distinguished:

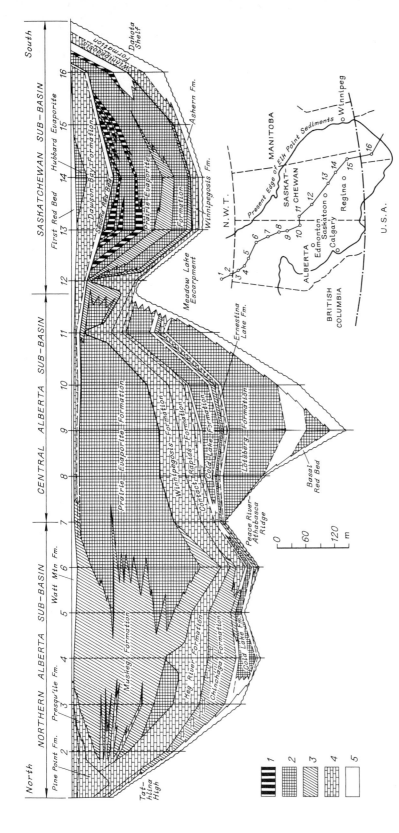

Fig. 54. Facies relationships in the Elk Point Basin. After Grayson et al. (1964), modified by Holter (1969). *1* potash, *2* halite, *3* anhydrite, *4* carbonate, *5* shale, calcareous or dolomitic

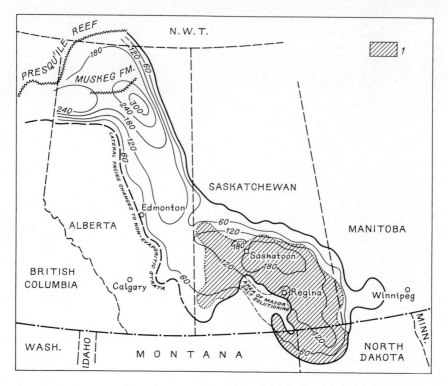

Fig. 55. Isopach map of the Prairie Evaporite. After Holter (1969). *1* areal distribution of potash. Isopach contours in meters

northern, confined to the Northern and Central Alberta sub-basins; and southern, lying inside the Saskatchewan sub-basin (Fig. 55). They are separated by a thinner zone, confined to the Meadow Lake Escarpment. The thickness in the northern zone is 300 m or more, and in the southern zone about 180 m.

The present north-eastern boundary of the Prairie and Muskeg Formations follows a pre-Cretaceous hiatus. The southern, western, and north-western boundaries are mainly sedimentary in nature; over most of the area, salt deposits are replaced by carbonate and red clayey carbonate rocks. In south-western Saskatchewan, on the Swift Current Platform, the absence of salt deposits is due to post-sedimentary processes, i.e., rock salt leaching. The Presqu'ile or the so-called Barrier Reef bounds evaporite distribution to the north-west.

In the north-west, the Muskeg Formation (Fig. 56) is composed of alternating carbonate and sulfate rocks. Near the Barrier Reef, the Zema area (McCamis and Griffith 1968), the salt-bearing Black Creek Formation can be distinguished between reefs made up of the Keg River Limestone. As a rule, it is not more than 80 m thick. The overlying part of the Muskeg Formation in the Zema area is salt-free. It can be divided into: the Lower Anhydrite Member, composed of microgranular anhydrite, intercalated with 25-m-thick dolomite; the Zema Member, dominated by the

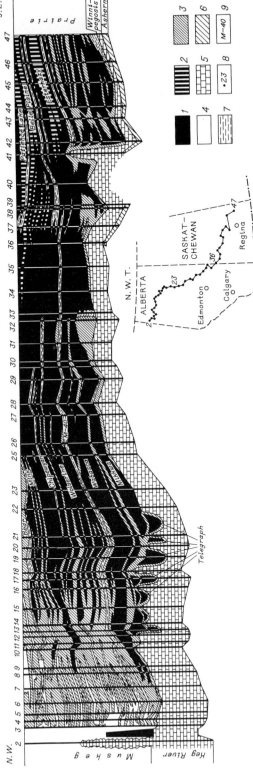

Fig. 56. Cross section of Muskeg and Prairie Formations. After Klingspor (1969). *1* halite, *2* potash, *3* anhydrite, *4* carbonate of Keg River and Winnipegosis Formations, *5* carbonate of Muskeg Formation, *6* shale and argillaceous carbonate of Ashern Formation, *7* clay of Muskeg and Prairie Formations, *8* location of sections and their number, *9* Muskeg 40 marker

dolomite, several to 23 m thick; the Upper Anhydrite Member with dolomite interbeds 25–135 m thick, and the Bistcho Member, composed of dolomite and limestone 20–30 m thick.

Farther south-east, the Muskeg Formation first is dominated by anhydrite and then by rock salt. There, the section can be divided into seven members, namely, Telegraph, Mikkwa, Wabasca, Wolverine, Chipewyan, Mink, and Bear (Klingspor 1969). In the middle part, easily discernible is a clayey marker bed, separating the formation into two halves, i.e., the Lower and Upper Muskeg, which allows the correlation of salt-bearing and sulfate-carbonate sections (Fig. 56).

The Telegraph Member fills in interreef zones, forming lenticular salt bodies. Its thickness reaches 100–110 m (Klingspor 1969). The lower 60-m-thick part consists of white and gray, medium- and coarse-grained, clayey rock salt, alternating with anhydritic varvite, which is often bituminous, especially at the base. Above is a 12–15 m thick unit of thin-bedded, microgranular, in places algal, dolomite alternating with nodular anhydrite. The upper part of the member is also composed of brownish-gray, medium-grained rock salt, intercalated with clayey-sulfate rocks, bituminous anhydrite, and dolomite. The thickness of this part does not exceed 30 m.

The Mikkwa Member lies as a veneer on many reefs of the Keg River and Winnipegosis Formations. The base of the member is composed of brown and dark brown, thin-bedded, microgranular anhydrite 3.0–3.5 m thick. Above it is medium-grained, white and brownish-gray rock salt with clayey interbeds, intercalated in the upper part with microgranular dolomite and anhydrite; the thickness is 12–15 m. The Mikkwa Member is 15–19 m thick.

Three m-thick layered anhydrite lies at the base of the Wabasca Member. The upper part of the member is composed of medium- and coarse-grained rock salt, in places clayey, white or gray, or red interbeds intercalated with anhydrite partings. The unit is about 40 m thick. The Muskeg 40 marker is made up of red and green mudstone, often saline and alternating with anhydrite. It is 2–3 m thick.

At the base of the Wolverine Member, 4–5-m-thick layered, microgranular, clayey white and bluish anhydrite occurs. An overlying section, 35–36 m thick, is composed of coarse- and medium-grained, layered, brown, and light brown rock salt in alternation with bluish-gray, microgranular anhydrite and saline, thin- and uniformly bedded mudstone. The member is 39–41 m thick.

The Chipewyan Member begins with a 5-m-thick, microgranular anhydrite bed with a dolomite interbed in the middle. The rest of the section is made up of clayey, medium- and coarse-grained rock salt with microgranular, nodular, and mosaic anhydrite interbeds. The total thickness of the member is 40–43 m.

The Mink Member is composed of rock salt alternating with anhydrite and dolomite. Anhydrite is mainly microgranular with mudstone partings, thin-bedded, and bituminous. The member is 26–27 m thick. The Bear Member is dominated by nodular, white, or dark brown anhydrite alternated with bituminous dolomite and breccia of dolomite and anhydrite fragments in argillaceous-anhydritic matrix. The member is 20 m thick.

In the Central Alberta and Saskatchewan sub-basins, equivalents of the Muskeg Formation are placed in the Prairie Formation (Prairie Evaporite). Figure 57 presents the subdivision of the formation proposed by different authors. In general, the Prairie

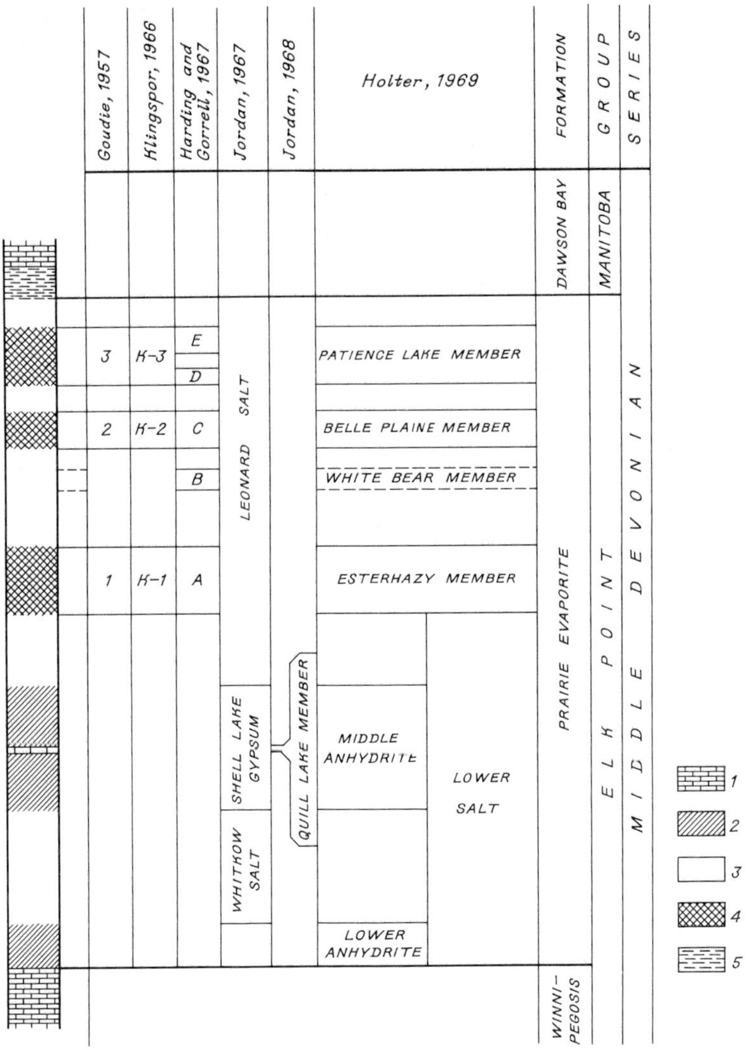

Fig. 57. Prairie Evaporite Formation composition. After Holter (1969). *1* carbonate, *2* anhydrite, *3* halite, *4* potash, *5* shale

Formation can easily be divided into two parts: the lower one, composed of rock salt with anhydrite beds and units; and the upper one, containing potash salts. The lower part is called the Lower Salt (Holter 1969). In the upper, potassium-bearing part of the section, four potash members can be distinguished, viz., Esterhazy, White Bear, Belle Plaine, and Patience Lake (Holter 1969, Worsley and Fuzesy 1979).

The Lower Salt is dominated by halite. Anhydrite with thin dolomite and rock salt interbeds often occurs at its base, known as the Lower Anhydrite. Its thickness does not exceed 12–15 m. There are also anhydrite beds in the middle Lower Salt. They are informally termed the Middle Anhydrite and are 30–35 m thick. Inside the Middle

Anhydrite the dolomite Quill Lake Member is recognized (Jordan 1968). The Lower Salt is 30–120 m thick. Greatest thickness was reported from areas north and northeast of Regina.

The Esterhazy Member comprises the lowermost potash beds of the Prairie Formation (Holter 1969). It is equivalent to: Zone 1 of Goudie; Zone A of Harding and Gorell (1967); and Zone K-1 of Klingspor (1966). The member is composed of rock salt with sylvite- and carnallite-bearing interbeds. The areal extent and thickness of the member are shown in Fig. 58. The maximal thickness of the White Bear Member is 8–10 m.

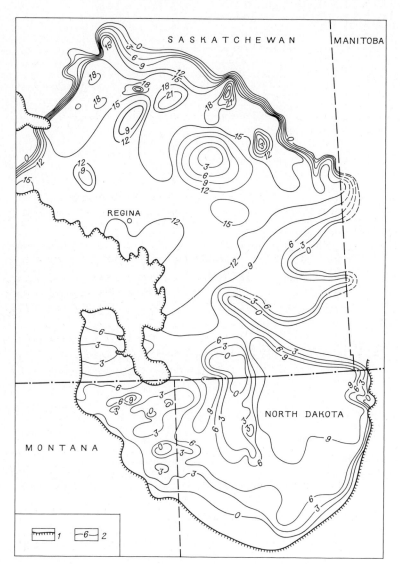

Fig. 58. Isopach map of the Esterhazy Member. Compiled from the data of Holter (1969), Anderson and Swinehart (1979), and Worsley and Fuzesy (1979). *1* limit of Prairie salt, *2* isopach contours, in meters

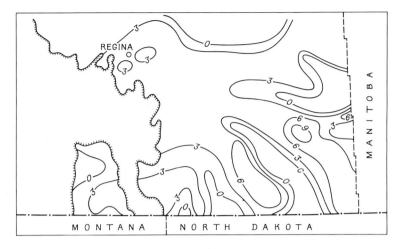

Fig. 59. Isopach map of the White Bear Member, south-eastern Saskatchewan. After Worsley and Fuzesy (1979). For legend see Fig. 58

It was distinguished as a separate potash member in southern Saskatchewan (Fig. 59) (Worsley and Fuzesy 1979). The member corresponds to Zone B of Harding and Gorell (1967). Potash beds contain mainly halite, sylvite, and carnallite.

The Bell Plaine Member comprises potash beds, placed by Goudie in Zone 2; by Harding and Gorrell (1967), in Zone C; and by Klingspor (1966), in K-2 Zone. The areal extent of the member is larger than that of underlying units (Fig. 60). Its thickness does not exceed 15–18 m. Above is an extensive rock salt bed 12–15 m thick.

The Patience Lake Member is most widespread (Fig. 61). Holter (1969) has assigned to it the uppermost potash beds of the Prairie Formation, formally placed in Zone 3 or in K-3 Zone (Klingspor 1966). In the Williston Basin, in North Dakota and Montana, U.S.A., equivalents of the Patience Lake Member are known as the Mountrail Member (Anderson and Swinehart 1979). The thickness of this potash member reaches 20–21 m. It is made up of rock salt with beds rich in potassic minerals: sylvite and carnallite. In the Saskatchewan sub-basin the Prairie Formation section is crowned by a rock salt member 3–12 m thick.

In Saskatchewan the top of the Prairie Formation plunges monoclinally from north-east to south-west from 600–3000 m or more, and respectively, potash members occur at depths of 800–900 and 2000 m. The main developed areas and prospects for potash salts in Saskatchewan are confined to a north-west-south-east striking zone where the depth to the top of the Prairie Formation is 800–1200 m.

In some southern regions of the West Canadian Basin, within the Dawson Bay Formation, the upper Elk Point Group contains the Hubbard evaporites (in Saskatchewan, the Dawson Bay Formation belongs to the lower Manitoba Group), which mainly occur in southern and south-eastern Saskatchewan (Fig. 62). There, they are composed of a relatively thin (not more than 20 m) rock salt unit alternating with anhydrite and finely laminated mudstone.

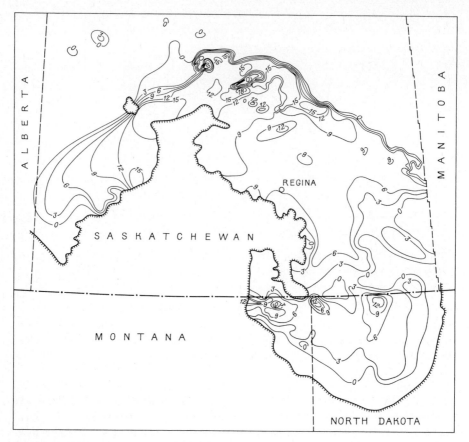

Fig. 60. Isopach map of the Belle Plaine Member. Compiled from the data of Holter (1969), Anderson and Swinehart (1979), Worsley and Fuzesy (1979). For legend see Fig. 58

In general, according to Pearson (1963), the Elk Point rock salt has the greatest thickness in eastern Central Alberta, where it exceeds 400 m (Fig. 63). In some regions of the Sasketchewan sub-basin, the net thickness reaches 180 m. The areal extent of rock salt in the Elk Point Group in the West Canadian Basin slightly exceeds 350,000 km^2. Apparently, a minimal salt volume may equal 40,000 km^3.

Among the Upper Devonian deposits evaporites are confined chiefly to the southern West Canadian Basin. In the lower Frasnian, they have been found at the level of the Beaverhill Lake Formation and its equivalents, viz., Davidson and Souris River Formations. Anhydrite in these formations appears south and south-east of Edmonton. Its proportion increases in south-eastern Alberta, Saskatchewan, and Montana. In southern Saskatchewan, rock salt beds appear in the upper Davidson Formation; they occupy an area of about 20,000 km^2 (Fig. 64). These salt-bearing deposits are assigned to the Davidson Evaporite (Lane 1964, Pearson 1963). It is mainly rock salt with thin lenticular mudstone interbeds; in the lower part, rock salt alternates with finely laminated brown anhydrite. The salt-bearing section is underlain by the

Fig. 61. Isopach map of Patience Lake and Mountrail Members. Compiled from the data of Holter (1969), Anderson and Swinehart (1979), and Worsley and Fuzesy (1979). For legend see Fig. 58

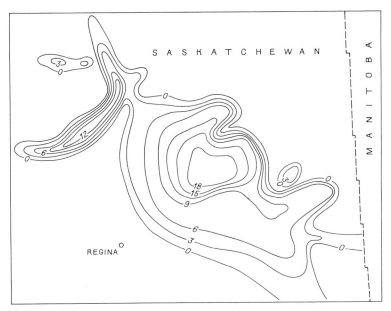

Fig. 62. Isopach map of the Hubbard Evaporite. After Pearson (1963). Isopach contours, in meters

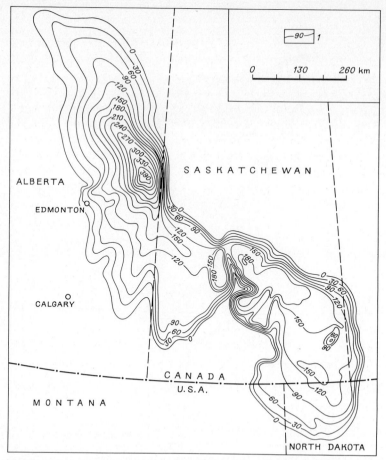

Fig. 63. Isopach map of total salt thickness in the Elk Point Group. After Pearson (1963). *1* isopach contours, in meters

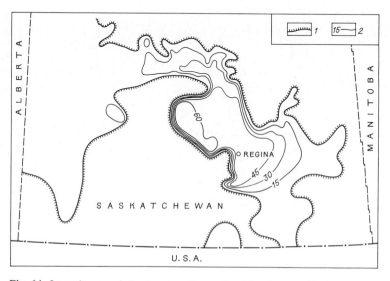

Fig. 64. Isopach map of the Davidson Evaporite. After Pearson (1963). *1* approximate edge of the Davidson Evaporite, *2* isopach contours, in meters

Sebkha Member (Price and Ball 1971), composed of alternating clayey limestone and dolomite with anhydrite lenses and interbeds about 0.6–1.5 m thick. Lenticular anhydrite interbeds and inclusions occur in underlying deposits of the Davidson Formation. The Davidson Evaporite is overlain by the Harris Formation (Lane 1964, Price and Ball 1971) made up of dolomite, anhydrite, limestone, and mudstone. Anhydrite interbeds in the lower Harris Formation are 1–4 m thick. The net thickness of rock salt in the Davidson Evaporite ranges from 10–15 to 60 m. The salt volume may equal 600 km^3.

Anhydrite is common in the upper Frasnian deposits that comprise the Duperow Formation and Woodbend Group, in the southern West Canadian Basin. This has been reported from southern Alberta, Saskatchewan, Montana, and North Dakota (Fig. 65).

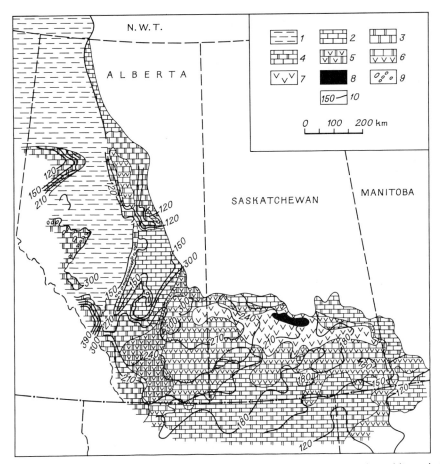

Fig. 65. Lithologic map also showing thicknesses of Woodbend Group deposits and its equivalent rocks (Duperow Formation and others) in West Canadian Basin. After Belyea (1964). *1* shale, *2* carbonate containing about 75% limestone, *3* carbonate containing more than 75% dolomite, *4* carbonate with equal amounts of dolomite and limestone, *5* dolomite with anhydrite accounting for 5%–20%, *6* dolomite and limestone with anhydrite accounting for 5%–20%, *7* dolomite and limestone with anhydrite accounting for more than 20%, *8* areal extent of rock salt, *9* isolated reef structures, *10* isopach contours, in meters

At this stratigraphical level rock salt was found only in southern Saskatchewan, where it builds up the thin Dinsmore Member (Kent 1967). The Dinsmore Member has been penetrated by a few boreholes in the vicinity of Saskatoon. Its thickness attains 60–65 m. The member is made up of rock salt with anhydrite and mudstone interbeds. It is underlain and overlain by anhydrite. The volume of salts is rather small.

Evaporites of the Birdbear Formation and the Winterburn Group are represented exclusively by sulfate rocks, namely, anhydrite, dolomite-anhydrite, and gypsum. Their distribution in the West Canadian Basin is shown in Fig. 66.

Sulfate and sulfate-carbonate rocks are widespread among the Famennian deposits of the southern West Canadian Basin. The uppermost Devonian salt-bearing unit was found in the lower Stattler Formation (Fig. 67). The thickness of rock salt beds varies between 10 and 30 m; the total thickness is 165–170 m. If the areal extent of rock salt is equal to 18,000–20,000 km^2, then its volume probably exceeds 150–200 km^3. The volume of rock salt in the West Canadian Basin may amount to 6.2×10^4 km^3.

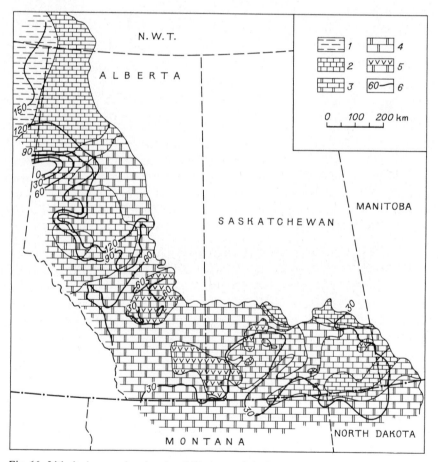

Fig. 66. Lithologic map also showing thicknesses of the Birdbear Formation, Winterburn Group, and equivalent rocks in West Canadian Basin. After Belyea (1964). *1* shale, *2* limestone, *3* limestone and dolomite, *4* dolomite, *5* dolomite and anhydrite, *6* isolines of thicknesses, in meters

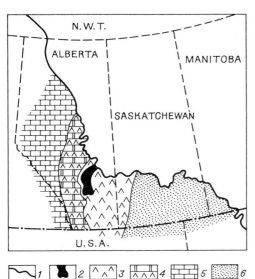

Fig. 67. Lithologic map of Stattler Formation deposits and equivalent rocks in Alberta and Saskatchewan. After Fuller and Porter (1969). *1* erosional edge of Stattler Formation and its equivalents. Depositional limit of: *2* salt, *3* anhydrite, *4* anhydrite and dolomite, *5* limestone, *6* red beds

Michigan Basin

The Michigan Basin lies between lakes Michigan and Huron and is bounded to the north by outcrops of Precambrian formations on the Canadian Shield, to the west, by the Wisconsin Arch, and to the east, by the Algonquin Arch. To the south-west, the Michigan Basin is separated from the Illinois Basin by the Kankakee Arch, and to the south-east, it is separated from the Appalachian Basin by the Findlay Uplift. The thickness of the Devonian sediments in the central part of the basin is about 1000 m. The Lower, Middle, and Upper Devonian strata can be distinguished in a composite section (Fig. 68). The Lower Devonian deposits, regarded there as the Garden Island and Bois Blanc Formations, fill in the most subsided central parts of the basin. The Garden Island Formation is dominated by light yellowish and brownish dolomitic sandstones with gray and dark gray dolomite interbeds and cherts. The formation does not exceed several meters. It correlates with the Oriskany Formation of the Appalachian Basin and belongs to the Siegenian (Lower Devonian) (Sanford 1967). In the Michigan Basin the Bois Blanc Formation represents the upper Lower Devonian. In central and north-western regions, it is chiefly composed of limestone, whereas dolomite is common to the west and south. The thickness of the formation varies between 30 m in marginal parts and 120 m in its central parts.

Eifelian deposits (Middle Devonian) over most of the Michigan Basin are subdivided (in ascending order) into the Amherstburg, Lucas, and Dundee Formations. The former two are often placed into the Detroit River Group. The Amherstburg Formation might be regarded as a basal unit of evaporitic deposits. It rests unconformably on the underlying strata and is generally represented by bluish and brownish limestone and dolomite, in places black, bituminous, and siliceous, among which interbeds of Sylvania quartz sandstone were recorded in some regions. The composition

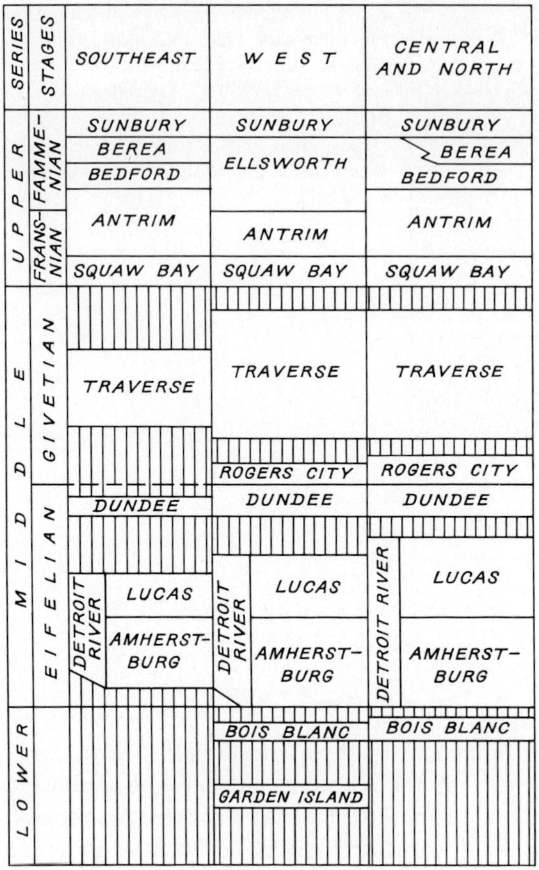

Fig. 68. Correlation chart of Devonian deposits in Michigan Basin. Compiled from the data of Sanford (1967)

of the Amherstburg Formation changes greatly across the area. In its most subsided central parts the basin is composed mainly of limestone; however, to the south-west and west dolomite dominates. To the south-east, the formation is entirely made up of dolomitic biogenic deposits, building carbonate reef structures, known as the Huronian Biostrome. The latter separates the area where the Amherstburg Formation is developed in the Michigan Basin from that of the lower Onondaga Group limestone in the Appalachian Basin. The Amherstburg Formation is 60–240 m thick.

The Lucas Formation comprises all the Devonian evaporite deposits of the Michigan Basin. In its northern and central parts, it is mainly composed of rock salt and anhydrite. At margins, evaporites pinch out, and the formation is composed of dolomite, replaced by limestone farther away from the center (Fig. 69). The thickness of the Lucas Formation ranges widely from 30 to 250–300 m. In subsided parts of the basin, up to eight rock salt beds 30 m thick may be distinguished (Landes 1951). They alternate with interbeds and units of anhydrite, gypsum, and, more rarely, dolomite. The thickest and most persistent dolomite interbed, several meters thick, was observed in the middle part of the formation; it divides the formation into two almost equal parts. This band is underlain and overlain by anhydrite. The thickest

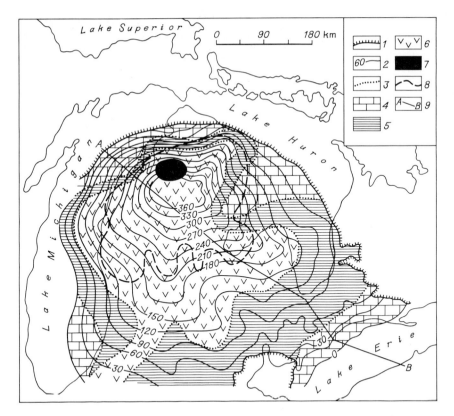

Fig. 69. Lithologic map also showing thicknesses of the Lucas Formation in Michigan Basin. After Sanford (1967). *1* limit of Lucas Formation outcrops, *2* isolines of thicknesses, in meters, *3* limits of lithologic complexes. Areas of essential development of: *4* limestone, *5* dolomite, *6* anhydrite, *7* salt; *8* depositional limit of salt, *9* line of section shown in Fig. 71

anhydrite units occur at the base and top of the formation. Rock salt of the Lucas Formation attains its greatest thickness, 130–135 m, in the northern part of the basin (Fig. 70). Rock salt lies at a depth of 400 m to the north and 1300 m in central regions. The areal extent of rock salt equals 25,000 km^2. Its volume may exceed 800 km^3. In the Michigan Basin rock salt pinches out towards the margin, and the entire Lucas Formation becomes sulfate and sulfate-carbonate.

The overlying Dundee Formation is made up of limestone and dolomite, its thickness being 30–120 m. The Rogers City Formation and the Traverse Group over most of the Michigan Basin are composed of limestone; it is only in the eastern and southeastern regions that the Traverse Group begins to be dominated by shale alternating with limestone. These deposits are similar in composition to the Hamilton Formation of the Appalachian Basin. Their thickness varies between 60 and 240 m. Frasnian and Famennian (Upper Devonian) deposits, distinguished as the Squaw Bay and Antrim Formations, are also dominated by carbonate rocks, viz., limestone and dolomite alternating with black and dark gray bituminous shale. It is only in the uppermost

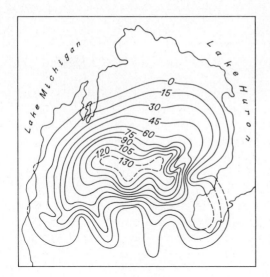

Fig. 70. Isopach map of Lucas Formation salt in Michigan Basin. After Landes (Lefond 1969). Isolines of thicknesses are given in meters

Upper Devonian that the Bedford, Berea, and Sunbury Formations become dominated by terrigenous rocks, viz., sandstone, siltstone, and mudstone, particularly widespread to the south-east. The total thickness of the Upper Devonian ranges from 100–300 m.

A brief discussion of the Devonian deposits in the Michigan Basin implies that evaporites of the Lucas Formation occur among the underlying and overlying carbonate deposits and replace them along the strike (Fig. 71).

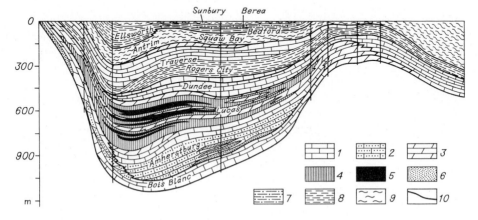

Fig. 71. Lithofacies profile of Devonian deposits in Michigan Basin. After Sanford (1967). *1* limestone, *2* arenaceous and silty limestone, *3* dolomite, *4* anhydrite, *5* salt, *6* sandstone, *7* siltstone, *8* mudstone, *9* shale, *10* limits of formations. For line of section see Fig. 68

Hudson and Moose River Basins

Description of basins given below is based on the evidence presented by Sanford and Norris (Sanford 1974, Sanford and Norris 1975).

Devonian deposits along with Silurian and Ordovician fill in a large syneclise in the Hudson Bay and in adjacent regions, situated on the Canadian Shield and bounded on all sides by Precambrian outcrops. This Paleozoic terrane belongs to the Hudson Platform (Fig. 72). Devonian deposits formed there within two epicontinental basins, namely, the Moose River and Hudson Basins are separated by the Henriette Maria Arch stretching from south-west to north-east. The Moose River Basin embraces the southern Hudson Bay and adjacent coastal regions. It is separated from the Michigan and Allegheny Basins to the south and south-west by the Fraserdale Arch. The Hudson Basin occupies the central Hudson Bay and the coastal part of the Tatnam Peninsula near the boundaries of north-eastern Manitoba and north-western Ontario. The basin is flanked by the Severn Arch and Bell Arch on the south-west and north-east, respectively.

In the Moose River and Michigan Basins Devonian deposits can be distinguished as separate formations showing good correlation and similar age. In ascending order they are: the Kenogami River, Stooping River, Kwataboahegan, Moose River, Murray Island, Williams Island, and Long Rapids Formations.

On the whole, the Kenogami River Formation is of Late Silurian/Early Devonian age. Only the upper part of the formation is assigned to the Early Devonian. In the

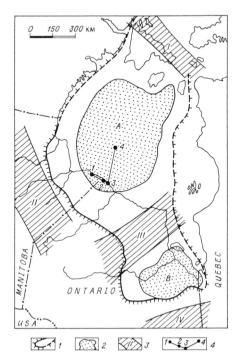

Fig. 72. Distribution of Devonian desposits in the Hudson Bay. Compiled from the data of Sanford and Norris (1975). *1* determined and inferred boundaries of Paleozoic deposits, *2* areal distribution of Devonian sediments (*A* Hudson Bay, *B* Moose River Basin), *3* arches and their numbers (*I* Bell, *II* Severn, *III* Henriette Maria, *IV* Freserdate), *4* boreholes and their numbers (*1* Kascattama No. 1, *2* Penn No. 2, *3* Penn No. 1, *4* Hudson Walrus A-71), and the profile shown in Fig. 73

Moose River Basin it is composed of light calcareous dolomite, oolitic dolomite, argillaceous dolomite, and by brecciform dolomitized limestone and silicified dolomite. Gypsum inclusions and lenses are common. The thickness of this part of the formation reaches 50–52 m in the Moose River Basin. In the Hudson Basin the upper Kenogami River Formation has been penetrated by three wells drilled on the Hudson Bay coast, namely, the Kaskattama No. 1, and the Penn No. 1 and 2 wells (Fig. 73).

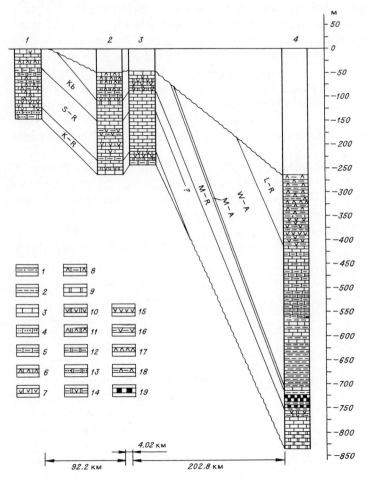

Fig. 73. Correlation of Devonian deposits along profile 1–4 (see Fig. 72). *1* siltstone, *2* mudstone, *3* limestone, *4* arenaceous limestone, *5* clayey limestone, *6* limestone with inclusions, lenses, and interbeds of gypsum, *7* limestone with inclusions, lenses, and interbeds of anhydrite, *8* clayey limestone with interbeds and lenses of gypsum, *9* dolomite, *10* dolomite with interbeds, lenses, and inclusions of anhydrite, *11* dolomite with interbeds, lenses, and inclusions of gypsum, *12* clayey dolomite, *13* clayey dolomite, locally arenaceous dolomite, *14* clayey dolomite with interbeds and lenses of anhydrite, *15* anhydrite, *16* anhydrite with mudstone interbeds and mudstone with inclusions and interbeds of anhydrite, *17* gypsum, *18* clayey gypsum and gypsinate mudstone, *19* rock salt. Figures above columns denote numbers of boreholes; for names see Fig. 72. Formations: *K-R* Kenogami River, *S-R* Stooping River, *Kb* Kwataboahegan, *M-R* Moose River, *M-A* Murray Island, *L-R* Long Rapids, *W-A* Williams Island

In the Kaskattama No. 1 well the section includes creamy to light gray fine-grained and aphanitic limestone with thin bands of light gray, argillaceous, silty, and sandy dolomite up to 10 m thick. The Penn No. 1 well includes, in ascending order:

	Thickness (m)
1. Light creamy, gray, and light greenish gray argillaceous dolomite	5.2
2. Light and red to brown micrograined limestone dolomitized to a variable degree with inclusions and bands of light gray anhydrite	15.0
3. Light gray, micro- and fine-grained oolitic dolomite limestone	5.2

The total thickness is 25.4 m.

The Penn No. 2 borehole penetrates the following section of the upper Kenogami River Formation, in ascending order:

	Thickness (m)
1. Light gray, strongly argillaceous dolomite	about 1.0
2. Brown thin-grained limestone	4.5
3. Light to brown dolomite with bands of bituminous dolomite with bands of anhydrite and crystals of pyrite	7.0
4. Pink dolomitized limestone	9.0
5. Light microgranular dolomitic limestone	6.0

The total thickness is 27.5 m.

The upper Kenograrni River Formation wedges out toward the central Hudson Basin. This part of the Kenogami River Formation is Late Gedinnian/Early Siegenian.

In the Moose River Basin, in the Albany River delta area, and along the lower reaches of the Stooping River at the northern margin of the basin, the Stooping River Formation contains limestone, dolomitic limestone, and dolomite. The limestone is usually calcarenitic, brown to grayish-brown, sometimes orange. The dolomitic limestone is brown, fine- and micro grained. Dolomite is dark to light brown, thin-crystalline. There are inclusions of gray and black chert in these rocks. Along the northern and north-western margins of the basin limestone dominates the Stooping River Formation, while the amount of dolomite decreases. In the center of the basin the lower part of the formation is dominated by dolomite and dolomitic limestone usually containing lenses of chert, as well as inclusions and veins of anhydrite. Limestone, locally silicified, is common in the upper part of the formation. Along the eastern margin the Stooping River Formation grades into limestone alternated with red sandy mudstone and green sandstone forming a separate Sextant Formation.

In the Hudson Basin the Stooping River Formation was penetrated by four deep wells. In the section of the Kaskattama No. 1 borehole the formation contains the following units, in ascending order:

	Thickness (m)
1. Light gray creamy thin-grained and aphanitic limestone	8.0
2. Light to dark gray aphanitic limestone, with inclusions of anhydrite	34.5
3. Light brown, blue to light gray medium-grained dense limestone with interbeds of gray argillaceous limestone	19.0

	Thickness (m)
4. Creamy aphanitic dolomite	8.2
5. Light to dark brown argillaceous, calcareous dolomite	4.5

The total thickness of the formation is 74.2 m.

In the Penn No. 1 and 2 wells limestone and dolomitic limestone dominate the lower part of the formation. In the upper part of the Penn No. 1 gray chert occurs in limestone, and creamy dolomite, argillaceous limestone, and gypsinate mudstone appear in the Penn No. 2 well. The thickness of the formation ranges there from 81–85 m. Equivalents of the Stooping Formation failed to be traced in the Hudson Walrus A-71 borehole due to their similarity to the overlying Kwataboahegan Formation; they form a single sequence.

Organic remains suggest Middle to Late Emsian age of the Stooping River Formation.

In the Moose River Basin, the Kwataboahegan Formation consists of dark brown, calcarinate, bituminous, massive, and thick-platy limestones more or less dolomitic. Carbonate banks and bioherms are common. Locally, limestone contains gypsum inclusions and selenite veins. The maximal thickness of the deposits is 75 m. In the Hudson Basin the Kwatoboahegan Formation also consists chiefly of light gray and light creamy limestone, commonly argillaceous and silicified, locally, gypsinate (Kaskattama No. 1 borehole). In more subsiding areas (Penn No. 1 and 2 wells) bituminous limestone and calcareous dolomite are widespread. In the central zone of the basin (the Hudson Walrus A-71 borehole) dolomite dominates the lower part of the section, and lenses of anhydrite occur up the section. The Kwataboahegan Formation is difficult to distinguish from the underlying Stooping River Formation. The thickness of the undifferentiated section is 76.5 m.

The Moose River Formation is represented by thin-grained and argillaceous limestone, dolomite, carbonate breccia, gypsum, and anhydrite. The maximal content of gypsum in outcrops was found in the belt extending from south-east to north-west 72 km long and about 16 km wide, i.e., from the Wakwaywkastic River to Cheepash River. Unusual land forms, creating gypsum mountains, are known there. There are many gypsum quarries in the Cheepash Basin. In the Moose River Basin the formation reaches 60 m in thickness. The thickness of gypsum in one of deep wells drilled in the Moose River Basin on the coast of Mike Island is 44 m. Brecciform carbonate rocks and breccias dominating the formation may have formed due to the leaching of gypsum beds. The data obtained from three deep boreholes suggest a highly variable composition of the Moose River Formation in the Hudson Basin. In two boreholes (Penn No. 1 and 2) the formation consists chiefly of thin- and medium-grained dolomitic and argillaceous limestone interbedded with dolomite. Rocks usually contain inclusions, lenses, and bands of gypsum and anhydrite, light gray flint, and minor red mudstone. The Hudson Walrus A-71 borehole drilled in the central Hudson Basin penetrated the following section, in descending order:

	Thickness (m)
1. White thin-grained calcarinate limestone	4.9
2. Light gray anhydrite	3.0
3. Light brown to gray anhydrite with inclusions of red mudstone	6.1

	Thickness (m)
4. Pale yellow to brown thin-grained calcarinate limestone with bands of red argillaceous limestone	3.0
5. Bright red mudstone	1.5
6. Interval poorly represented by a core consisting chiefly of rock salt	10.5
7. Orange to bright red mudstone and shale	10.0

The thickness exposed in the borehole is 39 m. The formation is of Middle Eifelian age.

The Murray Island Formation is easily recognized in the Devonian section of the Hudson and Moose River Basins. It is represented mainly by limestone and argillaceous and bituminous dolomite. The thickness of the formation varies from 6–20 m in the Moose River Basin and does not exceed 6–8 m in the Hudson Basin. The age of the formation is Upper Eifelian.

In the Moose River Basin, the Williams Island Formation, which includes Givetian (Middle Devonian) and Frasnian (Upper Devonian) deposits, is divided into the lower and upper units. The lower unit consists of greenish to gray shale and mudstone interbedded with yellow to brown sandstone, brown and gray gypsinate sandy mudstone, siltstone, and sandstone. Minor limestone and carbonate breccia are present. The lower unit reaches 45–50 m in thickness. The upper unit consists of thin- to medium-platy argillaceous limestone and light blue to gray calcareous mudstone. Gypsum and anhydrite occur locally up the section. Its thickness reaches 40–45 m. The total thickness of the formation is about 90 m.

In the Hudson Basin, the Williams Island Formation reaches 270–290 m in thickness. The formation there is composed of gray and pale yellow argillaceous and silty limestone, mudstone with inclusions and bands of gypsum and quartz sandstone. Red gypsinate mudstone and siltstone are common.

In both basins, the Devonian section is crowned by the Long Rapids Formation of Late Frasnian/Early Famennian age. It includes dark gray to black bimuninous shale and mudstone, and, in places, gray shale interlayered with dolomite and limestone. Purple red sandstone, brown to red mudstone interbedded with white gypsum is also present in the Hudson Basin. The rocks are usually saline and gypsinate there. In the Moose River Basin, the thickness of the formation does not exceed 85 m, and in the Hudson Basin it reaches 150 m.

In the Hudson Basin, salt deposits occupy the central part of the Hudson Bay proper (Fig. 74). The areal extent of evaporite rocks reaches 80,000 km^2, and that of rock salt is about 50,000 km^2. The average thickness of rock salt beds is about 10 m. Thus the salt content may amount to 5×10^2 km^3. Only sulfate rocks (gypsum and anhydrite) have been so far found in the Moose River Basin, and rock salt has not yet been found even in deep boreholes which are very few there. Therefore, rock salt is probably developed in the subsiding parts of the Moose River Basin where sulfate thickness increases.

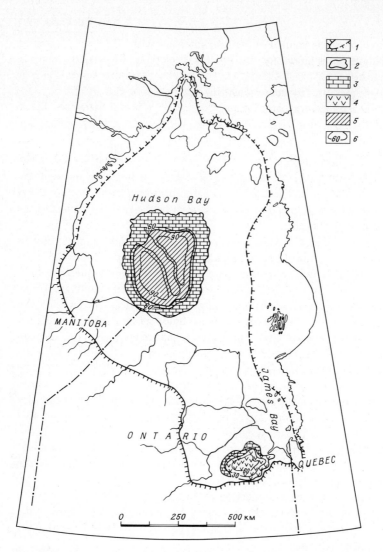

Fig. 74. Lithologic map and thicknesses of the Moose River Formation in the Hudson and Moose River Basins. After Sanford and Norris (1975). *1* determined and inferred boundaries of Paleozoic deposits distribution, *2* Moose River Formation. Areal distribution of: *3* limestone, clayey limestone, and dolomitized limestone, *4* gypsum, carbonate rocks, and carbonate breccias, *5* rock salt, gypsum, anhydrite, mudstone, and carbonate (salt zones), *6* lines of similar thicknesses, in meters

Adavale Basin

The Adavale Basin lies in southern Queensland, Australia (Fig. 75). It is very complex in structure. The major structural elements of the basin are: the Blackwater (Cooladdi, Quilpie, Langlo, Wanka, and Westgate trough embayments).

Salt-bearing deposits were reported from the eastern Adavale Basin, the Blackwater, Cooladdi troughs, and Langlo Embayment. They build up salt bodies, and have been penetrated by deep wells: namely, the Boree No. 1, Bonnie No. 1, Bury No. 1, Stafford No. 1, and Alva No. 1. Salt deposits are Middle Devonian in age. Their discussion is based on the data presented by Tanner (1967), Hill (1967), Galloway (1970), Brown et al. (1968), and Wells (1980).

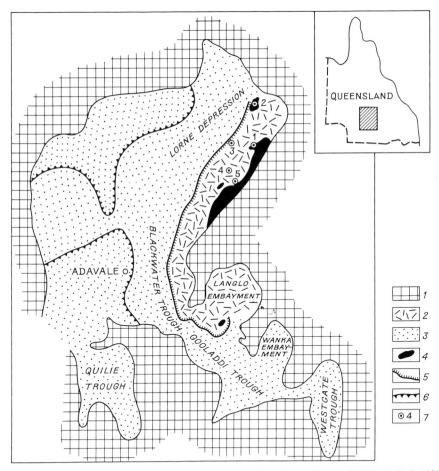

Fig. 75. Distribution of Boree Salt Member in Adavale Basin. After Tanner (1967) and Wells (1980). *1* Devonian absent (Permian or Mesozoic on basement), *2* areal extent of Devonian salt deposits, *3* areal extent of carbonate and terrigenous Devonian deposits, *4* salt bodies, *5* westward extent of Boree Salt Member deposition, *6* probable westward extent of Cooladdi Dolomite, *7* wells and their numbers (*1* Bury No. 1, *2* Boree No. 1, *3* Bonnie No. 1, *4* Alva No. 6, *5* Stafford No. 1)

In the eastern Adavale Basin the Devonian deposits penetrated by boreholes can be divided, in ascending order, into the Gumbardo, Log Creek, Etonvale, and Buckabee Formations.

The Gumbardo Formation is composed of andesite, andesitic tuff, arcoses, and conglomerates; the amount of the latter increases in the east. The formation is 750–800 m thick. It rests unconformably on the crystalline basement made up of Ordovician or Silurian granite, basalt, and phyllite (Tanner 1967). The formation may be regarded either as Early Devonian (Tanner 1967), or lower Middle Devonian in age (Galloway 1970).

The overlying Log Creek Formation comprises both terrigenous and carbonate deposits, that can be divided into the lower sandstone member, and the upper Bury Limestone Member. The formation is 178–418 m thick. The lower sandstone member often pinches out, whereas the thickness of the Bury Member increases up to 270–290 m. It is dominated by gray limestone alternated with dolomite and shale. The amount of dolomite increases up the section.

The Etonvale Formation can be divided into three members: the lower, D_3 or Boree Salt Member; the middle, Member D_2; and the upper, Member D_1. The Boree Salt Member was penetrated for the entire thickness by five wells. The Boree No. 1 well penetrated the member over the depth interval 1919.6–2426.5 m. The apparent thickness is 506.9 m. The upper part of the member (35.0 m) is made up of white, finely crystalline gypsum with dolomitic siltstone and shale and traces of anhydrite. Below it is rock salt with thin beds of shale and siltstone, often gypsinate; their thickness is 471.9 m. In the Bonnie No. 1 well, the Boree Salt Member is 100.6 m thick (depth interval: 2225.6–2326.2 m). It is composed of rock salt with rare clayey shale laminae.

The thickest section of the Boree Member (584.1 m) was penetrated by the Bury No. 1 well over the depth interval 1774.5–2358.6 m. Its upper part contains 23.8 m of anhydrite. Below the anhydrite deposits, shale and limestone were found, their thickness being 6.7 m. The remaining lower part of the member contains rock salt with traces of anhydrite and rare laminae and interbeds of shale. The Alva No. 1 well encountered the Boree Member at depths of 3096.1–3297.3 m (thickness is 201.2 m). The upper part, composed of halite with few thin shale interbeds can be distinguished there. Halite is pale brown, massive, and very finely crystalline. The lower part contains 11.5 m of anhydrites, overlying dolomite of the Bury Member. The Stafford No. 1 well penetrated salt deposits over the depth interval 2673.1–2749.6 m (thickness is 76.5 m).

In all three wells, Bonnie No. 1, Bury No. 1, and Boree No. 1, potassium minerals, mainly sylvite, were encountered in the Boree Member. Pods of sylvite were found in rock salt interbeds up to 10–15 cm thick (Wells 1980).

In all the wells, salt beds are overlain by Member D_2, composed mainly of sandstone and mudstone; its thickness reaches 260 m. An overlying Member D_1 also consists of sandstone and dolomitic shale.

Most workers assigned the Log Creek and Etonvale Formations to the Middle Devonian. The Upper Devonian deposits of the Adavale Basin are represented by thick (up to 1000 m) red sandstone, conglomerates, and shale, placed in the Buckabee Formation.

In the west evaporites pinch out, replaced by salt-free equivalents, and, according to Tanner (1967), they are composed of dolomite, known in the center of the basin as the Cooladdi Dolomite Member, assigned to the lower Etonvale Formation (Galloway 1970). The western limit of the dolomite distribution is shown in Fig. 75.

The data available allow us to outline the approximate area occupied by rock salt in the Adavale Basin. Salt deposits wedge out in the central Blackwater Trough and in the northern Gooladdi Trough. Rock salt forms a narrow belt 35–40 m wide, extending from north to south for 180–200 km. The area may cover 6000–8000 km^2. In the marginal eastern and northern parts of the basin, the thickness of the Boree Salt probably greatly increases owing to the numerous salt domes. Thus, one can assume that the total thickness of salt will be well in excess of 500 m, whereas an average minimal thickness for the entire area can be taken as equal to 50 m. Based on these values an approximate estimate for rock salt volume in the Adavale Basin will be 300–400 km^3.

Bonaparte Gulf Basin

Devonian salt-bearing deposits in the Bonaparte Gulf Basin, occupying coastal and shelf areas off northern Australia, were penetrated by two deep boreholes, the one drilled on the Pelican Island (Pelican Island No. 1), and the other in the Bonaparte Gulf, some 100 km from its south-western coast (Fig. 76). The basin is briefly discussed below on the basis of the data published by Wells (1980).

The Bonaparte Gulf Basin is bounded to the south-west by the Precambrian Kimberley block, and to the north-east by the Darwin block. The sides of the basin are cut by a fault system, and complicated by terraces subsiding towards a central zone, occupying the inner parts of the gulf. The basin is filled in by Cenozoic, Cretaceous, Jurassic, Triassic, Permian, Carboniferous, and Devonian deposits. Subsided areas may also contain Ordovician and Cambrian sediments. The presence of thick salt-bearing strata in the basin has long been inferred from geophysical data, primarily gravity and seismic, implying an extensive occurrence of diapiric intrusives similar to salt diapirs. These intrusives follow major faults. The depth to the tops of domes exceeds 1000–1200 m.

The assumption that these diapiric intrusives are salt domes was confirmed in 1971, when the Sandpiper No. 1 well intersected a cap rock zone of disturbed deposits 809 m thick (944–1753 m) and reached total depth after penetrating 139 m of halite (1753–1892 m). The cap rock is considered to be Upper Devonian to Lower Carboniferous, and the halite Upper Devonian (Famennian).

The Pelican Island No. 1 well was drilled to a depth of 1981.2 m. Over the depth interval 1791.3–1981.2 m (thickness is 189.9 m) it has encountered massive rock salt. The overlying deposits are Lower Carboniferous; the salt-bearing strata are considered Devonian.

The data available (Crist and Hobday 1973, Edgerly and Crist 1974, Laws and Kraus 1974, Laws and Brown 1976, Wells 1980) allow the assumption that Devonian salt-bearing deposits fill in the entire central, most subsided zone, of the Bonaparte Gulf Basin. Their areal extent may exceed 50,000 km^2. It is most likely that rock salt will be well over 100 m in thickness. In this case, its volume should amount to 5000 km^3.

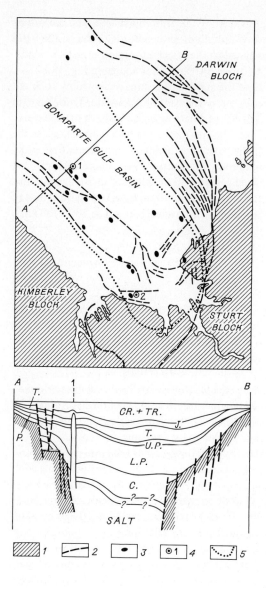

Fig. 76. Geological structure of south-eastern Bonaparte Gulf Basin. After Edgerley and Crist (1974). *1* land, *2* faults, *3* salt structures, *4* wells and their numbers (*1* Sandpiper No.1, *2* Pelican Island No. 1), *5* possible limit of salt

Devonian Evaporite Deposits in Other Basins

Devonian evaporites are at present reported not only from the above basins, but also from Minusinsk, Chulym-Yenisei, Kuznetsk, Teniz, Turgai, Ili, Moesian-Wallachian, Central Iowa, Illinois, Canadian Arctic Archipelago, Tindouf, Canning, Arckaringa, Carnarvon, Bancannia, and some other areas. Thus, within the Minusinsk Basin, sulfate rocks were found in the Middle Devonian deposits of the Abakan, Askyz, and Beysk Formations. In the former two formations gypsinate rocks with lenses and thin interbeds of gypsum and anhydrite occur among red sandstone and siltstone (the Abakan

Formation), or among pale yellow siltstone and marl (the Askyz Formation). The Beysk Formation contains rather thick beds of gypsum in limestone, marl, siltstone, and sandstone. The thickness of the Beysk gypsum-bearing formation varies between 50 and 110 m. There are some thick deposits of gypsum (Khamzas deposit) (see Fig. 37).

In the Chulym-Yenisei Basin, gypsum was found in the Beysk Formation. There, the thickness of gypsiferous strata reaches 55 m. Gypsum is mined in the Dankov deposit.

In the Kuznetsk Basin, gypsum bands were found along the eastern and southern margins among variegated and red, chiefly terrigenous deposits of the Podonin horizon of the Late Devonian. The Podonin sulfate is exposed in a number of deep boreholes, such as the Berdov, Ermakov, etc. Within the Teniz Basin, a sulfate-bearing sequence of Late Devonian age was penetrated by the Teniz test well down to 1430 m. Dolomite interbedded with anhydrite, up to 50 m thick, is exposed there (Ditmar 1966). In the Turgai Basin, sulfate rocks were found in the terrigenous-carbonate sequence of Famennian age (Litvin 1973). Gypsum bands were recorded in the central zone of the Borovsk Anticline. Evaporites in the Ili Depression were reported at Lake Sorbulak from red sandstone strata. There, gypsum forms lenses up to 10–12 m thick, extending for 3 km (Shcherbina 1954, Ivanov and Levitsky 1960). These strata are considered to be Late Devonian – Early Carboniferous.

A thick Middle Devonian sulfate-bearing sequence has been reported from the Moesian-Wallachian Basin (Patrulius et al. 1967, Polster et al. 1976). The sequence occupies vast areas in Bulgaria and Rumania, filling in the subsiding parts of the Varna and Wallachian Basins and the Tutrakan Trough. Its total thickness exceeds 1500 m. The thickest beds of anhydrite are associated with Givetian deposits where they are interbedded with clayey-carbonate and carbonate rocks, primarily with dolomite; sulfate rocks also occur in the Upper Eifelian deposits.

Halogenic deposits in the Central Iowa Basin, composed of anhydrite and gypsum, occur among Eifelian deposits (Middle Devonian). They build up the lower Wapsipinicon Formation of the Kenwood Sequence (Collinson et al. 1967). Anhydrite interbeds and units up to several meters thick in the Kenwood Sequence were observed over most of central and south-eastern Iowa. They alternate with clayey dolomite, limestone, and clay. Anhydrite and gypsum were also reported from Givetian deposits (Middle Devonian) of the Solon and Rapid Formations, the Cedar Valley sequence (Collinson et al. 1967). Thin anhydrite interbeds were also found in the northern Illinois Basin, in the Jeffersonville Formation (Collinson et al. 1967), assigned to the lower Eifelian.

In the Canadian Arctic Archipelago, gypsum makes up the Bird Fiord Formation on southern Ellesmere Island of Goose Fiord (McLaren 1963a, b, Kerr 1967b). This formation is composed of thin-bedded gray dolomite and sandy dolomitic marl, gypsum beds up to 2 m thick lying at the base of the section. In the middle part of the formation, gypsum accounts for about one third of the succession. In the upper part it occurs in the form of lenses. Gypsum-bearing deposits of the Bird Fiord Formation belong to the upper Middle Devonian. Gypsinate rocks are also known from the underlying Eids Formation (Kerr 1974).

In the Tindouf Basin, in north-western Africa, sulfate interbeds occur among the upper Middle Devonian and Famennian (Upper Devonian) (Hollard 1967).

Gypsum interbeds are developed among variegated sandy-clay deposits. The same deposits contain gypsum partings in some areas of the Dra'a Hammada and in the Anti-Atlas piedmont, including the Akka and Tata areas to the north.

In the Canning Basin, evaporites have been reported from the Tandalgoo red beds and the Mellinjerie Limestone (Wells 1980, Forman and Wells 1981). Most widespread are evaporites in the northern and central regions of the basin. In the Tandalgoo red beds and in the Mellinjerie Formation they have been penetrated, respectively, by the Sahara No. 1, Kidson No. 1, and Wilson Cliffs No. 1 wells.

Anhydrite among Devonian deposits of the Arckaringa Basin has been found in the black shale of the Cootanoorina well, and in Weedina No. 1 well, where it alternates with dolomite, dolomitic sandstone, and shale of probable Devonian age (Wells 1980). The Devonian salt-bearing deposits are probably developed in the Arckaringa Basin as well.

In the Carnarvon Basin, evaporites have been reported from the Gneudna Formation of the Yaringa No. 1 well (Wells 1980). Among Devonian deposits of the Bancannia Trough, western New Wales, evaporites occur among red bed sequences (Wells 1980).

Another region of Devonian salt occurrence should be mentioned. It embraces Central and Western Afghanistan, Pakistan, and South Iran. There, in the middle Harirud River near Rakha Village, Pakistan, as well as in Kerman area, Iran, Lower Devonian gypsum, anhydrite, and rock salt have been observed (Lapparent and Blaise 1966, Weippert and Wittekindt 1964, Durkoop et al. 1967, Morgunov and Rudakov 1972). Most workers determine the stratigraphic position of these evaporites to their occurrence below Eifelian deposits (Middle Devonian), but above Silurian sediments. This is also confirmed by the fact that pre-Middle Devonian conodonts were found in the salt-bearing sequence in Rakha area (Weippert and Wittekindt 1964). At the same time, this salt-bearing sequence is strongly deformed and may build up salt domes there (Durkoop et al. 1967). However, its age remains questionable. Rock salt in the Rakha area is Neogene rather than Lower Devonian (Jux and Schultz 1971). Salt-bearing deposits in the Kerman area, South Iran, regarded as Devonian, may also be much younger. The Lower Paleozoic, namely Cambrian age of these salts seems quite feasible.

Chapter VI
Carboniferous Salt Deposits

Carboniferous salt deposits have been recognized in the following basins: Chu-Sarysu, Mid-Tien Shan, Sverdrup, Williston, Paradox, Eagle, Maritime, and Saltville Basins. The present chapter also deals with the Amazon Basin, though recently some data have been published suggesting the Early Permian age of its salt deposits. Gypsum and/or anhydrite were reported from other basins of Europe, Asia, North and South America, Africa, and Australia (see Fig. 77).

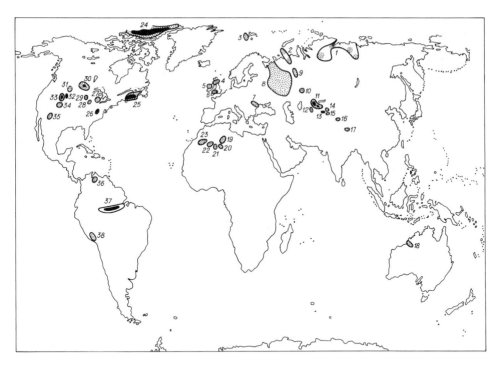

Fig. 77. Distribution of Carboniferous evaporites. Evaporite basins: *1* North Siberian, *2* Pechora-Novaya Zemlya, *3* Spitsbergen, *4* Northumberland, *5* North Ireland, *6* Central England, *7* Dobruja, *8* East European, *9* East Uralian, *10* Teniz, *11* Chu-Sarysu, *12* Chimkent, *13* Mid-Tien Shan, *14* Tyup, *15* Aksu, *16* Achikkul, *17* Lhasa, *18* Fitzroy, *19* Radames, *20* Illisie, *21* Ahnet, *22* Reggane, *23* Tindouf, *24* Sverdrup, *25* Maritime, *26* Saltville, *27* Michigan, *28* Illinois, *29* South Iowa, *30* Williston, *31* East Wyoming, *32* Eagle, *33* Paradox, *34* San Juan, *35* Orogrande, *36* Venezuela, *37* Amazon, *38* South Peruan. For legend see Fig. 3

Chu-Sarysu Basin

In Carboniferous time the Chu-Sarysu Basin repeatedly became a site of evaporite sedimentation. The salt deposits that developed there within the Dzhezkasgan, Tesbulak, and Kokpansor Depressions are Late Devonian or Early Carboniferous in age. They were discussed in Chapter V. A probable Early Carboniferous salt sequence known from the Tesbulak Basin, according to Sinitsyn and co-workers (1977), may be placed at the level of the Tournaisian.

The higher level of halogenic rocks in the Chu-Sarysu Basin is associated with Visean and Namurian deposits. Thus, the halogenic-carbonate sequence of Early-Middle Visean age, with bands of anhydrite and gypsum up to 1 m thick, is recognized in the Akkol Anticline, in the Talas Uplift, within the East Karakol Trough, in the Bestube Dome (Ditmar 1966, Aleksandrova 1971, Bakirov et al. 1971). Sulfate rocks in Upper Visean/Namurian deposits are present in the central and southern Chu Depression, in the Talas Uplift (Ditmar 1966, Bakirov et al. 1971). They appear to reach the north-western part of the Kirghiz Range where bands and rather thick gypsum and anhydrite beds are developed in Visean/Early Namurain sedimentary sequences of the Ulkunburul, Kishiburul, and Tekturmas Mountains (Shcherbina 1945, Ivanov and Levitsky 1960, Bakirov et al. 1971). The uppermost sulfate-bearing sequence is known as the Kyzylkanat sequence. It includes Middle/Upper Carboniferous terrigenous red beds and variegated deposits with lenses, inclusions, and veins of gypsum, and gypsinate rocks; in places thin interbeds of anhydrite and/or gypsum occur (Bakirov et al. 1971).

Mid-Tien Shan Basin

This basin is only tentatively regarded as a salt basin. It forms part of a single Carboniferous evaporite belt stretching sublatitudinally from the Chimkent Depression in the west to the Turuk Trough on the south slope of the Terskey-Alatau Range in the east (Fig. 78). Within this belt, halogenic rocks were found at different stratigraphic levels in Carboniferous deposits. In the Naryn Trough and adjacent mountain ranges evaporite deposits are generally Early/Middle Carboniferous; however, Late Carboniferous deposits occur as well.

The oldest sulfate rocks, primarily gypsum, occur as interbeds and lenses among shale and calcareous conglomerates and were found in Upper Tournaisian and Lower Visean deposits of the Dzhamandavan and Naryntau Ranges, i.e., in the south-western and eastern framing of the Naryn Depression. The thickness of the sulfate-bearing deposits ranges from 100–400 m in the Dzhamandavan Range and reaches 1000 m in the Naryntau Range (Poyarkov 1972). As early as 1958, Galitskaya described a 700–800 m thick Middle Visean sequence within the western framing of the Naryn Depression in the Kokiyrim Range; the deposits are gray marly limestone with bands and lenses of gypsum (Poyarkov 1972). North of the Naryn Depression, at different localities in the Moldotau Range, gypsum-bearing sediments are known among Visean

Mid-Tien Shan Basin

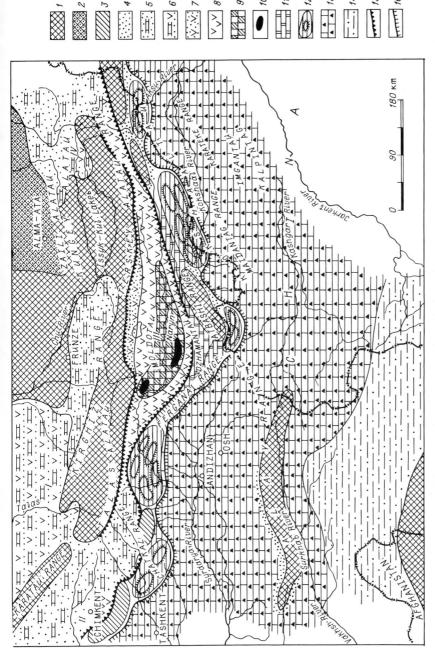

Fig. 78. Generalized lithologic map of Visean and Namurian (Lower Carboniferous) and Bashkirian (Middle Carboniferous) deposits of Tien Shan and adjacent areas. *1* land, *2* continental areas showing development of volcanics, *3* episodically existent islands. Areas of distribution of different sediments *(4–14)*: *4* terrigenous, *5* terrigenous and carbonate, *6* terrigenous, carbonate, and sulfate, *7* terrigenous, mostly red sulfate, *8* sulfate and carbonate, locally terrigenous, *9* possible areas of salt development, *10* inferred areas of salt distribution, *11* dolomite, *12* calcareous, probably reef, *13* siliceous-calcareous, *14* argillaceous and arenaceous, mostly gray, *15* zone of evaporite distribution in Tien Shan, *16* southern boundary of development of island areas and reef carbonate buildups. Evaporite basins: *I* Chu-Sarysu, *II* Chimkent, *III* Mid-Tien Shan, *IV* Tyup

and Namurian (Lower Carboniferous) deposits, as well as in the Bashkirian (Middle Carboniferous). Thus, in the Mount Dyudyunbel area, Lasovsky and Mozylev described a Namurian/Bashkirian section with a unit of alternating gray to yellow thin-bedded silicified limestone, black clay shale interbedded with gypsum, sandstone, and siltstone at the base. Some investigators reported the presence of bands and units of rock salt in addition to gypsum in the lower reaches of the Kokomeren River. Borisova and Egoshin have recorded rock salt lenses as large as 40 × 200 m among Carboniferous gypsum-bearing deposits on the left bank of the Kokomeren River at the mouth of its right tributary, Chon-Dobe, and a salt bed on the right bank of the Karachauli River, 8 km higher up the mouth. Luyk and Kovalev reported thin lenticules of halite in green to gray sandy-clay shales on the right bank of the Kokomeren River near its junction with the Kavyuk-Su River, 5 km north of the village of Kaa-Bulak.

Recently, Karas, Koroleva, Romanov, Churin, Zharikov, and Ermikova studied the Upper Paleozoic, Carboniferous deposits including those in the Naryn Depression and those of its framing. They reported the presence of sulfate and salt rocks in Bashkirian sedimentary sequences in the Bokaly and Karalarga Rivers Basins. The section along the Karalarga River, investigated by Zharikov, is of particular interest. It includes, in ascending order: (a) saline sandstone — 118 m; (b) sandstone and mudstone with seams and inclusions of rock salt in the upper and lower parts — 100 m; (c) mudstone and sandstone with inclusions of rock salt — 30 m; (d) saline sandstone with limestone interbeds at the top — 90 m. In the Bokaly River Basin, sulfate and possibly salt rocks have been assigned to the Chemanda Formation of the Bashkirian, though some investigators (Poyarkov 1972) assign it to the Visean and Namurian (Lower Carboniferous).

North of the Naryn Depression, gypsum is present in Visean to Namurian and Bashkirian deposits almost over the entire area embracing the Balykta and Sonkul Troughs. However, it occurs there mainly in terrigenous red beds. For example, in the Kavak Tau Mountains, on the north slope of the Moldotau Range, variegated gypsinate siltstone is present in the upper Karachauli Formation of Visean to Namurian age (Poyarkov 1972). Gypsiferous sediments also occur in the Aktaylyak Formation assigned to the Lower Bashkirian (Ektova and Belgovsky 1972). Additionally, along the south slope of the Dzhamandavan Range, gypsum interbeds were reported from the upper Kodzhagul Formation assigned to the Upper Carboniferous (Belgovsky 1972). Carboniferous sulfate rocks are also known east of the Naryn Depression in its continuation within the Turuk Trough (Poyarkov 1972).

Thus, Carboniferous evaporites are developed around the mountains framing the Naryn Depression. To the north-west, in the Kokomeren, Bokaly, and Karalarga Rivers Basins, rock salt is present. Recently the presence of Carboniferous rock salt was reported from the Naryn Depression as well. According to Karas and co-workers, the sequence is exposed in the Chellokkoin 6 borehole at depths of 2906–3150 m in Bashkirian deposits. The presence of rock salt was determined from geophysical data. New data and core samples described by G.A. Glazatova allow us to outline approximately six rock salt beds there. The four upper beds probably alternate with limestone and anhydrite, but the two lower and thickest beds contain, aside from anhydrite, numerous terrigenous bands, such as siltstone, sandstone, and gritstone. The upper part of the salt-bearing sequence is essentially sulfate-carbonate, and it is capped with limestone, overlain by effusive rocks at depths of 2308—2906 m. This sequence probably goes as far as the north-western framing of the Naryn Depression, where it crops out in the Kokomeren River Basin and in adjacent areas. At present, it is impossible to determine the extent of salt deposits in subsided zones of the Naryn Depression, but they may be developed in its interior. The age of the salt deposits is unclear. It is probably Visean, but some workers (Karas et al.) assign them to the Bashkirian (Middle Carboniferous).

The Mid-Tien Shan evaporite basin, presumed to be salt-producing, is rather extensive. It stretches from west to east for a distance of more than 400 km, the maximal width exceeding 100 km. Salt deposits have developed in the basin in an area of about 3000 km^2. If further investigations confirm the size of the basin, and if an average thickness of the salt strata is at least 200 m, and salt content amounts to 30%, then the volume of rock salt in the basin will account for about 150 km^3.

The zone of evaporite sedimentation including the Mid-Tien Shan evaporite basin occupied a certain paleogeographical position. On the south it was bounded by a zone of variable width consisting of an island chain; some islands existed throughout the Early to Middle Carboniferous, while others continually subsided. The Mid-Tien Shan evaporite basin proper was bounded on the south by an island chain situated within the Atbashi Range, and some other islands in the central Fergana Range. To the east some islands can be recognized within the Sarydzhaz Block, and to the west gently sloping islands can be outlined within the Chatkal Range. During Early to Middle Carboniferous time, lengthy reef and other carbonate bioherms formed between the islands. Thus, the zone including the islands and reef carbonate structures separated the belt of evaporite sedimentation from the open sea of normal salinity to the south. Calcosiliceous sediments accumulated chiefly in the open sea. It should be noted that a reef-insular zone was built up as a series of island arcs with the convex side facing south. One of these island arcs flanks the Mid-Tien Shan evaporite basin, and the other flanks the Chimkent Basin which also contained evaporites in Early Carboniferous time. The land was situated north of an evaporite sedimentation zone; terrigenous red beds or volcano-terrigenous deposits accumulated in intermontane basins developed on the mainland where the sea came from time to time. In some of these basins evaporite strata were accumulated.

Sverdrup Basin

This Carboniferous salt-bearing deposits have been recently recognized within the Canadian Arctic Archipelago Basin. Sulfate-bearing sequences, building up most diapir domes on Axel Heiberg, Amund Ringnes, Ellef Ringnes, Melville, and Ellesmere Islands, have been known there for many years (Blackadar 1963, Greiner 1963, McLaren 1963b, c, Norris 1963, Roots 1963a, b, c, Souther 1963, Thorsteinsson 1963, Tozer 1963, Christie 1964, Tozer and Thorsteinsson 1964). A wide distribution of sulfate-bearing sequences and their considerable thickness made many investigators believe that rock salt might also be present within the Canadian Arctic Archipelago in subsided parts of the Sverdrup Basin. This suggestion was confirmed in 1972 when the deep Hoodoo L-41 borehole was drilled in southern Ellef Ringnes

Island on the vault of the Hoodoo salt dome. Data obtained were first reported by Davies (1974a, b, 1975a, b). Evidence available suggests that evaporite deposits, including salt-bearing Carboniferous sediments, are present over much of the Sverdrup Basin. Thus, a new Paleozoic salt basin was recognized there and named the Sverdrup Basin. It will be briefly discussed below using data reported by Davies (1974a, b, 1975a, b), Thorsteinsson (1974), and Mayr (1975).

The Sverdrup Basin is situated within the Canadian Arctic Archipelago embracing the northern extremities of Prince Patrick and Melville Islands on the south-west, and stretching farther on the north-east up to north-western Ellesmere Island. Aside from these islands, the basin includes such large islands as Borden, Mackenzie King, Ellef Ringnes, Amund Ringnes, Cornwallis, Axel Heiberg, and many small islands (Fig. 79). The north-western boundary of the basin is defined only at two localities (in the extreme north-west of Ellesmere Island and on northern Axel Heiberg Island). In other areas the boundary disappears in the Arctic Ocean and is still unknown.

Carboniferous and Permian deposits are developed all over the Sverdrup Basin. Their thickness exceeds 2500 m. A great number of formations are recognized within

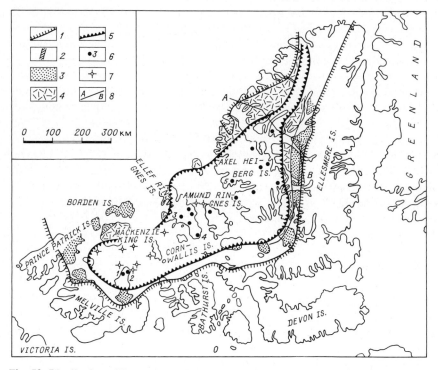

Fig. 79. Distribution of Permo-Carboniferous evaporites in Sverdrup Basin. After Thorsteinsson (1974) and Davies (1975a, b). *1* boundary of Sverdrup Basin, *2* near-shore terrigenous deposits, *3* near-shore terrigenous, sulfate-carbonate, and carbonate deposits, *4* predominantly sulfate and carbonate deposits, *5* distribution of salt deposits, *6* diapiric domes with outcrops of gypsum and anhydrite (*1* Barrow, *2* Cape Colqukoun, *3* Isachsen, *4* Hoodoo, *5* Southern Fiord), *7* inferred diapiric domes, *8* facies profile shown in Fig. 80

Sverdrup Basin

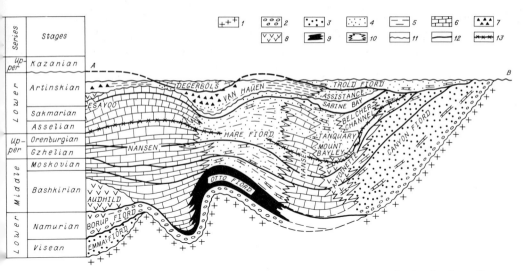

Fig. 80. Facies profile of Permo-Carboniferous deposits in Sverdrup Basin showing formations versus stratigraphic position. After Thorsteinsson (1974). *1* pre-Carboniferous deposits, *2* conglomerate, *3* sandstone, *4* siltstone, *5* mudstone and shale, *6* limestone and dolomite, *7* chert, *8* anhydrite and gypsum, *9* rock salt and anhydrite, *10* limits of formations, *11* unconformable occurrence of deposits, *12* stratigraphic boundaries, *13* Carboniferous/Permian boundary

the section (Fig. 80). The oldest sediments form the Emma Fiord Formation of the Visean (Lower Carboniferous) age. It consists mainly of sandstone and siltstone interbedded with gray and black calcareous mudstone. Conglomerate bands and coal lenses can be observed as well. The formation reaches 100–130 m in thickness. The above lying Borup Fiord Formation rests on the Emma Fiord Formation, and even Lower Ordovician and Cambrian rocks. The base of the formation is made up of red conglomerate; red and green-gray conglomerate and sandstone interbedded with siltstone, mudstone, and limestone that are predominantly gray, appear higher in the section. Some sections contain dolomite. The thickness varies from 120–350 m.

Mayr (1975) studied the Borup Fiord Formation in the Yelverton area on northern Ellesmere Island in detail and recognized three units: A, B, and C. Unit A consists of white sandstone and conglomerate with rare interbeds of red mustone. Unit B contains dark gray to black crystalline dolomite, black limestone, and calcareous mudstone. Gypsum and anhydrite interbedded with dark gray marl occur in the lower part of unit C in the Yelverton area. They are overlain by carbonate clay, massive dolomite, dark gray shale, and siltstone. Mayr suggested that unit C and probably unit B are to be compared with the Otto Fiord evaporite formation developed chiefly in the Sverdrup Basin interiors. The Audhild Formation is known only in the Kleybolt Peninsula of Ellesmere Island. It includes volcanogenic basic rocks between the Borup Fiord and Nansen Formation.

The Otto Fiord Formation contains most of the Carboniferous evaporite deposits in the Sverdrup Basin. The formation consists chiefly of medium- and thick-platy to massive anhydrite turning into gypsum at the surface. Anhydrite may amount to 80%

of the total thickness of the formation, and dark gray limestone together with minor mudstone account for the remaining 20%. Anhydrite of the Otto Fiord Formation builds up many salt domes. The thickness of the formation were estimated to be about 300–330 m.

After the drilling of the first Hoodoo L-41 deep borehole in the Hoodoo diapir on the south-eastern Ellef Ringness Island thick salt beds were recognized in the deep parts of the Otto Fiord Formation. After penetrating the top of gypsum and anhydrite at a depth of 331.5 m, the borehole penetrated the following rocks in descending order:

	Thickness (m)
1. Sulfate cap rock	12.0
2. Gypsum and anhydrite	23.0
3. Recrystallized limestone with stylolites	44.4
4. Anhydrite interbedded with limestone, dolomite, sandstone, and siltstone	201.0
5. Pure, essentially recrystallized rock salt. Inclusions of halopelites and seams of anhydrite were encountered in the rock salt core at a depth of 1200 m. Two beds of intrusive rocks of gabbroid composition were penetrated at a depth of 1890 m and 3210 m. A core of deformed anhydrite and carbonate rocks with abundant inclusions of rock salt was recorded from a depth of 3780 m. The anhydrite is thin-bedded. Such fossils as gastropod, mollusk, brachiopod, and ostracode were found in carbonate interbeds	3882.0

The well was abandoned at a depth of 4213.5 m in rock salt.

On the basis of these data and fossil occurrences, Davies (1975a, b) determined the age of the Otto Fiord Formation as Late Namurian/Early Moscovian. Rock salt is believed to be present at depth in other diapir domes of the Sverdrup Basin. The present depth of the salt base in the most downwarped parts of the basin reaches 7500 m. Salt deposits of the Otto Fiord Formation may occupy the entire interior zone of the basin (see Fig. 79).

Age equivalents of the Otto Fiord Formation in the marginal zones of the Sverdrup Basin are the upper Borup Fiord Formation and the lower Nansen Formation. The Otto Fiord Formation is overlain by the Hare Fiord Formation consisting chiefly of quartz siltstone, mudstone, limestone with various amounts of chert and quartz sandstone. The formation shows abrupt changes in composition over the entire area. The thickness varies widely from 300 to 1230 m. The age of the Hare Fiord Formation is Carboniferous to Permian ranging from the end of the Bashkirian (Middle Carboniferous) to the beginning of the Artinskian (Lower Permian).

The Nansen Formation contains limestone, in places with minor quartz. Limestone yields abundant fossils. Rocks are light to yellow-gray, in places green-gray in color. Chert lenses are encountered. The deposits are comparable with the Otto Fiord and Hare Fiord Formations. The age of the Nansen Formation varies from the Upper Namurian (Lower Carboniferous) to the Lower Artinskian (Lower Permian). The thickness ranges from 1200–2340 m.

The Canon Fiord, Antoinette, Mount Bayley, Tanquary, and Belcher Channel Formations are age equivalents to the Nansen Formation. The Canon Fiord Formation consists of variegated (gray, yellow, green, brown, and red) sandstone and siltstone; the amount of limestone, marl, and mudstone varies in different sections. Conglomerate occurs at the base of the formation. The formation reaches 1650 m in thickness. It ranges in age from the Bashkirian (Middle Carboniferous) to the Sakmarian (Lower Permian). The Antoinette Formation consists mainly of dark gray limestone with interbeds of siltstone, sandstone, and mudstone. Anhydrite and gypsum, as well as gypsinate limestone occur throughout the section. The thickness varies from 465–800 m. The age of the formation varies from the upper Middle Carboniferous (the end of the Moscovian) to the Late Carboniferous (Orenburgian). The Mount Bayley Formation is dominated by anhydrite and, in this respect it is similar to the Otto Fiord Formation. But in the Mount Bayley Formation, unlike the Otto Fiord Formation, anhydrite is interbedded with quartz siltstone and sandstone. The formation reaches 200–250 m in thickness. It is assigned to the Asselian (Lower Permian). The Tanquary Formation, up to 650 m thick, is represented by sandstone and limestone with interlayers of siltstone. The deposits are assigned to the Asselian, Sakmarian, and Lower Artinskian (Lower Permian). The Belcher Channel Formation consists of limestone with bands and lenses of sandstone and siltstone. Its thickness reaches 120 m. The age of the formation is Late Carboniferous/Early Permian (Gzhelian of the Upper Carboniferous to the Lower Artinskian of the Lower Permian).

Deposits of the Sabine Bay, Assistance, and Van Hauen Formations assigned to the Upper Artinskian are of continental origin. The Sabine Bay Formation, up to 200 m thick, is composed of light gray, yellow to orange, and red to brown quartz sandstone and conglomerate. The Assistance Formation consists of gray and yellow to brown sandstone and siltstone, up to 400 m thick. The Van Hauen Formation includes dark gray to black mudstone, siltstone, gray quartz sandstone, as well as dark gray to black cherts interbedded with silty schists. The thickness varies from 400–670 m.

The Lower Permian deposits are unconformably overlain by the Upper Permian sediments assigned to the Kazanian stage. They embrace the Trold Fiord and Degerböls Formations. The Trold Fiord Formation consists of gray, green, and brown sandstone interbedded with limestone. Its thickness varies from 30–300 m. The Degerböls Formation, up to 380 m thick, consists mainly of gray and yellow limestone.

Thus, in the Sverdrup Basin the evaporite deposits occur throughout the Carboniferous/Lower Permian section (from the Namurian, Lower Carboniferous, to Asselian, Lower Permian). Salt deposits range from the Late Namurian to the Moscovian, i.e., completely within the Early/Middle Carboniferous. Salt deposits form the Otto Fiord Formation. This rock salt unit appears to occupy 200,000 km^2. Its exposed thickness is 3882 m, and the total thickness exceeds 5000 m. However, great thicknesses are probably confined to salt domes, and normal thicknesses do not exceed 1500–2000 m. If an average thickness of rock salt over the area is 500 m, then rock salt in the Sverdrup Basin will amount to about 1×10^5 km^3. But in fact, the volume of rock salt should exceed this approximate estimate.

Williston Basin

Carboniferous salt deposits of the Williston Basin were discussed in great detail in many publications (Anderson and Hansen 1957, Edie 1958, Illing 1959, Pierce and Rich 1962, Macauley et al. 1964, Carlson and Anderson 1965, Haun and Kent 1965, Willis 1967, Proctor and Macauley 1968, Lefond 1969, Craig and Varnes 1979, Paleotectonic ... 1979, Gerhard et al. 1982). They occur among the Mississippian deposits whose thickness reaches 900 m in the central parts of the basin. The subdivision of the Mississippian deposits is shown in Fig. 81.

The Bakken Formation was distinguished at the base of the section in North Dakota and Montana, in the United States, and in southern Saskatchewan and Alberta, Canada. In North Dakota it consists of three members, viz., the lower, composed of black shale, mudstone; the middle, silty mudstone, and the upper, represented by black finely laminated mudstone. The thickness does not exceed 25 m.

In Manitoba, Saskatchewan, and Alberta the Bakken Formation is 30–35 m thick. At its base lies a shale unit that can be divided into three parts, the lower and the upper being composed of black shale, and the middle being composed of sandstone. Above lies a 10–12 m thick sandstone bed. The upper Bakken Formation also consists of black shale.

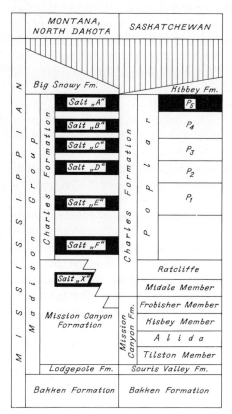

Fig. 81. Correlation chart of Mississippian deposits in Williston Basin showing stratigraphic position of salt horizons. Compiled from the data of Anderson and Hansen (1957), Macauley et al. (1964), Carlson and Anderson (1965), and Proctor and Macauley (1968)

In the central Williston Basin, North Dakota, the overlying mainly carbonate and evaporite section is known as the Madison Group. In ascending order three formations, Lodgepole, Mission Canyon, and Charles, can be distinguished within the group. The Lodgepole Formation is dominated by sliced argillaceous or siliceous light and dark gray limestone. Farther south its equivalents are known under the name of the Souris Valley Formation. The latter can be divided into three members. The basal member consists of shale and biogenic limestone about 6 m thick. The middle 100–200 m thick member is composed of limestone. Shale dominates the upper 30 m thick member.

As to the Mission Canyon Formation, in the central Williston Basin it consists of massive and thick platy brownish and light yellowish-gray, in places clayey or silicified limestone, locally oolitic and partly dolomitized. To the north and north-east, in southern Canada, anhydrite appears in the section. There, it is subdivided into four members: Tilston, Alida, Kisbey, and Frobisher. The base of the Tilston Member is made up of oolitic and algal limestone. Above there are lenticular anhydrite and clayey siltstone interbeds varying in thickness from 3–12 m. The Tilston Member is 30–60 m thick. The Alida Member is dominated by oolitic limestone intercalated with clayey and silty limestone; its thickness is 5 m. The Kisbey Member consists of siltstone and clayey limestone about 10 m thick. In southern Saskatchewan, the Frobisher Member is composed of oolitic and biogenic limestone, clayey and sandy dolomitic limestone, dolomite, and anhydrite. The thickness is 50–55 m. In North Dakota, in central areas of the basin, rock salt interbeds indexed horizon X were distinguished at the level of the Frobisher Member but within the upper Mission Canyon Formation.

The Charles Formation encompasses, in fact, all the salt deposits. In central areas, it is 200–210 m thick. It is composed of rock salt, anhydrite, dolomite, and limestone. In North Dakota, six salt horizons were distinguished within the formation; in ascending order they are indexed by letters A, B. C, D, E, and F. In southern Saskatchewan and in Manitoba, the Charles Formation can be divided into three members, viz., Midale, Ratcliffe, and Poplar. Anhydrite is present in all of them. So, the Frobisher Anhydrite (overlying the member of the same name of the Mission Canyon Formation) is easily discernible in the Midale Member. The upper Midale Member is built up of limestone. It is 26–27 m thick. The Ratcliffe Member is an alternation of anhydrite, dolomite, and limestone interbeds. Anhydrite wedges out south-eastward and carbonate dominates the entire section. The thickness of the Ratcliffe Member is 25–35 m. The greatest thickness (120 m) was recorded for the Poplar Member. It consists of anhydrite, dolomite, clayey limestone, and mudstone. Five evaporite beds indexed in ascending order by the letters P_1, P_2, P_3, P_4, and P_5 were recognized in the upper Poplar Member, each bed containing anhydrite not more than several meters thick. Bed P_5 has the thickest anhydrite. In some areas of Saskatchewan, rock salt was reported from bed P_5. The correlation between evaporite beds of the Poplar Member and salt horizons of the Charles Formation in North Dakota remains uncertain. Bed P_5 is assumed to occupy the same level as that of salt horizon A in the central Williston Basin. In North Dakota, lower salt horizons pinch out north- and north-eastward.

East of Williston Basin, in central Montana, the Madison Group is built up of carbonate rocks. In southern Montana and adjacent regions of Wyoming, anhydrite

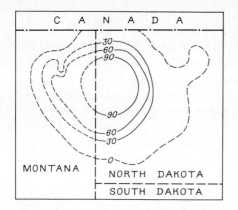

Fig. 82. Aggregate thickness of Mississippian salt beds in Williston Basin. After Anderson et al. (Pierce and Rich 1962). Isolines of thicknesses (known and inferred), in meters

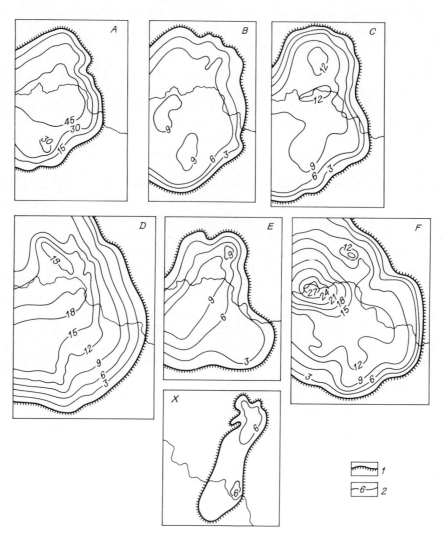

Fig. 83. Map of thicknesses of Mississippian salt horizons *A, B, C, D, E, F,* and *X* in North Dakota, U.S.A. After Anderson and Hansen (1957). *1* limit of salt, *2* isolines of thicknesses, in meters

appears among deposits of the Madison Group. Carbonate breccia, limestone, and dolomite are common there. Anhydrite interbeds and inclusions occur throughout the entire section.

In the Williston Basin, the upper Mississipian deposits are placed into the Big Snowy Group. It incorporates three formations, viz. Kibbey, Otter, and Heath. The Kibbey Formation in turn can be divided into three members, namely, the lower, gray mudstone-shale, the middle, limestone, and the upper, sandstone. The Otter Formation is dominated by green-gray and gray shale with light gray limestone interbeds. The Heath Formation is composed of gray and black shale. The Big Snowy Group totals 120 m and above.

Rock salt is known from the central Williston Basin, namely, North Dakota, Montana, and, partly, Saskatchewan (Fig. 82). The total thickness of the rock salt does not generally exceed 100 m. Upper salt horizon A has the greatest thickness (up to 45 m) (see Fig. 83).

The thickness of horizon F and horizon D is not more than 30 m and 20 m, respectively. The thickness of all the remaining salt horizons is 10–12 m. In the Williston Basin a depth above sea level to the top of the upper salt beds ranges from 450–2100 m (Fig. 84). The volume of rock salt, if its areal extent is taken to be 80,000 km^2, with an average thickness of 30 m, might exceed 2.4×10^3 km^3.

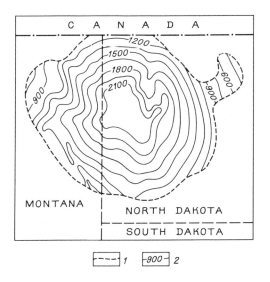

Fig. 84. Map showing depth below sea level to top of highest salt of Mississippian age in Williston Basin. After Kunkel (Pierce and Rich 1962). *1* limit of salt, *2* isolines of depths to top of highest salt, in meters

Maritime Basin

Carboniferous evaporites have long been known in some areas of New Brunswick and Nova Scotia, Canada. There are large gypsum deposits; they have been mined since the 1830's. As early as 1917, rock salt was found there in the Mississippian sedimentary strata.

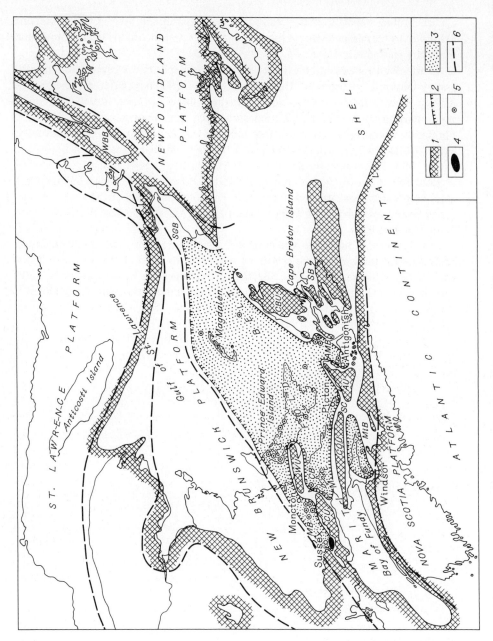

Fig. 85. Distribution of Carboniferous deposits in Maritime Provinces. Compiled from the data of Gussow (1953), Hamilton (1961), Belt (1965), Kelley (1970), Poole et al. (1970), Howie and Barss (1975), and Anderle et al. (1979). *1* Carboniferous deposits absent, *2* limit of salt deposits, *3* known and possible extent of salt deposition, *4* areal distribution of potash salts, *5* deep boreholes drilled in evaporites, *6* boundaries of major tectonic elements, *MB* Moncton sub-basin. *CB* Cumberland sub-basin, *MiB* Minas sub-basin, *SC* Stellarton Cap, *AMB* Antigonish-Mabou sub-basin. *SB* Sydney sub-basin, *SGB* St. George's sub-basin, *WBB* White Bay sub-basin, *CU* Caledonian Upland, *WU* Westmoland Upland, *COU* Cobequid Upland, *AU* Antigonish Upland, *CBU* Cape Breton Upland

Carboniferous sedimentary basins had encompassed almost the entire Maritime Fold Belt, adjacent regions of the New Brunswick Platform, and Atlantic continental shelf (Fig. 85). Sedimentary strata were deposited in a series of connected troughs or intermontane basins; in some of them evaporites accumulated as well. They were repeatedly discussed in literature (Gussow 1953, Bancroft 1957, Bell 1960, Hamilton 1961, Crosby 1962, Greiner 1962, Boyle 1963, Howie and Cumming 1963, Collins 1964, Belt 1965, Evans 1965, 1967, Withington 1965, Cooper 1966, Benson 1967, 1970a, b, Kelley 1967, 1970, Shaw and Blenchard 1968, Lefond 1969, Taylor 1969, Howie and Barss 1975, Anderle et al. 1979).

The best known are deposits from the Moncton, Cumberland, Minas, Antigonish-Mabou sub-basins. Figure 86 shows different proposed subdivisions and correlations for Carboniferous sedimentary strata of the above sub-basins.

The Horton Group was distinguished at the base of the section. In the Moncton sub-basin it can be divided into three formations, namely, Mamramcook, Albert, and Moncton. The Mamramcook Formation incorporates terrigenous red rocks known also as the Lower Red Beds. Red conglomerates with sandstone, siltstone, and shale interbeds were observed in the lower part of the formation. Siltstone and purple-red, brick-red, and brownish-red shale with green and greenish-gray interbeds dominate the upper part. In places siltstone and shale are intercalated with gray and reddish-gray limestone. The thickness varies from 120–2100 m.

The Albert Formation encompasses gray deposits lying between the Lower Red Beds (Mamramcook Formation) and the Upper Red Beds (Moncton Formation). It is composed of dark gray and gray shale, locally calcareous, alternated with gray and brown-gray clayey and limy sandstone and siltstone as well as gray and light gray limestone. The composition of the formation is nonpersistent across the area. In some parts of the Moncton sub-basin it is built up mainly of light and dark gray quartz sandstone, often bituminous, intercalated with bituminous sandstone and calcareous shale rare with dark gray and buff limestone. The other areas are dominated by dark gray clayey-limy biogenic shale and yellowish limestone with interbeds of finely laminated gray siltstone. The proportion of dark gray, finely laminated bituminous shale increases in the west. In the Pollet River and Sussex areas, the Albert Formation contains shale, siltstone, sandstone, and conglomerates. Farther north, shale, siltstone, and some gray limestone dominate. To the north-west, buff-gray dolomite builds up almost the entire section. On the Stony Creek oil and gas field, and in the lower Petitcodiac River, interbeds of halogenic rocks, including rock salt, were found in the upper Albert Formation. The thickness of the entire formation ranges from 150 m to 1500–1600 m.

On the Stony Creek field, the Albert Formation is subdivided into four members. In ascending order they are: Dawson Settlement, Frederick Brook, Hiram Brook, and Gautreau (Greiner 1962).

The Dawson Settlement Member is made up of light gray limy, quartz, and feldspar-quartz sandstone intercalated with dark gray limy, sandy shale, totalling 400 m. The Frederick Brook Member is dominated by gray and dark gray bituminous and limy shale with interbeds of clayey limestone. To the west, sandstone and siltstone interbeds appear whereas clayey limestone and dolomite were observed to the north and north-east; they are 100–300 m thick. The Hiram Brook Member consists of

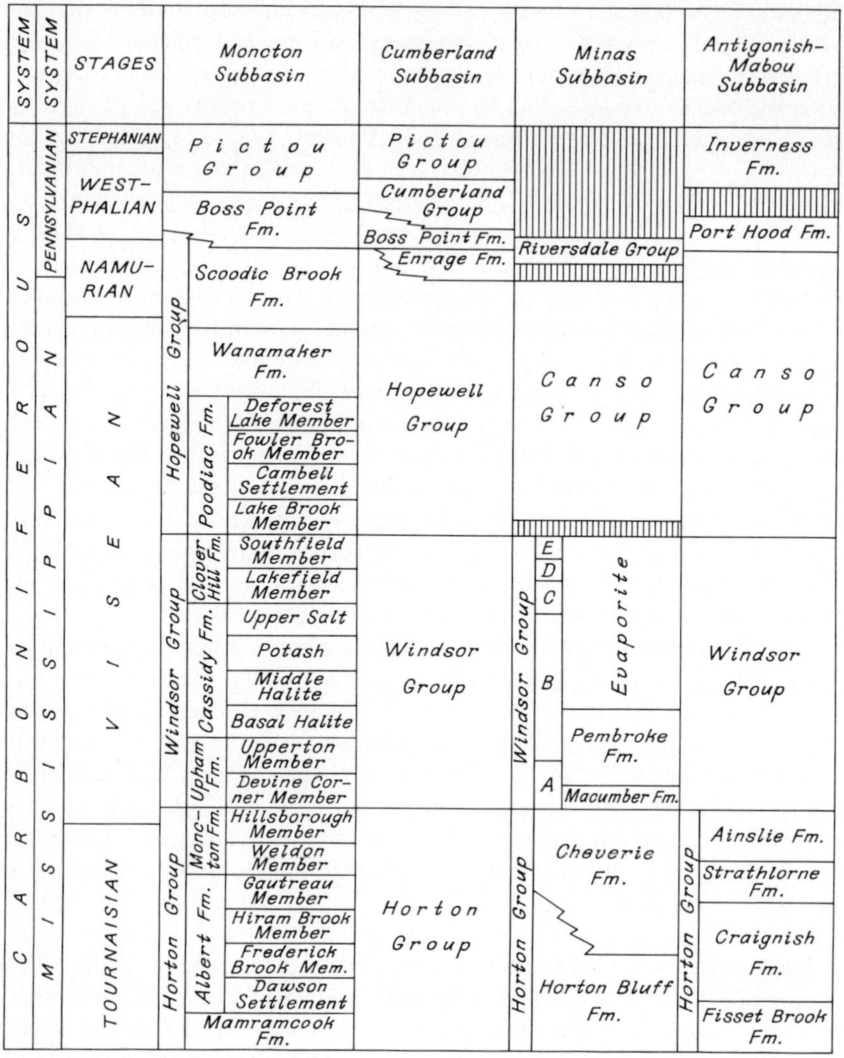

SYSTEM	SYSTEM	STAGES	Moncton Subbasin	Cumberland Subbasin	Minas Subbasin	Antigonish-Mabou Subbasin
CARBONIFEROUS	PENNSYLVANIAN	STEPHANIAN	Pictou Group	Pictou Group		Inverness Fm.
		WEST-PHALIAN	Boss Point Fm.	Cumberland Group		
				Boss Point Fm.	Riversdale Group	Port Hood Fm.
		NAMURIAN	Scoodic Brook Fm.	Enrage Fm.		
	MISSISSIPPIAN	VISEAN	Hopewell Group / Poodiac Fm. — Wanamaker Fm. / Deforest Lake Member / Fowler Brook Member / Cambell Settlement / Lake Brook Member	Hopewell Group	Canso Group	Canso Group
			Clover Hill Fm. — Southfield Member / Lakefield Member		Windsor Group — E / D / C / B / A — Evaporite / Pembroke Fm. / Macumber Fm.	Windsor Group
			Cassidy Fm. — Upper Salt / Potash / Middle Halite / Basal Halite	Windsor Group		
			Upham Fm. — Upperton Member / Devine Corner Member			
		TOURNAISIAN	Moncton Fm. — Hillsborough Member / Weldon Member / Gautreau Member	Horton Group	Horton Group — Cheverie Fm. / Horton Bluff Fm.	Horton Group — Ainslie Fm. / Strathlorne Fm. / Craignish Fm. / Fisset Brook Fm.
			Albert Fm. — Hiram Brook Member / Frederick Brook Mem. / Dawson Settlement			
			Mamramcook Fm.			

Fig. 86. Stratigraphic subdivisions and correlatives of Carboniferous sequence in Maritime Basin. Compiled from the data of Gussow (1953), Bell (1960), Hamilton (1961), Crosby (1962), Greiner (1962), Boyle (1963), Belt (1965), Kelley (1970), Howie and Barss (1975), and Anderle et al. (1979)

siltstone, sandstone, and shale, often limy and bituminous, and about 600 m thick. Gray and dark gray laminated clayey and sandy dolomite are known to the northeast. To the west, sandstone dominates and conglomerates appear in the section. Two beds of white gypsinate limestone were reported from the upper part of the member in some western regions.

The Gautreau Member is marked by halogenic deposits. On the Stony Creek field, they were penetrated by deep boreholes on the Hillsborough Syncline, on either side of the Petitcodiac River (Gussow 1953). The member consists of rock salt intercalated with anhydrite and shale. Its peculiar feature is the presence of glauberite which occurs as inclusions in shale, rock salt, and anhydrite, or makes up separate interbeds (Hamilton 1961, Evans 1970). A variable thickness of rock salt (70–488 m) (Gossow 1953, Hamilton 1961) is due to salt tectonics. Depth of occurrence varies from 210–530 m.

The Gautreau Member salt deposits occur in the Moncton sub-basin and are confined to the Albert Basin (Gussow 1953, Hamilton 1961); its northern and southern limits lie respectively 2.5 km north of Hillsborough and 20 km south of Moncton. Rock salt has not been observed everywhere and occurs in the form of large lenses or diapirs outlined mainly from gravity data. The probable areal extent of the Albert rock salt does not exceed 100 km^2 and its volume, at a mean thickness of 100 m for the area, accounts for 10 km^3.

The lower Moncton Formation consists of red siltstone, shale, sandstone, and conglomerate, 1500–1600 m thick (Weldon Member) and its upper part is made up of red sandstone and conglomerate (Hillsborough Member) up to 1000 m thick.

In other areas of the Maritime Basin, no evaporites have been found in the Horton Group as yet. It is represented by red, gray, and brownish arkosic and feldspathic conglomerate, red and gray sandstone, siltstone, and shale with ferrugeinous limestone bands. The thickness varies over a wide range of 500 to 3500–4000 m.

In the Maritime Basin, evaporites are predominantly associated with the Windsor Group. In the type section of the Windsor area, Nova Scotia, the group can be divided into five subzones marked A, B, C, D, and E. The first two subzones are confined to the lower Windsor, and the other three to the upper Windsor.

In the northern and central Moncton sub-basin the Windsor Group corresponds to subzones A, B, and C of the type section (Gussow 1953). Subzone A starts with laminated black bituminous dolomite and limestone overlain by sandy, dolomitic anhydrite which is also laminated, light in color, with rare red shale interbeds. The thickness of subzone A is 10–15 m, locally 20–30 m. Subzone B is subdivided into four members, namely: (1) limestone; (2) red shale with gypsum bands; (3) massive gypsum; and (4) red shale, sandstone, and conglomerate. The limestone member, 6–7 m thick, includes dark gray limestone with limy grayish-green and buff sandstone intercalations. The second 95 m thick member is composed of white and pink gypsum with red shale bands. The third member consists of massive white and gray gypsum, locally with red shale; its thickness exceeds 30 m. The fourth member is built up of red shale up to 55–60 m thick. The thickness of subzone B totals 180–192 m.

At Hillsborough, three members are discernible in subzone B. The lower member contains laminated clayey, dark to light gray, green and pink limestone, 45 m thick. The second member, 95 m thick, consists of red shale with limestone lenses. The third member is represented by massive light gray to white gypsum, 36 m thick. In the Hillsborough environs, the thickness of subzone B totals 180–200 m.

In the north-western Moncton sub-basin, subzone B section contains only two members. The lower part consists of brownish-gray and grayish-pink, locally grayish-white limestone; the upper part is made up of gray and white massive gypsum and anhydrite. As a rule, the thickness of limestone does not exceed 12–20 m, while that of gypsum reaches 300 m.

In the south-westernmost Moncton sub-basin, potash salt is known from the Windsor Group (Anderle et al. 1979). Salt has been found 25 km south-west of Sussex over an area of 200 km^2 within the Marchbank Syncline. There, the Windsor Group comprises the Upham, Cassidy Lake, and Clover Hill Formations.

The Upham Formation is made up of light-colored laminated to massive limestone which forms the Devine Corner Member, and overlying massive to laminated anhydrite known as the Upperton Member.

The Cassidy Lake Formation embraces all the salt-bearing deposits. It is subdivided, in ascending order, into the following members: Basal Halite, Middle Halite, Potash, and Upper Salt.

The Basal Halite Member consists of very light, gray to clear, medium to coarsely crystalline halite that characteristically contains dark argillaceous halite bands of interstitial gray-green anhydrite claystone. The argillaceous bands become more numerous in the upper segment of the member. White nodules containing 0.5 mm of danburite occur at the lower contact. The thickness varies from 1.5 m in the southern marginal parts to 150 m in the center.

The Middle Halite Member contains light brown, medium-grained crystalline halite, which is characterized by dark gray to dark brown argillaceous halite bands consisting of interstitial gray-green anhydrite claystone and red-brown claystone. The thickness varies from 25–80 m.

The Potash Member can be divided into two submembers: the sylvinite bed above and the lower gradational bed below. The latter contains orange or light brown halite with red-rimmed crystals of sylvine; its thickness ranges from 3–18 m. The sylvinite bed is characterized by sylvine mineralization greater than 15% KCl, consisting entirely of sylvinite and halite with no potassium-magnesium salts. This bed is mainly composed of brick-red to blood-red sylvine crystals, 2–5 mm in diameter, in a matrix of dark brown to dark orange, finely crystalline halite. The sylvinite bed is not homogeneous but is interrupted by barren halite and low-grade sylvinite beds, 2.5– 30 cm thick. The bed reaches 30 m in thickness. The thickness of the Potash Member varies from 18–45 m.

The Upper Salt Member is very distinctive. It is characterized by orange, brown, colorless, and argillaceous halite beds; red sylvinite beds and laminae; claystone and gray anhydrite laminae; and by the presence of certain borate minerals. The member is subdivided very generally into three parts. The upper part consists of light brown to light orange, finely crystalline halite with disseminated gray-green claystone, red sylvinite beds and laminae, and gray-green claystone laminae with numerous borate clusters and laminae. The middle part of the member is characterized by light brown to light orange and colorless, finely crystalline halite with clayey halite beds, red sylvine beds, and laminae with a maximum of 24 anhydrite laminae (1 mm to 1 cm), and by minor claystone laminae. Borate crystals occur as described in the interval above. The lower part consists of orange-colored finely crystalline halite with gray anhydrite laminae. The thickness of the Upper Salt Member reaches 60 m.

The Clover Hill Formation consists of the Lakefield and Southfield Members. The Lakefield Member contains gray massive to bedded anhydrite about 8 m thick. The Southfield Member consists of gray-green claystone 3–15 m thick.

In the Cumberland sub-basin, the Windsor Group also contains salt deposits associated with subzone B. They are distributed in the extreme south of New Brunswick, embracing the coastal regions of Cumberland, Meringoin Peninsula, Shepody Bay, and its western coast. Salt lenses are thought to occur there within the Meringoin Anticline, and Locher Lake and Tantreta Faults. An areal dimension of rock salt was reliably established from gravity lows. Rock salt occurs at considerable depths, viz., more than 900–2300 m. Only in the apical parts of anticlines does rock salt lie at smaller depths; however, a salt unit is disturbed there by salt tectonics.

In the Cumberland area (northern Nova Scotia), halogenic deposits are known from the upper Windsor Group. According to Bell's data (Evans 1970) the following section lies above limestone of zone A:

Lower zone B

1. Gypsum, anhydrite, and rock salt with red sandstone, siltstone, marl, limestone, and limy shale bands . . . . over 150 m
2. Red shale, marl, siltstone, and sandstone . . . . 240 m

Upper zone B and zone C

3. Gray limestone, gray limy shale, red shale and marl . . . . 38 m

Zone D

4. Red sandstone, siltstone, marl, shale with limestone bands and lenses . . . 120 m
5. Gypsum. . . . . about 7.5 m

In the more easterly regions of Nova Scotia, the Windsor Group is usually not divided. However, rock salt units and beds occur there in some areas. For example, at Malagash, rock salt was drilled as early as 1917 at a depth of 26 m. At Pugwash, a rock salt bed was found at a depth of 189 m. At Nappan, rock salt strata were penetrated by deep holes over a depth interval of 277–500 m, and one of the holes was drilled to a depth of 1466.1 m but did not penetrate the entire sequence. In all the above localities, the Windsor Group contains, along with rock salt, also gypsum, anhydrite, dolomite, limestone, shale, and siltstone interbeds.

In the Malagash, Pugwash, and Cumberland areas, carnallite and sylvinite were found in rock salt of the Windsor Group (Evans 1970). At Malagash, pink and yellowish-green sylvine crystals occur in lenses as thick as 1–2 m. Red carnallite makes up separate zones.

At Pugwash, polyhalite and rinneite bands and inclusions were observed along with sylvine and carnallite; however, the latter two remain dominant potash minerals (Evans 1967). They are either scattered among terrigenous rocks or form sylvine-carnallite-halite lenses. There are also segments built up of carnallite breccia where carnallite occurs in a mixture with clay and marl; the thickness of carnallite breccia ranges from a few centimeters to 3 m.

In the Minas sub-basin, in the Horton-Windsor area, the Windsor Group is divided into three parts. The Macumber Formation, the Pembroke Formation, and evaporite strata are recognized in the lower, middle, and upper parts, respectively (Crosby 1962). Some colleagues propose subdividing the upper evaporite into three members locally (Boyle 1963). The Macumber Formation is made up of buff and light gray, sandy, well laminated limestone about 10 m thick. The Pembroke Formation contains carbonate breccia locally interbedded with red limy sandstone and shale. The thickness of the formation does not usually exceed 30 m.

Evaporites of the Windsor Group are widespread in the Horton-Windsor area. At many localities, gypsum and anhydrite crop out. The halogenic deposits are usually built up of limestone, shale, gypsum, anhydrite, and rock salt interbeds. Shale, occurring mainly in the upper part of the sequence, is brick-red and greenish-gray in color. Limestone is often biogenic, locally oolitic. Gypsum and anhydrite are most common among halogenic rocks. They are mined in gypsum pits situated in the Windsor, Cheverie, and Walton areas. Gypsum is locally 300 m thick. In the Horton-Windsor area, rock salt beds were penetrated by a number of boreholes over a depth interval of 100–150 m. Boyle (1963) believes that in the Windsor-Cheverie-Walton area evaporites occur in the middle Windsor Group, and that the upper part of the group consists of two sequences. The lower evaporite overlying sequence is known as the Tennycape Formation. It consists of red and green shale interbedded with siltstone, gypsum, and anhydrite; its thickness totals 180–185 m. The upper sequence contains limestone, dolomite, and sandstone, up to 450 m thick.

In the Antigonish-Mabou sub-basin, the Windsor Group deposits are very common. In the Antigonish area, they consist of grayish-red and greenish-gray siltstone and shale interbedded with greenish-gray to dark gray crystalline, argillaceous, and oolitic limestone, gypsum, and anhydrite. Near the village of South Antigonish Harbour, a rock salt unit containing gypsum and anhydrite interbeds is 300 m thick. Somewhat farther east, a rock salt bed was penetrated at a depth of 600 m (Benson 1970a, b).

On Cape Breton Island, the Windsor Group is composed of massive red siltstone, shale, sandstone, limestone, and dolomite thin bands, as well as gypsum, anhydrite, and rock salt (Kelley 1967). The proportion of different rocks in the group is as follows: red terrigenous rocks – 70%; gypsum and anhydrite – 17%; rock salt – 6%; limestone and dolomite – 7%. Carbonate bands occur throughout the entire section. Salt deposits were discovered when deep oil test wells were drilled on the north-western side of Ainslie Lake and at Mabou. One of the wells penetrated rock salt over a depth interval of 395.8–1700.5 m; another well entered rock salt at a depth of 1365.5 m, however at a depth of 2093.6 m it did not penetrate the entire section. The Imperial Port Hood No. 1 well penetrated rock salt over the depth interval of 1856.2–2019.3 m; the rock salt is light in color, interbedded with shale, anhydritic siltstone, anhydrite, dolomite, and argillaceous limestone. The other sixteen rock salt beds were encountered over the depth interval 2019.3–2944.4 m. The Mabou No. 1 well, drilled 5 km south of Mabou, encountered rock salt at a depth of 658.4 m and penetrated it for 26.2 m. Rock salt seems to occur in the central Windsor Group. On Cape Breton Island, the thickness of the Windsor Group reaches 750 m in salt-free sections and exceeds 2000 m in salt-bearing sections.

In the St. George's sub-basin, western Newfoundland, thick gypsum units were found among equivalents of the Windsor Group distinguished there as the Codroy Group (Rose et al. 1970).

Geophysical and deep drilling data suggest that salt deposits of the Windsor Group occupy much of the north-western Maritime Basin in the Gulf of St. Lawrence area (Howie and Barss 1975). They were penetrated there by the HB Fina Northumberland Strait F-26 and Sarep HQ Brion Island No. 1 wells.

The Windsor Group is ubiquitously overlain by thick terrigenous deposits where the Canso and Hopewell Groups were recognized in the lower part, and the Riversdale, Cumberland, Pictou Groups and their equivalents in the upper part. They are composed of red and gray shale, siltstone, sandstone, and conglomerate, their total thickness ranging from 2 to 4–5 km.

The data available allow an approximate estimate on the areal distribution of the Windsor Group rock salt in the Maritime Basin. It may reach 55,000 km^2. If the average thickness of rock salt over the area is 200 m, then its volume will amount to about 1.2×10^4 km^3. A total volume of rock salt with allowance made for that of the Albert Formation may exceed 1.3×10^4 km^3.

Saltville Basin

Carboniferous salt-bearing deposits in eastern North America have been reported in the south-western Appalachian Basin. Evaporites are confined to the Maccrady Shale of Mississippian age and distributed in a narrow belt stretching from south-west to north-east along the Saltville Thrust. The following discussion is based on data published by Withington (1965), Cooper (1966), and DeWitt (1979).

The Maccrady Shale consists chiefly of red and yellow shale with sandstone and dolomite interbeds containing gypsum, anhydrite, and rock salt in the upper section. The thickness varies from 20–30 to 250–600 m. Evaporites occur in a 1.5–2 km wide belt following the Saltville and Pulaski Thrusts from the border of Tennessee and Virginia States in the south-west to the town of Blacksburg, Montgomery County, Virginia, in the north-east, for a distance of 200 km (Fig. 87). Halogenic formations are likely to occur in northern Tennessee and in western Virginia, where gypsum exposures are known from Hagan.

In Virginia, the Mississippian deposits are better known in the area between Plasterco and Locust Cove, on the Greendale Syncline. The central part of the syncline is filled in by Mississippian deposits drilled by a number of deep boreholes and mined for gypsum. Along the Saltville Thrust, Cambrian deposits have overthrusted Mississippian strata.

The Chattanooga Shale, made up of thin lamellar shale, siltstone, and sandstone, lie at the base of Mississippian deposits. They are overlain by the Price Sandstone characterized mainly by weathered rusty quartzose sandstone and conglomerate. Siltstone, sandstone, and black shale bands with single coal seams can be observed in the upper part. The Price Sandstone exceeds 140–150 m in thickness.

The Maccrady Shale is easily traced on the Greendale Syncline, though it is subject to substantial facies change. To the north-west, the thickness of the Maccrady Shale measures 40–45 m. Most of the formation is built up of red and green medium-platy sandstone alternating with red shale. Some 10 m below the top, rare limestone bands can be distinguished. The uppermost part consists of red, green, and gray shale.

The thickness of the Maccrady Shale increases to 75 m to the north-east. In this segment of the Greendale Syncline, chestnut to brown siltstone (30 m) occurs at the base, dolomite and carbonate shale (7–8 m) in the center, and red, gray, and green

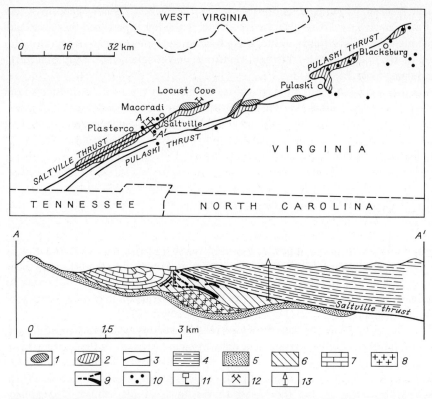

Fig. 87. Distribution of Maccrady Shale evaporites in south-west Virginia and section across Greendale Syncline and Saltville Thrust. After Withington (1965). *1* outcrops of Maccrady Shale, *2* areal distribution of evaporites inferred from salt occurrences, *3* faults, *4* Cambrian and younger deposits, *5* Price Sandstone, *6* Maccrady Shale, *7* Little Valley Limestone, *8* rock salt, *9* gypsum and anhydrite, *10* salt pits, *11* gypsum pits, *12* evaporite pits or drill holes, *13* deep well

plastic shale (30—37 m) at the top. In the Plasterco, Saltville, and Locust Cove areas, the plastic shale member contains gypsum and anhydrite. Massive gypsum beds in the Maccrady plastic shale member were found near Blacksburg. There, the Maccrady Shale can be divided into three members. They are: (1) the lower red sandstone and clayey silt; (2) the middle clayey fossiliferous dolomite; (3) the upper red and green plastic shale with variegated anhydrite and gypsum beds.

The thickness of the Maccrady Shale increases to the south-east. For instance, in the Saltville District, the thickness reaches 450—500 m. Rock salt beds up to 30 m thick appear there; gypsum and anhydrite gain in thickness as well. Evaporites, along with dolomite and limestone, build up the upper member of plastic shale. Salt deposits are known at present only in a limited area. They are thought to stretch over a distance of 3.5—4 km beneath the Saltville Thrust (Cooper 1966).

On the south-eastern Greendale Syncline, the Maccrady Shale, despite a sharp increase in the thickness of the upper plastic shale member, retains its three-member structure; sandstone and carbonate can be followed in the lower and middle parts, respectively. The thickness of rock salt totals 250 m. Twelve to 20 rock salt beds can be recognized there.

In the Locust Cove District, the plastic shale member contains gypsum beds up to 15 m thick; some of them are being mined. The thickness of the Maccrady Shale reaches 380 m. Rock salt has not been found as yet. To the south, the thickness of the Maccrady Shale decreases to 100 m.

The Little Valley Limestone overlying the Maccrady Shale consists of yellowish-brown to buff limestone, limy shale, and sandstone. Between Plasterco and Locust Cove, their thickness measures no less than 150 m. Locally, clayey dolomite, gypsum, and rock salt beds are present in the lower part (Cooper 1966). Above them, there is the lower Hillsdale Limestone, made up of biogenic shale and siltstone overlain by black siliceous, locally algal limestone. The thickness is 70–75 m.

Even higher Mississippian deposits in the Greendale Syncline form the Ste Genevieve Limestone, the Taggard Formation, and the Girkin Limestone, all dominated by limestone. The overlying Fido Sandstone, the Cove Creek Limestone, and the Pennington Formation, developed in the adjacent areas of Virginia and West Virginia, are characterized chiefly by variegated shale, siltstone, and sandstone with rare limestone bands.

In the Appalachian Basin, the Mississippian deposits are unconformably overlain by the thickest red coal-bearing formations of Pennsylvanian age.

The Maccrady Evaporite occurs beneath the Saltville and Palaski Thrusts. In the area between Plasterco and Locust Cove, the thrust dips south-eastward at an angle of $20°–60°$. These evaporites have been penetrated by boreholes beneath the thrust. For example, 1.5 km south-east of Plasterco, a borehole encountered evaporites at a depth of 731.5 m.

Near the thrust, the Maccrady Shale contains inclusions and blocks of gypsum, anhydrite, rock salt, limestone, and sandstone. Such blocks occur among red and green marl and shale. According to Withington (1965), only the Maccrady Shale was dislocated along the thrust. The overlying Little Valley Limestone forms a local syncline there with gently lying north-western and overturned south-eastern limbs. The underlying Price Sandstone plunges gently south-eastward beneath the Saltville Thrust (see Fig. 87). According to Cooper (1966), the Mississippian deposits form an overturned syncline, whose south-eastern limb is displaced by the Saltville Thrust.

The salt-bearing formation proper is built up of rock salt interbedded with anhydrite, dolomite, limestone, and shale. Rock salt in normal succession is usually bounded by zones of tectonic salt breccia. Alternating rock salt and salt-free rocks can be observed only in some undeformed lenses. Most of brecciform zones contain up to 60%–75% of rock salt.

The distribution of salt strata in the Appalachian Basin and the Saltville Basin have not been outlined as yet. According to Cooper (1966), an areal extent of evaporites in the Saltville Basin reaches about 5000 km^2. Withington (1965) shares the opinion that an areal extent of the Mississippian evaporites in the Appalachian Basin is much greater than that known at present. The studies of salines indicate the presence of salt strata in Virginia and adjacent areas not only along the Saltville Thrust, but near the Palaski Thrust as well; they may continue north-eastward into West Virginia. Their occurrence at a great distance from the Palaski Thrust is not inconceivable.

These sulfate deposits may be of Carboniferous age despite the fact that some researchers (Rodgers 1970) assign them to the Middle Cambrian. Considering the aforesaid, an areal extent of evaporites in the Saltville Basin may essentially exceed 5000 km². For now it seems infeasible to estimate the volume of rock salt in the Saltville Basin.

Paradox and Eagle Basins

Paradox Basin and most of the Eagle Basin lie on the Colorado Plateau. The Paradox Basin, embracing the bordering states of Utah, Colorado, Arizona, and New Mexico, is bounded to the south by the Defiance Uplift; to the west, by the Circle Cliffs Uplift and San Rafael Swell; and to the east and norht-east, by the Uncompahgre Uplift. The south-eastern Paradox Basin is open towards the San Juan Basin, from which it is separated by the Four Corners Platform. The Eagle Basin is situated in Colorado. It is bounded to the south-west and north-east, respectively, by the Uncompahgre

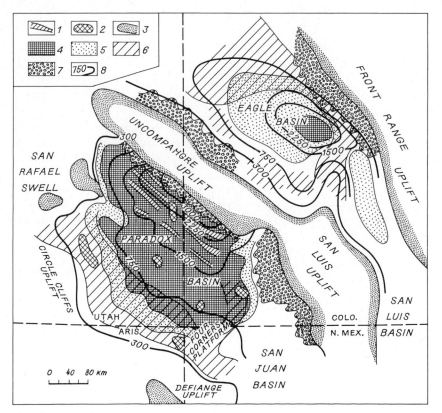

Fig. 88. Isopach-facies map, Pennsylvanian System of Paradox and Eagle Basins. After Peterson and Hite (1969). *1* salt anticline, *2* Tertiary intrusive, *3* Pennsylvanian absent, *4* salt, *5* anhydrite, *6* shelf carbonate, *7* arkose, *8* isolines of thicknesses, in meters

Uplift, the Front Range Uplift. The south-eastern Eagle Basin is connected with the San Luis Basin. Both the Paradox and the Eagle basins were subject to intense downwarping in Pennsylvanian time. Over 3000 m of sediments were accumulated there. Coarse-grained terrigenous deposits were formed along old uplifts, whereas in the central areas of the basin, rather thick salt strata were accumulated. Sulfate-carbonate and carbonate rocks were deposited (Figs. 88, 89) in the south-western and south-eastern Paradox Basin, and in the north-western Eagle Basin. Piedmont slopes of troughs, adjacent to the Uncompahgre and Front Range uplifts, are the most subsided parts.

The Paradox Basin has been widely discussed in such papers as those of Wengerd and Strickland (1954), Herman and Sharps (1956), Herman and Barkell (1957), Wengerd and Matheny (1958), Hite (1960, 1961, 1968), Elston et al. (1962), Wengerd (1962), Ohlen and McIntyre (1965), Baars et al. (1967), Mattox (1968), Peterson (1968), Peterson and Hite (1969), Mallory (1975), McKee (1975). The subdivision of Pennsylvanian deposits is given in Fig. 90.

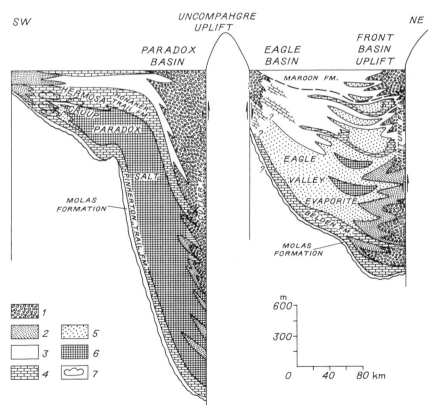

Fig. 89. Cross section showing gross facies relations across Paradox and Eagle Basins. After Peterson and Hite (1969). *1* conglomerate and arkose, *2* sandstone, red shale, and limestone, *3* red shale and sandstone, *4* limestone and dolomite, *5* anhydrite and shale, *6* halite, anhydrite, and black shale, *7* mounds

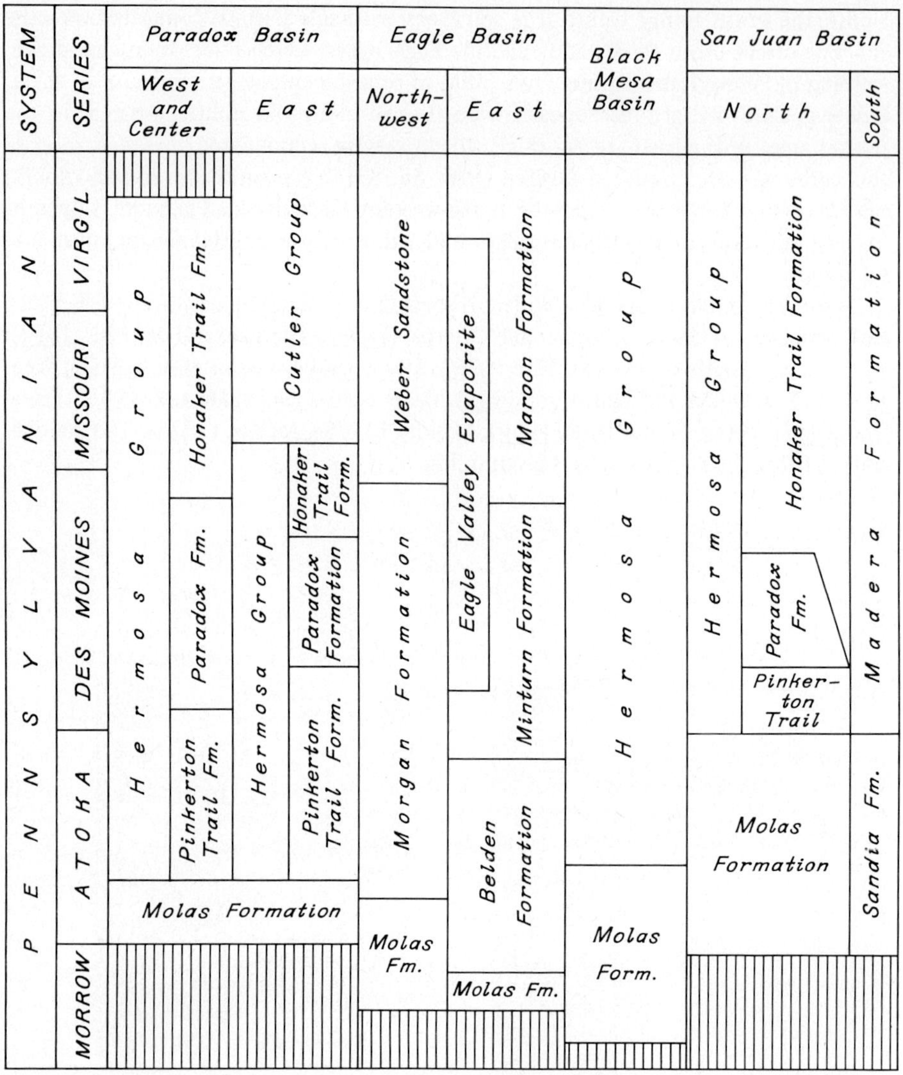

Fig. 90. Correlation of Pennsylvanian deposits of Paradox, Eagle, Black Mesa, and San Juan Basins. Compiled from the data of Wengerd and Matheny (1958), Peterson et al. (1965), Ohlen and McIntyre (1965), Baars et al. (1967), Peterson and Hite (1969), and Mallory (1971, 1975)

At the base, the Molas Formation is distinguished, made up of red shale and siltstone with gray limestone interbeds in the upper part; its thickness does not exceed 60–70 m. Overlying deposits are known as the Hermosa and Cutler groups. The latter occurs only in the eastern part of the basin, whereas the Hermosa Group, comprising halogenic deposits, is developed both in the eastern, central, and western regions. The Hermosa Group is divisible in ascending order into three formations, viz., Pinkerton Trail, Paradox, and Honaker Trail Formations (Wengerd and Matheny 1958, Ohlen and McIntyre 1965, Baars et al. 1967, Peterson 1968, Peterson and Hite 1969).

Pinkerton Trail Formation is predominantly made up of dolomite and black mudstone. It is 60–65 m thick.

Paradox Formation comprises all the salt-bearing deposits. In zones of the highest salt saturation its thickness is over 1200 m. In apical parts of salt anticlines, disturbed by salt tectonics, especially to the north-east, the thickness locally is twice as large, i.e., 2500–3000 m and even 4000 m. The Paradox Formation is composed of rock salt alternating with potash salts, anhydrite, dolomite, and black shale. Towards the margin, salt deposits pinch out and are replaced first by sulfate-carbonate, and then by carbonate deposits.

Red shale marker horizons are easily discernible among salt-bearing deposits. They extend into a salt-free section and allow tracing a facies change pattern. Thick black shale interbeds bind the sections, composed of carbonate, anhydrite, and rock salt. Each section marks a major sedimentary cycle. Five major cycles were recognized in the section; each cycle was given a name, namely, Alkali Gulch, Barker Creek, Akah, Desert Creek, and Ismay cycles. Their thickness ranges from 30 m in carbonate facies to 100 m and more in salt-bearing facies. These major cycles comprise two or more separate evaporite cycles. Twenty-nine of these halite-bearing evaporite cycles, each ranging from a fraction of a meter to several tens of meters in thickness have been identified in the deep part of the basin (Fig. 91). From the base of the evaporite cycle upwards, the sequence is: (A) silty, calcareous black shale, (B) silty dolomite, (C) anhydrite, (D) halite with or without potash salts; (C) anhydrite, (B) silty dolomite, and (A) black shale (Peterson and Hite 1969).

Potash salts are confined to the following salt beds or evaporite cycles, which are numbered 5, 6, 9, 13, 16, 18, 19, 20, 21, 24, and 27 (Hite 1961). In general, potash salts are composed of sylvinite and carnallite rocks. Minor polyhalite, kieserite, and rinneite have also been observed. Potash salts are very common. For instance, in salt bed 19, whose structure is shown in Fig. 92, they occur in a band which extends for more than 175 km. On the north-western Moab Valley Anticline, where salt-bearing strata occur in an undisturbed succession, the potash salt thickness in bed 19 equals 190–200 m. There, they are composed mainly of alternating carnallite rocks and halite. However, a mineral composition of potash salts changes along the strike, i.e., carnallite rocks are often replaced by sylvinite, as in the area between the Cane Creek, Seven Mile, and Salt Valley anticlines.

Potash salts in the Paradox Basin occupy an area over 15,000 km^2 (Fig. 93). Potash salts occur at depths between 770 and 6350 m. Potassic horizons, accessible for mining, are commonly confined to the apical parts of such anticlines as Cane Creek, Moab Valley, and others, where a depth to the top of rock salt ranges from 480–730 m, and the upper rock salt horizons lie at a depth of 800 and 1800 m. Potash salts of salt beds 5 and 9 can be mined there.

On anticline apexes, strongly disturbed by salt tectonics, a depth to the top of rock salt is often less than 200 m; relatively thick and high-grade potash deposits in salt beds 18, 19, 20, and 21 also occur at a much smaller depth.

In general, the Paradox Formation is represented by the multiple alternations of interbeds of rock salt and potash salts, anhydrite, dolomite, and mudstone. Sulfate rocks, viz., anhydrite and gypsum, building up separate units, traceable over an extensive area, lie at the base and the top. The lowest anhydrite unit, which marks

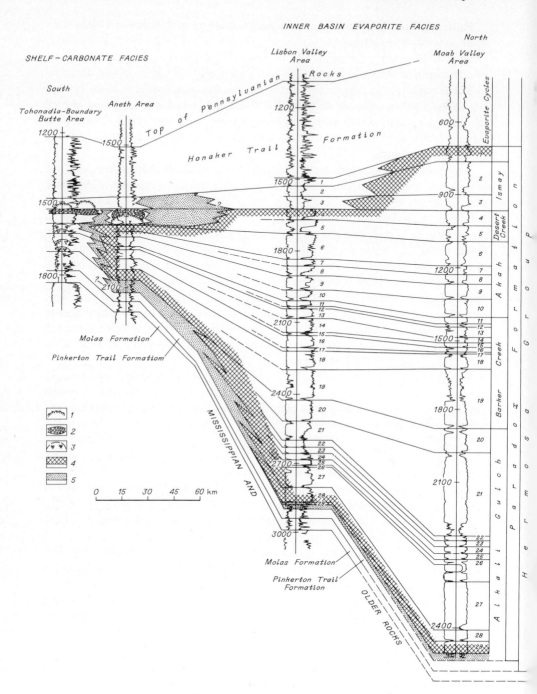

Fig. 91. Correlation of radioactivity logs across Paradox Basin showing relationship between shelf-carbonate facies and inner-basin evaporite facies. After Peterson and Ohlen (Peterson and Hite 1969). *1* algal plate mound, *2* "Leached oolite", *3* Paradox bioherm, *4* boundary of halite facies, *5* boundary of anhydrite facies

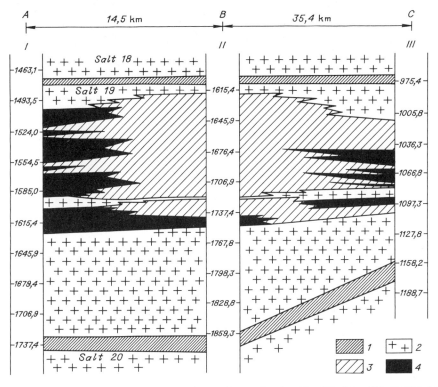

Fig. 92. Potash-bearing cycle 19 of the Paradox Formation (see Fig. 93). After Hite (1961). *I* Cane Creek Anticline, Texas Gulf Producing Well 1-x, *II* Seven Mile Anticline, Deli-Taylor Oil Corp. Well 2, *III* Salt Valley Anticline, Defense Plant Corp. Reeder 1. *1* shale-anhydrite marker, *2* halite, *3* halite and sylvite, *4* halite and carnallite

the beginning of the Paradox Formation section, rests on the Pinkerton Trail Dolomite, and is overlain by salt deposits of the Alkali Gulch cycle. It represents a basal member of the cycle. The upper anhydrite unit, lying at the top of the Ismay cycle, crowns the section of the Paradox Formation. This unit is overlain by the Honaker Trail Formation. To the south, to the south-west, and to the north-west, the extent of the formation decreases due to pinching-out and facies change of its lower and upper horizons. However, in each particular section, at its base and top, the under- and overlying anhydrite units are present, but their stratigraphic position differs greatly from that of similar units in subsided parts of the basin. The general succession of salt-bearing deposits in the Paradox Formation is clearly seen in Figs. 89, 91, and 94.

To the south, south-east, and north-west, they are replaced by carbonate deposits; all of them, as a rule, belong to the non-subdivided Hermosa Group. North-east, near the Uncompahgre Uplift, the upper evaporites are replaced by terrigenous deposits of the Cutler Group.

As a whole, rock salt of the Paradox Formation covers an area not less than 30,000 km^2 (see Fig. 93). A total thickness of salt-bearing rocks, occurring between

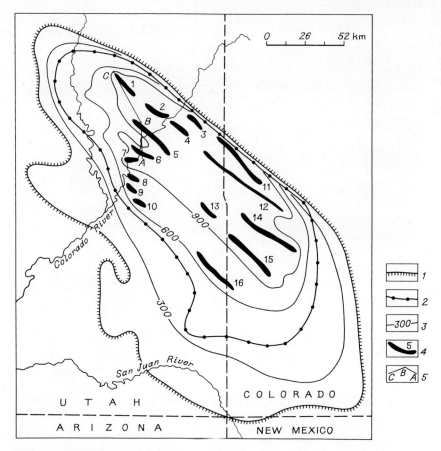

Fig. 93. Distribution or rock and potash salts in Paradox Basin. Compiled from the data of Hite (1961), and Pierce and Rich (1962). *1* approximate limits of the salt of the Paradox Formation, *2* approximate limits of major potash zones of the Paradox Formation, *3* isopach contours of rock salt, in meters, *4* salt anticline (*1* Salt Valley, *2* Cache Valley, *3* Fisher Valley, *4* Castle Valley, *5* Seven Mile and Moab Valley, *6* Cane Creek, *7* Shafer Dome, *8* Lockhart Dome, *9* Rustler Dome, *10* Gibson Dome, *11* Sinbad Valley, *12* Paradox Valley, *13* Lisbon Valley, *14* Gypsum Valley, *15* Dolores, *16* Boulder Knoll), *5* section shown in Fig. 92

the top of the uppermost and the base of the lowermost rock salt beds ranges there from 150–200 to 900 m. The greatest thicknesses of salt in the basin are inferred for a band, adjacent to the Uncompahgre Uplift, which may reach 2000 m and more. Single boreholes have penetrated salt deposits for 4000 m and have not encountered the basement. Salt saturation of sections near the areas, where rock salt pinches out, accounts for not less than 10% of a total thickness, varying in inner parts of the depression between 30% and 99% (Pierce and Rich 1962). A mean salt saturation of sections, calculated by 30 boreholes, drilled on salt anticlines, according to Shoemaker et al. (Pierce and Rich 1962), equals 72.4%. With allowance made for these data, the following estimates were taken to determine salt volume in the Paradox Basin:

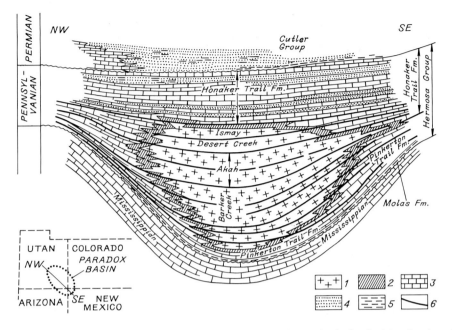

Fig. 94. Salt deposits of Paradox Basin. After Baars et al. (1967). *1* salt, *2* anhydrite, *3* carbonate, *4* sandstone, *5* shale, *6* black shale marker beds

areal extent, 25,000 km²; a mean thickness of salt-bearing deposits, 600 m; salt saturation, 60%. In this case, the volume of salts in the basin will equal 9×10^3 km³.

Over most of the basin, Paradox Formation is overlain by the Honaker Trail Formation. The lower Honaker Trail Formation is dominated by dolomite, shale, and siltstone. Its middle part contains limestone and siltstone, whereas sandstone is common in the upper part.

In the south-western marginal parts of the basin, the Honaker Trail Formation forms a single sequence with the Pinkerton Trail Formation; and so they are placed into the non-subdivided Hermosa Group. To the north-east, the proportion of terrigenous material greatly increases in the Honaker Trail Formation, while that of carbonate rocks gradually decreases and, finally, the entire formation wedges out, being replaced along the strike by coeval deposits of the Cutler Group. The Honaker Trail Formation is 300 m thick.

The Cutler Group is composed of conglomerate, arkose gritstone, and sandstone, red and green mudstone, and shale; minor carbonate rocks and black mudstone were recorded as well. The thickness reaches several thousand meters. Pennsylvanian deposits in the Eagle Basin are less well known than those of the Paradox Basin. They have received the most study by Mallory (1966, 1971, 1975), and Peterson and Hite (1969), whose data served as a basis for the following description.

In the northern and western parts of the basin, the Molas, Morgan Formations, and Weber Sandstone are distinguished in the Pennsylvanian section, and, in the eastern part, Molas, Belden, Minturn, and Maroon Formations are singled out. Evaporites developed in the central part of the basin are grouped into the Eagle Valley Evaporite (see Fig. 90).

Molas Formation consists predominantly of red shale, siltstone, and sandstone. Carbonate rocks, placed to the east and center in the Belden Formation, and in the north and west, assigned to the lower Morgan Formation, occur up the section. The Belden Formation, along with carbonates, represented by limestone and dolomite, accounting for more than 20%, contains dark gray and black shale, as well as mudstone and sandstone; approaching the Front Range Uplift, the amount of terrigenous material increases. The lower Morgan Formation is predominantly composed of limestone and dolomite.

Molas and Belden Formations, and the lower Morgan Formation contain no evaporites. They appear higher in the section, starting with the so-called evaporite interval. The lower boundary of the evaporite interval is drawn to the east at the lower Minturn Formation, in a subsided central zone of the basin, at the base of the Eagle Valley Evaporite, and to the north, in the middle Morgan Formation. The upper boundary of the evaporite interval is within the lower Maroon and the upper Morgan Formations.

Minturn Formation is dominated by terrigenous rocks, its thickness exceeding 2700 m. It comprises red and gray conglomerate, gritstone, sandstone, siltstone, and shale. Lenticular interbeds of gypsum and anhydrite appear in the lower part, at the level of the evaporite interval. In some regions, reef carbonate units are developed. Anhydrite is common in the upper Minturn Formation. This part of the section is built up of carbonates, placed into the Jacque Mountain Limestone Member, extending for a long distance and paralleling the south-western boundary of the Front Range Uplift. To the west, limestone is replaced by anhydrite.

In the central zone of the Eagle Basin, the Eagle Valley Evaporite is predominantly made up of rock salt, anhydrite, gypsum, gypsinate mudstone, variegated shale, siltstone, and sandstone, as well as gray and black mudstone, and shale. The composition of the formation varies greatly across the area. In the Eagle Valley, light gypsinate mudstone and siltstone occur along with gypsum interbeds and 30 cm-thick bands of silicified dark gray limestone, consisting locally of red shale and siltstone. Farther south, massive and thick-platy gypsum appears in the upper part of the formation. At Eagle, the upper Eagle Valley Evaporite is composed of clayey anhydrite, black and gray shale and dark-colored siltstone; while its lower part consists of saline mudstone, siltstone, and rock salt. Not far from Cattle Creek, the upper part of the section, penetrated by a number of deep holes, is composed of gray and brownish anhydrite alternating with light siltstone, sandstone, and dark shale about 650 m thick; its lower part is dominated by rock salt.

Eagle Valley Evaporite contains also potash salt interbeds. They have been reported from two boreholes east of the Eagle River Anticline, in the Alkali Trough. At depths of 1097–1220 m, two potash salt beds were penetrated, respectively, 1.8 and 2.1 m thick. The potash salts are composed of sylvite and carnallite rocks.

Salt-bearing deposits pinch out to the east, north-east, and north-west. At the Front Range Uplift, evaporite is replaced by terrigenous rocks of the Minturn and Maroon formations. In the north, they grade into thinner deposits of the Morgan Formation. The Eagle Valley Evaporite ranges in thickness from 300–2700 m.

In the north-westernmost part of the basin, the uppermost Morgan Formation is composed of limestone alternating with gray sandstone. Gypsinate and anhydrite sandstone appear in the section approaching the evaporitic shelf.

In north-eastern Eagle Basin, the Pennsylvanian section is crowned by the Maroon Formation, made up of red conglomerate, gritstone, sandstone, siltstone, and shale. It is over 1000–1200 m thick. The lower part of the formation contains interbeds of gypsinate mudstone and siltstone. Its upper part is dominated by sandstone and siltstone. Coarse-grained rocks in the formation are most common at the Front Range Uplift. To the north-west, the upper part of the Maroon Formation is marked by facies change, namely, the appearance of white, cream-colored and light gray Weber Sandstone. The thickness of the latter is not more than 250 m.

Thus, in the Eagle Basin, salt-bearing deposits are confined to the lower Eagle Valley Evaporite. They were reported from central parts of the basin, in the Alkali Trough, and from the Cattle Creek Anticline area. Salt strata pinch out rapidly and grade into terrigenous deposits towards the mountain flank and into sulfate, sulfate-carbonate, and carbonate deposits to the north-west (Fig. 95).

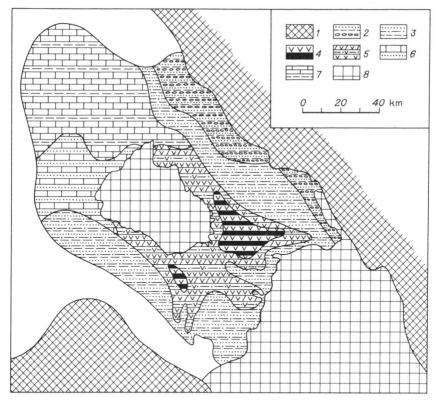

Fig. 95. Lithology of the lower part of the Eagle Valley Evaporite and equivalents in the Eagle Basin. After Mallory (1971). *1* Uncompahgre and Front Range Uplifts, *2* conglomerate, grit, sandstone, siltstone, and shale, *3* grit, sandstone, siltstone, and shale, *4* halite, gypsum, and anhydrite, *5* gypsum, anhydrite, and gypsiferous and anhydritic mudstone and siltstone, *6* interbedded limestone and sandstone, *7* limestone, *8* outcrop of Eagle Valley Evaporite and Precambrian rocks

The thickness of rock salt beds varies between 30–40 and 100–120 m. In some regions, it greatly increases due to the manifestations of salt tectonics. As a whole, the thickness of the evaporite interval deposits ranges from 300–2700 m and above. According to Mallory (1971), a total thickness of evaporites (rock salt, anhydrite, and gypsum) exceeds 300 m. An average thickness of rock salt in a salt-producing zone may reach 300 m. The areal extent of rock salt equals 5000 km^2. Thus, its volume will amount to 1.5×10^2 km^3.

Amazon Basin

Salt deposits were found in the Amazon Basin in 1954, when two deep holes were drilled: one in the vicinity of Nova Olinda, in the lower Madeira River area; and the other, in the lower Tapajoz River area. Both boreholes penetrated thick rock salt strata. At present, numerous boreholes, drilled there, have pentrated the entire evaporite section, enabling the study, in detail, of the composition, structure, and spatial distribution pattern of evaporites. Only some of these data have been published (Szatmari et al. 1979); therefore, the brief characterization of the Amazon evaporite basin given below is based essentially on the earlier evidence (Oliveira 1956, Petri 1958, Link 1959, Morales 1959, Harrington 1962, Loczy 1963, 1964, Benavides 1968, Lefond 1969).

The Amazon evaporite basin lies in northern Brazil. It extends in a west-easterly direction, from the Atlantic Ocean to the upper Amazon River, for more than 2000 km, and it is 300–600 km wide.

The emplacement of the basin took place in the Early Paleozoic. Paleozoic sedimentary strata crop out within two belts 30–50 km wide, which extend from east to west and are parallel to the flow of the Amazon River. Paleozoic strata plunge down the center and there they are unconformably overlain by Meso-Cenozoic formations. The Amazon Basin can be divided into three major regions: (1) the western, Upper Amazon Depression; (2) the central, Mid-Amazon Depression, and (3) the eastern, Lower Amazon Depression. To the west, the Upper Amazon Depression is separated by the buried Iquitos Arch from the Cis-Andian Trough; to the east, the Purus Swell detaches it from the Mid-Amazon Depression. The Gurupa Swell can be distinguished between the Mid-Amazon and the Lower Amazon Depressions. Evaporites have been reported from the Mid-Amazon and the Upper Amazon Depressions (Fig. 96). Hitherto, evaporites were assumed to occur among Pennsylvanian deposits (Petri 1958, Morales 1959, Benavides 1968, Bigarella 1975, The Encyclopedia ... 1975); however, recently their probable Lower Permian age was inferred owing to the fact that they are underlain by the Upper Carboniferous (Stephanian) limestone and sandstone and are overlain by the Upper Permian to Lower Triassic red shales (Szatmari et al. 1979). At the same time, the lower part of the evaporites which are intercalated with fossiliferous limestone, probably belong to the Late Carboniferous, and the entire evaporite succession can be regarded as Permo-Carboniferous.

The Monte Alegre Formation, composed mainly of gray sandstone with siltstone and mudstone interbeds, can be distinguished at the base of the Pennsylvanian section.

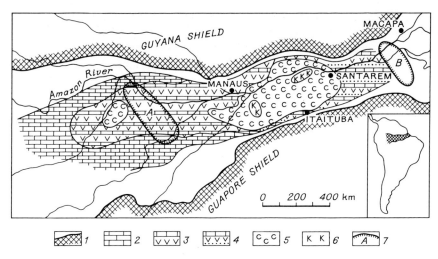

Fig. 96. Lithology of the Nova Olinda, Amazon Basin. Compiled from the data of Szatmari et al. (1979). *1* outcrops of Precambrian rocks, *2* limestone, *3* interbedded limestone, dolomite, and anhydrite, *4* interbedded limestone, dolomite, anhydrite, sandstone, and siltstone, *5* halite, gypsum, anhydrite, and carbonate rocks, *6* potash, *7* swells (*A* Purus, *B* Gurupa)

Its thickness ranges from 60–250 m. Above it lies the Itaituba Formation, made up of limestone, black mudstone, and gray sandstone with anhydrite interbeds whose numbers increase up the section. Limestone often contains a diverse fauna, mainly brachiopods, gastropods, and corals. This formation is 200–300 m thick. The section is crowned with the Nova Olinda Formation, made up of sandstone, siltstone, mudstone, dolomite, limestone, anhydrite, gypsum, and rock salt. Salt-bearing rocks dominate the upper part of the formation. The rocks are red, violet, green, or yellow. The Nova Olinda Formation varies in thickness from 800 m at the northern flank of the basin to 1200 m in the axial zone, and to 650 m along its southern flank.

Evaporite deposits are cyclic in structure. In the central and western Amazon Basin, the cycles in the lower Nova Olinda Formation are composed of carbonate rocks and anhydrite. In the middle and upper parts of the formation, halite is the dominant evaporite cycle. Limestone and anhydrite are restricted to the basal part of each cycle, becoming thinner in each succeeding one. All in all, seven major cycles can be distinguished in the section (Szatmari et al. 1979). Cycle VII contains potash salts, represented by sylvinite. The presence of thick diabase beds is a peculiar feature of the evaporite sequence.

In general, salt deposits are confined to the upper Nova Olinda Formation. They fill in the most subsided parts of the Amazon Basin, underlying central areas of the Mid-Amazon Depression and eastern Upper Amazon Depression. Eastwards, rock salt pinches out and salt-bearing deposits grade into sulfate and sulfate-carbonate; the latter in turn become replaced by carbonate. In the extreme eastern regions of the basin and near northern and southern margins, the proportion of terrigenous rocks abruptly increases in the section. Such features of a facies change of deposits suggest that the Amazon evaporite basin was a deep embayment with the sea water inflow from the open ocean, situated along the western margin of the present South American continent (Harrington 1962, Benavides 1968, Szatmari et al. 1979).

Salt-bearing deposits in the Amazon Basin cover an area of about 100,000 km^2. If an average total thickness of rock salt over the entire area is 200 m, then rock salt volume will amount to 2×10^4 km^3.

Carboniferous Evaporite Deposits in Other Basins

A great number of Carboniferous evaporite basins containing only sulfate rocks such as gypsum and/or anhydrite have been recently recognized. There are 29 sulfate basins, namely the North Siberian, Pechora-Novaya Zemlya, Spitsbergen, Northumberland, North Ireland, Central England, Dobruja, East European, East Uralian, Teniz, Chimkent, Tyup, Aksu, Achikkul, Lhasa, Fitzroy, Radames, Illisie, Ahnet, Reggan, Tindouf, East Wyoming, San Juan, Orogrande, South Iowa, Illinois, Michigan, Venezuela, and South Peruvian basins.

Within the North Siberian Basin, in which evaporites persisted not only in the Devonian, but also during the Early Carboniferous, there are halogenic rocks (exclusively gypsum and anhydrite) of Carboniferous age in the north-western Siberian Platform at Norilsk, in the Khatanga, Kyutingda, and Kempendyai Troughs (Matukhin and Menner 1974, Vaag and Matukhin 1982). West of Norilsk, sulfate rocks are exposed in the Tundrin Formation section containing several anhydrite beds associated with two lower beds. Anhydrite is interbedded with gray dolomite, dolomitic marl, and dolomitic limestone in the first bed, while in the second overlying bed anhydrite occurs among red marl (Matukhin and Menner 1974). The areal extent of the Tundrin sulfate-bearing sequence is still uncertain. In the Khatanga Trough, interbeds of gypsum and anhydrite occur in the carbonate sequence. Menner (1965) places them in the central part; nevertheless, their position in the section is still uncertain. In the Kyutingda Trough, situated in the north-eastern Siberian Platform in the Kyutingda River valley and bounded by the Olenek and Daldyn Uplifts in the north-east and south-west, respectively, sulfate rocks are associated with the Tournaisian deposits and assigned to the gypsum-bearing sequence recognized there (Kuteinikov and Syagaev 1962, Fradkin 1964, 1967, Menner 1965, Biterman et al. 1970, Vaag and Matukhin 1982). Its lower part is composed of interbedded dolomitic limestone, dolomite, and gypsum, and the upper part of dolomitic marl, clay, and gypsum. East of the trough, gypsum bands are rather persistent; their thickness varies from 1–10 m. The thickness of the entire gypsum-bearing sequence ranges from 50–90 m. In the Kempendyai Trough, evaporites (gypsum and anhydrite) occur only in the lowermost Carboniferous, in the lower Kurunguryakh Formation (Fradkin 1964, 1967, Nakhabtsev and Fradkin 1970). They are interbedded there with dolomite and tuffaceous siltstone. The thickness of the sulfate-bearing part of the Kurunguryakh Formation is 90 m.

The Pechora-Novaya Zemlya evaporite basin is recognized by the presence of two sulfate sequences; one sequence is known in the Pechora Basin; it runs from the Usa River mouth to the Ilyich River mouth, and the other is known on Novaya Zemlya along the western coast of the Makarov Bay and on Alevastrovyi Island.

Early Carboniferous age is proposed for both sequences. They are usually assigned to the Visean, and partly to the Namurian (Atlas ... 1969). In the Pechora Basin, anhydrite and gypsum are interbedded with marl, dolomitic limestone, and dolomite; the thickness of the sulfate-carbonate sequence varies from 100–150 m (Chermnykh 1962, 1966, Nikolaev and Sivkov 1982). On Novaya Zemlya, the thickness of the gypsum-bearing sequence is about 30 m (Rogozov 1970).

The Spitsbergen Basin was the site of sulfate accumulation during Carboniferous and Permian time. Two gypsum-bearing sequences have been recognized there; the lower sequence is assigned by Harland (1964) to the Bashkirian (Middle Carboniferous). Some geologists (Sasipatrova 1967) consider it to be Visean. The lower gypsum-bearing sequence is confined to western Spitsbergen.

Within the East European Basin, gypsum and anhydrite are known by the Carboniferous deposits in many regions. The distribution pattern is inadequately known, though the Carboniferous deposits are sufficiently well-studied; however, we are unable to determine with certainty the number of sulfate sequences developed in the basin; for the time being they are approximately outlined. The oldest sulfate unit can be identified as the Ozersk sequence. It is developed in the Moscow Syneclise, as well as in adjacent regions of the Volga-Uralian area (Tokmovo, Tatar, and Zhiguli-Pugachev Arches, Ryazan-Saratov Trough). A high portion of sulfate rocks in the sequence is reported from the southern Moscow Syneclise in the Tula District. It consists of gypsinate banded dolomite, sulfate-dolomitic rocks, and gypsum. In the Novomoskovsk, Venev, Domnin, and Zaraisk Districts, the number of gypsum beds in the section increases, with a continuous gypsum unit, 15–23 m thick in its upper part. The maximal sulfate saturation is reported from the area near the town of Laptevo. The thickness of the Ozersk sequence reaches 65–70 m (Birina et al. 1971). In the north, the sequence is essentially sulfate-dolomite. These sections can be observed near Podolsk, Monino, Rybinsk, Lyubim, and Soligalich. The thickness varies there from 20–62 m. In the northwestern Moscow Syneclise, the Ozersk sequence contains rather thick (about 10 m) gypsum deposits. Within the Tatar Arch, the sequence becomes essentially carbonate, but anhydrite inclusions occur locally in its central area (Rauzer-Chernousova et al. 1967). Lenses and seams of gypsum and anhydrite are encountered in some sections in the Zhiguli-Pugachev Arch (the Krasnaya Polyana-1 borehole). The Ozersk sulfate sequence is rather widespread in the East European Basin. Many investigators assign it to the Zavolzhsk horizon of the Tournaisian.

The overlying sulfate-bearing sequence, developed almost throughout the East European Carboniferous evaporite basin, occurs at the level of the Oka and Serpukhov superhorizons of the Visean, and at the level of the Protvin horizon of the Namurian (Lower Carboniferous). The sequence is tentatively named the Oka-Serpukhov sequence. It is present in the central and western parts of the Moscow Syneclise (districts of Nepeitsevo, Tutaev, Shar'ya, Lyubin, Soligalich, Kashin, Krasnyi Kholm), where it is composed mainly of dolomite and dolomitized limestone with gypsum inclusions, lenses, and bands (Birina et al. 1971); the sequence extends into the Gorky-Volga Region (districts of Oparino, Kotelnich) and into the Upper Kama Basin (the Glazov key borehole), where bands and inclusions of thin anhydrite occur in terrigenous-carbonate and carbonate variegated deposits; the sequence is reported from the northern and southern parts of the Tatar Arch (Golyshurma and Baitugan areas) and from the Melekes Depression where sulfate rocks occur in the form of inclusions and veinlets of anhydrite throughout the carbonate section of the Oka and Serpukhov superhorizons; the sequence is known in the Zhiguli-Pugachev Arch where numerous inclusions of gypsum and anhydrite occur in dolomite and dolomitized limestone (Rauzer-Chernousova et al. 1967, Efremov et al. 1982). The same Oka-Serpukhov sulfate sequence seems to continue in the northern East European Basin and farther into the Basin of the middle reaches of the Severnaya Dvina River where Barkhatova (1963) reported the presence of anhydrite bands in dolomites tentatively assigned to the Devyatino Formation. The upper horizons of this sulfate sequence locally include deposits of the Namurian Protvin horizon; this can be observed in the zone of Vyatka folding, in the Upper Kama Basin, in the Tokmovo and Tatar Arches, and in the Sergiev-Abdulin Depression (Rauzer-Chernousova et al. 1967). The thickness of the Oka-Serpukhov sequence varies over a wide range of 50–60 m to 350 m.

The third sulfate sequence is recognized at the level of the Kashira, Podolsk, and Myachkov horizons of the Moscovian (Middle Carboniferous). It can be named the Kashira-Myachkov sequence. It lies within the Moscow Syncelise, in the zone of Vyatka folding, in the Upper Kama Basin, in the Cherdyn region of the Cis-Uralian Depression, and in the Ryazan-Saratov Trough; the sequence probably remains sulfate-bearing in some areas of the Tatar and Tokmovo Arches, and it primarily contains carbonate rocks with bands, lenses, and inclusions of gypsum and anhydrite. Sulfate rocks become more extensive in the Shar'ya, Soligalich, Chukhloma, and Lyubim areas (Barkhatova 1963, Rauzer-Chernousova et al. 1967, Nalivkin and Sultanaev 1969, Shik 1971, Birina et al. 1971). The thickness of the Kashira-Myachkov sequence is about 200 m in the most sulfate-bearing sections.

The fourth sulfate sequence is associated with the Kasimov superhorizon of the Gzhelian (Upper Carboniferous). It is clearly defined in the Moscow Syneclise where its thickness ranges from 40–70 m. South-western and western boundaries of sulfate rocks in the Kasimov sequence stretch from the town of Kasimov roughly to Moscow and the town of Krasnyi Kholm. North-east and east of this boundary, minor gypsum bands, 0.3–0.5 m thick, appear in the sequence consisting chiefly of limestone and dolomite. Then, north-west of the boundary, running through the towns of Kovrov, Gavrilov Yam, and Tutaev, the sequence becomes highly gypsinate, and the thickness of separate gypsum beds reaches 0.5–1 m (Goffenshefer 1971). Rather reliable data suggesting that the Kasimov sulfate sequence continues eastward and north-eastward into the zone of Vyatka folding and into the upper Kama Basin, as well as toward the Tokmovo and Tatar Arches, have been reported (Rauzer-Chernousova et al. 1967). The sequence possibly stretches northward into the Timan area and into the Vychegda River Basin (Barkhatova 1963).

The overlying fifth sulfate sequence includes deposits of the Klyazma horizon belonging to the Gzhelian (Upper Carboniferous). It is known all over the East European Carboniferous evaporite basin. In the Moscow Syneclise, its thickness reaches 60–80 m. The sequence is chiefly composed of dolomitized limestone and dolomite with gypsum bands, the thickness of which varies from 1 to 3–4 m (Goffenshefer 1971). Within the Vyatka folding zone, in the Upper Kama Basin and in the Cherdyn region of Cis-Urals, the Klyazma sequence remains sulfate-bearing carbonate sequence; however, the portion of sulfate rocks there decreases and they form rare bands, lenses, and inclusions. The thickness of the sequence ranges from 70—100 m. In the area of the Tokmovo and Tatar Arches, the Melekes and Sergiev-Abdulin Depressions, lenses and bands of gypsum and/or anhydrite are common in the Klyazma sequence among dolomitized limestone, calcareous dolomite, and dolomite. The section is not altered within the Zhiguli-Pugachev and Orenburg Arches. Two anhydrite units, up to 50 m thick each, were reported (Orekhovo, Kuleshovka, and Neklyudovo Districts) (Grachevsky et al. 1969). The Klyazma sulfate-bearing sequence probably reaches the eastern margin of the Caspian Depression where sulfate rocks were found in the Upper Gzhelian (Dalyan and Posadskaya 1972). It should be noted that the Kasimov and Klyazma sulfate sequences, recognized as separate members can be differentiated with some certainty only within the Moscow Syneclise. In other regions of the East European Basin, particularly in the Volga-Uralian area, they are hardly distinguishable. Only one sulfate-bearing sequence may be present in this region at a level of the Gzhelian.

The uppermost sixth sulfate-bearing sequence can be arbitrarily recognized as the Noginsk sequence among the Carboniferous deposits of the East European Basin. It is situated at the level of the Orenburgian stage of the Upper Carboniferous and becomes most distinct in the Moscow Syneclise where, north of the boundary, running through the town of Murom, Sudogda, Kolchugino, and Zagorsk, inclusions, lenses and bands of gypsum and anhydrite, 2—4 m thick, appear among carbonate rocks (Goffenshefer 1971). The same sequence may have been reported from the Timan area, and from the Vychegda and Pinega Rivers Basin where bands of gypsum and anhydrite, 0.7—1 m or even 5.7 m thick, are known locally in the Orenburgian deposits (Barkhatova 1963). The Noginsk sulfate sequence may stretch into the Upper Kama Basin and the Tokmovo Arch where anhydrite bands were recognized among the Orenburgian deposits (Rauzer-Chernousova et al. 1967).

Within the Northumberland Basin in Scotland, sulfate rocks are present in Visean calcareous sandstone. Inclusions and bands of gypsum occur among dolomite, mudstone, and sandstone over a small area (George 1958, 1963).

The distribution of sulfate-bearing deposits in the Northern Ireland Basin is restricted. They are associated with the Roscunish argillaceous sequence assigned to the Upper Visean (Padget 1953, Oswald 1955, Caldwell 1959, George 1963, Dixon 1972). West et al. (1968) have recognized the sulfate-bearing sequence in the Northern Ireland Basin as a separate Aghagrania Formation subdivided into units. Gypsum beds are associated there with the lower Meenymore member and with the uppermost Corry member. Large gypsum lenses are locally 2.2 m in size, but usually gypsum is present as intercalations, inclusions, and veinlets.

The Central England Basin has a limited areal extent as well. The Visean anhydrite series is exposed there in boreholes near Hathern, Leicestershire (Llewelling and Stabbins 1968).

The Dobruja Basin is recognized from the presence of sulfate rocks (anhydrite) in the Tournaisian and Visean deposits exposed within the Dobruja Depression in Moldavia (Kaptsan et al. 1963, Kaptsan and Safarov 1965b, 1969). The deposits were penetrated there by three boreholes (R-30, R-31, and R-32). Anhydrite beds occur among limestone and calcareous dolomite. The thickness of the carbonate sulfate-bearing sequence is 176 m. It is believed to continue southward.

In the East Urals evaporite basin, the number of sulfate-bearing sequences is unknown as the basin has not yet been outlined and it is roughly determined as a single region of halogenic sedimentation. Along the eastern slope of the Urals, individual outcrops of the Carboniferous sedimentary successions may have accumulated in isolated basins, and several basins of evaporitic sedimentation may have existed there at different periods of time. Three widely spaced outcrops of sulfate rocks are presently known there. The largest gypsum rock outcrop is located near Magnitogorsk (Ivanov and Levitsky 1960). A sulfate sequence consisting of marl, calcareous sandstone, and conglomerate with two gypsum units was outlined there as early as 1932. The thickness of the lower unit varies from 27–99 m, and that of the upper one from 17–20 m. The second terrane of Carboniferous gypsum-bearing deposits is located along the eastern slope of the South Urals in the Bagaryak district (Ivanov and Levitsky 1960). Gypsum beds, the thickness of which varies from 5–10 m to 15–37 m, are associated with variegated sequence represented by conglomerate, sandstone, limestone, and marl. Their total thickness is 300 m. The third terrane of sulfate rocks is known north of the Magnitogorsk terrane on the eastern slope of the Central Urals (Pronin 1969). Red beds with thin layers and lenses of gypsum occur here.

The Chimkent evaporite basin is a part of the Carboniferous evaporite belt stretching into the Mid-Tien Shan Basin. The basin is not well known. Sulfate rocks were reported from the Tournaisian deposits. The thickness of the gypsum-bearing sequence ranges from 25–60 m. It occurs within a carbonate series including tuff and tuff porphyrite bands (Shcherbina 1945, Ivanov and Levitsky 1960).

In the Teniz Basin, sulfate rocks show a limited distribution. They are known in the upper Kirey Formation where limestone bands with large crystals of gypsum and gypsinate sandstone and siltstone occur among red conglomerate, gravelstone, and sandstone. The proposed age of this part of the Kirey Formation is Bashkirian (Litvinovich 1972).

The Tyup Basin including the Tyup and, in part, Tekes Troughs of the North Tien Shan is situated east of Issyk-Kul Lake between Kirghizia and Kazakhstan. Sulfate rocks occur as gypsum bands in the Tyup and Chaarkuduk Formations of the Bashkirian (Middle Carboniferous) (Chabdarov and Sevostianov 1971, Ektova and Belgovsky 1972).

The Aksu Basin is situated along the north-western margin of the Tarim Depression in Sinkiang (China). Sulfate rocks (gypsum) are associated there with the Middle Carboniferous Kurukusum Group developed in the Aksu River Basin in the Asgan-Bulag-Taga area. They are observed among variegated sandstone and marl. The thickness of the sulfate-bearing sequence exceeds 300 m (Regional Stratigraphy . . . 1960).

The Achikkul Basin has been tentatively recognized. It embraces the northern foothills of the Przhevalsky Range between Ayagkumkul and Achikkul Lakes. The basin includes the Arktag sequence of red coarse-grained sandstone with limestone interbeds and gypsum lenses. The age of the sequence is uncertain and regarded as Permo/Carboniferous. Sulfate deposits in the Achikkul Basin and in other basins of Central Asia are probably Middle Carboniferous (Regional Stratigraphy . . . 1960).

The Lhas Basin was named arbitrarily. Sulfate rocks are known from the Pando Formation consisting of dark gray and green shales, quartzites, and limestones with gypsum lenses. The formation is 2000–2500 m thick. It is rather widespread south of the Nyenchhen-Thanglha. The location of gypsum in the Pando section is not yet clear. The entire formation is assigned to the Carboniferous. As to the stratigraphic level of sulfate rocks, it is similar to that of the northern evaporite basins of Central Asia (Regional Stratigraphy . . . 1960).

The Fitzroy Basin, situated in western Australia, includes the Anderson sulfate-bearing formation. It is represented by sandstone, siltstone, and mudstone interbedded with dolomite and anhydrite. The formation is 1500–2000 m thick. The age is Late Carboniferous (Westphalian/Lower Stephanian) (Thomas 1959, Veevers and Wells 1961, Brown et al. 1970, Wells 1980).

Small Rhadames, Illisie, Ahnet, Reggane, and Tindouf Basins are situated in northwestern Africa, mainly in the Algerian Sahara (Aliev et al. 1971). In the Rhadames Basin, sulfate rocks were recorded in the Upper Namurian where unit E is composed chiefly of alternating clay and dolomite with rare intercalations of white anhydrite and some siltstone and fine-grained sandstone. Within the Illisie Basin, two sulfate-bearing sequences are known: the lower El Adeb Larach and the upper Tigentourin sequences. In both sequences, gypsum bands occur among terrigenous-carbonate variegated deposits. In the Ahnet Basin, gypsum and anhydrite beds occur along the western slope of Djebel Berg Uplift in the uppermost horizons of the Visean deposits consisting mainly of limestone and mudstone. In the Reggane Basin, a gypsum-bearing unit is distinguished at the Visean D level; it consists of white and pink gypsum and anhydrite interbedded with clay, marl, dolomite, and limestone; the thickness of the complex ranges from 120–207 m. Another sulfate-bearing sequence is determined in the same basin at the level of the Upper Namurian. It is represented by variegated clay, sandstone, and limestone with lenses and thin layers of gypsum. The Late Visean Quarkziz sulfate sequence is widespread in the Tindouf Basin. It is divided into three units: the lower limestone unit consists of limestone and dolomite interbedded with variegated mudstone (black, red, green, and creamy), sandstone, and white anhydrite, up to 356 m thick; the central clayey-anhydrite unit is composed of alternating anhydrite and mudstone; anhydrite locally is 45 m thick, the total thickness amounts to 111.5 m; the upper limestone unit, 102 m thick, contains thin layers of anhydrite. In the Tindouf Basin, anhydrite is known not only in the central part, but in the southern part as well, where horizons of sulfate rocks among the Upper Visean deposits are tens of meters thick. They belong to the Ain El Barka Formation.

In the Illinois Basin, significant amounts of bedded gypsum and anhydrite occur in the lower part of the St. Louis Limestone in Illinois, Indiana, and Kentucky, U.S.A. Lesser amounts of these evaporites as nodular masses, geode fillings and veinlets are also found in the St. Louis, Salem, and Harrodsburg Limestones, in some units of the Borden Group or Formation, and in the Fort Rayne Formation (McGregor 1952, Saxby and Lamar 1957a, b, Bond et al. 1968, Sable 1979). The lower part of the St. Louis Limestone, less than 6 to more than 60 m thick, is referred to as the evaporite unit or evaporite-bearing zone. In this unit, gypsum and anhydrite beds are 0.3–6 m thick, they are intercalated with limestone, dolomitic limestone, dolomite, and minor mudstone.

In the South Iowa Basin, anhydrite and gypsum are restricted to the St. Louis Limestone and Pella Formation (Johnson and Vondra 1969). In the Michigan Basin, gypsum and anhydrite beds have been reported in the Michigan Formation of Late Mississippian age (Vary et al. 1968, Wanless and Shideler 1975, Cohee 1979). As many as eight beds of gypsum have been reported in the Michigan Formation in south-western Michigan. In the central part of the basin, anhydrite beds are alternated with shale, dolomite, and sandstone. The total thickness of gypsum beds ranges between several and 10–12 m, mounting to 30 m in the center of the basin.

In the Orogrande Basin, situated in southern New Mexico and western Texas, west of the Pedernal Uplift, the Late Carboniferous (Missourian/Virgilian) upper part of the Magdalena Formation is sulfate-bearing. Gypsum beds among carbonate rocks and marls are known in the El Paso region (Crosby and Mapel 1975), but the thickest gypsum beds (3–12 m) were reported from the uppermost sequence in the south of the basin in New Mexico. They are also interbedded there with carbonate rocks (Bachman 1975). The San Juan Basin is situated south-east of the Paradox Basin. They are separated by a zone of carbonate biogenic buildups, where the evaporites of the Paradox Formation thin out. Outside this zone, gypsum-bearing mudstone interbedded with dense dolomite and limestone appears at the same level, but this time in the San Juan Basin. They form the Paradox Formation. The East Wyoming Basin is only tentatively recognized because it is not confined to a certain sedimentary basin, but traceable over much of a shelf zone which in Pennsylvanian time occupied eastern and north-eastern Wyoming and adjacent areas of South Dakota and Nebraska (Maughan 1975, Prichard 1975, Paleotectonic . . . 1975). Sulfate rocks (gypsum and anhydrite) occur there among terrigenous-carbonate deposits forming the Pennsylvanian Minnelusa Formation. They develop in eastern Wyoming and stretch into central South Dakota.

Sulfate-bearing sequences of Pennsylvanian age are known in the Venezuela and South Peruvian basins in South America. In the Venezuela Basin, gypsum is associated with the Palmarito Formation, and in the South Peruvian Basin, gypsum is assigned to the Tarma Group (Benavides 1968).

Chapter VII
Permian Salt Deposits

During the Permian, a period of immense salt accumulation, several vast evaporite basins, such as the East European, Central European, Midcontinent, and Peru-Bolivian Basins were formed. Salt deposits are also known from the following basins: Alpine, Moesian, Chu-Sarysu, Supai, and North Mexican (Fig. 97). Fourteen Permian basins with gypsum and/or anhydrite occurrences were recognized: Darvaza, Karasu-Ishsay, Dobruja, Rakhov, Mecsek, Dinarids, North Italian, Spitsbergen, East Greenland, Sverdrup, Rio Blanco, Parnaiba (Maranhão), Arabian, and Browse Basins.

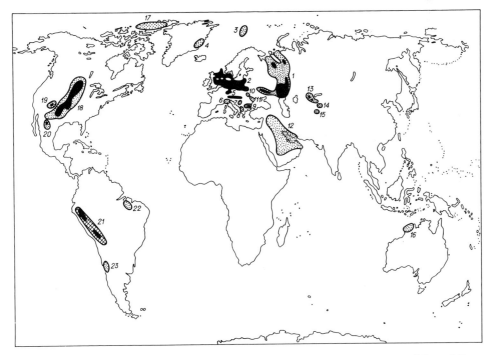

Fig. 97. Distribution of Permian evaporites. Evaporite basins: *1* East European, *2* Central European, *3* Spitsbergen, *4* East Greenland, *5* Alpine, *6* North Italian, *7* Mecsek, *8* Dinarids, *9* Moesian, *10* Rakhov, *11* Dobruja, *12* Arabian, *13* Chu-Sarysu, *14* Karasu-Ishsay, *15* Darvaza, *16* Browse, *17* Sverdrup, *18* Midcontinent, *19* Supai, *20* North Mexican, *21* Peru-Bolivian, *22* Parnaiba (Maranhão), *23* Rio Blanco. For legend see Fig. 3

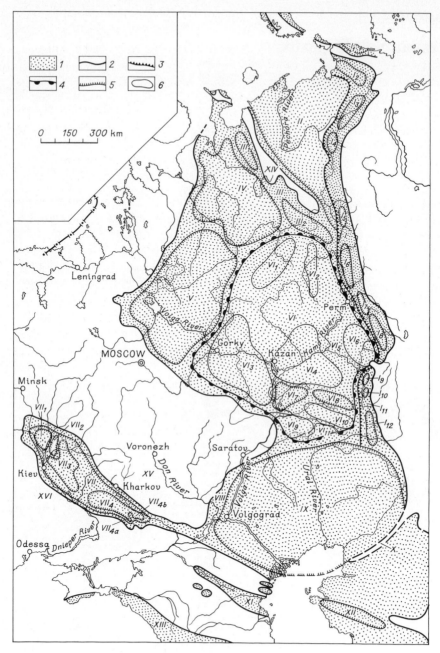

Fig. 98. Distribution of Permian deposits in East Europe. Compiled by the author together with Merzlyakov. *1* areal extent of Permian deposits, *2* limit of Permian deposits, *3* westward extent of Cis-Uralian Trough, *4* limit of Volga-Ural Anteclise, *5* limits of syneclises and troughs, *6* limits of positive structures (uplifts and inliers). *I* Cis-Uralian Trough and component depressions and basins (I_1 Kara Depression, I_2 Korotaikha Depression, I_3 Kosju-Rogovo Depression, I_4 Bolshesognino Basin, I_5 Upper Pechora Depression, I_6 Solikamsk Depression, I_7 Chusovaya Depression, I_8 Yuryuzan-Sylva Depression, I_9 Sim-Inzer Depression, I_{10} Voskresensk-Shikhany Depression, (cont. p. 205)

East European Basin

The East European Basin is one of the largest Paleozoic evaporite basins with evaporite deposits over an area of 1.5 million km^2. They occur in the Mezen Depression, Moscow Syneclise, Volga-Ural Anteclise, central and southern Cis-Uralian Trough, and the Caspian and Dnieper-Donets Depressions (Fig. 98).

Permian deposits have been investigated well, yet incompletely. To the north, they are not well-known, while in the Dnieper-Donets Depression and the central and eastern areas, particularly in the Volga-Ural area and the Cis-Uralian Trough, where numerous Permian type sections are situated, they have been thoroughly investigated. Due to complex deformation by salt tectonics, an adequate study of the structure and spatial distribution pattern of the involved salt deposits in the Caspian Depression has not been completed.

The entire East European Basin and its separate parts have often been discussed. Many investigators pertain to stratigraphy and lithology of Permian deposits. The following publications are referred to when discussing the East European Permian Basin: Ivanov (1932, 1965), Rykovskov (1932), Kovanko et al. (1939), Ruzhentsev (1948), Forsh (1951, 1955, 1976, 1979), Lyutkevich et al. (1953), Dubinina (1954), Chochia (1955), Fiveg (1955a, b, 1959, 1973a–e), Lyutkevich (1955, 1963, 1969, 1975), Burkovskaya (1956a, b), Nesterenko (1956, 1958, 1970, 1978), Soloviev (1956), Vakhrameeva (1956), Eventov (1957, 1962), Makarova (1957, 1959), Rybakov (1958, 1960, 1962), Borozdina (1959), Aizenshtadt et al. (1960), Ivanov and Levitsky (1960), Vakhrameeva and Gorkun (1960), Aizenshtadt and Pinchuk (1961), Brailovsky (1961a, b), Golubtsov (1961, 1966), Gordievich and Sanarov (1961), Lapkin (1961a, b, 1964, 1966), Latskova (1961, 1967, 1981), Levenshtein (1961), Sokolova et al. (1961), Svishchev (1961), Graifer et al. (1962), Khatyanov (1968a, b), Kopnin (1962, 1963, 1965, 1977), Kovanko (1962), Lipatova (1962), Urusov et al. (1962, 1967), Zamarenov (1962, 1970), Aizenshtadt and Gershtein (1963), Avrov and Kosmacheva (1963), Barkhatova (1963), Bondarenko (1963), Gavrish (1963, 1964), Kel'bas (1963), Martynov and Khnykin (1963), Pakhtusova (1963a, b), Porfiriev (1963), Rotai (1963), Shmelev (1963), Zoricheva (1963, 1966), Arkhipov (1964), Banera (1964), Gordon-Yanovsky and Bespalov (1964), Aizenshtadt and Gorfunkel (1965), Bobrov and Korenevsky (1965), Butkovsky et al. (1965), Lapchik (1965), Tikhvinsky (1965, 1967, 1970a, b, 1971, 1973, 1974, 1976a, b), Tikhvinsky et al. (1967), Varlamov (1965), Arabadzhi et al. (1966), Sokolova (1966), Stepanov and

---

◄ I_{11} Sterlitamak-Meleuz Depression, I_{12} Sakmara Depression,), *II* Pechora Basin, *III* Timan (Pritimansky) Trough (*III$_1$* Safonovo Trough, *III$_2$* Vychegda Trough), *IV* Mezen Depression, *V* Moscow-Syneclise, *VI* Volga-Ural Anteclise (*VI$_1$* Kotelnich Uplift, *VI$_2$* Komi-Permyak Uplift, *VI$_3$* Tokmovo Arch, *VI$_4$* Tatar Arch, *VI$_5$* Birsk Saddle, *VI$_6$* Bashkir Arch, *VI$_7$* Melekes Basin, *VI$_8$* Sergiev-Abdulin Basin, *VI$_9$* Zhiguli-Pugachev Arch, *VI$_{10}$* Orenburg Arch, *VI$_{11}$* Buzuluk Depression), *VII* Dnieper-Donets Depression (*VII$_1$* Pripyat Depression, *VII$_2$* Chernigov Uplift, *VII$_3$* Nezhin Depression, *VII$_4$* Poltava Depression, *VII$_{4a}$* Kalmius-Torez Trough, *VII$_{4b}$* Bakhmut Trough), *VIII* Volga Monocline, *IX* Caspian Depression, *X* South Emba Uplift, *XI* Kuma-Manych Trough, *XII* Mangyshlak Trough, *XIII* norther slope of the Caucasus Range, *XIV* Timan Uplift, *XV* Voronezh Anteclise, *XVI* Ukrainian Shield

Forsh (1966), Ignatiev (1967), Tikhvinskaya (1967), Tikhvinskaya and Chepikov (1967), Bobrov et al. (1968), Budanov and Molin (1968), Butkovsky and Omelchenko (1968), Ermakov et al. (1968, 1977), Fiveg and Banera (1968), Gelfand and Karpenko (1968), Ivanov and Voronova (1968, 1972, 1975), Lobanova (1968), Movshovich (1968, 1970, 1977), Oborin (1968), Supronyuk et al. (1968), Vasiliev (1968), Ashirou and Efremov (1969a,b), Atlas ... (1969), Blizeev and Vishnyakov (1969), Bobukh et al. (1969), Bulekbaev et al. (1969), Ivanov et al. (1969, 1977), Korenevsky et al. (1968), Seiful-Mulyukov and Sheremetiev (1969), Sementovsky (1969), Shcherbakov and Volgina (1969), Vainblat (1969), Vinogradova and Oshchepkov (1969), Volchegursky et al. (1969), Blizeev (1970, 1971a, b), Blom and Vavilov (1970), Khursik et al. (1970), Korenevsky (1970, 1973, 1981), Diarov (1971a, b, 1972, 1974, 1977), Diarov and Dogalov (1971), Diarov and Dzhumagaliev (1971), Dneprov and Koltypin (1971), Ermakov (1971), Gorbatkina and Strok (1971), Khavin (1971), Pinchuk (1971), Svidzinsky (1971a, b), Zholtaev (1971), Zhuravlev and Svitoch (1971), Bazhenova (1972), Dalyan and Posadskaya (1972), Odoleev (1972), Shafiro (1972, 1975, 1977), Slivkova et al. (1972), Tikhvinsky and Blizeev (1972), Zamarenov et al. (1972), Zhuravlev (1972), Zhuravlev et al. (1972), Gorbov (1973), Shafiro and Sipko (1973), Shchedrovitskaya (1973), Oshakpaev (1974), Ustritsky (1975), Bogatsky et al. (1977), Ermakov and Grebennikov (1977), Golubev (1977), Kapustin et al. (1977), Kolotukhin et al. (1977), Kopnin and Zueva (1977), Kopnin et al. (1977), Krichevsky et al. (1977), Krylova et al. (1977), Kudryashev and Myagkov (1977), Levenshtein et al. (1977), Morozov (1977), Muldakulov and Marchenko (1977), Sapegin (1977), Svidzinsky et al. (1977), Tikhvinsky et al. (1977), Tretyakov (1977), Trofimova and Efremov (1977), Azizov and Tikhvinsky (1978), Derevyagin et al. (1979, 1981a, b), Merzlyakov (1979), Grebennikov and Ermakov (1980), Kopnin and Kukharchuk (1980), Kopnin and Maloshtanova (1980), Morozov et al. (1980), Zharkov et al. (1980), Bocharov and Khalturina (1981), Derevyagin (1981), Diarov et al. (1981), Golubev and Brovchenko (1981), Komissarova (1981), Kopnin and Korotaev (1981), Kovalsky (1981), Levina et al. (1981), Makarov (1981), Sapegin and Yanin (1981), Tretyakov and Sapegin (1981), Tumanov (1981), Turkov and Shudabaev (1981).

Permian deposits occur as a wide continuous belt from the Barents Sea to the Caspian Sea. To the east, they are bounded by the pre-Permian outcrops in the Urals; to the west, they taper out near the Baltic Shield and Voronezh Anticlise. To the south-west, Permian deposits form an embayment in the Dnieper-Donets Depression. Permian sedimentary strata are uncommon in the southern areas of the Soviet Union, occurring in general outside the East European Basin. They were reported from the Kuma-Manych and Mangyshlak Troughs, Northern Caucasus, and southern Crimea. The distribution of Permian strata and the main structural elements of the East European Basin are shown in Fig. 98.

The thickness of Permian strata vary over a wide range (Fig. 99). A gradual increase in total thickness is observed from west to east toward the Urals. Thicknesses of 2000–2500 m and 5000–6000 m to the south and north of the Cis-Uralian Trough, respectively, were reported. In the Caspian Depression, original thicknesses are unknown due to the influence of salt tectonics on Permian deposits. There are a large number of diapirs, some of which are shown in Fig. 99. The original thickness of the

East European Basin

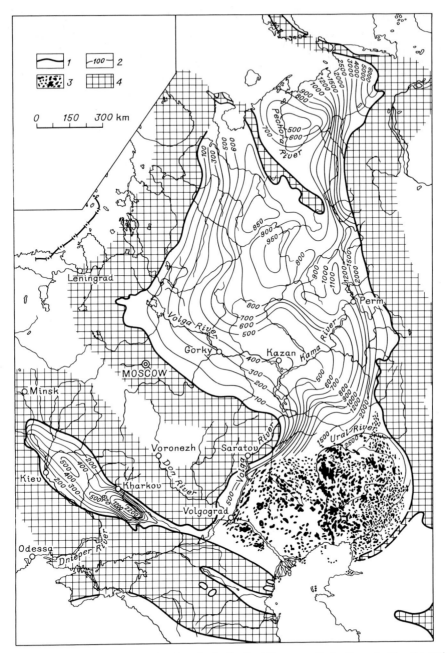

Fig. 99. Isopach map of Permian deposits in the East European Basin. Compiled by the author together with Merzlyakov. *1* limit of Permian deposits, *2* isopach contours, *3* salt domes, diapirs, and anticlines with higher thicknesses of salt deposits, *4* Permian absent

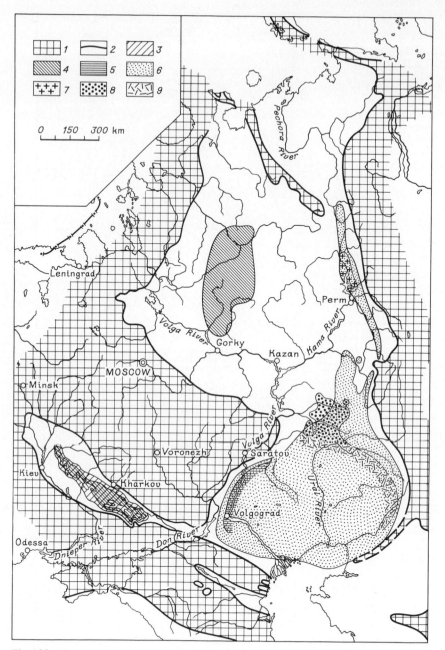

Fig. 100. Distribution of salt deposits in East European Basin. Compiled by the author together with Merzlyakov. *1* Permian absent, *2* limit of Permian deposits. Depositional limit of: *3* Asselian, *4* Sakmarian, *5* Artinskian, *6* Kungurian, *7* Ufimian, *8–9* Kazanian (*8* known, *9* probable)

Permian sedimentary strata is inferred to be 7—8 km in the deeper central parts of the depression (Vasiliev 1968, Zhuravlev et al. 1972). In the south-eastern Dnieper-Donets Depression, Permian strata attain 2000—2200 m.

In the East European Basin, evaporites occur throughout the Permian section (Figs. 100 and 101). Salt-bearing deposits were recognized at six stratigraphic levels: Asselian, Sakmarian, Artinskian, and Kungurian (Lower Permian); and Ufimian and Kazanian (Upper Permian). The oldest Asselian salt deposits were found only in the Dnieper-Donets Depression. Sakmarian rock salt is believed to occur in two areas: (1) the Dnieper-Donets Depression, where the Kramatorsk Salt Formation may occur at the level of the Sakmarian (Korenevsky et al. 1968); however, it is conceivably the upper part of the formation that may be assigned to the Lower Sakmarian (Lapkin 1961a, b, 1966); (2) the Dvina-Sukhona Basin, contiguous parts of the Mezen Depression, Moscow Syneclise, and Volga-Ural Anteclise. Artinskian salt strata are believed to occur in the north-western and western peripheral zones of the Caspian Depression, in the southern Saratov District, and in the Volgograd District (Latskova 1961, 1967, Fiveg and Banera 1969, Shafiro 1972). Kungurian salt deposits are very common in the East European Basin. They were discovered in many parts of the Cis-Uralian Trough (Upper Pechora, Solikamsk, Chusovaya, Yuryuzan-Sylva, Sim-Inzer, Voskresensk-Shikhany, Sterlitamak-Meleuz, and Sakmara Depressions), in the south-eastern Volga-Ural Anteclise, and throughout the Caspian Depression. Ufimian rock salt was found over the entire area of the Solikamsk Depression (Kopnin 1962, Ivanov 1965). Kazanian salt deposits are known from the southern Volga-Ural Anteclise and the northern and eastern Caspian Depression (Zhuravlev 1972, Zhuravlev et al. 1972).

It should be emphasized that the stratigraphic position of many salt sequences has not yet been determined. Therefore, the above assignments are only tentatively accepted at present. Some of the most disputable problems will be dwelt upon in subsequent sections.

Asselian salt deposits (Fig. 102) were found in the Dnieper-Donets Depression, where they form the Nikitovo and Slavyansk Formations (Nesterenko 1956, 1978, Levenshtein 1961, Bobrov and Korenevsky 1965, Korenevsky et al. 1968, Korenevsky 1970, 1973, 1981, Levenshtein et al. 1977). The salt deposits are underlain by the Kartamysh Formation. The lower boundary of the Asselian is drawn at the base of this formation. The lower Kartamysh Formation is composed of red shale, siltstone, and sandstone interbedded with carbonate rocks making up marker horizons. The upper part of the formation consists of variegated, terrigenous rocks with marker dolomitic interbeds. The Kartamysh Formation contains twelve markers labeled $Q_1 - Q_{12}$, in ascending order. In some places of the depression, anhydrite bands, beginning from marker band Q_8, are present in the upper part of the section.

The Nikitovo Salt Formation includes predominantly sand-clayey rocks with bands and beds of carbonate rocks and rock salt units. Carbonate beds form distinct marker horizons labeled, in ascending order: R_1, R_1^1, R_2, R_3, R_4, and R_4^1. Salt units are generally designated by formal names.

The Nikitovo Formation starts with carbonate bed R_1 represented by dolomitic limestone, locally with anhydrite interbeds up to 2—3 m thick. Above is a terrigenous unit of sandstone and siltstone with rare anhydrite bands; in the northern Bakhmut Trough, the lower part of the terrigenous unit contains a 22—25 m rock salt bed

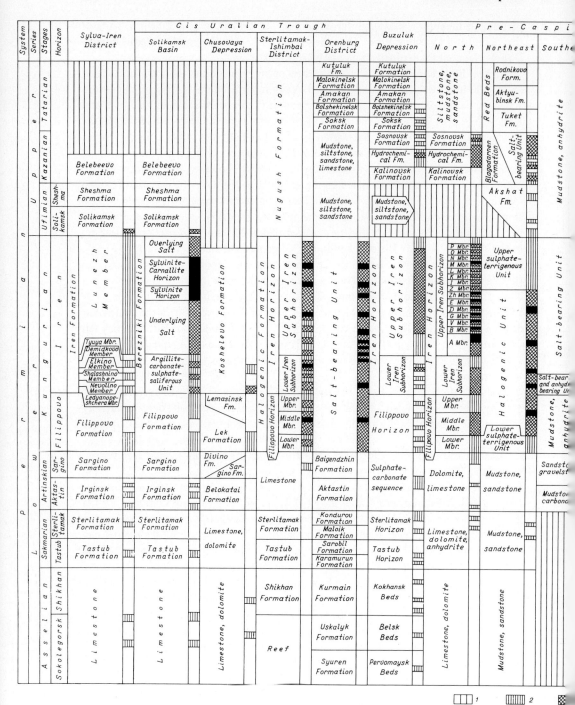

Fig. 101. Correlation chart of Permian deposits in East European Basin. Compiled by the author together with Merzlyakov. *1* deposits absent, *2* anhydrite and gypsum, *3* salt, *4* potash, *5* bischofite rocks

East European Basin 211

...er Dome	Chelkar Dome	Elton Dome	Volga Monocline	Bakhmut Trough	Tokmov Arch	Dvina-Sukhona Region	Upper Pechora Depression	Horizon	Stages
...ndstone, ...lomerate	Red Beds	Red Beds	Red Beds	Koreneu Formation / Shebelinsk Formation (Dronov Group)	Carbonate-terrigenous sequence / Transition Beds / Podluzhniki / Opoki / Shikhany	Vyatka Horizon / Severodvinsk Horizon / Sukhona Formation / Nizhneustie Formation (Urzhum Horizon)	p_2^c	Tatarian	
...dstone, ...stone			Gray mudstone, limestone	Peresazhsk Formation	Seryi Kamen / Podboi / Sloistyi Kamen / Yadrenyi Kamen / Dolomite, limestone, mudstone	Dolomite, gypsum / Limestone, dolomite		Kazanian	
			Red mudstone, siltstone		Sandstone, siltstone, mudstone, dolomite, anhydrite	Red Beds (Vikhtovsk Fm.)	p_2^b / p_2^a	Sheshma / Soli-kamsk	Ufimian
...ying Anhydrite / ...lying Salt / ...r Halite Mbr. / ushaktau Mbr. / ...ite-Halite Mbr / ... Anhydrite / ...Halite Mbr. / ...yhalite-...inite Mbr. / Lower ...drite-Hali-te Mbr. / ...aitau ...nation	Anhydrite / Upper Halite zone / Upper Subzone / Potash Horizons 1-3 / Main Anhydrite / Lower Subzone / Potash Horizons 1-7 / Lower Halite zone (Salt-bearing Unit, Potash zone)	Overlying Anhydrite / Upper Halite Horizon / Carnallite-Halite Member / Lower Anhydrite-Halite Mbr. / Anthraconite Horizon / Cover Salt / Ore Member / Transition Member / Underlying Salt / Lower Halite Horizon (Potash Horizon)	Anhydrite / Marker VII / Marker E / Marker VI / Marker D₁ / Marker D / Marker V / Marker C / Marker IV / Marker B / Marker III / Marker A₂ / Marker IIa / Marker A₁ / Marker II / Marker A / Marker I (7 Rhythm / 6 Rhythm / 5 Rhythm / 4 Rhythm / 3 Rhythm / 2 Rhythm / 1 Rhythm) (Cycles IX-X / Cycle VIII / Cycles VI-VII / Cycles IV-V / Cycle III / Cycles I-II)		Anhydrite, gypsum, dolomite (Verkhnekuloi Formation)	Anhydrite, gypsum, dolomite, mudstone / Filippovo Formation	Red Beds / Iren Formation / Filippovo	Iren / Filippovo	Kungurian
					Anhydrite, dolomite	Gypsum, anhydrite, dolomite	Limestone	Aktas-tin	Artinskian
				Sylvite-carnallite Horizon / Sylvite-bearing Horizon / Belbasov Horizon (Kramatorsk Fm.)	Limestone, dolomite / Anhydrite, gypsum, dolomite	Salt series / Dolomite, anhydrite	Mudstone, limestone	Sterli-tamak / Tastub	Sakmarian
				Krasnoselsk Beds / Nadbryantzevo Bed / Bryantzevo Bed / Podbryantzevo Bed / Karfagen Beds / Tor Bed / Svyatogorsk Bed / Kamensk Bed / Spivak Bed (Slavyansk Fm. / Nikitovo Fm. / Kartamysh Formation) (Carbonate sequence / Sulphate-carbonate sequence)	Dolomite, anhydrite / Dolomite, anhydrite	Dolomite, anhydrite / Dolomite, limestone	Limestone	Shikhan / Sokolegorsk	Asselian

■ 4   ⫽ 5

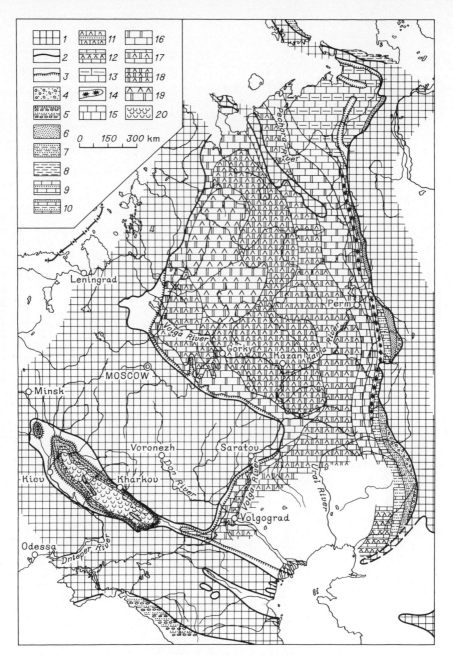

Fig. 102. Lithologic map of Asselian deposits in East European Basin. Compiled by the author together with Merzlyakov. *1* Permian absent, *2* limit of Permian deposits, *3* limit of Asselian. Depositional limits of: *4* conglomerate, gravelstone, sandstone, *5* sandstone, mudstone, siltstone, and conglomerate, *6* mainly sandstone, *7* mudstone, siltstone, and sandstone, *8* mainly mudstone, shale and siltstone, *9* terrigenous and carbonate rocks, *10* sandstone and mudstone interbedded with carbonate rocks, *11* terrigenous-carbonate rocks interbedded with anhydrite, *12* limestone, mudstone, and anhydrite, *13* limestone and argillaceous limestone, *14* carbonate reef structures, *15* mainly limestone, *16* limestone and dolomite, *17* limestone and dolomite with inclusions and interbeds of anhydrite (gypsum), *18* dolomite with interbeds and inclusions of anhydrite (gypsum), *19* dolomite interbedded with anhydrite, *20* salt

known as the Spivak Salt Bed. The terrigenous unit is overlain by carbonate horizon R_1^1, 0.3–1.3 m thick. Above lies another terrigenous unit made up essentially of shale, siltstone, several anhydrite interbeds, and thin rock salt beds referred to as the Kamensk Bed; its thickness does not exceed 11 m. This unit is overlain by carbonate marker horizon R_2 composed of dolomite and limestone 7–8 m thick. Even higher lies the Svyatogorsk Salt Bed developed along the northern flank of the Bakhmut Trough. Its thickness ranges from 6–12 m. Tapering out, the member is replaced by sulfate-terrigenous rocks (shale and anhydrite). The Svyatogorsk Salt Bed is overlain by carbonate rocks and close marker horizons R_3 and R_4 with a thickness totaling 14–21 m. In the northern Bakhmut Trough, carbonate horizon R_4 is overlain by three rock salt beds grouped into the Tor Salt Bed. The two lower beds range from 15–30 m in thickness; the upper bed is 10–12 m thick. Anhydrite occurs between, above, and under the rock salt beds. The upper Nikotovo Formation is composed of shale and siltstone with anhydrite interbeds; its thickness varies from 10–15 to 35 m. A thin carbonate band marked R_4^1 is found in some places.

Thus, the Nikitovo Formation is represented entirely by alternate beds and units of rock salt, as well as terrigenous and carbonate rocks. The total thickness of the formation ranges from 75 to 240–250 m. Two rhythms of salt accumulation are recognized in the formation corresponding to the Spivak Salt Bed and the Tor Salt Bed (Korenevsky et al. 1968).

In the Bakhmut Trough, the Slavyansk Formation consists mainly of rock salt and anhydrite interbedded with carbonate and terrigenous rocks. Carbonate and anhydrite marker horizons are marked S; a number is assigned to each horizon. Thick rock salt units are designated by formal names.

The most conspicuous carbonate marker horizons are S_2, S_3, and S_4 in the lower, central, and upper parts of the section, respectively. A series of rather thick (up to 10–15 m) rock salt beds known as the Karfagen Beds lies below. The Podbryntzevo Bed with a variable thickness of 13–40 m is easily distinguished above horizon S_2. Carbonate marker horizon S_3 is overlain by the Bryantzevo Bed; its thickness is estimated to reach 40–43 m. Up the section, marker horizon S_4 is underlain by the Nadbryntzevo Bed 12–42 m thick. Four rock salt beds can be distinguished above horizon S_4; they compose the upper Slavyansk Formation and are grouped under the name of the Krasnoselsk Beds. Their thickness ranges from 6–10 to 25–30 m. Many anhydrite interbeds, as well as thinner rock salt and terrigenous rock beds can be distinguished in the Slavyansk Formation along with the known rock salt beds and carbonate marker horizons.

The thickness of the Slavyansk Formation varies from 400–600 m in the Bakhmut Trough and the south-eastern Dnieper-Donets Depression; it reaches 400 m in the Kalmius-Torez Trough, and ranges from 40–140 m in the north-western Dnieper-Donets Depression.

The Asselian salt series tapers out along the flanks of the Dnieper-Donets Depression. Sulfate-carbonate-terrigenous, sulfate-terrigenous, and terrigenous deposits marking the near-shore zones of sedimentation are developed in this area.

No Asselian salt has been found as yet over the remaining part of the East European Basin, although sulfate, sulfate-carbonate, and carbonate rocks are widespread.

Their distribution shows a conspicuous, regular pattern. The entire central part of the basin is occupied by sulfate and sulfate-carbonate rocks. They dominate the northwestern area including the Mezen Depression and the Moscow Syneclise, as well as the north-western Volga-Ural Anteclise (Lyutkevich 1955, Makarova 1957, Pakhtusova 1963a, Zoricheva 1966, Tikhvinsky 1967, 1970b, Gorbatkina and Strok 1971, Korenevsky 1973, Tikhvinsky et al. 1977, Merzlyakov 1979). The Asselian is characterized by dolomite interbedded with anhydrite and gypsum.

The proportion of anhydrite in the section decreases when traced south- and south-eastward towards the central parts of the Volga-Ural Anteclise. The Sokolegorsk horizon is dominated by dolomite with minor gypsum and anhydrite. Sulfate rocks are very common in the Shikhan horizon (Tikhvinsky 1970b).

The dolomite, anhydrite, and gypsum terrane is bounded on all sides by the area containing limestone and the above mentioned rocks. A unit of similar lithology was found in the Komi-Permyak Uplift, the Upper Kama Depression, the eastern and southern Volga-Ural Anteclise, and to the west, in the Moscow Syneclise (Makarova 1957, Tikhvinsky 1967, 1970b, Gorbatkina and Strok 1971). Anhydrite and gypsum occur as inclusions or bands; they are especially numerous in the Upper Asselian, in particular, among deposits of the Shikhan horizon.

A lithological unit composed of limestone and dolomite with anhydrite (gypsum) inclusions and bands may well continue into the Caspian Depression. This was reported from the northern and western flank zones of the depression (Rybakov 1958, Svishchev 1961, Latskova 1967, Urusov et al. 1967, Movshovich 1970 and others). The occurrence of sulfate rocks in the Asselian terrigenous-carbonate deposits was also recorded along the south-eastern flank of the depression (Zamarenov 1970).

In the western marginal part of the East European Basin, the Asselian deposits are usually represented by limestone and dolomite. A similar zone is recognized to the east.

An abrupt facies change is typical of the Asselian deposits developed along the eastern side of the East European Platform and in the Cis-Uralian Trough. From west to east, three narrow zones each marked by a peculiar lithology of rocks can be recognized. The western zone extends along the eastern side of the platform and is dominated by limestone. To the east, almost throughout the entire western margin of the Cis-Uralian Trough, a zone of thick carbonate deposits composed of biogenic limestone, locally forming reef structures, is found. Farther east, terrigenous rocks appear in the section; on the eastern slope of the Cis-Uralian Trough, they completely replace carbonate deposits. Terrigenous strata are composed of siltstone and sandstone or gravelstone and conglomerate in some western areas. Asselian terrigenous deposits were also recorded in the north-eastern and eastern Caspian Depression and the Urals, in the Aktyubinsk area (Zamarenov 1970, Dalyan and Posadskaya 1972).

Hence, the distribution of deposits differing in composition suggests that during the Asselian age, the East European Basin was landlocked to the west and east. Salt accumulated in an embayment cutting deep inland and situated within the Dnieper-Donets Depression. The volume of rock salt which accumulated there can be estimated from the following data: the areal extent of rock salt of the Nikitovo Formation exceeds 20,000 km^2; the average thickness of the formation equals approximately 200 m; and salt content is 40%. Thus, the volume of rock salt of the Nikitovo Formation will run as high as 1.6×10^3 km³. The areal distribution, the average thickness, and salt content of the Slavyansk Formation may be considered at least 30,000 km², 200 m, and 50%, respectively. The volume of rock salt in the Slavyansk Formation would equal 3.0×10^3 km³. Hence, in the Dnieper-Donets Depression, Asselian rock salt may exceed 4.6×10^3 km³.

The distribution pattern of the Sakmarian (Lower Permian) deposits in the East European Basin is roughly the same (Fig. 103). Two Sakmarian terranes of salt accumulation can be outlined. Besides the Dnieper-Donets Depression, rock salt is also known from the Severnaya Dvina-Sukhona Rivers Interfluve in the contiguous areas of the Mezen Depression, Moscow Syneclise, and Volga-Ural Anteclise.

Salt deposits of a probable Sakmarian age compose the Kramatorsk Formation in the Dnieper-Donets Basin. According to the data of Korenevsky and co-workers (1968), the formation consists chiefly of rock salt (85% of the total volume). Anhydrite beds account for 5–6% of the volume; their thickness, usually does not exceed 1 m; and only four beds reach 3–7 m. Variegated anhydrite contains abundant halite crystals. Interbeds of potash salts, namely sylvinite, carnallite, carnallite-halite, sylvite-carnallite, and carnallite-sylvite rocks occur in the upper part of the formation. They form two horizons: the lower, Privoleno sylvinite horizon; and the upper, Chasov-Yar sylvinite-carnallite horizon. Six lithological units, each starting with terrigenous and anhydrite interbeds and ending chiefly with rock salt, can be distinguished in the formation (Bobrov and Korenevsky 1965).

Horizon T_1 known as the Belbasov Motley Bed occurs ubiquitously at the base of the Kramatorsk Formation and at the base of the first unit. It consists of fulvous, saline siltstone and sandstone with halite inclusions and rock salt interbeds up to 3 m thick. An anhydrite interbed and locally pelitomorphic limestone called "waffle" are discernible at the top. The horizon ranges in thickness from 8–40 m. The upper part of the lower unit is represented mainly by white and transparent rock salt including four beds of contaminated halite; the lower bed is fulvous and composes an intervening horizon T_1^1. Phenocrysts of polyhalite and laminae of kieserite-halite, carnallite-kieserite, and carnallite are discernible in places. The thickness of the salt bed and of the entire first unit is estimated to be 60–130 m and 80–150 m, respectively.

The second unit starts with the anhydrite horizon T_2 and ends with a bed of white and transparent rock salt 50–70 m thick. Its thickness ranges from 50–80 m. In some areas of the north-western Dnieper-Donets Depression, sylvinite interbeds occur in rock salt. Siltstone-anhydrite marker horizon T_3 (3–7 m) lies at the base of the third lithological unit; its upper part is represented by a thick rock salt sequence (100–200 m) with pink and fulvous rock salt interbeds which form intervening markers labelled, in ascending order, T_3^1, T_3^2, T_3^3, T_3^4, and T_3^5. A bed of variegated sylvinite (t_3^1) up to 3.35 m thick can be distinguished locally between beds T_3^3 and T_3^4. The thickness of the third unit ranges from 100–120 m.

Marker horizon T_4 represented by cellular anhydrite 2.2–3.8 m thick forms the base of the fourth lithological unit. Up the section, rock salt with anhydrite interbed T_4^1 overlain by silvite-halite (7.5 m) and variegated sylvinite (5.5 m) beds can be seen. The halite of the fourth unit and of the entire bulk of sediments is estimated to be 25–60 m and 30–65 m thick, respectively. Potash-bearing parts of the third and fourth units are grouped into the Privoleno sylvinite horizon up to 100 m thick.

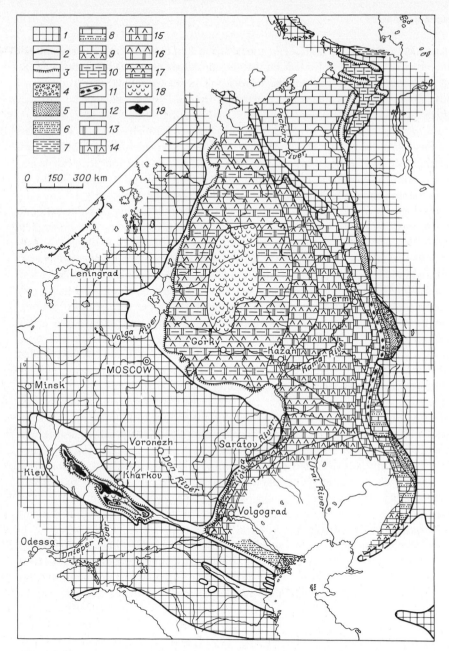

Fig. 103. Lithologic map of Sakmarian deposits in East European Basin. Compiled by the author together with Merzlyakov. *1* Permian absent, *2* limit of Permian deposits, *3* limit of Sakmarian deposits. Depositional limits of: *4* conglomerate, gravelstone, and sandstone, *5* mainly sandstone, *6* mudstone, siltstone, and sandstone, *7* mainly mudstone, shale, and siltstone, *8* sandstone and mudstone interbedded with carbonate rocks, *9* limestone, mudstone, and anhydrite, *10* argillaceous limestone, *11* carbonate reef structures, *12* mainly limestone, *13* limestone and dolomite, *14* limestone and dolomite with inclusions and interbeds of anhydrite (gypsum), *15* dolomite with interbeds and inclusions of anhydrite (gypsum), *16* dolomite interbedded with anhydrite, *17* anhydrite (gypsum) interbedded with dolomite and minor mudstone, *18* salt, *19* potash

The fifth unit starts with the Torez variegated horizon T_5 composed of fulvous and buff, saline siltstone and shale interbedded with rock salt (20–40 m). The upper, larger part of the section, about 75 m, contains rock salt impregnated by sylvite, carnallite, langbeinite, and eight potash salt beds composed of sylvinite, carnallite, carnallite-halite, sylvite-carnallite, and carnallite-sylvinite rocks with halite-langbeinite and polyhalite seams. These potash rocks are assigned to the Chasov-Yar sylvinite-carnallite horizon, which is about 60 m thick. Two or three anhydrite-polyhalite interbeds occur in the upper part of the section. The thickness of the unit locally exceeds 100 m. The base of the upper sixth lithological unit contains the anhydrite horizon T_6 (1.4–2.0 m) overlain by white rock salt interbedded with fulvous, saline siltstone, up to 20 m thick.

The total thickness of the Kramatorsk Formation reaches 600 m and 400 m in the Bakhmut and Kalmius-Torez Troughs, respectively; it decreases to 200 m and 100–150 m in the north-western Dnieper-Donets Depression. Near the flanks, the upper horizons of the formation recede from the section and its thickness gradually decreases and tapers out completely. Simultaneously, the formation varies in composition. Terrigenous rocks become dominant. The minimal volume of rock salt in the Kramatorsk Formation is estimated at 1×10^3 km^3 with regard to its areal extent of 10,000 km^2 and average thickness of 100 m.

In the Dvina-Sukhona Interfluve, salt deposits compose the Verkhnekuloi Formation; they were penetrated by a number of deep wells (Shangal, Solvychegodsk, Koryazhema, and other wells). The Shangal well penetrated five rock salt beds 0.1–3.7 m thick. In the central part of the Sukhona River Basin, the salt series consists of 5–7 rock salt beds 9–10 m thick; the total thickness amounts to 15–35 m; salt saturation accounts for 35–40%. The Solvychegodsk and Koryazhema wells penetrated three rock salt interbeds each, with thicknesses varying from 0.7–2.8 m in the former and amounting, respectively, to 1.4 m, 4.6 m, and 9.6 m in the latter. Salt saturation accounts for about 38% (Ivanov and Levitsky 1960).

The Sakmarian salt sequence in the Belbazh rock salt deposit, Gorky District, has received the most attention (Tumanov 1981). It contains six rock salt beds (P_0, P_I, P_{II}, P_{III}, P_{IV}, and P_V) separated by anhydrite beds labeled P_{0-1}, P_{1-2}, P_{2-3}, P_{3-4}, and P_{4-5}. In the most complete sections, salt content reaches 32–48%. Rock salt bed P_0 is 1.65 m thick. Bed P_I up to 9.8 m thick, is composed of clayey rock salt with minor anhydrite. Bed P_{II}, 9–11 m thick, is represented by clean white rock salt. Rock salt beds P_{III} and P_{IV} are 13 m and 2.5 m thick, respectively. They are separated by an anhydrite band which thins out to the west. This salt sequence is underlain by anhydrite and dolomite units 25–36 m thick. The salt deposits are overlain by a 50–70 m anhydrite sequence composed of interbedded anhydrite, gypsum, and dolomite.

Ivanov (Ivanov and Levitsky 1960) reported that salt deposits of the Verkhnekuloi Formation may occupy an area 300 km long and 200 m wide or even larger. At the present time, they are known to extend south to north for almost 600 km in a 150–250 km wide belt. The areal extent of rock salt exceeds 10,000 km^2. The volume of rock salt may be as high as 1×10^2 km^3, assuming an average thickness of 10 m over the entire area.

A considerable amount of sulfate deposits have been accumulated outside the zones of salt accumulation of the East European Basin. They are widespread in the

Mezen Depression, Moscow Syneclise, Volga-Ural Anteclise, and Caspian Depression. In the Moscow Syneclise, anhydrite occurs throughout the Sakmarian section along with dolomite, minor marl, and clay (Gorbatkina and Strok 1971). Anhydrite also dominates the north-eastern Volga-Ural Anteclise (Makarova 1957, Tikhvinsky 1967, 1970b). Eastwards, the proportion and thickness of anhydrite and gypsum decreases; the section is first dominated by dolomite, then limestone appears. Carbonate strata with interbeds and inclusions of anhydrite and gypsum are widely developed in the central Volga-Ural Anteclise (Porfiriev 1963, Tikhvinsky 1967, 1970b), and to the south, on the Zhiguli-Pugachev and Orenburg Arches (Rybakov 1960, Svishchev 1961, Tikhvinsky 1967).

Porfiriev (1963) reported that to the east, the sulfate-carbonate deposits are replaced by a zone of limestone and dolomite, and then by a narrow zone of limestone developed at the western extremity of the Cis-Uralian Trough. Farther east, a zone of biohermal carbonate deposits with numerous reef complexes was recognized (Nalivkin 1950, Trofimuk 1950, Porfiriev 1963). East of this zone, terrigenous rocks appear; they dominate the central and eastern Cis-Uralian Trough.

Sakmarian sulfate-carbonate deposits were also reported from the western and northern flanks of the Caspian Depression (Urusov et al. 1962, 1967, Movshovich 1970, Korenevsky 1973, Derevyagin et al. 1981). Carbonate formations occur in the north-eastern and eastern areas of the depression, while terrigenous deposits lie along the margins (Ruzhentsev 1948, Zamarenov 1970, Dalyan and Posadskaya 1972). A sulfate-carbonate succession is developed to the south-east (Dalyan and Posadskaya 1972, and others). Sakmarian terrigenous formations are believed to occupy the southwestern part of the Caspian Depression (Lapkin 1961a, b, and others). The composition of the Sakmarian deposits in the central, lower parts of the depression are as yet unknown, although limestone seems to occur there (Forsh and Goryacheva 1979).

The net volume of Sakmarian rock salt in the East European Basin is estimated at 1.1×10^3 km^3.

The Artinskian deposits in the East European Basin embrace mainly sulfate-carbonate and carbonate rocks (Fig. 104). Evaporites are restricted chiefly to the eastern areas: in the Moscow Syneclise, in the central, northern, and southern Volga-Ural Anteclise, and in the north-western Caspian Depression. The latter is characterized by the presence of rock salt in the upper part of the section (Latskova 1961, 1967, Shafiro 1972). Salt deposits have not been studied in detail there. Rock salt bands, which are intercalated with anhydrite and dolomite, are estimated to be a few meters thick. The distribution and volume of the Artinskian rock salt is still unknown.

The spatial distribution of the Artinskian lithological units is characterized by the same west-east change of deposits, which is also typical of the Sakmarian and Asselian deposits. It is in this direction that the sulfate-dolomite strata grade into dolomite-limestone deposits with interbeds and phenocrysts of anhydrite and gypsum. Farther away there is a limestone zone bounded to the east by a zone of argillaceous limestone with reef structures. Terrigenous rocks including conglomerate are widely developed in the central and eastern Cis-Uralian Trough. They are also known from the eastern and south-western Caspian Depression.

In the East European Basin, the thickest salt strata with potash salt horizons were recognized in the Kungurian deposits (Fig. 105).

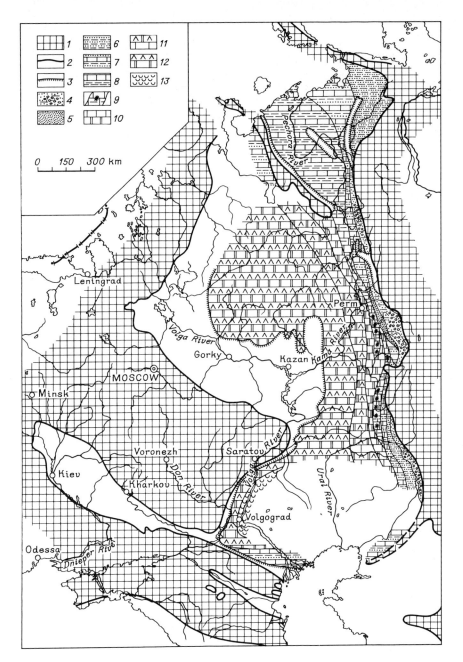

Fig. 104. Lithologic map of Artinskian deposits in East European Basin. Compiled by the author together with Merzlyakov. *1* Permian absent, *2* limit of Permian deposits, *3* limit of Artinskian deposits. Depositional limits of: *4* conglomerate, sandstone interbedded with mudstone, *5* mainly sandstone, *6* mudstone, siltstone, and sandstone, *7* sandstone and mudstone interbedded with carbonate rocks, *8* limestone, clay, and mudstone, *9* limestone, argillaceous limestone, and carbonate reef structures, *10* limestone and dolomite, *11* limestone and dolomite with inclusions and interbeds of anhydrite (gypsum), *12* dolomite with interbeds and inclusions of anhydrite (gypsum), *13* rock salt

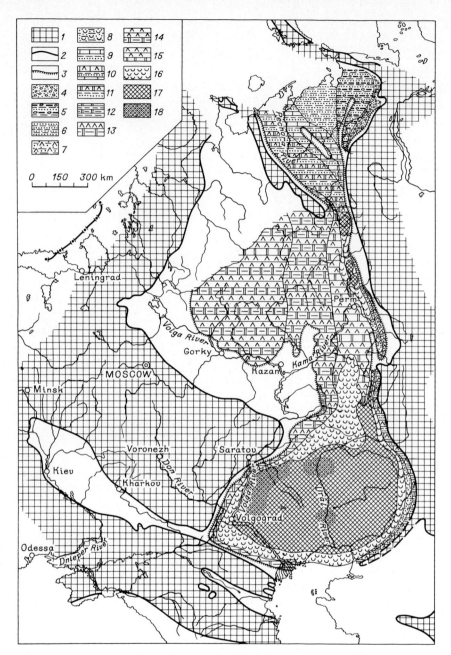

Fig. 105. Lithologic map of Kungurian deposits in East European Basin. Compiled by the author together with Merzlyakov. *1* Permian absent, *2* limit of Permian deposits, *3* limit of Kungurian deposits. Depositional limits of: *4* terrigenous coarse-grained rocks, minor siltstone, and mudstone, *5* terrigenous coaly deposits, *6* mudstone, siltstone, and sandstone, *7* the same as *6*, with inclusions and interbeds of gypsum (anhydrite), *8* terrigenous salt deposits, *9* sandstone and mudstone interbedded with limestone, *10* sandstone, siltstone, and mudstone with limestone interbeds and occasional inclusions of anhydrite (gypsum), *11* sandstone, siltstone, and mudstone interbedded with dolomite and anhydrite (gypsum), *12* clayey dolomite and dolomitic marl, *13* the same as *12*, interbedded with anhydrite, *14* alternation of anhydrite, clayey dolomite, and marl, *15* dolomite and anhydrite, *16* salt, *17* potash, *18* bischofite rocks

The Kungurian type sections are situated on the Ufa Plateau and in the Sylva-Iren parts of Kama areas at Perm. The Filippovo Formation composed of dolomite and limestone is assigned to the lower Kungurian. The upper Kungurian embraces the Iren Formation consisting of seven members, in ascending order:

1. Ledyanopeshchera Member (30–34 m) — Gypsum and anhydrite with dolomite bands in the lower part.
2. Nevolino Member (6–8 m) — Dolomite and dolomitized limestone.
3. Shalashnino Member (12–35 m) — Gypsum and anhydrite with inclusions and seams of dolomite and clay.
4. Elkino Member (2–8.5 m) — Dolomite.
5. Demidkovo Member (18–35 m) — Gypsum and anhydrite with dolomite veinlets and laminae.
6. Tyuya Member (4–14 m) — Dolomite and limestone.
7. Lunezh Member (46–74 m) — Gypsum and anhydrite with seams and phenocrysts of dolomite and clay.

Correlation with this section allowed the distinction of the Filippovo and Iren Formations in the remaining East European Basin, not only in the contiguous parts of the Cis-Uralian Trough and Volga-Ural Anteclise, but farther west in the Moscow Syneclise, and to the south, in the Caspian Depression. At present, it is widely known that in all the above areas, salt series are confined only to the Filippovo and Iren horizons (Kungurian). This opinion is confirmed chiefly by the fact that in the Chusovaya and Solikamsk Depression, nearest to the type section, rock and potash salt strata are associated with the Filippovo and in particular, with the Iren Formations. In the Solikamsk Depression, this problem was resolved unambiguously.

In the Cis-Uralian area at Perm, the Filippovo horizon is marked by the carbonate, carbonate-sulfate, carbonate-sulfate-clayey, and conglomerate-sandy types of the section (Graifer et al. 1962, Ivanov 1965). The section of the carbonate type is typical of the eastern margins of the platform. The section of the carbonate-sulfate type occurs farther east and covers most of the Solikamsk Depression. Here, the Filippovo Formation is divided into four members: (1) the lower anhydrite, 10–45 m; (2) the lower carbonate, 5–37 m; (3) the upper anhydrite, 8–38 m; and (4) the upper carbonate, 5–24 m. The total thickness of the formation is estimated at 165–190 m. The carbonate-sulfate-clayey and the conglomerate-sandy types of the section are developed in the eastern Solikamsk Depression. Equivalents of the Filippovo Formation are assigned to the Lek Formation (Graifer et al. 1962), which contains sandstone and shale with anhydrite interbeds at the base; and sandstone, shale, and marls interbedded with dolomite and limestone at the top.

Graifer assigned the salt series of the Solikamsk Depression to the Berezniki Formation. It can be placed entirely at the level of the Iren horizon. Ivanov recognized the following members, in ascending order: (1) the Clay-Carbonate Sulfate-Salt Member; (2) the Underlying Salt Member; (3) the Potash Member; (4) the Overlying Salt Member; and (5) the Upper Transition Member. This section is correlative with the type sections of the Iren horizon, namely, the Clay-Carbonate Sulfate-Salt Member is considered to be an equivalent of the six lower members of the Iren Formation, from the Ledyanopeshchera Member to the Tyuya Member inclusive. It is composed of marl, shale, dolomite, minor limestone, anhydrite, rock salt, siltstone, and sandstone.

Rock salt also seems to occur at the level of the Shalashnino and Demidkovo Members. The thickness of the Clay-Carbonate Sulfate-Salt Member ranges from 150–300 m.

The remaining three overlying members of the salt series in the Solikamsk Depression lie at the level of the seventh, Lunezh Member of the Iren Formation.

The Underlying Salt Member is subdivided into three horizons (Ivanov and Voronova 1975). The lower horizon is composed of units, beds, and interbeds of rock salt and terrigenous-chemogenic rocks represented by clay(marl)-anhydrite-dolomite, clay(marl)-gypsum-anhydrite, and dolomite-anhydrite varieties, minor siltstone, and sandstone. The number of units and beds of terrigenous-chemogenic rocks ranges from 1–5 m and vary from 0.5–43 m in thickness. Rock salt beds are 1.8–38.5 m thick. The total thickness of the horizon is 35–130 m. The middle horizon consists mainly of rock salt with no noticeable interbeds of terrigenous-chemogenic rocks. The thickness of the horizon varies over a range of 80–130 m to 420–440 m. The upper horizon contains rock salt with some sylvite inclusions at the top; its thickness ranges from 8–66 m. The net thickness of the Underlying Salt Member is estimated to be 50 m and less in zones where it wedges out and 500–515 m in the central part of the depression, averaging 300–350 m.

The Potash Member of the Solikamsk Depression has been studied most intensely (Ivanov 1932, 1965, Fiveg 1948, 1955a, b, 1959, 1973d, Dubinina 1954, Vakhrameeva 1956, Kopnin 1965, Fiveg and Banera 1968, Ivanov and Voronova 1975, Golubev 1977, Sapegin 1977, Kopnin and Kukharchuk 1980, Kopnin and Maloshtanova 1980, Golubev and Brovchenko 1981, Kopnin and Korotaev 1981, Sapegin and Yanin 1981, Tretyakov and Sapegin 1981). The Potash Member is represented by alternate potash- and rock salts. In general, it is subdivided into two horizons – the lower sylvinite and the upper sylvinite-carnallite horizons. The thickness of the former ranges from 7–8 to 30–40 m, with an average of 20–22 m. Four sylvinite beds, separated by rock salt beds, were recognized in the lower horizon, in ascending order: Krasnyi[6] III, Krasnyi II, Krasnyi I, and Bed A. The sylvinite-carnallite horizon embraces nine potash beds labeled B, V, G, D, E, Zh, Z, I, and K, which are interbedded with rock salt beds. The thickness of a horizon varies from 20–115 m, 60 m on the average.

The following description of potash beds is based on data published by Ivanov and Voronova (1975). Krasnyi III sylvinite can be divided into three beds: Kr III^a, Kr III^b, and Kr III^V. The lower Kr III^V bed consists of two, red sylvinite bands 2–6 m thick with a 7 cm thick rock salt band separating them. Bed Kr III^b, 1.35 m thick, composed of two, buff-pink sylvinite bands, with a light gray rock salt band with traces of sylvite between them. Bed Kr III^a is composed of red, fine-grained sylvinite and gray, yellowish-gray, fine-, medium-, and micrograined rock salt with thin (1–4 mm to 1–1.5 cm) clay-anhydrite and clayey interbeds. The thickness of sylvinite and rock salt interbeds is 1–7 cm and that of the bed is 0.7–1.25 m.

Krasnyi II sylvinite contains seven bands; the odd bands are rich in sylvite, while even bands are dominated by halite. The thickness is 2.2–6.0 m. Band 7, occurring at the base of Krasnyi II sylvinite, is represented by alternate interbeds of red and

---

6 Krasnyi is the Russian translation of red

pinkish-red sylvinite and yellowish-gray to gray rock salt with thin (1–3 mm) clayey and clay-anhydrite laminae. Band 6 consists of gray to yellowish-pink, and medium- to micrograined rock salt with clayey to clayey-anhydrite interbeds 1–3 mm to 1–1.5 cm thick and rare interbeds (1–3 cm) of red sylvinite. Band 5 consists of alternate red and pink sylvinite interbeds, 0.5–8 cm thick, and light gray and yellowish-pink rock salt of the same thickness. Band 4 is made up of dark gray to gray, medium-grained, locally orange-pink and pink-gray, fine- and micrograined rock salt with clay interbeds (4 cm thick) with traces of sylvite in the lower part. Band 3 is marked by alternate red and pink sylvinite with gray to yellowish-gray rock salt. The thickness of sylvinite interbeds and rock salt is respectively, 1–10 cm and 2–6 cm. Band 2 is dominated by gray, pinkish-gray, and orange-pink rock salt with sylvite inclusions and patches and carbonate-clayey interbeds. Band 1 consists of alternating red sylvinite interbeds 1–10 cm thick and gray rock salt 0.5–8 cm thick.

Krasnyi I sylvinite, 1.15 m thick, is marked by the alternation of red, pink, white, and yellowish-pink sylvinite with gray, yellowish-gray and light-blue to blue rock salt containing thin carbonate-clayey interbeds. Sylvinite A is often subdivided into Bed A′, and Bedded sylvinite A. Bed A′ is rather thin (0.25 m) and is marked by the intercalation of red and orange-red sylvinite with buff-gray and blue-gray rock salt. In general, there are three sylvinite interbeds about 3–5 cm thick and two halite interbeds 1–3 cm thick. Bedded sylvinite A is characterized by a frequent alternation of sylvinite and rock salt. Sylvinite interbeds are dark red, pink, pale pink, and white, while those of rock salt are gray, yellowish-pink, bluish-gray, and blue. In all there are 40 beds, each consisting of the following bands: (1) saliferous clay, 1–3 mm thick; (2) halite, 3.5 cm thick; and (3) sylvinite, 0.5–6 cm thick. The thickness of Bedded sylvinite A attains 1.2–1.3 m. Potash B has a complex composition. In general, it is motley sylvinite, however, in some areas the bed consists of carnallite rock intercalated with rock salt. The thickness varies from 1.4–2.8 m. Motley sylvinite is usually massive and contains milky-white sylvite grains with buff-red rims, as well as grains of colorless, pink, and blue halite. The carnallite rock is orange-red, buff-red, pinkish-yellow, and medium- to coarse-grained.

Potash V is marked by the alternation of carnallite rock of various shades of red and light gray to yellowish-gray halite with interbeds of saliferous clay. In some areas, the bed consists of motley sylvinite and gray to orange-yellow rock salt. The thickness varies from 2.8–6.6 m. Potash G is mostly composed of carnallite rock intercalated with rock salt. Motley sylvinite is very scarce. Carnallite rock is sealing wax red, orange, orange-pink, yellow, and lemon yellow. The bed is 2.5–7.7 m thick. Potash D is the alternation of sealing wax red and orange-red carnallite rock, gray to pinkish-gray rock salt locally with interbeds and lenses of motley sylvinite. The thickness varies from 2.75–8.1 m. Potash E is the alternation of buff-red and orange-red carnallite rock with gray and white rock salt. In places, the bed consists of motley sylvinite. When the bed is carnallite or sylvinite in composition, its average thickness is 7.4 m and 3.1 m, respectively. Potash Zh is composed of carnallite rock with rock salt interbeds and laminae of saliferous clay. The thickness does not exceed 1 m. Potash Z consists either of carnallite rock or motley sylvinite. Its average thickness is 0.5–0.75 m. Over most of the Solikamsk Depression, Potash I consists of two carnallite bands, one is orange or brown-red and the other is lemon-yellow; they are

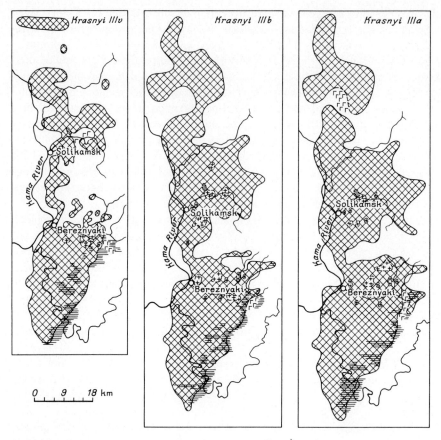

Fig. 106. Lithofacies maps of potash beds Krasnyi III^v, III^b, III^a, II, and A. After Ivanov and Voronova (1975). *1* red sylvinite, *2* red sylvinite II, represented by salt in lower beds, *3* subterranean erosion of potash salt. Salt rocks with the following content of carbonate-clayey material: *4* increased (5–10%), *5* high (10–20% and over), *6* intrastratal salt, *7* rock salt replacing potash salt in marginal zones, *8* erosional edge of the bed (*a* known, *b* probable)

separated by light gray rock salt. Locally, carnallite rocks are replaced by motley sylvinite composed of milky-white sylvite and blue halite grains. The average thickness of the bed is 1.65 m. Potash K is composed mainly of alternate carnallite rock, rock salt, and saliferous clay interbeds about 1 m thick.

The composition and facies change pattern of the potash beds in the Solikamsk Depression are shown in Figs. 106 and 107.

The Overlying Salt Member is composed of light gray, pink, yellow, and fine- to micro-grained halite with clayey and clay-anhydrite interbeds (up to 6–10 cm). The thickness ranges from 15–36 m. The Upper Transition Member is represented by alternate rock salt beds and marl. It consists of two parts, namely, the lower rock salt and the upper marl and rock salt. The Upper Transition Member is 40–55 m thick.

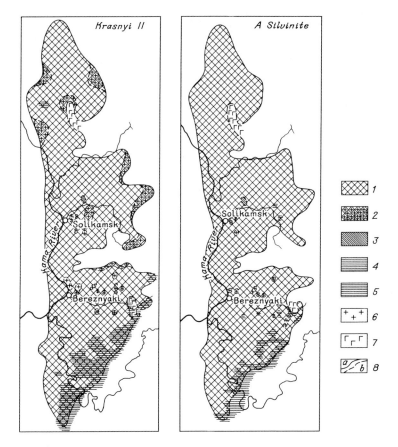

Fig. 106 (continued)

For a more detailed description of the Berezniki Formation of the Solikamsk Depression, the reader is referred to the publications of Golubev (1977), Kudryashov and Myagkov (1977), Kopnin and Kukharchuk (1980), Kopnin and Maloshtanova (1980), Golubev and Brovchenko (1981), Kopnin and Korotaev (1981).

The areal extent of potash salt in the Solikamsk Depression is 4500 km^2 (Fiveg and Banera 1968), while according to Ivanov it does not exceed 3100 km^2. The Berezniki Formation is 800–900 m and 400–500 m thick, respectively, in the subsided parts and on the flanks. Rock salt in the Kungurian deposits of the Solikamsk Depression amounts to approximately 4.5×10^3 km^3, when an areal extent of 9000 km^2 and an average thickness of 500 m is assumed (Fiveg and Banera 1968). According to Ivanov, the volume of rock salt in the Solikamsk Depression equals 3.7×10^3 km^3.

In the Upper Pechora Depression, the Kungurian salt deposits are divided into the Filippovo and Iren Formations. Chochia (1955) inferred a possible occurrence of some Filippovo units in the Kolva Vishera Region and adjacent northern areas along the upper Pechora River. These units were also recognized in the Upper Pechora Depression. The lower anhydrite unit is 16–41 m thick; the lower carbonate unit,

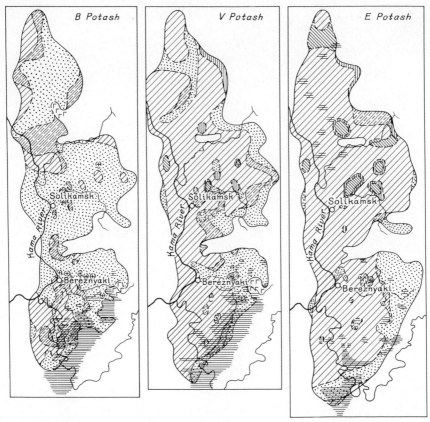

Fig. 107. Lithofacies maps of potash beds B, V, E, K, Z. After Ivanov and Voronova (1975). *1* carnallite rock, *2* motley sylvinite, *3* alternation of carnallite rocks and motley sylvinite, *4* sylvinite cap, *5* subterranean erosion of potash salt bed. Salt rocks with the following content of carbonate-clayey material: *6* increased (5–10%), *7* high (10–20% and over), *8* intrastratal rock salt, *9* rock salt replacing potash salts in marginal zone, *10* erosional edge of the bed (*a* known, *b* probable)

composed of dolomitized limestone and minor dolomite, attains 10–57.5 m; the upper anhydrite and upper carbonate units are, respectively, 6–97 m and 3.5–29 m thick (Ivanov et al. 1959). The total thickness of the Filippovo Formation is 150 m.

Ivanov and co-workers (1959) provided data for the Iren Formation in the western Upper Pechora Depression, where it can be divided into seven units. Units 1, 3, and 5, composed dominantly of anhydrite, are respectively, 23–130 m, 27–247 m, and 16–120 m thick. Their thickness increases eastward. Units 2 and 4 are represented by dolomite; the thickness of unit 2 varies from 0–21 m, while that of unit 4 is 3–20 m. The lower part of units 6 and 7 is composed of terrigenous rocks, while their upper part consists of anhydrite and gypsum. To the east, salt-bearing deposits up to 400–435 m thick appear in the upper part of unit 7 (Ivanov and Voronova 1968).

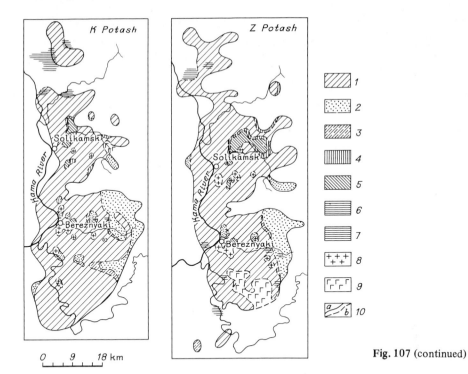

Fig. 107 (continued)

In the Upper Pechora Depression, the salt deposits are divided into three members: (1) the Underlying Salt Member, 125–260 m, locally 370 m; (2) Potash Member, 9–55 m; and (3) Overlying Salt Member, 4–35 m (Ivanov 1965). The Potash Member is represented by carnallite rocks and sylvinite intercalated with rock salt. In the Upper Pechora Depression, salt deposits occupy an area of 6600 km^2 (Fiveg and Banera 1968). Their average thickness is apparently 150 m. In this case, the volume of rock salt will be approximately 1×10^3 km^3.

In the Chusovaya and Yuryuzan-Sylva Depressions, rock salt was reported from the Ledyanopeshchera and Shalashnino Members of the Iren horizon (Ivanov 1965). Rock salt of the Ledyanopeshchera Member is 40–80 m thick. The Shalashnino Member contains two rock salt beds: the upper bed 45 m and the lower bed 20 m separated by a 20 m anhydrite unit. Southward, the thickness of rock salt decreases to 2–9 m. Rock salt has not been found in the upper Iren horizon of the Chusovaya Depression. Since the Chusovaya and Yuryuzan-Sylva Depressions are not well-known, an approximate estimate for the rock salt cannot be given.

The Filippovo and Iren horizons are distinguished in the Volga-Ural Anteclise area (Blizeev 1970, 1971a, b, Tikhvinsky 1971, Tikhvinsky and Blizeev 1972). The Filippovo horizon can be divided into the Lower and Upper Iren subhorizons. The former comprises the lower six members of the classical Kungurian section from the Ledyanopeshchera to the Tyuya Member consists of the lower, middle, and upper members. The Iren horizon inclusive, while the latter corresponds to the Lunezh Member and is subdivided into the Lower and Upper Lunezh Beds. When traced from north to south and south-east, rock salt bands and beds begin appearing in all horizons of the Kungurian as they approach the Caspian Depression and the Cis-Uralian Trough. Halite interbeds first appear in the Upper Lunezh Beds in some northern areas of Bashkiria; they then occur in the Lower Lunezh Beds. Farther south, rock salt bands are first observed in the Nevolino and Elino Members and then in the Tyuya Member. The Lower Iren subhorizon also becomes anhydrite- and salt-bearing. Locally, rock salt occurs in the Filippovo horizon.

The lower member of the Filippovo horizon contains up to 16 m of dolomite on the south-eastern slope of the Tatar Arch. South-eastward, anhydrite and gypsum inclusions and interbeds appear in the section. In southern Bashkiria and the Orenburg District, anhydrite interbeds attain 4–6 m and the thickness of the member reaches 18–20 m. In the southern Cis-Uralian Trough and south-western Caspian Depression, rock salt units up to 16 m appear in the section; the thickness of the member increases to 55–60 m.

On the south-eastern slope of the Volga-Ural Anteclise, in eastern Tataria, the Kuibyshev, and north-eastern Orenburg Districts, the lower part of the middle member is composed of clayey dolomite with inclusions and interbeds of gypsum, while its upper part is composed of anhydrite with dolomite laminae. The thickness of the member ranges from 0–25 m. On the south-western slope of the Ufa Plateau, the member is represented by dolomite with limestone interbeds; anhydrite and gypsum interbeds appear in some western and north-western sections. The thickness of the member is 29–45 m. On the south-eastern monoclinal side of the Russian Platform (Orenburg District, Bashkiria, south-eastern Kuibyshev District), the middle member is dolomitic in composition, its thickness increasing southward and south-eastward from 27–30 m to 76–80 m. In the western Cis-Uralian Trough, rock salt bands begin to appear as they grade into thick beds and units. In the central areas of the trough, in the Sim-Inzer, Sterlitamak-Meleuz, and Sakmara Depressions, the member is represented by alternate rock salt and anhydrite. The salt saturation accounts for 70–85%. The thickness increases to 200–240 m. A potash salt horizon up to 10–20 m thick was distinguished in the upper part of the middle member in the most subsided zone of the above depressions. It is composed of sylvinite and carnallite rocks; to the south, it contains polyhalite rocks.

On the north-eastern slope of the Tatar Arch and Birsk Saddle, the upper member is composed of dolomite and clayey dolomite. On the south-eastern slopes of the Tatar and Zhiguli-Pugachev Arches, the proportion of clayey dolomite increases. The member is 12–14 m thick. To the east and south-east, anhydrite interbeds appear in the middle part of the member and its thickness is 16 m. Farther south-east, anhydrite interbeds are 13–18 m thick; the upper member totals 30–34 m. In the southern Cis-Uralian Trough on the western slope, 25–34 m thick rock salt beds are intercalated with anhydrite and dolomite. Here, the thickness of the member is 63–67 m. In the southern and central parts of the trough, rock salt attains 70–80 m and the thickness of the entire member increases to 85–90 m and in places to 116–200 m. Thus, salt deposits of the Filippovo horizon occur mainly in the most subsided parts of the Cis-Uralian Trough and in the northern Caspian Depression. In the former region, rock salt is observed in all three members of the Filippovo horizon and its total thickness attains 250—300 m. On the northern flank of the Caspian Depression, the middle member becomes salt-bearing.

Over most of the Volga-Ural Anteclise, six members are easily distinguished within the Lower Iren subhorizon: the Ledyanopeshchera Anhydrite Member (2.5—10 m); Nevolino Dolomite Member with rare interbeds and inclusions of anhydrite (5—9 m); Shalashnino Anhydrite Member (0.5—1.5 m to 6—7 m, locally 16 m); Elkino Dolomite Member with anhydrite inclusions (5—8 m); Demidkovo Dolomite Member with dolomite interbeds (22—34 m); and Tyuya Dolomite Member with rare anhydrite interbeds (5—6 m to 10—12 m). The total thickness of the section varies from 45—60 m to the north to 70 m to the south. Sulfate rocks account for 50—70% of the section. The Lower Iren subhorizon retains this composition up to the south-eastern sides of the Russian Platform. Within the Buzuluk Depression and Cis-Uralian Trough, the rocks become mainly salt-bearing in composition. For instance, rock salt dominates the southern part of the Cis-Uralian Trough (south of the town of Sterlitamak), Caspian Depression, and it also occurs in a narrow bent zone running north-west of the syneclise along the eastern and northern flanks of the Buzuluk Depression. This subhorizon is represented by alternate rock salt and anhydrite units. Its thickness ranges from 70 to 118—120 m, locally reaching 180—220 m. The lower Ledyanopeshchera Anhydrite Member and the uppermost Tyuya Member, which are composed of anhydrite and clayey dolomite interbeds totaling 20 m, are easily discernible in the section.

On the western side of the Ufa Plateau, the Lower Lunezh Beds are composed of carbonate rocks, dominantly limestone interbedded with dolomite. Anhydrite starts to appear in the section when traced westward and south-westward. The deposits are 51—54 m thick. Anhydrite with interbeds of clayey anhydrite and clay up to 50—55 m thick are developed in western and central Bashkiria. Anhydrite with one or two (2—4 m) dolomite interbeds in the lower part of the section occurs on the south-eastern slope of the Tatar Arch and within the Zhiguli-Pugachev Arch. Rock salt interbeds appear in the upper part of the section. In the Lower Lunezh Beds, rock salt is present on the south-eastern side of the platform, in the northern Caspian Depression, and in the southern Belaya Trough, Cis-Uralian Trough. Two lithological units — halite-anhydrite and anhydrite-halite — replacing each other along the strike are distinguished. The former is restricted to a transitional band between dolomite-anhydrite and anhydrite-halite terranes and consists of anhydrite with one or two rock salt beds up to 25 m thick. The thickness of the unit varies from 44—119 m, increasing south- and south-eastward. The salt saturation amounts to 30—40%. The second unit is dominated by rock salt with some anhydrite interbeds. Its salt saturation accounts for 70—80%. Potash salts, namely, sylvinite and carnallite rocks are observed in the upper part of the anhydrite-halite unit. This unit is developed within the Buzuluk Depression, in the southern Cis-Uralian Trough. Its thickness ranges from 120—220 m on the flanks of the Caspian Depression to 370 m in the central Cis-Uralian Trough.

The Upper Lunezh Beds, which are composed of gypsum and anhydrite with dolomite interbeds in the lower part and rock salt lenses in the upper part, are developed in the narrow band stretching along the eastern and south-eastern slopes of the Tatar Arch, as well as in the Birsk Saddle area. Their thickness is 53—60 m. South-east and eastward, the thickness of rock salt beds and units increases greatly. The Upper

Lunezh Beds represented by alternate rock salt units (40–130 m) and anhydrite (10–25 m) are known from the south-eastern side of the Russian Platform and from the Buzuluk, Sterlitamak-Meleuz, Sakmara (Cis-Uralian Trough), and northern Caspian Depressions. A general increase in thickness is marked from north to south, namely, 200 m at the latitude of the town of Ufa, 330 m in West Sterlibashevo field, 478 m in Sorochinsk field, and 1137 m in Chinarevo field. An increase in thickness from 217–1640 m in the same direction is also observed in the Cis-Uralian Trough. Both on the platform and in the Cis-Uralian Trough, the most complete sections contain 6–8 rock salt units; their number increases to 14 in the flank zone of the Caspian Depression, where they are intercalated with anhydrite units. Polyhalite rocks and sylvinites were found in seven salt units within the Cis-Uralian Trough.

The Kungurian salt sequence of the Caspian Depression was studied in more detail on its flanks, namely, in the Volga Monocline, along its northern limb, as well as in the north-eastern and eastern parts. In the center of the depression, where salt deposits were strongly deformed, they are known only from some salt structures, e.g., the Inder, Chelkar, and Elton Domes, which were specially drilled for potash salts or other mineral resources. At present however, the salt series was reflected only in local stratigraphic charts, which do not allow a reliable correlation.

The Kungurian section of the Volga Monocline is one of the standard sections in the Caspian Depression. Shafiro (1972) recognized 16 marker beds and units (see Fig. 101). Eight of them contain dolomite-anhydrite and anhydrite interbedded with rock salt (markers I, II, IIa, III, IV, V, VI, VII). The remaining markers (A, A_1, A_2, B, C, D, D_1, E) are composed of polyhalite-halite (A, A_1, A_2, B, D), bischofite or carnallite-bischofite (C, D_1), carnallite-halite or sylvite-halite (E) rocks. The Kungurian salt deposits of the Saratov-Volgograd area attain 1000 m and consists of seven rhythms, each starting with dolomite-anhydrite and terminating with salt rocks.

Rhythm 1. Marker I is composed of dolomite-anhydrite 15–20 m thick. Up the section, it grades into a halite unit with minor polyhalite – Marker A.

Rhythm 2. Marker II is a 10–15 m dolomite-anhydrite bed replaced on the flank by halite-anhydrite rocks. Marker A_1 is a 15–25 m polyhalite-halite unit. Marker IIa is a 1–2 m dolomite-anhydrite bed. Marker A_2 is a 20–40 m polyhalite-halite unit.

Rhythm 3. Marker III is a 40–50 m unit consisting of two dolomite-anhydrite beds (5–10 m and 15–20 m) separated by rock salt (5–20 m). Eastward, the latter is gradually replaced by anhydrite. In some sections of the flank zone, the lower anhydrite bed is replaced by strongly anhydritized rock salt. Marker B is a 25–50 m halite-polyhalite unit.

Rhythm 4. Marker IV is a 5–10 m dolomite-anhydrite bed with minor rock salt. Marker C is a 10–30 m bischofite bed with carnallite seams.

Rhythm 5. Marker V is a 10–15 m dolomite-anhydrite bed. Marker D is a 30–45 m polyhalite-halite unit. Marker D_1 is a 10–35 m carnallite-bischofite bed.

Rhythm 6. Marker VI is a 20–35 m dolomite-anhydrite bed with rock salt bands. Marker E is a 20 m unit of rock salt interbedded with carnallite or sylvinite.

Rhythm 7. Marker VII is a 10–20 m dolomite-anhydrite bed with rock salt bands.

Ermakov (1971), Ermakov and Grebennikov (1977), Grebennikov and Ermakov (1980) proposed a slightly different subdivision of the Kungurian salt deposits in the Volga Monocline. They distinguish 11 rhythmic units, grouped into six cycles. The first cycle encloses four rhythmic units, each beginning with sulfate-carbonate or sulfate rocks and crowned by rock salt, with minor polyhalite and rare sylvite and carnallite inclusions. The fourth, uppermost unit, contains potash salt interbeds with considerable thickness in places. The other cycles are correlated, respectively, with five upper rhythms according to the scheme of Shafiro. In fact, the schemes differ only in the subdivision of the lower salt-bearing deposits, whereas their upper parts are recognized in a similar way. The stratigraphic position of bischofite horizons is determined unambiguously. The major marker horizons are represented by the same anhydrite and dolomite-anhydrite beds. According to Banera and Gorbov (Gorbov 1973), the Kungurian salt sequence of the Volga Monocline can be divided into ten cycles, whereas Tikhvinsky (1974) believes that in the part of the section assigned to the Upper Iren subhorizon, like in the central and northern Caspian Depression, 13 cycles can be distinguished, grouped into the Ulagan, Elton, Chelkar, and Inderbor beds (Tikhvinsky 1976a). Derevyagin and co-workers (1981b) distinguished 11 rhythmic units in the section.

Recently, rhythmic units in the Volga Monocline area have been given individual names. Thus, Pisarenko, Makarov and Derevyagin (Derevyagin et al. 1981b) gave the following names to rhythmic units, in ascending order: Volgograd, Balyklei, Karpensk, Volga, Lugovo, Pogozhye, Antipovka, Pigarev, Dolina, Eruslan, and Overlying Anhydrite.

Occurring at the base of the Volgograd rhythmic unit is an up to 55 m thick dolomite and anhydrite bed containing large proportions of magnesite and polyhalite. The upper part of the rhythmic unit is composed of alternate dolomite, anhydrite, and polyhalite rocks with rock salt. Its thickness is 120–170 m. The Balyklei rhythmic units, 150–220 m thick, is dominated by rock salt; however, dolomite, dolomite-anhydrite, and anhydrite beds, intercalated with carnallite and carnallite-halite rocks up to 3–8 m thick, also occur. The lower Karpensk rhythmic unit is composed of anhydrite, ranging in thickness from 10–20 m to 25–35 m. Lying above are polyhalite-halite rocks with bischofite and bischofite-carnallite interbeds. Their thickness is 100–130 m, locally 215–350 m. The Volga rhythmic unit is characterized by a small thickness (20–55 m). In its lower part, clayey dolomite-anhydritic beds, containing white, rounded inclusions of borate minerals can be distinguished; this section is 10–30 m thick. Up the section, rock salt is interbedded with anhydrite, 20–45 m in thickness. The Lugovo rhythmic unit is composed at the base of anhydrite and dolomite-anhydrite (15–20 m); up the section, it grades into rock salt with polyhalite and polyhalite and halite rocks containing potash salt and bischofite interbeds (20–50 m). The thickness of the unit in some regions increases to 100–130 m and even 140–180 m. The Pogozhye rhythmic unit begins with a 5–15 m thick anhydrite bed overlain by salt with interbeds of clayey anhydrite, potash salts (sylvinite and carnallite rocks, 2–7 m thick), bischofite, and bischofite-carnallite rocks. The thickness of the upper part of this rhythmic unit ranges from 40–170 m. In ascending order, its section consists of: an anhydrite-dolomite bed (10–25 m); rock salt (25–40 m); polyhalite-carnallite and polyhalite-carnallite-halite rocks with kieserite and sylvite

(15–50 m). In the upper part, a 20–50 m thick bischofite bed containing carnallite and sylvite can be distinguished. The Pigarev and Dolina rhythmic units are composed of anhydrite beds with dolomite at the base; their thickness varies between 5 and 30 m. The upper half is composed of rock salt interbedded with sylvite-halite, carnallite-halite, and sylvite-carnallite-halite rocks; the thickness is 10–60 m. The thickness of the Pigarev and Dolina rhythmic units is, respectively, 50–100 m and 50–70 m. The thickness of the Eruslan rhythmic unit varies between 45 and 200 m. A 8–15 m thick anhydrite bed lies at the base and up the section, rock salt alternates with anhydrite. The lower part of the Overlying Anhydrite rhythmic unit is composed of intercalated anhydrite and rock salt, while its upper part is composed of anhydrite, dolomite, mudstone, siltstone, and sandstone. The thickness reaches 80 m.

Easily distinguished within the Volga Monocline are two bischofite horizons confined to the Pogozhye and Antipovka rhythmic units. They are known as the Narimanovo and Gorodishche horizons (Zharkov et al. 1980). The Narimanovo horizon has been studied in detail. In its lower part, there is a zone 15.14 m thick of carnallite rocks and rock salt with minor bischofite-carnallite and bischofite rocks. The middle part of the horizon is dominated by bischofite rocks, 29.16 m thick. It is characterized by colorless, lucid, light, and dark gray bischofite with inclusions, in places numerous, of milky-white carnallite and yellowish and pale-pink, semi-opaque bischofite rocks. Quite peculiar is a 15.7 m thick unit in the middle of a bischofite zone, composed of bluish, smoky bischofite and bischofite-carnallite rocks. It contains black, bituminous clay selvages and seams (up to 1 cm) with admixed halite having an odor of hydrosulphur. The upper Narimanovo horizon (18.1 m thick) comprises, along with bischofite, also bischofite-halite, halite-carnallite-bischofite, and halite-carnallite rocks. This part is distinguished as a carnallite-bischofite zone. Only this section contains both anhydrite-kieserite-bischofite rocks and bischofite with admixed anhydrite and kieserite and bischofite with anhydrite; thin-bedded rocks of kieserite-bischofite composition occur as well.

On the Elton Dome, some Kungurian salt deposits were studied in detail by Svidzinsky (1971a, b), who recognized four horizons; the Lower Halite horizon (I), the potash horizon (II), the Anthraconite horizon (III), and the Upper Halite horizon (IV).

The Lower Halite horizon (I) is composed of uniform rock salt with seasonal anhydrite seams. The apparent thickness is 469 m. The horizon consists of three members: (1) the member marked by nonpersistent rhythmic intercalation and composed of rock salt with anhydrite laminae, grouped into rhythmic units; the number of laminae is variable (1–2 to 7–11), with rare polyhalite and celestine inclusions; the apparent thickness of the member is 198 m; (2) the member consisting of a few interbeds and composed of rock salt with isolated anhydrite seams (1–2 cm) forming 5–10 cm thick rhythms, locally with admixed kieserite and polyhalite; the thickness varies from 111–155 m; and (3) the member represented by a rhythmic intercalation and composed of rock salt in the form of anhydrite seams in groups of 3–5, which are distributed rather uniformly throughout the section; isolated inclusions of kieserite, rare sylvite, and carnallite occur as well; the thickness is about 200 m.

In the potash horizon (II), three producing potash salt beds, dominantly sylvinite and carnallite in composition, can be distinguished. Together with separating rock salt beds, they are placed into a Producing unit. Below, the Transisional Member and the Underlying Salt Member, and, above, the Overlying Salt Member can be recognized.

The Underlying Salt Member is composed of coarsely crystalline rock salt with minor, seasonal anhydrite seams. Minor, admixed kieserite, sylvinite, carnallite, and nonpersistent carnallite interbeds 2–3 m thick were also found. The thickness of the member varies between 43 and 246 m.

The Transitional Member is represented by rock salt with inclusions and nonpersistent interbeds of sylvinite and carnallite rocks. Locally, the thickness attains 63 m.

The Producing unit is composed of alternated potash salt beds assigned to the lower (1), middle (2) and upper (3) producing beds and rock salt beds. The thickness of the latter is variable. From the lower to the middle producing beds, it varies from 8.4–162.8 m, whereas between the middle and upper beds, it varies from 8–82 m. When the middle producing bed is absent, the thickness of rock salt between the remaining potassic beds increases to 100–240 m. The lower producing bed is 7.56–48 m thick. Two bands, a sylvinite band (4.45–40 m) with an admixed polyhalite, syngenite, kieserite, and carnallite and a carnallite band (19.7 m) are easily discernible. The middle producing bed can be traced on the north-eastern and south-western flanks of the Elton Uplift. It also consists of two bands: the lower one, represented by carnallite rock (1.45–8.24 m) and the upper one, made up of sylvinite rock (0.87–21 m). The upper producing bed has a 1.65–7.67 m sylvinite band at the base and a 0.82–11.27 m carnallite rock band at the top.

The Overlying Salt Member is 16–128 m thick. It is made up of rock salt with seasonal anhydrite seams and inclusions of sylvite, carnallite, minor kieserite, and syngenite.

The Anthraconite horizon (III) was named for nonpersistent interbeds of anthraconite, a finely crystalline calcite impregnated by petroleum bitumen. The horizon is composed of limestone, dolomite, rock salt, and halopelite. Within the horizon, three members can be recognized, in ascending order: (1) halopelite, 2.3–7.4 m; (2) anhydrite-dolomite-halite, 3.8–28 m; and (3) limestone-dolomite-anhydrite, 3.8–24.5 m. The total thickness ranges from 9.9–60 m.

The Upper Halite horizon (IV) consists of rock salt 800 m thick. This section contains four potassium-bearing beds, numbered 5, 6, 7, and 8, which are confined to the middle carnallite-halite member. The latter is underlain by the lower anhydrite-halite member and is overlain by the upper anhydrite-halite member. Potassium-bearing bed 5 is mainly carnallite, locally sylvite in composition. Bed 6 contains admixed carnallite; bed 7 is dominated by carnallite; and bed 8 contains carnallite-kieserite rocks.

The total thickness of the studied salt-bearing part of the section on the Elton Dome reaches 1700 m.

Kungurian salt deposits on the Cherkar Dome were also studied in detail. Korenevsky (Korenevsky and Voronova 1966) distinguished, in ascending order: the lower halite zone, a potassium-bearing zone, and the upper halite zone. The potassium-bearing zone can be divided into the lower carnallite-sylvinite, anhydrite-halite, carnallite-halite, anhydrite-carnallite-halite, and the upper sylvinite subzones. Diarov (1971a, b) distinguished the main anhydrite horizon in this section. He subdivided the potassium-bearing zone into two subzones: the lower and the upper subzones, each containing bischofite beds. Seven and three potassium-bearing horizons were recognized in the lower and upper subzones, respectively.

On the Inder Dome, the salt sequence can be divided into the Sutpaitau, Kzyltau, Kurgantau, and Totdzhal Formations.

The lower Sutpaitau Formation is composed of rock salt interbedded with anhydrite and minor carnallite, kieserite, sylvite, polyhalite, boracite, and ascharite inclusions. Sylvite interbeds were observed as well. The apparent thickness of the formation is approximately 1000 m. The Kzyltau Formation can be divided into three members: the lower, anhydrite-halite member (100–150 m); the middle, polyhalite-sylvinite member (50–200 m); and the upper, halite member (100–200 m).

The Kurgantau Formation can be divided into four members: (1) the Anhydrite Member (or the Main Anhydrite Member after Diarov); (2) the Görgeyte-Halite Member; (3) the Shushaktau Member, embracing the upper potassium-bearing zone of the section; and (4) the Upper Halite Member. The Main Anhydrite Member is of stratigraphical importance, being easily discernible on many domes in the central part of the Caspian Depression and so allowing a correlation of salt sequences (Diarov 1971a). The member consists of two combined anhydrite beds, namely, the lower and the upper beds, respectively, 2–3 m and 10–20 m in thickness. These beds are separated by a 3–6 m rock salt unit. The total thickness of the member amounts to 15–30 m. The Görgeyte-Halite Member is made up of rock salt interbeds with görgeyte-halite rocks (5–50 m) and contains inclusions and seams of potash and potash-magnesium salt. It ranges in thickness from 150–250 m.

The Shushaktau Member is about 200 m thick. Its base contains a lower potash-boron horizon, composed of carnallite-kieserite, polyhalite-sylvinite, and sylvinite rocks with inclusions of preobrazhenskite, boracite, kaliborite, hydroboracite, and other rocks; the horizon is 5–30 m thick. Above lies rock salt interbedded with black clay and thinly laminated anhydrite-clayey rock and bands of sylvite-polyhalite-kainite and kaliborite-halite rocks in the upper part; this unit is 4–10 m thick. The thickness of the salt member totals 100–150 m. The upper potash-boron horizon is made up of sylvinite-polyhalite, kainite-sylvinite, sylvinite, and minor carnallite rocks with inclusions of kaliborite, hydroboracite, preobrazhenskite, boracite, and other rocks. The thickness of the horizon is 5–25 m. The Upper Halite Member is composed of rock salt with seasonal anhydrite seams and is impregnated by polyhalite, carnallite, hydroboracite, and preobrazhenskite. The thickness of the member varies between 30 and 200 m.

The lower and upper parts of the Totdzhal Formation are represented, respectively, by rock salt (overlying rock salt) and anhydrite (overlying anhydrite). The formation is 300–500 m thick. The thickness of the studied salt-bearing part of the section on the Inder Dome totals 2100 m. The average thickness of potash salt is 61 m.

Tikhvinsky (1974, 1976a, b) and Azizov and Tikhvinsky (1978) suggested a possible subdivision of the Kungurian salt deposits in the Caspian Depression into 11 potassium-bearing horizons, grouped into three potassium-bearing zones. The lower zone is confined to the Ulagan Beds and encompasses three potassium-bearing horizons; the middle zone, consisting of five potassium-bearing horizons, is assigned to the Elton Beds; the upper zone occurs in the Chelkar Beds and comprises three potassium-bearing horizons.

Along the eastern and north-eastern margins of the Caspian Depression, the Kungurian salt deposits gradually become terrigenous. There, rock salt pinches out and the section is composed of anhydrite interbedded with clay, sandstone, and marl; then terrigenous rocks become dominant. In the Emba River Basin, the Kungurian deposits are represented by three sequences: the lower, sulfate-terrigenous; the middle, salt-bearing; and the upper, sulfate-terrigenous.

In the Cis-Urals, in the Aktyubinsk area, two types of Kungurian sections can be recognized, namely, the Aleksandrovsk and Aktyuba types. The former is characterized by the presence of sulfate-carbonate-terrigenous rocks; this type of a section is developed to the east and north-east. The second type is represented by rock salt and minor potash salts (Zamarenov 1970).

Despite the vague structure of the Caspian Depression sections, the given characteristics are well confirmed due to available evidence. The correlation of sections and their relation to the Volga Monocline standard section, on the one hand, and the type section of the Volga-Ural Anteclise and the Kama Basin, at Perm, on the other hand, are presently still considered ambiguous. Fig. 101 exemplifies such a correlation.

The distribution of sylvinite, carnallite, polyhalite, kieserite, and bischofite rocks in the Kungurian deposits within the Caspian Depression is shown in Fig. 108.

Estimates of the thickness of the Kungurian salt deposits occurring in the Caspian Depression have been given by many investigators (Aizenshtadt and Gershtein 1963, Arabadzhi et al. 1966, Fiveg and Banera 1968, Movshovich 1968, Zhuravlev and Svitoch 1971, Zhuravlev 1972). The original thickness of the salt series in the central Caspian Depression was estimated at 4000–5000 m. Toward the margins, it decreases to 1000–2000 m. According to Movshovich, rock salt in the depression should amount to 9.2×10^5 km^3. This value exceeds the estimate obtained by Fiveg and Banera (7.1×10^5 km^3), which could be considered as the minimal volume. Estimations of other investigators were much larger.

The overall volume of the Kungurian rock salt in the East European Basin is determined as follows: Fiveg and Banera (1968) believe that 0.7×10^5 km^3 of salt was deposited on the Volga-Ural Anteclise and in platform areas, adjacent to the Caspian Depression to the north; their estimate of the depressions of the Cis-Uralian Trough was 0.2×10^5 km^3. According to these figures, rock salt in the Kungurian deposits will amount to approximately 1.0×10^6 km^3.

The distribution pattern of the Ufimian and Kazanian (Upper Permian) deposits of various composition within the East European Basin is shown in Fig. 109.

Rock salt in the Ufimian deposits was reported from the Solikamsk Depression, some adjacent areas of the Cis-Uralian Trough, and the eastern part of the East European Platform (Kopnin 1965, Ivanov 1965). It is confined to the lower Solikamsk Formation, which is distinguished as a clay-marly or salt-marly sequence. Its thickness ranges from 30–40 to 150–160 m.

It is composed of dark and dark gray clay and marl with rock salt interbeds 5–25 m thick, mainly in the lower and middle parts of the section. In places, there are several (up to 3–4) rock salt interbeds and units alternated with mirabilite and sylvinite laminae. Rock salt occurs within a belt 5–10 km wide, extending along the Kama River from the Vishera River mouth to the north and to the Berezniki Region to the south.

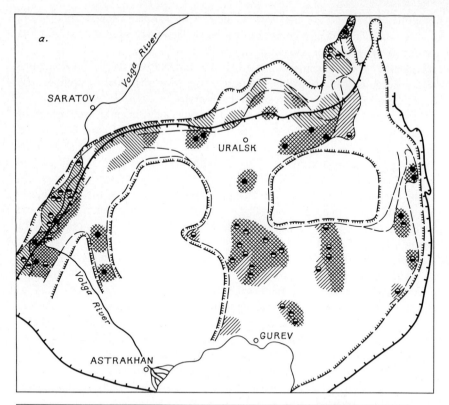

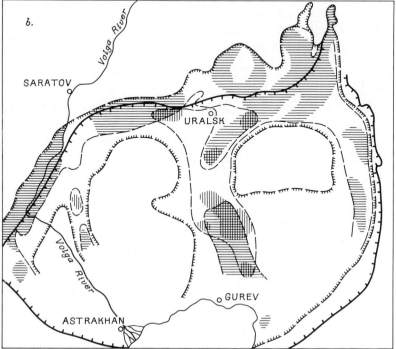

Fig. 108a, b

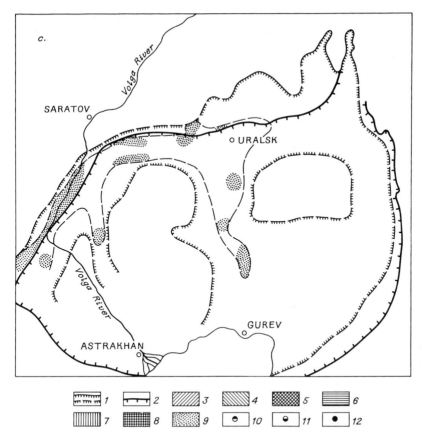

Fig. 108a–c. Distribution of sylvinite and carnallite rocks (a), polyhalite and kieseritic rocks (b), and bischofite rocks (c) in Caspian Depression. After Tikhvinsky (1976a, b). *1* limit of potash deposits, known and probable, *2* distribution of salt structures. Depositional limit (*solid line* known, *dotted line* probable) of: *3* sylvinite, *4* carnallite rock, *5* sylvinite and carnallite rocks, *6* polyhalite rock, *7* kieseritic rock, *8* polyhalite and kieseritic rocks, *9* bischofite rock. Thick beds inferred from core material of: *10* sylvinite, *11* carnallite rock, *12* sylvinite and carnallite rocks

No rock salt has been reported from the other regions of the East European Basin, where the recognition of Ufimian deposits seems to be quite reliable. Some investigators have indicated the presence of rock salt in the Ufimian deposits on the eastern flank of the Caspian Depression (Dalyan and Posadskaya 1972). However, their Ufimian age is debatable (Zhuravlev 1972).

Anhydrite and gypsum are widespread in the Ufimian deposits. They occur as inclusions, interbeds, and locally as rather thick units in carbonate and terrigenous-carbonate deposits on most of the Moscow Syneclise, Volga-Ural Anteclise, and the northern and western Caspian Depression. Some sections were described in detail in many publications (Rybakov 1962, Urusov et al. 1962, Tikhvinskaya and Chepikov 1967, Nalivkin and Sultanaev 1969, etc.). Thus, it is noted only that the amount of sulfate rocks gradually increases from west to east as the Urals is approached.

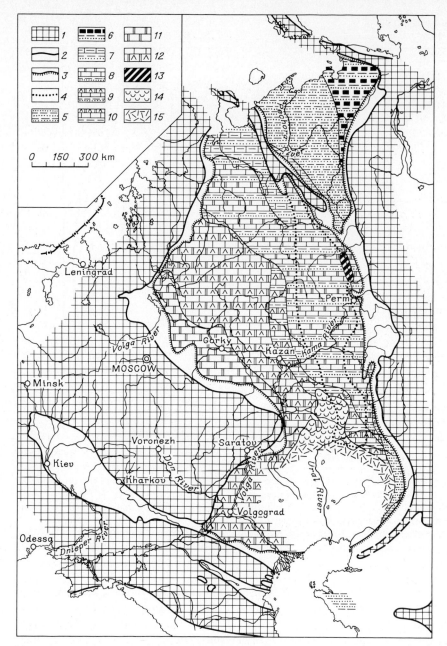

Fig. 109. Lithologic map of Ufimian and Kazanian deposits in East European Basin. Compiled by the author together with Merzlyakov. *1* Permian absent, *2* limit of Permian deposits, *3* limits of Ufimian and Kazanian deposits, *4* limit of area where complete sections of Ufimian and Kazanian deposits are present (in the remaining part either lower or the entire Ufimian absent and in eastern areas Upper Kazanian absent). Distribution: *5* mainly terrigenous rocks, *6* terrigenous, coal-bearing rocks, *7* sandstone, siltstone, limestone, and clayey limestone, *8* sandstone, mudstone, limestone, and dolomite, *9* the same rocks with interbeds and inclusions of anhydrite (gypsum), *10* mudstone, limestone, and dolomite with interbeds and inclusions of anhydrite (gypsum), *11* dolomite, *12* limestone and dolomite interbedded with anhydrite, *13* rock salt in Ufimian deposits, *14–15* rock salt in Kazanian deposits (*14* known, *15* probable)

Kazanian salt deposits were reported from the southern Volga-Ural Anteclise, within the Buzuluk Depression. Forsh (1951, 1955, 1976, 1979) described them in detail in a number of publications and compiled bed-by-bed schemes for the Kazanian of the eastern Russian Platform, in the third volume of the *Atlas of Lithologic-Paleogeographic Maps of the USSR*. The reader is referred also to the publications of Rybakov (1960), Svishchev (1961), Kovanko (1962), and Sementovsky (1969).

The salt-bearing series is distinguished as a Hydrochemical Formation which occurs in the lower Upper Kazanian deposits at the level of the Isakly Beds. It is underlain by a terrigenous-carbonate unit assigned to the Oboshin Beds; below is a carbonate member of the Kalinovsk Formation (Lower Kazanian).

An anhydrite unit made up of anhydrite with dolomite seams can be distinguished at the base of the salt-bearing series. In the Buzuluk area, it is 20—25 m thick. Lying above is a rock salt unit; its thickness in the Samara River Basin ranges from 0—220 m. Salt deposits are confined to a narrow trough extending from south to north, from the middle and upper Samara River to the upper Bolshoi Kinel River. The trough filled by rock salt, is rather sinuous in outline. Its western boundary follows a line joining the towns of Buzuluk and Buguruslan; while the northern and north-eastern boundary follows the right side of the Bolshoi Kinel River near Alekseevo and Matveevka Villages; the eastern boundary lies near the town of Sharlyk and west of the Samara River. The salt sequence is overlain by anhydrite with dolomite interbeds; their thickness near Buzuluk reaches 40—45 m.

The trough, filled by salt deposits of the Hydrochemical Formation, is open toward the Caspian Depression. In this direction, the thickness of salt deposits increases. In the Sorochinsk area, it reaches 200—220 m, whereas near Tashly Village it amounts to 280—300 m. The stratigraphic position of salt-bearing deposits also changes south- and south-eastwards. Salt deposits were found also in the overlying Sosnovsk Formation and according to Zhuravlev and co-workers (1972), in the underlying Kalinovsk Formation as well.

Kazanian salt deposits on the northern flank of the Caspian Depression are extremely widespread. They were penetrated on the western side by the Orekhovo No. 1 well, where the thickness of anhydrite and salts of the Hydrochemical Formation attains 35 m. South-west of Tashly Village, Kazanian salt deposits up to 460 m thick were penetrated by a G-2 well on the Pavlovo Salt Dome. Farther south-west on the Tsyganov Dome, rock salt interbedded with terrigenous rocks up to 83 m thick was also found in the Upper Kazanian. The above data led Zhuravlev and co-workers (1972) to the well-grounded assumption that the Upper Kazanian salt deposits occur in much of the Caspian Depression up to its eastern margins. This conclusion is confirmed since in some parts of the Cis-Urals, Aktyubinsk area, the salt-bearing sequence is also developed in the Kazanian deposits. It was penetrated on the Kumsai and Lugovskoe uplifts. It is assumed that in other regions (Ashchi, Nikolaevskoe, Severnyi Loktybai, Muyunkum), rock salt beds occur in the Upper Permian deposits. Their thickness may increase westwards.

No rock salt has been reported from the Kazanian sedimentary strata from other areas of the East European Basin. In the Moscow Syneclise, Mezen Depression, and Volga-Ural Anteclise areas, anhydrite and gypsum comprising sulfate-carbonate, sulfate-carbonate-terrigenous, and sulfate-terrigenous deposits are very common.

At present, it is difficult to estimate the volume of rock salt in the Ufimian and Kazanian deposits of the East European Basin. The Ufimian salt deposits have a limited extent and a small volume. The volume of the Kazanian salts may exceed 8×10^3 km^3, if the assumed rock salt areal extent in the Buzuluk and Caspian depressions is 80,000 km^2 and average thickness is about 100 m.

Upper Kazanian salt deposits occupy the highest position in the Permian section of the East European Basin. The overlying Tatarian deposits are represented mainly by red beds and variegated, terrigenous rocks with inclusions and lenses of gypsum and anhydrite.

It should be noted that most of the rock salt in the salt deposits of the East European Basin is associated with the Kungurian (Lower Permian) deposits. The Kungurian period was marked by enormous salt accumulation and by formation of salt strata unique in composition and containing bischofite deposits. During other epochs, salt accumulation took place only in certain regions of the basin, and compared to Kungurian age, was limited. The entire bulk of Permian salts in the East European Basin exceeds 1.02×10^6 km^3.

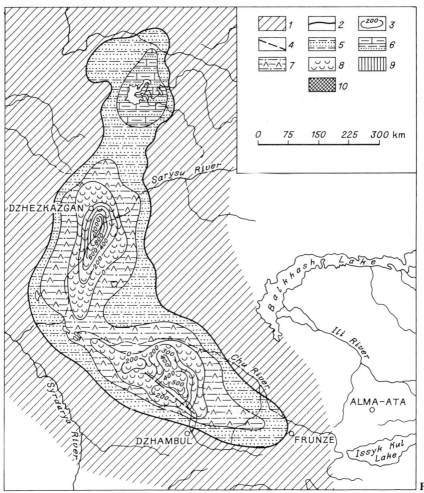

Fig. 110

Chu-Sarysu Basin

M.A. ZHARKOV and G.A. MERZLYAKOV

Permian salt deposits are developed within most of the Chu-Sarysu Depression (Fig. 110). They were reported from its north-western part, the Dzhezkazgan Depression; and to the south, on the Ulanbel-Talas Uplift and in the Chu Depression. The following description is based mainly on the investigations of Ditmar (1962a, b, 1963, 1965, 1966), Bakirov (1965a, b, 1977), Shakhov (1965a, b, 1968), Bakirov and Belyashov (1966), Kumpan (1966), Ditmar and Tikhomirov (1967), Gulyaeva et al. (1968), Gabai (1969, 1977), Orlov et al. (1969), Dobretsov and Kumpan (1973), Sinitsyn et al. (1977), and Merzlyakov (1979).

Fig. 110. Lithologic map of Permian deposits in Chu-Sarysu Basin. Compiled from the data of Ditmar and Tikhomirov (1964), Orlov et al. (1969), and Bakirov and Belyashov (1966). *1* Permian absent, *2* limit of Permian deposits, *3* isopach contours of salt units, *4* faults. Depositional limits of: *5* sandstone, siltstone, and mudstone, *6* clayey limestone, marl, siltstone, and sandstone, *7* sandstone and siltstone with interbeds and inclusions of gypsum and anhydrite, *8* salt. In cores: *9* anhydrite and gypsum rocks, *10* salt

Salt deposits, reported from the Dzhezkazgan Depression, are confined to the Zhidelisai and Kingir Formations. In the southern Chu-Sarysu Depression, salt-bearing deposits belong to the Karakyr, Sorkol, and Tuzkol Formations. Their age position and correlation are shown in Fig. 110.

The Zhidelisai salt sequence is 280–350 m thick; in the Dzhezkazgan Depression it attains 800 m. The sequence is made up of mudstone, cross-bedded siltstone, and crimson red, brown, and pink-gray sandstone alternated with beds and interbeds of rock salt. The latter is pink or red-brown, often folded, and rich in terrigenous material. There are gypsum and anhydrite inclusions and interbeds. The Kingir Formation is mainly gray. Its lower part is composed of gray marl and clayey limestone alternated with lucid, gray rock salt beds. Inclusions and interbeds of gypsum and anhydrite occur throughout the section. Glauberite forms scattered impregnations, crystals, intergrowths, rare interbeds, and lenses. Orlov and co-workers (1969) emphasized that the salt sequence of the lower Kingir Formation is 170–190 m thick. Rock salt accounts for 22%–75% of the section. The areal extent of salt is no less than 1500 km^2. Along the periphery of the depression, salt deposits are replaced by terrigenous-carbonate, and then by terrigenous rocks. In the Zhidelisai Formation, rock salt accounts for 40%–72% of the section. The areal extent of these salt strata is about 10,000 km^2. On the basis of this data, the volume of the Permian rock salt in the Dzhezkazgan Depression was estimated to be 2.8×10^3 km^3.

Salt deposits, occurring in the Chu Depression and confined to the Karakyr, Sorkol, and Tuzkol Formations, are chiefly made up of red-brown gypsinate and anhydritized siltstone, mudstone, and sandstone.

The salt sequence of the Tuzkol Formation, occurring in the south-western Chu-Sarysu Depression, is composed mainly of red siltstone, mudstone, and pink-gray, fine-grained sandstone alternated with thin rock salt beds. Terrigenous rocks are often saline and contain inclusions of halite crystals. Interbeds and lenses of gypsum and anhydrite occur throughout the section; those of limestone being much rarer. Glauberite crystals and aggregates were found in the lower part of the section. The thickness is over 400 m. The areal extent of this salt sequence exceeds 30,000 km^2 and the volume exceeds 3×10^3 km^3, assuming an average total thickness of 100 m. The total volume of rock salt in the Chu-Sarysu Basin is no less than 5.8×10^3 km^3.

Moesian Basin

The Moesian salt basin is situated in eastern Bulgaria within the Moesian Plate (see Fig. 140). Salt-bearing deposits were recorded at Provadiya Village, near the Sofia-Varna railway (Spasov and Yanev 1966, Lefond 1969, Yanev 1969, 1970a, b, 1971, Garetsky 1972). They comprise a salt stock about 850 m long and about 400 m wide. Recently, new data have become available which enable a more detailed description of the basin (Polster et al. 1976, Zhukov et al. 1976, Yanev 1981).

The evaporite series are exposed in numerous, deep boreholes in the vicinity of Varna, Provadiya, Mirovo, Bezvoditsa, Vetrino, Zlatar, Totleben, Kaliakra, Turgovishte, Elenovo, Chereshovo, and in adjacent southern regions of Rumania. According to

Yanev (1969, 1970a, b, 1971, Zhukov et al. 1976), two types of section can be distinguished in the Upper Permian: salt-bearing and argillaceous, terrigenous sulfate-bearing. The salt-bearing series was studied in the greatest detail using drilling data from the vicinity of Mirovo Village (OP-1) and Bezvoditsa (P-75) wells. There is a salt-bearing unit in the lower part of the section; in the upper part, salts dominate. The former is composed of thin-bedded, gray and ash-gray mudstone with bands of siltstone, marl, and rock salt. Thin dolomite layers are very common, as well as carbonate breccia with intercalations and lenses of anhydrite. The thickness of the lower unit varies from 150 m (Bezvoditsa) to 890 m (Mirovo). The salt-bearing unit is represented mainly by pale, yellow-to-white, and pink-to-red rock salt with alternate mudstone, anhydrite, and dolomite. According to Zhukov et al. (1976), primary succession is obscured by salt tectonics. As a result, the sequence is composed mainly of brecciated rocks with mudstone, marl, and dolomite fragments enclosed in the salt matrix. Lenses of gypsum and glauberite are often recorded, as well as carnallite veinlets and selvages. The thickness of the salt-bearing unit is over 2140 m and 380 m near Mirovo and Bezvoditsa, respectively.

Salt deposits in the Moesian Basin are traced from Mirovo and Bezvoditsa Villages almost to Omurtag, i.e., 100 km from east to west. The salt zone is surrounded by deposits of argillaceous-terrigenous sulfate type, represented mainly by siltstone, mudstone, and more rarely, dolomite. Yanev distinguished three units among them. The lower unit is composed of green-gray to dark gray, purple-brown to brownish-red, thin-bedded mudstone, alternated with siltstone, anhydrite, and dolomite. Anhydrite often occurs as pockets and lenses. The thickness of the unit is about 640 m. It is correlated with all the salt-bearing deposits of the Mirovo and Bezvoditsa area. The overlying member is composed of massive mudstone and siltstone. The rocks are bright red, brick-red to brownish-red, more rarely dark and greenish-gray. All the rocks are calcareous to a greater or lesser extent; inclusions, lenses, and thin layers of anhydrite are often recorded. The thickness of the unit varies from 160–1093 m. The third unit crowning the Permian section is characterized by variegated rocks; it is arenaceous-argillaceous in composition; its thickness varying from 54–136 m.

Within the Varna Depression, equivalents of the salt-bearing deposits are represented mainly by a sulfate-carbonate sequence. According to Yanev, it can be distinguished as the third, independent type of sequence occurring east of this salt accumulation area. It probably marks a transition from a salt basin to the open sea of normal salinity, which, in his opinion, was in the place of the present Black Sea (Zhukov et al. 1976, p. 124).

An area of evaporite accumulation within the Moesian Basin was flanked on its west by an intensely subsiding depression of the East Cis-Balkan area, where thick terrigenous strata were being accumulated. Red siltstone and deltaic sandstone are about 2 km thick. The available data indicate that the Moesian salt basin was bounded by land in the north, north-west, and west, where source areas were located on the North Bulgarian Uplift and in South Dobruja. No data are as yet available on the presence of any source areas at the southern boundary. It is most likely that the salt basin was also connected in the south with a sea of normal salinity.

The salt-bearing deposits over the Moesian Depression are underlain by thick, Lower Permian, terrigenous red beds. In places, inclusions and lenses of anhydrite are sporadically observed in the upper part of the red beds.

In the Moesian salt basin, the determined areal extent of salt deposits is 3000 km^2. Assuming an average rock salt thickness for this area of 300 m, then the minimal volume will amount to 9×10^2 km^3.

Central European Basin

Permian salt deposits are widespread in central West Europe. They form a belt extending from east to west for more than 1600 km; the width ranges from 300–600 km. They occupy most of Lithuania, Poland, G.D.R., West Germany, Holland, Denmark, England, and the North Sea. The area covered by Permian evaporites exceeds 700,000 km^2. Salt strata have been found both in Lower Permian (Rotliegende) and Upper Permian (Zechstein) sedimentary sequences. Zechstein salt formations are very common and their occurrence helps to outline the Central European Basin.

The Lower Permian deposits (Rotliegende) consist chiefly of red terrigenous and volcanic-terrigenous beds, while the Upper Permian strata (Zechstein) are represented mainly by evaporites. Permian salt deposits have been studied thoroughly and described by many investigators (Tinnes 1928, Ahlborn 1934, 1955, Lotze 1938, 1957a, b, 1958a, b, Smith 1928, Smith 1950, Dunham 1948, 1967, Stewart 1949, 1951a, b, 1953, 1954, 1963a, b, Fleck 1950, Raymond 1953, Brinkmann 1954, Ahlborn and Richter-Bernburg 1955, Dietz 1955, Herrmann and Richter-Bernburg 1955, Kühn 1955, Richter-Bernburg 1955a, b, c, 1957, 1959a, b, 1960, 1972a, b, 1981, Roth 1955a, b, 1972, Herrmann 1956, Fabian 1957, 1958, Trusheim 1957, 1960, 1964, Braitsch (1962, Hartwig 1958, Hoppe 1958, 1960, Jung 1958a, b, 1966, 1968a, b, Schulze 1958, 1960, 1962, Teichmüller 1958, Wolburg 1958a, b, Borchert 1959, 1972, Brucren 1959, Tokarski 1959, Wilkening 1959, Ekiert 1960, Hecht 1960, Ivanov and Levitsky 1960, Knak 1960, Langbein and Seidel 1960, Poborski 1960, 1964, 1969, 1970, Schreiber 1960, Werner et al. 1960, Kölbel 1961, Seidel 1961, 1965, 1966, Vala 1961, Dittrich 1962, 1966, Jankowski and Jung 1962, 1964a, b, 1966, Krason 1962, 1964, Löffler and Schulze 1962, Marr 1962, Schwandt 1962, Stolle 1962, 1967, Tomaszewski 1962, 1981, Walger 1962, Braitsch and Herrmann 1963, Kerkmann 1966, Mötzung 1963, Pawłowska 1963, 1964, 1970a, b, c, Pawlowska and Poborski 1963, 1970, Sannemann 1963, Borchert and Muir 1964, Fuchtbauer 1964, 1972, Gottesmann 1964, Kłapciński 1964, 1967, Quester 1964, Sorgenfrei and Buch 1964, Ullrich 1964, Dadlez and Dembowska 1965, Donovan and Dingle 1965, Fuchtbauer and Goldschmidt 1965, Konstantynowicz 1965, Langbein 1965, 1973, Meyer 1965a, b, Podemski 1965, 1968, 1972, Rentzsch 1965, Schettler 1965, 1972, Szaniawski 1965, 1966, Tonndorf 1965, Wienholz and Wirth 1965, Andreas et al. 1966, Baar 1966, Eastwood 1966, Haase 1966, Havlena 1966a, b, Jahne 1966, Jahne and Pielert 1966, Jungwith and Sellert 1966, Katzung 1966, 1968, 1970, 1972, 1975, Konitz 1966, Lützner 1966, Merz 1966, Münzberger et al. 1966, Oettel and Voitel 1966, Philipp 1966, Plumhoff 1966, Sokolowski 1966, 1970, Sorgenfrei 1966, Steiner 1966, Tasler 1966, Dadlez 1967a, b, Döhner 1967, 1970a, b, 1976, D'Ans 1967, Herrmann et al. 1967, Heybroek et al. 1967, Hodenberg and Kühn 1967, Hofrichter 1967, Hoyningen-Huene 1967, Kent 1967a, b, 1968, 1980,

Korich 1967, Kosmahl 1967, Kröll and Nachsel 1967, Meinhold and Reinhardt 1967, Rost and Schimanski 1967, Salski 1967, Shurawlew 1967, Simon 1967, Bartenstein 1968, Döhner et al. 1968, Helmuth 1968a, b, Hinz 1968, Jung and Lorenz 1968, Suveizdis 1968, Brunstrom and Walmsley 1969, Dadlez and Marek 1969, Meier 1969, 1975, 1981, Hemman 1970, 1972, Heynke and Zänker 1970, Jahne et al. 1970, Jurkiewicz 1970, Kaczmarek 1970, Kent and Walmsley 1970, Knoth 1970, Milewicz 1970a, b, 1971, 1981, Milewicz and Pawłowska 1970, Paech 1970, Poborski and Pawłowska 1970, Reichel 1970, Reichenbach 1970, Schirrmeister 1970, Schwab 1970, Seifert 1970, 1972, Siedlecke 1970, Smith 1970, 1972, 1981, Wyżykowski 1970a, b, Kühn and Dellwig 1971, Lorenc 1971, Pokorski 1971, 1981, Arthurton and Hemingway 1972, Glennie 1972, Ivanov and Voronova 1972, Marek and Znosko 1972, Stołarczyk 1972, Bush et al. 1973, Falke 1974, Glushko et al. 1974, 1976, Charysz 1975, Dickenshtein et al. 1975, 1976, Döhner and Elert 1975, Flügel 1975, Gurary 1975, Lyutkevich 1975, Neumann and Schön 1975, Taylor and Colter 1975, Watson and Swanson 1975, Ziegler 1975, 1978a, b, 1980, Stolle and Döhner 1976, Kijewski et al. 1979, Smith and Crosby 1979, Korenevsky et al. 1980, 1983, Woods 1979, Antonowicz and Knieszner 1981, Antsupov et al. 1981, Depowski et al. 1981, Harwood 1981, Korenevsky and Poborsky 1981, Lützner et al. 1981, Nemec 1981, Nemec and Porębski 1981, Paul 1981, Roniewicz et al. 1981, Solak and Zołnierczuk 1981, Wijhe 1981, Wagner et al. 1981). The following description of Permian deposits situated in the Central European Basin is based on the above works.

In central and western Europe, the Lower Permian deposits fill in large troughs and depressions emplaced on the Hercynian, Caledonian, and Precambrian folded basement. The depressions were locally restricted by mountain structures. In Early Permian time, sedimentation was accompanied by a rather intense tectonic activity. This led to the accumulation of a lithologically variable complex of terrigenous, coal, volcanic, and salt deposits.

Such a major structure as the Middle European Depression was formed during the Lower Permian (Katzung 1972); it extends from the North Sea on the west to central Poland on the east (Fig. 111). The depression includes a large number of basins filled by thick Rotliegende strata, e.g., the English Basin, North-West German Basin, Polish Trough, Zielona Gora Depression, Laskowice Graben, and Podlasie Trough.

The Middle European Depression is bounded by the Fenno-Scandian High and the Byelorussia Massif to the east and north-east, by the Caledonian structures of Britain and Scandinavia to the north-west and west and by the London-Brabant, Rhenish, and Bohemian Massifs to the south and south-west.

Narrow and deep depressions and troughs were formed south and south-west of the Middle European Depression in the Lower Permian. Some of them were then deep embayments of some major troughs (Český-Budějovice, Boskovice, Central Bohemian, Erzgebirge, and Mühlhausen-Infeld Troughs and the Weser Depression), while others formed a system of troughs extending as a single continuous belt south-westward (Saale, Oos, Saar-Werra, Wittlich, Burgundy, Baden, and Schramberg Troughs).

Lower Permian sedimentary strata are locally rather thick. In the deepest depressions and troughs, thicknesses exceed 1000 m and in places reach 1500 m and even 2000 m (Fig. 112). The greatest thicknesses were recorded in the north-western Oos

Fig. 111. Distribution of Lower Permian (Rotliegende) deposits in central West Europe. Compiled from the data of Vala (1961), Suveizdis (1963), Trusheim (1964), Havlena (1966b), Katzung (1966, 1968, 1970, 1972, 1975), Plumhoff (1966), Kent (1967a, b, 1968, 1980), Hoyningen-Huene (1967), Brunstrom and Walmsley (1969), Rutten (1969), Kent and Walmsley (1970), Milewicz (1970a, 1981), Milewicz and Pawłowska (1970), Pawłowska (1970a), Garetsky (1972), Richter-Bernburg (1972a), Bush et al. (1973), Dickenshtein et al. (1975), Ziegler (1973, 1978a, b, 1980), Antsupov et al. (1981), Lützner et al. (1981), and Pokorski (1981). *1* areal distribution of Lower Permian (Rotliegende) deposits, *2* troughs, basins, and depressions and their numbers (*I* Ulster-Manx-Furness Basin, *II* Clyde-Ouse Trough, *III* Central North Sea Basin, *IV* North Danish Basin, *V* Bamble Trough, *VI* English Basin, *VII* North-West German Basin, *VIII* West Netherland Basin, *IX* Channel Basin, *X* Nibus Basin, *XI* Wittlich Trough, *XII* Burgundy Trough, *XIII* Baden Trough, *XIV* Schramberg Trough, *XV* Oos Trough, *XVI* Saar-Werra Trough, *XVII* Weser Depression, *XVIII* Mühlhausen-Ilfeld Trough, *XIX* Saale Trough, *XX* Bitterfeld-Torgau-Mühlberg Trough, *XXI* Erzgebirge Trough, *XXII* East Franconian Trough, *XXIII* Česky-Budějovice Trough, *XXIV* Boskovice Trough, *XXV* Central Bohemian Trough, *XXVI* Döhlen Trough, *XXVII* Nisa-Bóbr Trough, *XXVIII* Laskowice Graben, *XXIX* Zielona Gora Depression, *XXX* Poznan Graben, *XXXI* Polish Trough, *XXXII* Podlasie Trough, *XXXIII* Stupsk Basin, *XXXIV* Peri-Baltic Depression), *3* highs, massifs, plateaus, uplifts, and their numbers (*1* Grampian High, *2* Pennine High, *3* Irish Massif, *4* Cornish Platform, *5* London-Brabant Massif, *6* Armorican Massif, *7* Rhenish Massif, *8* Hunsrück Uplift, *9* Morvan-Vosges Uplift, *10* Spessart-Ruhla Swell, *11* Upper Donau Uplift, *12* Nürnberg Uplift, *13* East Thüringia Uplift, *14* Halle-Wittenberge Uplift, *15* Flechtingen Uplift, *16* Lausitz Uplift, *17* Sudetes High, *18* Radom-Lublin Plateau, *19* Wolsztyn Dissected Highland, *20* Marury Plateau, *21* Chełmo Lowland, *22* Pomeranian Highland, *23* Fenno-Scandian High, *24* Ringkøbing-Fyn High, *25* Mid-North Sea High)

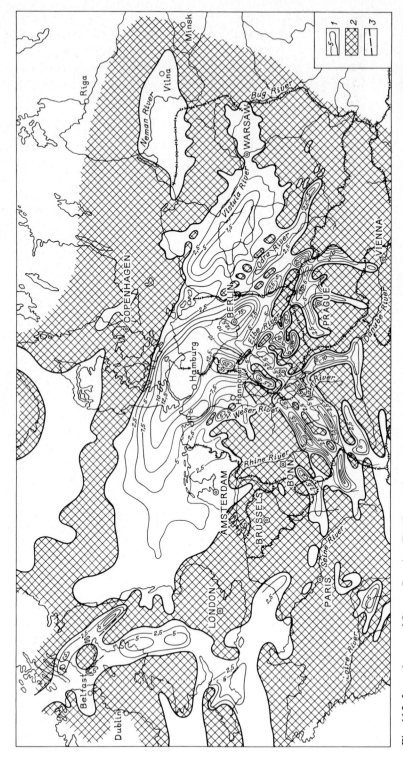

Fig. 112. Isopach map of Lower Permian (Rotliegende) deposits in central West Europe. Compiled from the data of Trusheim (1964), Havlena (1966b), Milewicz (1970a, 1981), Milewicz and Pawłowska (1970), Garetsky (1972), Katzung (1972), Richter-Bernburg (1972a), Dickenshtein et al. (1975), Ziegler (1978a, b, 1980), Pokorski (1981), and others. *1* isopach contours, in hundreds of meters, *2* Lower Permian absent, *3* lines of drastic change in thickness related to faults

Trough, in the Werra River Basin, in some southern parts of the Middle European Derpession, in particular, north of the Flechtingen Uplift, and between the Lausitz Uplift and the Sudetes High, where the Rotliegende exceeds 2000 m. In the Saar-Werra Trough, in some areas of the North-West German Basin, in the Boskovice Trough, and in the Central Bohemian Trough, the Lower Permian formations attain a thickness of 1000—1500 m.

The deposits of Rotliegende are characterized by a drastic facies change (Fig. 113). They are divided into Autunian (Lower Rotliegende) and Saxonian (Upper Rotliegende). The Lower Rotliegende consists of red terrigenous deposits (conglomerate, sandstone, siltstone, and shale) with coal-bearing and bituminous gray rocks and acid and basic effusives and tuffs. The Upper Rotliegende consists mainly of red sandstone, siltstone, and shale. Evaporite deposits are confined to the Upper Rotliegende.

In Europe, a Lower Permian salt sequence is known only in the western Middle European Depression, where it fills lower zones of the North-West German and English Basins. The area covered by rock salt in the Rotliegende deposits was contoured in Fig. 113 from the data presented chiefly by Plumhoff (1966), Kent (1967a, b, 1968), Brunstrom and Walmsley (1969), Kent and Walmsley (1970), Katzung (1972), Richter-Bernburg (1972a), and Ziegler (1978a, b, 1980). The boundaries of the Lower Permian area of salt accumulation have not yet been outlined, although they have been rather roughly drawn from geophysical data and data from occasional boreholes drilled in Rotliegende salt series of diapirs.

On some diapirs in the lower Elbe River, salt rocks (Haselgebirge) composed of coarsely crystalline halite with red clay inclusions lie at the base of the exposed section. Their thickness exceeds 400 m. They are overlain by 15 m of red sandstone and marl; above are Zechstein deposits with conglomerate and copper slate (Kupferschiefer) at the base (Hofrichter 1967).

In the lower Elbe River, the thickness of the Lower Permian salt sequence exceeds 1000 m. There it composes elongate double-salt swells. The lower part of the sequence, consisting chiefly of rock salt, is overlain by a thin, red clay bed followed by Haselgebirge, where anhydrite, marl, and sandstone interbeds occur (Trusheim 1960, Plumhoff 1966, Zhuravlev 1972, Glushko et al. 1974).

Salt deposits of Rotliegende were penetrated in the north-eastern North-West German Basin on Fehmarn Island, Baltic Sea. The Fehmarn Z-1 well penetrated to 925 m. The upper part of the sequence, 757 m thick, is composed of siltstone and clay with numerous rock salt interbeds. Red, fine-grained sandstone, 75 m, lies below. Conglomerate and quartz porphyry, 93 m, occur at the base. Lower Permian salt deposits were penetrated by numerous wells in the North Sea.

Some investigators (Kent and Walmsley 1970) believe that the Rotliegende salt sequence, despite incomplete penetration, can attain an average of 1500 m. The areal extent of the salt deposits currently exceeds 80,000 km^2. Assuming an average thickness of rock salt for the entire area of 100 m, even in this case, the volume of rock salt accumulated during the Lower Permian will run as high as 8×10^3 km^3. However, actual salt volume may be an order of magnitude higher than the above estimate.

Zechstein (Upper Permian) salt deposits cover an area of about 700,000 km^2 (Fig. 114) and fill in the following basins and depressions: Polish-Lithuanian Depression, Central Polish Depression, North-East German Basin, Altmark Depression,

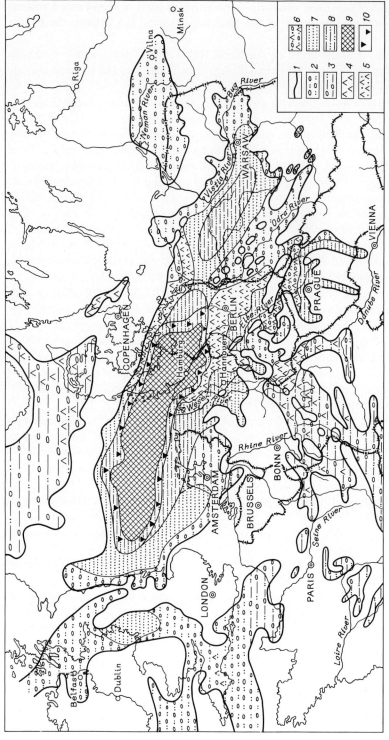

Fig. 113. Lithologic map of Lower Permian (Rotliegende) deposits in central West Europe. Compiled from the data of Schreiber (1960), Trusheim (1964), Katzung (1970, 1972, 1975), Knoth (1970), Milewicz (1970), Milewicz and Pawłowska (1970a, 1981), Pawłowska (1970), Reichel (1970), Schwab (1970), Garetsky (1972), Richter-Bernburg (1972a), Dickenshtein et al. (1975), Ziegler (1973, 1978a, b, 1980), Lithofacies-Paleogeographical Atlas . . . (1978), Pokorski (1981), and others. *1* limit of Lower Permian, *2* conglomerate, gritstone, sandstone, *3* conglomerate, sandstone, siltstone, mudstone, *4* effusive formations, *5* terrigenous-effusive deposits, *6* coarse-grained, mainly conglomerate-sandy deposits with effusive rocks, *7* mainly sandstone, *8* sandstone, siltstone, mudstone, *9* salt deposits, *10* anhydrite

Central European Basin

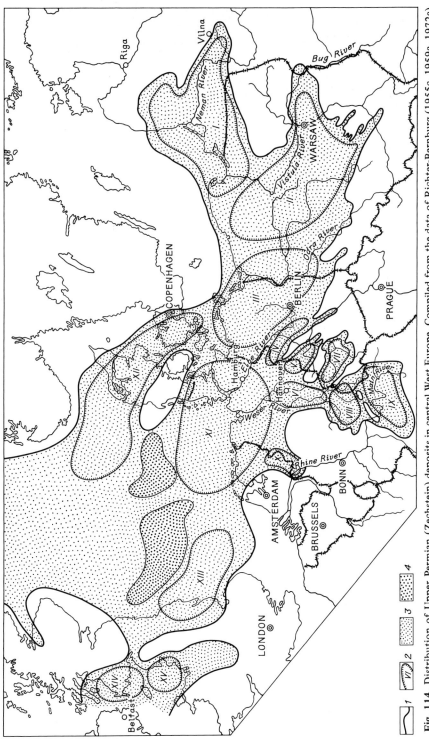

Fig. 114. Distribution of Upper Permian (Zechstein) deposits in central West Europe. Compiled from the data of Richter-Bernburg (1955c, 1959a, 1972a), Fabian (1958), Teichmüller (1958), Hoppe (1960), Poborski (1960, 1970), Löffler and Schulze (1962), Stewart (1963b), Suveizdis (1963), Trusheim (1964), Seidel (1965), Sorgenfrei (1966), Kent (1976b), Bartenstein (1968), Hinz (1968), Jung (1968b), Podemski et al. (1968), Brunstrom and Walmsley (1969), Meier (1969), Pawlowska (1970b, c), Pawlowska and Poborski (1970), Arthurton and Hemingway (1972), Ziegler (1973, 1978a, b), Smith (1981), Wagner et al. (1981), and others. *1* limit of Upper Permian (Zechstein) deposits, *2* limit of troughs and their numbers (*I* Polish-Lithuanian Depression, *II* Central Polish Depression, *III* North-East German Basin, *IV* Altmark Depression, *V* Subhercynian Depression, *VI* Weser Depression, *VII* Thüringia Depression, *VIII* Werra Fulda Depression, *IX* Frankonia Basin, *X* Lower Rhine Basin, *XI* North-West German Basin, *XII* Norwegian-Danish Basin, *XIII* English Basin, *XIV* Solway Firth Basin, *XV* Manx-Furness Basin), *3* areas covered by Upper Permian (Zechstein) deposits, *4* arches and swells in North Sea

North-West German Basin, Norwegian-Danish Basin, and English Basin; thus, forming a single zone of subsidence which, like that of the Lower Permian, can be called the Middle European Depression. Zechstein salt strata are common in the Lower Rhine Basin, Weser Depression, Subhercynian Depression, Thüringia Depression, Werra Fulda Depression, and Franconia Basin. Anhydrite and gypsum were recorded in the Solway Firth and Manx-Furness Basins.

Zechstein salt deposits reach 1000–1500 m (Fig. 115). These estimates are available for the Central Polish Depression and the North-East German Basin. The total thickness of salt deposits probably increases westward into the North-West German Basin and into the North Sea. However, several hundreds of diapir structures developed in these areas make it difficult to accurately determine the true thickness of salt deposits; it may range from 1500–2000 m. Lesser thicknesses of Zechstein deposits were recorded in the Lower Rhine Basin, in the Weser, Subhercynian, Thüringia, Werra Fulda Depressions, and Franconia Basin. In general, the total thickness does not exceed 600–800 m.

A rather detailed stratigraphic subdivision of the Zechstein deposits is available for the Central European Basin (Fig. 116). Richter-Bernburg (1955b) divided Zechstein into four parts: Zechstein 1 (Werra Series), Zechstein 2 (Stassfurt Series), Zechstein 3 (Leine Series), and Zechstein 4 (Aller Series). Another, uppermost series, Zechstein 5 (Ohre Series) has since been recognized in the central North-East German Basin, in the Altmark, and Central Polish Depressions (Wirth 1967, Jung 1968b, Reichenbach 1970). Therefore, five Zechstein series are recognized.

Such a stratigraphic subdivision is based on cycles of salt accumulation (Richter-Bernburg 1955a, b). Each cycle starts with terrigenous, generally, clayey deposits overlain by carbonate and sulfate (anhydrite) rocks followed by a sequence represented by rock salt and in places, by potash salt; the cycle is crowned by the Upper Anhydrite.

In the Central European Basin, five salt sequences, namely (in ascending order), the Werra-Steinsalz, Stassfurt-Steinsalz, Leine-Steinsalz, Aller-Steinsalz, and Ohre-Steinsalz, have been recognized in the Upper Permian composite section. Minute subdivisions within the series have letter indexes: T – for clay, Ca – for carbonate, A – for anhydrite, Na – for rock salt, and K – for potash salt. Numbers are added to the letter indexes to denote the position of beds in a particular cycle of Zechstein: 1 – for Zechstein 1, 2 – for Zechstein 2, 3 – for Zechstein 3, 4 – for Zechstein 4, and 5 – for Zechstein 5. Thus, indexes T1, Ca1, Na1, and K1 denote clayey, carbonate, anhydrite beds, and rock salt and potash bands of Zechstein 1 (Werra Series); indexes T2, Ca2, A2, Na2, and K2 denote respective interbeds of Zechstein 2 (Stassfurt Series), etc. In cases of more minute subdivision, interbeds are suffixed with letters of the Greek alphabet and in some cases with initial letters of local geographic names (e.g., Kaliflöz Thüringen of the Werra Series is labeled K1Th; Kaliflöz Hessen of the same series is denoted by K1H; Kaliflöz Ronnenberg and Kaliflöz Riedel are labeled K3Ro and K3Ri, respectively).

Type sections of individual Zechstein series, which were used for stratigraphic subdivision, are situated in different parts of the Central European Basin. Thus, the type localities of the Werra Series are the Werra Fulda Depression and the Lower Rhine Basin; the type localities of the Stassfurt Series are the Subhercynian and Thüringia

Fig. 115. Isopach map of Upper Permian (Zechstein) deposits in central West Europe. Compiled by Zharkov and Merzlyakov from the data of Richter-Bernburg (1955a, c, 1972a), Schauberger (1955), Trusheim (1957, 1960, 1964), Teichmüller (1958), Wolburg (1958a, b), Wilkening (1959), Hoppe (1960), Löffler and Schulze (1962), Suveizdis (1963), Seidel (1965), Sorgenfrei (1966), Heybroek et al. (1967), Kent (1967a), Meinhold and Reinhardt (1967), Simon (1967), Bartenstein (1968), Hinz (1968), Jung (1968b), Podemski et al. (1968), Brunstrom and Walmsley (1969), Dadlez and Marek (1969), Meier (1969), Zhuravlev (1972), Dickenshtein et al. (1975), Taylor and Colter (1975), Ziegler (1973, 1977, 1978a, b), and Lithofacies-Paleogeographical Atlas ... (1978). *1* limit of Upper Permian (Zechstein) deposits, *2* isopach contours, in hundreds of meters; *3* salt domes, diapirs, and anticlines, *4* uplifts and zones with decreased thickness or complete absence of salt, *5* Upper Permian (Zechstein) absent

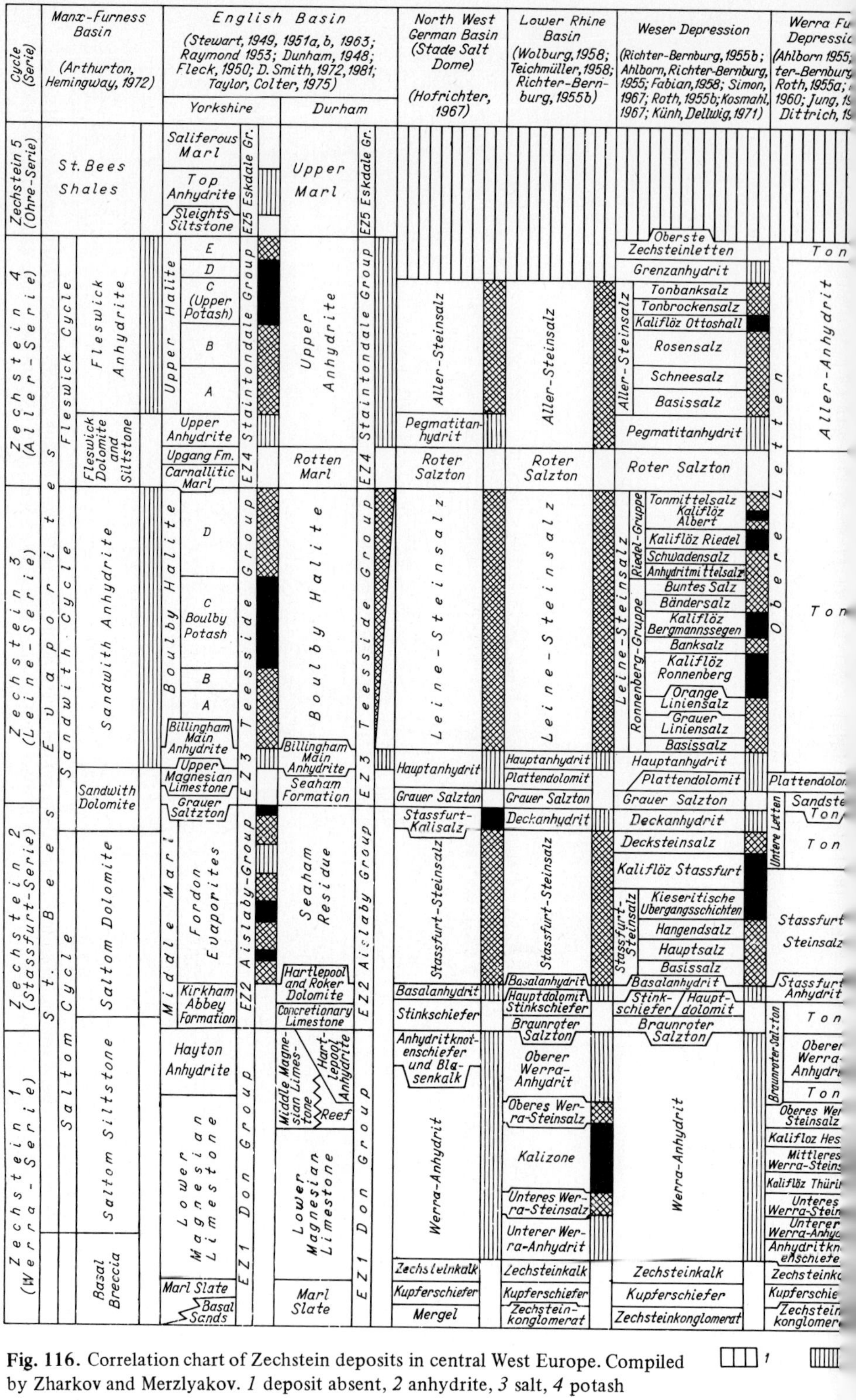

Fig. 116. Correlation chart of Zechstein deposits in central West Europe. Compiled by Zharkov and Merzlyakov. *1* deposit absent, *2* anhydrite, *3* salt, *4* potash

	Thüringia Depression (Seidel, 1965; Stolle, 1967; Merz, 1966; Jung, 1968b; Hoyningen-Huene, 1967)	Altmark Depression (Löffler, 1962; Jung 1968b; Hoyningen-Huene, 1967; Reichenbach, 1970)	North-East German Basin (Hoyningen-Huene, 1967; Jung, 1968b; Hemmann, 1970; Reichenbach, 1970; Schirrmeister, 1970)	Central Polish Depression (Poborski, 1960; Wagner e.a. 1981)	Polish-Lithuanian Depression (Vala, 1960; Suveizdis, 1963; 1975, 1976)		Cycle (Serie)
	Anhydrit	Anhydrit		Boundari Anhydrite			Zechstein 5 (Ohre-Serie)
	Schluffstein	Schluffstein		Top Youngest Halite			
	Ohre-Steinsalz	Ohre-Steinsalz	PZ4b	Anhydrite			
	Anhydrit	Anhydrit		Upper Red Pelite			
	Schluffstein	Schluffstein					
Oberste Steinletten	Oberste Zechsteiletten						Zechstein 4 (Aller-Serie)
Grenzanhydrit	Grenzanhydrit	Grenzanhydrit	Grenzanhydrit	Upper Youngest Halite			
Tonbrockensalz		Tonbrockensalz					
Kaliflez Ottoshall	Aller-Steinsalz	Schneesalz	Anhydritmittelsalz Tonflockensalz Kristallsalz Schneesalz	Aller-Steinsalz	Upper Pegmatite Anhydrite Lower Youngest Halite Lower Pegmatite Anhydrite		
Basissalz	Basissalz	Basissalz		Lower Red Pelite			
Pegmatitanhydrit	Pegmatitanhydrit	Pegmatitanhydrit	Pegmatitanhydrit				
Roter Salzton	Roter Salzton	Roter Salzton	Roter Salzton				
Tonflockensalz Schwadensalz Anhydritmittelsalz Orangeaugensalz and Bändersalz Kristallsalz	Leine-Steinsalz	Tonflockensalz and Schwadensalz Anhydritmittelsalz Kristallsalz	Ronnenberg-Gruppe Riedel-Gruppe	Tonflockensalz Schwadensalz Anhydritmittelsalz Kaliflöz Ronnenberg I Übergangsschichten Kaliflöz Ronnenberg III Kristallsalz	Leine-Steinsalz	Ronnenberg-Gruppe Riedel-Gruppe	Oberes Leine-Steinsalz Kaliflöz Ronnenberg Unteres Leine-Steinsalz	PZ3	Younger Halite Younger Potash			Zechstein 3 (Leine-Serie)
Liniensalz	Liniensalz	Liniensalz					
Basissalz	Basissalz	Basissalz				Halite	
Hauptanhydrit	Hauptanhydrit	Hauptanhydrit	Main Anhydrite				Zechstein 2 (Stassfurt-Serie)
Plattendolomit		Plattendolomit	Platy Dolomite	Galindas Fm.	Limestone and Dolomite		
Grauer Salzton	Grauer Salzton	Grauer Salzton	Gray Pelite				
Deckanhydrit	Deckanhydrit	Deckanhydrit	Screening Anhydrite		Upper Anhydrite		
Decksteinsalz	Decksteinsalz	Decksteinsalz	Screening Older Halite	Mudstone			
Kaliflöz Stassfurt	Kaliflöz Stassfurt	Kaliflör Stassfurt	Older Potash				
Stassfurt-Steinsalz	Stassfurt-Steinsalz	Stassfurt-Steinsalz	Older Halite	Aistmark Formation	Halite		
Basalanhydrit	Basalanhydrit	Basalanhydrit	Basal Anhydrite		Basal Anhydrite		
Hauptdolomit Stassfurt-Ton	Stinkschiefer	Stinkschiefer	Main Dolomite		Zhalgiryay Formation		
Oberer-Dolomit Oberer Werra-Anhydrit Oberer Werra-Ton			Upper Anhydrite		Mudstone		Zechstein 1 (Werra-Serie)
					Upper Anhydrite		
Oberes Werra-Steinsalz	Werra-Steinsalz	Werra-Steinsalz	Oldest Halite		Halite		
Unterer Werra-Anhydrit Unteres Werra-Steinsalz Mittlerer Werra-Ton Unterer Dolomit Zechsteinkalk			Lower Anhydrite	Pregol Formation	Lower Anhydrite		
Zechsteinkalk	Zechsteinkalk	Zechsteinkalk	Zechstein Limestone		Novoakmyansk Fm.		
Kupferschiefer Limestone Bank	Kupferschiefer	Kupferschiefer	Copper Shale		Sosnava Fm.		
Zechsteinkonglomerat	Zechsteinkonglomerat	Sandstein	Sandstein		Kalvar Fm.		

Depressions; and those of the Leine and Aller Series in the Weser Depression. The type sections of the Ohre Series are situated in the North-East German Basin and in the Altmark Depression.

Zechstein deposits of the Werra Fulda, Weser, Thüringia, Subhercynian Depressions, and the Lower Rhine Basin have been studied in great detail. A composite section of the German Zechstein is as follows. Basal Zechsteinkonglomerat (C1) lies at the base of Zechstein deposits. Above lies Kupferschiefer (T1) overlain by Zechsteinkalk (Ca1). The halogenic part of the Werra Series starts in the Werra Fulda Depression with Anhydritknotenschiefer (CaA1) overlain by Unterer Werra-Anhydrit (Alu)[7]. Unteres Werra-Steinsalz (Na1α), Kaliflöz Thüringen (K1Th), Mittleres Werra-Steinsalz (Na1β), Kaliflöz Hessen (K1H), and Oberes Werra-Steinsalz (Na1γ) lie above this area. The section is capped by Oberer Werra-Ton (T1r)[8] and Oberer Werra-Anhydrit (A1r, A1β, or A1o).

The most complete section of the Stassfurt Series in the Thüringia and Subhercynian Depressions contains, in ascending order: Ton 2 (Stassfurt Ton); Hauptdolomit (Ca2); Basalanhydrit (A2); Stassfurt-Steinsalz (Na2), which is divided into anhydrite, polyhalite, and kieserite zones; Kaliflöz Stassfurt (K2); Decksteinsalz (Na2r); and Deckanhydrit (A2r).

In many areas of the Central European Basin, the Leine Series starts with Grauer Salzton (T3) overlain by Plattendolomit (Ca3) and Hauptanhydrit (A3). The third Zechstein cycle ends with Leine-Steinsalz (Na3), which is divided in the Weser, Subhercynian, and Altmark Depressions and in the North-East German Basin into the lower, Ronnenberg Group and the upper, Riedel Group. In the Weser Depression, the lower, Ronnenberg Group consists of two potash beds — Kaliflöz Ronnenberg (K3 Ro) and Kaliflöz Bergmannssegen (K3 Be); in the Altmark Depression, they are represented by Ronnenberg II and Ronnenberg I. In the Weser Depression, the upper Riedel Group is also composed of two potash beds: the lower, Kaliflöz Riedel (K3 Ri) and the upper, Kaliflöz Albert (K3 Ab). In the Leine Series, the potash beds are underlain and overlain by rock salt; subdivisions and indexes are presented in Fig. 116.

The fourth Zechstein cycle (Aller Series) has a band of Roter Salzton (T4) at the base. It is overlain by Pegmatitanhydrit (A4) followed by Aller-Steinsalz (Na4). Kaliflöz Ottoshall (K4 Ot) is known to lie locally in the Weser and Subhercynian Depressions. The Aller Series ends with Grenzanhydrit (A4r) and locally, when the overlying Ohre Series is absent, with Oberste Zechsteinletten (4r). Zechstein 5 (Ohre Series) contains, in ascending order: Schluffstein (T5), anhydrite (A5), rock salt (Na5), Schluffstein (T5r), and anhydrite (A5r).

Thus, the composite section of German Zechstein consists of five rock salt beds and eight potash beds: two beds (Thüringen and Hessen) occur in the Werra Series; one bed (Stassfurt) — in the Stassfurt Series; four beds (Ronnenberg, Bergmannssegen, Riedel, and Albert) — in the Leine Series; and one bed (Ottoshall) — in the Aller Series. The latter has a limited distribution. Correlative marker horizons are: Kupferschiefer (T1), Zechsteinkalk (Ca1), Hauptdolomit (Ca2d), Stinkschiefer (Ca2st), Basalanhydrit (A2), Grauer Salzton (T3), Plattendolomit (Ca3), Hauptanhydrit (A3), Roter Salzton (T4), and Pegmatitanhydrit (A4).

---

7 In some other areas, the interbed is denoted A1α
8 Denoted also T1γ

The correlation of Zechstein deposits of the central and southern central European Basin with salt series developed in its eastern and western parts poses some problems, although detailed stratigraphic standards and a number of marker horizons are known. To the west — in the English Basin and to the east — in the Central Polish and Lithuanian Depressions, local stratigraphic units have been distinguished in Zechstein deposits.

In the Central Polish Depression, Zechstein deposits have been divided into four sequences known as the Oldest Halite, Older Halite, Younger Halite, and Youngest Halite. Poborski (1960), Werner et al. (1960), and Wagner and co-workers (1981) showed that these sequences correspond to all the five series of German Zechstein. This conclusion was based on the following: Copper Shale and Zechstein Limestone is overlain by the Oldest Halite; the dolomite lying under the Older Halite is the equivalent of the Hauptdolomit (Ca2d) of the Stassfurt Series; the Gray Pelite underlies the Younger Halite and is correlated with the Grauer Salzton (T3); the Lower Red Pelite and the Lower Pegmatite Anhydrite which occur in the middle part of the Younger Halite are correlated, respectively, with Roter Salzton (T4) and Pegmatitanhydrit (A4) of the Aller Series; and the Upper Red Pelite, which lies in the middle part of the Youngest Halite, is the equivalent of the Schluffstein (T5) of the Ohre Series. The above correlation enabled the recognition of equivalents of all the five series of German Zechstein in deposits of Polish Zechstien. They are marked PZ1, PZ2, PZ3, PZ4a, and PZ4b. The minute subdivision is shown in Fig. 116.

When traced eastward toward the Central Polish Depression margins, salt deposits thin out and red pelite begins to dominate the upper Younger Halite and the Youngest Halite. Peculiar clay-saline rocks similar to Haselgebirge are common. These equivalents of salt deposits are known as Brown and Red Zubers.

In the Polish-Lithuanian Depression, the position of Permian salt deposits in a section can be best determined using the correlation of Polish and German Zechstein. With some certainty, the Pregol Formation may be assigned to the Werra Series; the Aistmark Formation may be assigned to of the Stassfurt Series; and the Galindas Formation may be considered as the equivalent of the Leine Series (Suveizdis 1963, 1975, 1976). A gradual enrichment of the upper Zechstein deposits in red stones (Brown and Red Zubers) to the east suggests that in Lithuanian red beds may occur throughout the section. It is probable that the Suduv Formation, which is composed of red and brown marl and siltstone with gypsum lenses and inclusions, may be their equivalent (Vala 1961).

The correlation of German and English Zechstein has been studied recently in detail (D. Smith 1972, 1981, Taylor and Colter 1975). In the English Basin, Zechstein deposits are divided into five cycles marked EZ1, EZ2, EZ3, EZ4, and EZ5. The first cycle of English Zechstein encompasses deposits of the Don Group, which is comparable with the Werra Series. The second cycle corresponds to the Aislaby Group; it is correlated with the Stassfurt Series. The third cycle embraces the Teesside Group and is correlated with the Leine Series. The fourth cycle includes the Staintondale Group; it is considered an equivalent of the Aller Series. The fifth cycle of English Zechstein, known as the Eskdale Group, is correlated with the Ohre Series. The English Zechstein section contains three salt sequences, each associated with the upper parts of the Aislaby, Teesside, and Staintondale Groups. The lower sequence is assigned to the Fordon Evaporite; the middle sequence is assigned to the Boulby Halite; and the upper, to the Upper Halite. Potash horizons are present in the upper two salt sequences, namely, in the Boulby Potash and Upper Potash. For a more detailed subdivision of the Zechstein deposits in the English Basin, the reader is referred to Fig. 116.

In the western Central European Basin, in the Manx Furness Basin, Zechstein deposits form the St. Bees Evaporite (Arthurton and Hemingway 1972). It consists of three cycles: the Saltom, Sandwith, and Fleswick Cycles. Siltstone occurring in the lower Saltom Cycle may be, with all probability, placed at the level of the Werra Series; overlying dolomite of the same cycle may be correlated with the Stassfurt Series; and the Sandwith and Fleswick Cycles may be roughly considered as equivalents of the Leine and Aller Series, respectively. A comprehensive discussion of each Zechstein cycle is given below.

Zechstein 1 (Werra Series)

Composition of the Werra Series deposits in the Central European Basin is schematically shown in Fig. 117. Rather thick (more than 50–100 m) rock salt strata of the lower Zechstein cycle are common in the Werra Fulda Depression, Franconia Basin, Thüringia Depression, Lower Rhine Basin, Weser Depression, Norwegian-Danish Basin, Central Polish, and Polish-Lithuanian Depressions. Salt deposits are developed at the Central European Basin margins, while anhydrite and carbonate with thin rock salt bands occur in the interior parts of the North-East German and North-West German Basins and the Subhercynian and Altmark Depressions.

The lower Zechstein cycle is characterized by the deposition of gypsum (anhydrite) walls bordering to the north, areas of salt accumulation in the Lower Rhine Basin, in the Werra Fulda, and Thüringia Depressions (Richter-Bernburg 1955a, c, 1957, 1959a, 1972a, b, 1981, Seidel 1965, Dittrich 1966, Münzberger et al. 1966, Jung 1968a, b, and others). Anhydrite walls can be also followed along the northern margin of the basin in north-eastern Mecklenburg, GDR, as well as to the south, in Oberlausitz (Münzberger et al. 1966, Jung 1968a, b). Anhydrite walls bound lower parts of the Central Polish Depression and flank the Polish-Lithuanian Depression on the west and south. At the Central European Basin margins, the Werra deposits are composed of sulfate-carbonate and carbonate rocks to the north, east, and south. To the west, in the English Basin, dolomite with anhydrite interbeds is common. Predominantly terrigenous deposits of the Werra Series fill the southern Franconia Basin and Manx Furness Basin. Potash salts of the Werra Series have been reported from three areas of the Central European Basin: (1) Werra Fulda Depression; (2) Lower Rhine Basin; and (3) Polish-Lithuanian Depression.

The Werra Series has received the most study in the Werra Fulda Depression. A great number of potash pits and wells, which were drilled in the entire section, are located in the Werra and upper Fulda Basins. Twenty-two potash pits are being mined in the Werra and Fulda potash districts. The best known are: Marx-Engels I, II; Springen I, II, III, IV, and V; Alexandershall; E. Thälmann I, II, and III; Hattorf; and Neuhof-Ellers. Potash beds have been studied thoroughly in these areas.

A composite section of the Werra Series in the Werra Fulda Depression (Table 2) consists of Zechsteinkonglomerat (C1), Kupferschiefer (T1), Zechsteinkalk (Ca1),

Central European Basin 259

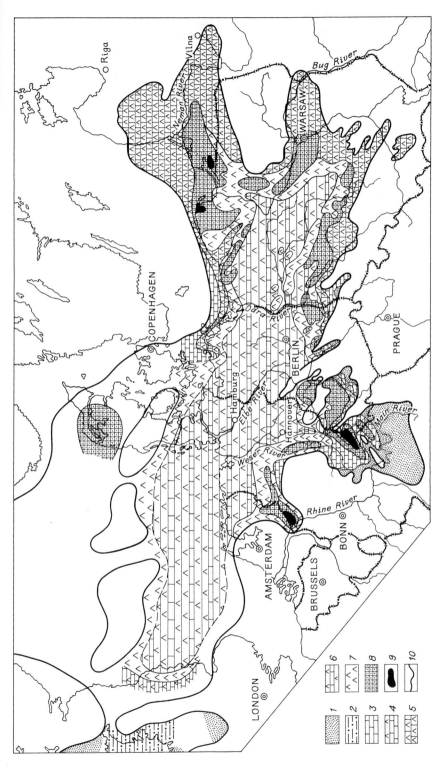

Fig. 117. Lithologic map of Zechstein 1 deposits (Werra Series) in Central European Basin. Compiled by Zharkov and Merzlyakov from the data of Stewart (1949, 1951a, b), Richter-Bernburg (1955a, 1951a, b, 1972a. b, 1981), Teichmüller (1958), Wolburg (1958a, b), Hoppe (1960), Ivanov and Levitsky (1960), Löffler and Schulze (1962), Suveizdis (1963, 1975), Trusheim (1964), Seidel (1965), Hoyningen-Huene (1967), Bartenstein (1968), Hinz (1968), Jung (1968a, b), Brunstrom and Walmsley (1969), Poborski (1969, 1970), Arthurton and Hemingway (1972), Smith (1972, 1981), Stołarczyk (1972), Wagner et al. (1981), and others. Depositional limits of: 1 terrigenous rocks, 2 mainly siltstone, 3 carbonate rocks, 4 carbonate rocks interbedded with anhydrite, 5 carbonate rocks with approximately equal amount of anhydrite, 6 mainly anhydrite with minor rock salt and carbonate rocks, 7 anhydrite (zones of anhydrite walls), 8 rock salt (shown in areas where its thickness exceeds 50 m), 9 potash salt, 10 limit of Upper Permian (Zechstein) deposits

Table 2. Stratigraphic subdivision and thickness of the Werra Series in the Werra and Fulda potash districts. (Adapted from Ahlborn 1955, Roth 1955a, Hoppe 1958, 1960, Dittrich 1966, Jahne et al. 1970)

Cycle, series	Strata	Thickness (m)	
		Werra Basin	Upper Fulda Basin
Zechstein 2 (Stassfurt Series)	Braunroter Salzton (T1r + A1r + T2)	10	10
Zechstein 1 (Werra Series)	Oberes Werra-Steinsalz (Na1 γ)	100 (75–120)	50
	Kaliflöz Hessen (K1H)	2–15	2
	Mittleres Werra-Steinsalz (Na1β)	60 (50–70)	55
	Kaliflöz Thüringen (K1Th)	2–10	2.5
	Unteres Werra-Steinsalz (Na1α)	100 (70–130)	90
	Unterer Werra-Anhydrit and Anhydritknotenschiefer (CaA1 + A1u)	6–8	10
	Zechsteinkalk (Ca1)	6–8	5–25
	Kupferschiefer (T1)	0.3–1	0.5–1
	Zechsteinkonglomerat (C1)	1–2	about 5

Unterer Anhydrit (A1), Unteres Werra-Steinsalz (Na1α), Kaliflöz Thüringen (K1Th), Mittleres Werra-Steinsalz (Na1β), Kaliflöz Hessen (K1H), Oberes Werra-Steinsalz (Na1γ), and Braunroter Salzton. According to Dittrich (1966), Merz (1966), and Jung (1968b), the lower part of the Braunroter Salzton was assigned to the Werra Series, though many investigators place it into the Stassfurt Series (Ahlborn 1955, Roth 1955a, Hoppe 1958, 1960, Jahne 1966, Jahne et al. 1970).

In the Werra Basin, a salt sequence of the lower Zechstein cycle totals 250–300 m in thickness. When traced toward the east, north-east, and north, the thickness decreases and completely wedges out near Thüringer Wald. This is caused by leaching and postsedimentary erosion of rock and potash salt. The leaching zone is well-defined in the north-eastern Werra potash district. To the south-east, the outlines of rock salt and potash salt are original sedimentary features (Trusheim 1964, Dittrich 1966, Haase 1966). Salt leaching zones and related subsidences and negative structures are also developed in the interior parts of the salt-bearing area. On these negative structures, the Werra rock salt strata, including potash beds, were completely washed out. Diabase is common in Zechstein deposits in the Werra Basin (Roth 1955a, Hoppe 1958, 1960).

Unteres Werra-Steinsalz (Na1α) in the Werra potash district is divided into four horizons: Na1α$_1$, Na1α$_2$, Na1α$_3$, and Na1α$_4$ (Jahne et al. 1970). Kaliflöz Thüringen (K1Th) has a complex structure and displays marked lithological variation. For the western Werra Basin, Roth (Hoppe 1960) proposed the subdivision of Kaliflöz Thüringen (K1Th) into three zones: lower, sylvinite-kieserite zone; middle, sylvinite-halite zone; and upper zone, which displays a variation in lithology across the area and is composed of either Umwandlungssylvinit (recrystallized secondary sylvinite), rock salt, Trümmercarnallitit (clastic carnallitic rocks), or Hartsaltz. The lower zone was subdivided into the Untere kieseritreiche Gruppe and Obere kieseritreiche Gruppe. The middle zone contains three rock salt beds: 1, 2, and 3. Furthermore, a so-called dreigeteilte Steinsalzbank (three-partite rock salt band) lies between rock salt beds 1 and 2.

Farther south-east in the Werra area, the Kaliflöz Thüringen was also divided by Konitz (1966). Nine horizons have been recognized down the E. Thälmann potash pit, beneath the zone displaying lithological variation. They are:

Horizon 1 — Kieseritreiche Hartsalzbank . . . . 0.3 –0.5 m
Horizon 2 — Sylvinreiche Hartsaltzbank. . . . . 0.2 –0.3 m
Horizon 3 — Steinsalzbank 1 . . . . 0.1 –0.2 m
Horizon 4 — Geschichtete Hartsalzbank 1 . . . . 0.1 –0.25 m
Horizon 5 — Zwischenlage. . . . . 0.01–0.1 m
Horizon 6 — Geschichtete Hartsalzbank 2 . . . . 0.1 –0.25 m
Horizon 7 — Dreigeteilte Steinsalzbank . . . . 0.1 –0.15 m
Horizon 8 — Geschichtete Hartsalzbank 3 . . . . 0.15–0.20 m
Horizon 9 — Ungeschichtete Sylvinitlage

Above these horizons either Trümmercarnallitit or Umwandlungssylvinit appears. Toward the west and south-west, Hartsalz horizons, beginning with horizon 9 and followed by lower horizons including horizon 1, gradually thin out. The Kaliflöz Thüringen is composed of carnallite and minor sylvinite rocks. However, basal Hartsalz bands (lower horizon 1) are almost ubiquitous. Elsewhere, carnallite rocks are underlain by a band of unlaminated sylvinite (horizon 9), which is not a pure stratigraphic, but rather a peculiar lithological unit marking a lithological boundary between carnallite rocks and Hartsalz. In the Kaliflöz Thüringen, carnallite rocks are known mainly from Dietlas and Menzengraben potash districts, where the central, subsided zone of the sedimentary basin seems to be located.

According to Oettel and Voitel (1966) and Jahne and co-workers (1970), the Kaliflöz Thüringen has a somewhat different composition in the northern Werra potash district. With respect to the areas of predominant Hartsalz development, this potash bed is subdivided into six zones, each starting and ending with bands of clay. These zones consist of one or several rhythms showing the following succession: rock salt → sylvinite or Hartsalz → rock salt. Zones 2, 4, 5, and 6 are represented mainly by one rhythm, while zones 1 and 3 consist of four and three rhythms, respectively. In the northern Werra potash district, Steinsalzbank 1 and Steinsalzbank 2, as recorded by Konitz and Roth farther south and south-west, are also easily discernible. They occur in zones 2 and 4, respectively. Steinsalzbank 1 is characterized by the presence of three clay-kieseritic bands and one double clay-kieseritic band at the top and at the base, respectively; occasional red sylvine crystals occur as well. Two indistinct clay-kieserite partings several millimeters thick lie in the middle part of Steinsalzbank 2. An essential Hartsalz composition was recorded in the Kaliflöz Thüringen along the margins of the northern Werra District and in apical parts of some anticlines. The section becomes almost pure carnallitic in composition towards the center of the basin. Facies change similar to that recorded on the south takes place from the margins to the center of the basin. In the northern Werra potash district, carnallite and transitional varieties dominate, in general, over Hartsalz in the Kaliflöz Thüringen.

In the north-eastern Werra Fulda Depression, Mittleres Werra-Steinsalz (Na1β), which separates the Kaliflöz Thüringen from the Kaliflöz Hessen, is sudivided into

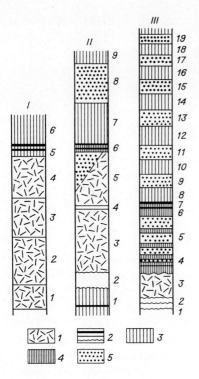

Fig. 118. Subdivision of Kaliflöz Hessen in Werra River Basin. After Jahne and Pielert (1966). *I* after Hoppe, *II* after Roth and Weber, *III* after Jahne and Pielert. *1* Hartsalz, *2* clayey and halopelitic interbeds, *3* recrystallized (secondary) salt, *4* primary salt, *5* carnallite rocks. Numerals at columns denote units described in text

four horizons marked $Na1\beta_1$, $Na1\beta_2$, $Na1\beta_3$, and $Na1\beta_4$ (Konitz 1966, Jahne et al. 1970).

In the north-eastern Werra Fulda Depression, the Kaliflöz Hessen (K1H) was divided by Hoppe (1960) into six bands, in ascending order (Fig. 118):

1. Wurmsalz, locally depleted Hartsalz, or kieseritic halite with three wavy bands of clay .... 0.5 –2.0 m
2. Hartsalz LO_3. Lower part enriched in kieserite and depleted in sylvite. Locally represented by Flockensalz .... 0.7 –1.2 m
3. Hartsalz LO_2. Middle part depleted in sylvite .... 0.4 –0.8 m
4. Hartsalz LO_1. Upper part sylvite-rich .... 0.25–0.6 m
5. Rock salt with two, thin, string-like bands of clay several tens of millimeters thick each. The bands are directly overlain by a discernible Tonlöser, which separates the Kaliflöz Hessen into two parts: the lower, essentially Hartsalz and the upper, composed of either rock salt or rock salt with carnallite or carnallite rocks .... 0.3 –0.4 m
6. Rock salt

In the Werra Basin, the Kaliflöz Hessen was subdivided by Roth and Weber into eight horizons (see Fig. 118, II). The Hartsalz section of the bed was further subdivided into three parts: the lower or Wurmpartie, the middle or Flockensalzpartie, and the upper.

1. Basal rock salt.
2. Lower or vermicular part made up of rock salt and minor Hartsalz with two vermicular bands of clay at the base and at the top.
3. Middle or flocculent part represented by sylvinite and kieseritic Hartsalz.
4. Overlying double band of clay, separating the middle and upper parts of the Hartsalzlager.
5. Upper part in some regions is represented by carnallite rocks, in others by Hartsalz with carnallite inclusions or with partings of carnallite rocks. This part crowns the section of the Hartsalzlager of Kaliflöz Hessen, which occurs in its lower part. Above appears a so-called Begleitlager made up mainly of rock salt and carnallite rock.
6. Thin Tonlöser is easily discernible across the area and separates the underlying Hartsalzlager from the overlying Begleitlager.
7. Rock salt band recognized at the base of the Begleitlager.
8. Overlying carnallite part of the Begleitlager, which crowns the Kaliflöz Hessen. Above appears the Oberes Werra-Steinsalz.

In the Werra potash district, the Kaliflöz Hessen was thoroughly studied by Jahne and Pielert (1966). Seventeen horizons were recognized and given letter indexes (see Fig. 118, III):

1–2. Hartsalz with three bands of clay (Horizon L_1) .... 0.2–0.3 m
3. Hartsalz or Flockensalz (L_2) .... 1.1–1.4 m
4. Rock salt containing various amounts of sylvine and three bands of clay (L_3). .... 0.2–0.3 m
5. Carnallite rock with two salt bands (L_4) .... 0.8–1.2 m
6. Rock salt with occasional string-like bands of clay in the middle part and the base (L_5). The unit terminates the lower part of Kaliflöz Hessen. .... several centimeters
7. Tonlöser separating the lower Hartsalz part of Kaliflöz Hessen from the upper, predominantly halite-carnallite Begleitlager .... several centimeters
8. Rock salt (H_1) .... up to 1.0 m
9. Carnallite rock (BC_1) initiating the Begleitlager. .... 0.8–1.2 m
10. Rock salt (H_2) .... 0.4–0.7 m
11. Carnallite rock (BC_2). .... 0.8–1.2 m
12. Rock salt (H_3) .... 0.4–0.7 m
13. Carnallite rock (BC_3). .... 0.5–1.0 m
14. Rock salt (H_4) .... 0.4–0.7 m
15. Carnallite rock (BC_4). .... 0.3–0.6 m
16. Rock salt (H_5) .... 0.4–0.8 m
17. Carnallite rock (BC_5). .... 0.2–1.2 m
18. Rock salt (H_6) .... 0.2–0.4 m
19. Carnallite rock (BC_6). .... over 0.2 m

The above data show that in the Werra potash district, the Kaliflöz Hessen consists of two well-defined parts: the lower part represented chiefly by Hartsalz of various composition and minor carnallite rock at the top; and upper part, carnallite-halite in composition, which is recognized as the Begleitlager. Both parts of the Kaliflöz Hessen have suffered marked facies changes. Jahne and Pielert have thoroughly studied facies changes of the lower part of the Begleitlager (horizons L_2-L_5).

In the north-eastern Werra Fulda Depression, the Oberes Werra-Steinsalz was divided by Dittrich (1966) and Jahne and co-workers (1970) into two parts ($Na1\gamma_1$ and $Na1\gamma_2$).

Bed $Na1\gamma_1$ consists of gray, essentially medium crystalline rock salt with clay-anhydrite partings spaced about 10 cm apart. Approaching the top, they partly scatter and turn into clay-anhydritic patches. A halite interbed with kieserite up to 3 m thick, labeled $Na1\gamma_{kie}$, can be recognized at the base. The thickness is 15—25 m.

Bed $Na1\gamma_2$ is made up of reddish, coarsely crystalline rock salt with cubic clay-anhydrite inclusions. Toward the top, the proportion of anhydrite decreases and that of clay increases. An anhydritic marker horizon labeled $Na1\beta_{2a_1}$, 10—20 cm thick, lies at the base. The latter is overlain by another anhydritic band $Na1\gamma_{2a_2}$ up to 10 cm thick. The bed consists of three parts dominated by anhydritic and clayey partings. These parts are marked $Na1\gamma_{2a}$, $Na1\gamma_{2at}$, $Na1\gamma_{2t}$. The thickness is 60—75 m.

In the north-eastern Werra Fulda Depression, the Werra Series ends with brown to red saline clay (Braunroter Salzton), the lower and upper parts of which, according to Dittrich (1966), Merz (1966), and Jung 1968b) are assigned to the lower Zechstein cycle and to the Stassfurt Series, respectively. The lower part of the brown to red saline clay is subdivided into clayey band (T1r), 4—8 m thick, and the Oberer Werra-Anhydrit (A1r), 0.4—0.8 m. Dittrich (1966) pointed out that when salt deposits wedge out near the zones of ancient rises, the Oberer Werra-Ton (T1r) unconformably rests in places on older beds, which thin out at the rises.

In general, the Werra Series section is the same in the south-western Werra Fulda Depression, in the upper Fulda River Basin, i.e., in the Fulda potash district (Richter-Bernburg 1955c, Roth 1955a, Kühn and Dellwig 1971). All the stratigraphic units discussed for the Werra potash district can be recognized there as well. The Kaliflöz Thüringen and Kaliflöz Hessen, as well as the Unteres Werra-Steinsalz, Mittleres Werra-Steinsalz, and Oberes Werra-Steinsalz are easily discernible in the section. However, each unit has a smaller thickness (Roth 1955a). Wurmsalz and Flockensalz occur in Kaliflöz Hessen. The Begleitlager represented mainly by carnallite rocks is persistent at the top of the bed; however, predominant rocks are sylvite-kieseritic in composition.

Potash salts in the Werra Fulda Depression are distributed only in its interior parts. At the depression flanks, namely, approaching the Spessart Swell on the south-east, Thüringer Wald on the north-east, and the Rhenish Massif on the west, potash beds thin out and grade into rock salt. A salt sequence then omits from the Werra Series section. Near the ancient rises, the thickness of carbonate rocks of Zechsteinkalk (Ca1) and Unterer Werra-Anhydrit (A1) sharply increases. Anhydrite walls and reef structures, where rock salt wedges out, are common. The anhydrite walls are unconformably overlain by Oberer Werra-Ton (T1r). All these peculiarities were discussed in detail by Dittrich (1966).

South-west of the Spessart Swell, in the Franconia Basin, the Werra Series consists exclusively of rock salt; potash salt has not been found. In the marginal south-western and southern areas of the basin, red terrigenous, essentially clayey deposits appear in the Werra Series. Anhydrite interbeds can be locally traced before tapering out. A "pelite" facies (Richter-Bernburg 1955c) or so-called "Franconish facies", which seems to be the equivalent of the Werra Series, is developed there (Trusheim 1964, Dittrich 1966).

The thickness of the Werra Series salt deposits in the Werra Fulda Depression and in the Franconia Basin is shown in Fig. 119. The thickness reaches 250–300 m and exceeds 100 m, in the former and in the latter, respectively. Their areal extent can be estimated at 7000 km^2. The average thickness of rock salt for the entire area can be approximated at 100 m. The volume of the Werra Series rock salt in the Werra Fulda Depression and in the Franconia Basin may amount to 7×10^2 km^3.

The Werra Series has been studied in detail in the Thüringia Depression. According to Seidel (1965), the thickness of units composing the series vary over the following limits:

Zechsteinkonglomerat (C1)	0– 10 m
Kupferschiefer (T1)	0– 1 m
Zechsteinkalk (Ca1)	0– 25 m
Unterer Dolomit (Ca1u)	0– 40 m
Unterer Werra-Anhydrit (A1u)	0–150 m
Werra-Steinsalz (Na1)	0–300 m
Oberer Werra-Anhydrit (A1o)	0–130 m
Oberer Dolomit (Caα0)	0– 15 m

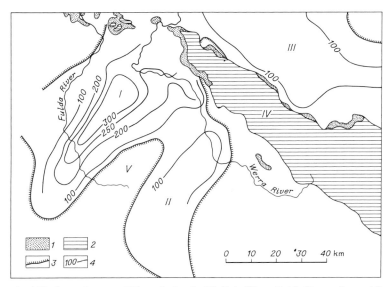

Fig. 119. Isopach map of Werra-Steinsalz (Na1) in Werra Fulda Depression and Franconia Basin. After Dittrich (1966). *1* outcrops of Zechstein deposits, *2* outcrops of pre-Zechstein deposits, *3* depositional limit of rock salt, *4* isopach contours of rock salt, in meters. Structural elements: *I* Werra Fulda Depression, *II* Franconia Basin, *III* Thüringia Depression, *IV* Thüringer Wald, *V* Spessart Swell

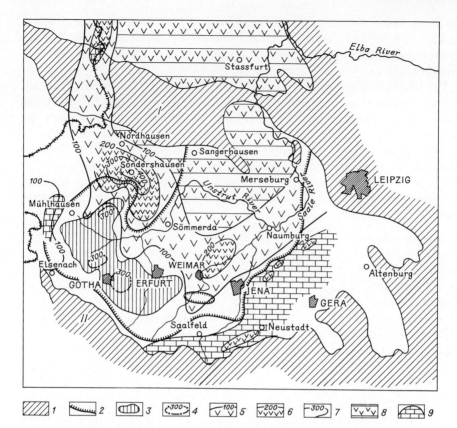

Fig. 120. Lithologic map also showing thicknesses of Werra Series deposits in Thüringia Depression. After Münzberger et al. (1966). *1* outcrops of pre-Zechstein deposits (*I* Harz, *II* Thüringer Wald), *2* depositional limit of salt, *3* area where salt exceeds 200 m in thickness, *4* isopach contours of salt 300 m thick, *5* area where anhydrite ranges from 100–200 m in thickness, *6* area where anhydrite exceeds 200 m, *7* isopach contours of anhydrite, *8* area where anhydrite 40–80 m thick is interbedded with salt laminae, *9* area where light marginal carbonates over 10 m thick and dark carbonate less than 10 m occur

The composition and change pattern of thickness of the lower Zechstein cycle in the Thüringia Depression are given in Fig. 120. The Werra salt deposits fill the central Thüringia Depression, but the maximal thicknesses of salt deposits were reported from the south-western margin, in the area between the towns of Erfurt, Gotha, and Mühlhausen. North- and north-eastward, toward the central, deeper parts of the depression, the thickness of rock salt drastically decreases; in the lower Unstrut River, it does not exceed 100 m; farther north-east, it reduces to 10–15 m; and in places, salt wedges out completely. In the central part of the depression, in an arcuate belt extending from Nordhausen on the north-west to Weimar on the south-east, and farther north-east to Naumburg, a zone of the greatest thickness for anhydrite was recorded; here, anhydrite forms walls flanking the area of maximal salt accumulation to the north-east. At the depression margins, the thickness of anhydrite and that of the entire Werra Series decreases. In the south-eastern Thüringia Depression, carbonate deposits dominate.

Copper slate or marl are widespread in the depression. They are black or buff, bituminous rocks. To the north-east, in the Sangerhausen and Mansfeld Troughs, clay marked by lead-zinc or copper mineralization occurs in the copper slate horizon. To the south-west, west, and south-east, copper slate is locally difficult to distinguish from overlying limestone; copper slate contains limestone lenses and forms a single basal bed.

The Zechsteinkalk occurs as a specific facies across most of the central Thüringia Depression (Fig. 121, I). It is represented by dark gray, brownish-gray, and black limestone with lenticules and inclusions of fine-cloddy ("pearly") anhydrite and pyrite nodules. In the extreme western, south-western, southern, and south-eastern

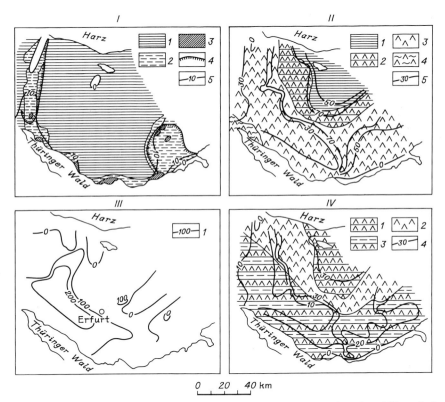

Fig. 121. Lithologic map showing thicknesses of deposits of the main units of Werra Series in Thüringia Depression. After Seidel (1965). *I* Zechsteinkalk (Ca1) and Dolomit (Ca1αu): *1* areal distribution of typical Zechsteinkalk deposits, *2* areal distribution of limestone and dolomite, *3* reefs made up of dolomite, *4* areal distribution of Kupferschiefer, *5* isopach contours, in meters. *II* Unterer Anhydrit (A1u): *1* thin-laminated and wavy-laminated (flaser) anhydrite, *2* wavy-laminated anhydritic varvite with thin-laminated anhydrite, *3* wavy-laminated (flaser) anhydrite, *4* knotted thin-laminated anhydrite typical of Werra district, *5* isopach contours, in meters. *III* Werra-Steinsalz (Na1): *1* isopach contours, in meters. *IV* Oberer Werra-Anhydrit (A1r): *1* thin-laminated anhydrite (varvite), *2* wavy-laminated (flaser) anhydrite, *3* wavy-laminated (flaser) anhydrite with Oberer Werra-Ton, *4* isopach contours, in meters

parts of the depression, dolomite appears in the section along with limestone; generally, dolomite composes the upper part of a horizon. Reefs made up of gray and yellow to gray limestone and dolomite up to 50 m thick are developed there.

The Unterer Werra-Anhydrit in the Thüringia Depression displays a marked variation in composition (Fig. 121, II). In the subsided parts, particularly to the northeast, thin-laminated, shaly anhydrite interbedded with wavy-laminated flaser anhydrite is common. To the west, south, and south-east, the zone is flanked by an area of dominant, laminated anhydrite. In the southern Thüringia Depression, clayey interbeds appear, which occur chiefly as Oberer Werra-Ton (T1γ), which rests on Werra-Steinsalz (Na1). The Werra series is capped by upper, brown-gray to buff, locally oolitic dolomite, common, generally, in the central deeper parts of the depression.

In the Thüringia Depression, the Werra-Steinsalz covers an area no less than 40,000–45,000 km^2. The thickness of salt deposits varies over wide limits. Over an area of about 1200 km^2, it measures about 200 m. In this area, the volume of rock salt totals 2.4×10^2 km^3. The remainder of the volume may be estimated, assuming an average thickness of 10 m for an area of 40,000 km^2, at 4×10^2 km^3. Hence, the total volume of the Werra salt in the Thüringia Depression will account for 6.4×10^2 km^3.

In the Subhercynian Depression, the thickness of the Werra Series decreases mainly at the cost of rock salt. The Zechsteinkonglomerat, Kupferschiefer, and Zechsteinkalk can be distinguished at the base of the series. However, a large volume of the series is accounted for by the overlying Werra-Anhydrit (Table 3). The Unterer Werra-Anhydrit, 28–34 m, overlying Werra-Steinsalz about 10–15 m, and Oberer Werra-Anhydrit 20–36 m are discernible in the southern and central parts of the depression. The Oberer Werra-Anhydrit terminates the section of the lower Zechstein cycle. In some northern areas, rock salt is absent from the Werra Series. Here, the Unterer Werra-Anhydrit merges with the Oberer Werra-Anhydrit, thus forming a single unit made up entirely of anhydrite (Jung 1958a, b, 1968a).

In the north-western Subhercynian Depression, in adjacent areas of the Weser Depression, and north-west of the Flechtingen Uplift, the Werra Series contains no rock salt (Philipp 1966). The thickness of basal deposits varies across the area from 0.02–3.5 m. They are composed of marl and clay, chiefly calcareous or dolomitized, locally with anhydrite inclusions. Kupferschiefer (T1) is represented by dark gray and gray to black, silty, micaceous, locally pyritized shale 0.1–0.5 m. Zechsteinkalk (Ca1) is easily discernible and has a thickness of 3.0–6.35 m. It consists of gray, dolomitized, in places argillaceous, anhydritic limestone. The upper part of the series is composed of anhydrite 43.5–55 m thick. Dolomite, in the form of individual, tapered interbeds and inclusions, lies at the base and at the top of the anhydrite unit.

In general, in the Subhercynian Depression, the Werra-Steinsalz covers an area not less than 50,000 km^2. The average thickness of salt over the area can be approximated at 10 m. In this case, the volume of rock salt in the depression will total 5×10^2 km^3.

In the Weser Depression, the Werra Series is salt-bearing only to the west. In the eastern and north-eastern parts of the depression, anhydrite has a greater thickness (90–100 m) and builds up entire walls (Herrmann and Richter-Bernburg 1955, Herrmann 1956, Herrmann et al. 1967, Richter-Bernburg 1972a, 1981). To the west, rock salt with a thickness exceeding 200 m appears in the section. A salt sequence

Table 3. Stratigraphic subdivision and thickness of the Werra Series in the north-eastern Thüringia, Subhercynian, and Altmark Depressions (Adapted from Jung 1958a, Schulze 1958, Löffler and Schulze 1962)

Horizon	Thickness (m)								
	North-eastern Thüringia Depr.		Subhercynian Depression						Altmark Depression
	Unstrut River Basin	Tautschenthaler Swell	Aschersleben area	Bernburg area	Stassfurt area	Schönebeck area	Allertal Graben		Calvörde Region
Oberer Werra-Anhydrit (A1o)	36.0	25.0	20.0–22.0	20.0	20.0–29.0	20.0	38.0–40.0		45.0
Werra-Steinsalz (Na1)	10.5–11.5	0.0–12.0	2.0–9.0	10.0–11.0	6.0–15.0	3.0	no salt		no salt
Unterer Werra-Anhydrit (A1u)	39.0	30.0	28.0–33.0	34.0	30.0	30.0			
Zechsteinkalk (Ca1)	2.5	5.5	6.0–7.0	4.0–10.0	4.0–10.0	5.0	4.0–10.0		3.0
Kupferschiefer (T1)	0.4	0.3–0.4	0.4	0.4–0.5	0.1–0.3	0.4	0.3		0.3
Zechsteinkonglomerat (C1)	0.45	0.0–3.0	2.0	3.0	2.0	4.0–5.5	1.0		5.0

occurs chiefly west of the Weser River over an area of 2500 km². The volume of rock salt measures approximately 5×10^2 km³.

In the Lower Rhine Basin, the Werra deposits fill a rather narrow trough extending farther south into the Rhenish Massif (Teichmüller 1958, Wolburg 1958a). The trough, following Werra-Steinsalz (Na1), extends for 120–130 km from south to north with a width varying from 40–70 km. A salt sequence of the lower Zechstein cycle is developed only in the central part of the trough; toward the flanks, it thins out and gives way to anhydrite deposits. In the central part of the southern Lower Rhine Basin, potash salt is common (Fig. 122).

The following is a composite section of the Werra Series in the study area, in ascending order: Zechsteinkonglomerat (C1), Kupferschiefer (T1), Zechsteinkalk (Ca1), Unterer Werra-Anhydrit (A1α), Unteres Werra-Steinsalz (Na1α + β), Kalizone (K1), Oberes Werra-Steinsalz (Na1γ), and Oberer Werra-Anhydrit (A1β). Such a complete section is typical of the central, deeper part of the basin, where it was penetrated by many deep wells. Here, the Werra Series has a thickness of 375.7 m. The thickness of rock salt is estimated at 100 m.

Farther south, the thickness of salt increases and exceeds 200 m near the town of Wesel. Between Borthot and Kamp-Lintfort, salt drastically thins out. The Kalizone wedges out across the Rossenrei area; in the Kamp-2 well, the entire salt sequence

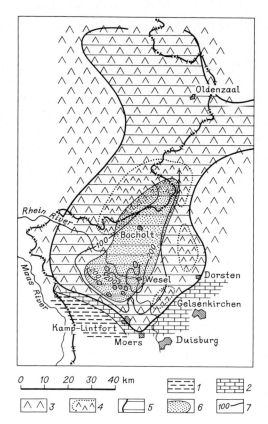

Fig. 122. Lithologic map of Werra Series deposits in Lower Rhine Basin. After Teichmüller (1958) and Wolburg (1958a, b). *1* red clay and fine-grained sandstone, *2* gray and brown to red marl, limestone, and dolomite with organic fossils; local carbonate reef buildups, *3* anhydrite, *4* anhydrite walls, *5* salt, *6* potash, *7* isopach contours of rock salt, in meters

does not exceed several tens of meters. However, the thickness of the Unterer Werra-Anhydrit essentially increases; here, a terminal anhydrite wall bordering the area of salt accumulation to the south is developed. Lithological variations can be observed in logs of the Kamp-1, Fridrih-Henrih-57, and Kamp-4. Marl can be encountered in the first borehole, while in the second, it dominates the section. Marl is an equivalent of the entire halogenic part of the Werra Series in the central parts of the basin. In the Kamp-4 well, interbeds of dolomite and dolomitized limestone appear among marl. Farther south and south-west, terrigenous rocks dominate the section, red rocks appear, then clay and sandstone become common and mark a coastal marginal zone of the basin. These lithological variations were described in detail by Teichmüller (1958).

A change in composition and thickness of underlying horizons shows the same trend. In the central Lower Rhine Basin, Zechsteinkalk (Ca1) is represented chiefly by dolomitic marl. Farther south, it is replaced by marly limestone and dolomite; their thickness gradually increases from 5–6 m to 25–50 m. Dolomite occurs along the flank; its thickness is 75 m and more. The Kupferschiefer thins out southward. Deposits vary similarly when traced south-eastward. There the salt sequence wedges out at an anhydrite wall; then anhydrite wedges out near a zone composed of carbonate deposits of Zechsteinkalk (Ca1), which gains in thickness due to reef structures.

North of the central Lower Rhine Basin, first the Kalizone and then rock salt omit from the section. North of Tubbergen, the Unterer Werra-Anhydrit and Oberer Werra-Anhydrit merge. In the North-West German Basin, the lower Zechstein cycle is represented mainly by a salt-free section. In the Lower Rhine Basin, the Kalizone is bounded to the north by thick Unterer Werra-Anhydrit, which builds up an anhydrite wall (Teichmüller 1958, Richter-Bernburg 1972a, b, 1981).

The volume of Werra Series salt in the Lower Rhine Basin, assuming an areal extent of 11,000 km^2 and an average thickness of 50 m, totals 5.5×10^2 km^3.

In the North-West German and North-East German Basins and in the Altmark Depression, the Werra Series embraces a thin complex of sediments, which, in general, is salt-free or contains thin rock salt bands. The series crops out on some salt domes and along the margins of the basins in the Elbe Basin, in the northern coastal regions of the GDR, and to the south, in the Altmark Depression. Some of these sections are described below. For example, on the Stade Salt Dome, the Zechstein is represented by the following units (Hofrichter 1967):

Mergelsteine (C1) –	0.9 m
Kupferschiefer –	0.35 m
Zechsteinkalk –	4.0 m
Werra-Anhydrit –	40–60 m
Anhydritknotenschiefer and Blasenkalk –	1.25 m

The Werra Series is 47.5–67.5 m thick. At the south-western margin of the North-West German Basin, in the eastern Netherlands, anhydrite reaches 300 m in thickness and forms a wall (Wijhe 1981).

In the extreme south-west of the Altmark Depression, in the Calvörde Region, the Werra Series has a thickness of 50–52 m. The upper 45 m, where no salt has been found, is composed of anhydrite (Löffler and Schulze 1962, Münzberger et al. 1966, Jung 1968b).

In the North-East German Basin, the Werra salt deposits were reported from its most south-eastern areas at Niederlausitz, GDR, and adjacent areas of Poland, near the towns of Gubin and Wschowa.

At Niederlausitz, rock salt exceeds 100 m in thickness. Here, all the horizons of the series known from the more westerly areas of the Zechstein basin can be traced (Jung 1968a, b). To the north-west and probably to the north, the area covered by rock salt is bounded by an anhydrite wall where rock salt thins out. North of the wall, in the interior subsided parts of the North-East German Basin, rock salt is either absent from the lower Zechstein cycle or it is very thin (Jung 1968a). The same lithological variations as those of the Werra Series in the Lower Rhine Basin take place in the interior parts of the Central European Basin.

In the Nysa-Łuzycka and Bobr Rivers Interfluve, Poland, and the Zary Pericline, and in the North Sudetic Trough, the Werra Series (lower Polish Zechstein Cycle PZ1) was penetrated by many wells (Milewicz 1971). In a section just above basal sandstone or conglomerate about 5 m thick lies Copper Shale overlain by Zechstein Limestone, Basal Anhydrite, and rock salt; the series is crowned by the Upper Anhydrite. The thickness of these units in the northern areas of the Nysa-Łuzycka and Bobr Interfluve is presented in Table 4.

Table 4. Stratigraphic subdivision and thickness of Werra deposits in the Gubin area. Adapted from Milewicz (1971)

Stratigraphic unit	Thickness (m)				
	Gubin	Sekowice	Zeklin	Dichów	Żarkow
Upper Anhydrite	38–50	48–57	30–37	28–52	
Oldest Halite	78–343	61–80	52–247,	39–180	100
Basal Anhydrite	21–159	124–167	58–160	53–122	
Zechstein Limestone	3–6	4–6	3–5	3–8	85
Copper Shale	0.5–1	0.5–1	0.5–1	0.5–1	0.6

Salt deposits of Cycle PZ1 are developed to the north, between Gubin and Lubsko. South-east and south of Lubsko, rock salt thins out completely. The Basal Anhydrite is more common to the north. Anhydrite has a maximal thickness (over 300 m) south of the area of the thickest rock salt, where an anhydrite wall can be recognized. The wall seems to join a similar wall outlined in the Niederlausitz area. Southward, in the area where rock salt thins out, the Upper Anhydrite merges with the Basal Anhydrite and the entire section becomes anhydritic. In the Żarkow area, their total thickness exceeds 100 m. Farther south, anhydrite is replaced by carbonate deposits which occur both in the Żary Pericline and in the North Sudetic Trough. The thickness increases to 85 m. Reef structures are common (Milewicz 1971). In the far southern areas of the North Sudetic Trough, the Werra Series consists almost entirely of red sandstone.

In the south-western Central Polish Depression, deposits of Cycle PZ1 are completely exposed east of the Gubin-Lubsko area, in the vicinity of Wschowa (Poborski

1960). The Basal Conglomerate, Copper Shale, Zechstein Limestone, Lower Anhydrite 125 m, Rock Salt 130 m, and Upper Anhydrite 32 m are present there.

Cycle PZ1 has been studied in some detail in the Fore-Sudetic Monocline, in the Lubin-Legnica area, and the Sieroszowice area (Podemski 1965, Lorenc 1971, Tomaszewski 1981). A salt sequence was penetrated north of Sieroszowice and near Kasimirow; it is almost 89 m thick. It is underlain and overlain by the Lower Anhydrite 35 m and the Upper Anhydrite of the same thickness, respectively. South-west, the total thickness of anhydrite increases to 80–100 m. In this direction, the salt sequence thins out and the Upper Anhydrite merges with the Lower Anhydrite. They are underlain by a carbonate sequence ranging in thickness from 80–120 m. The Copper Shale is very common there. It is represented by dolomitic, dolomite-marly, and bituminous varieties (Konstantinovich 1972). Sandstone interbeds can also be encountered.

In the lower Zechstein cycle (PZ1), rock salt has been traced from Lubin farther south-east into the Nida Basin (Antonowicz and Kniesner 1981, Wagner et al. 1981). In the south-eastern Central Polish Depression, in the Nida Trough, a complete section of the cycle was penetrated by the Pagow IG-1 borehole (Jurkiewicz 1970). The following stratigraphic units were penetrated over a depth interval of 2660–2547.2 m, in adcending order:

1. Shale and siltstone interbedded with sandstone   3   m
2. Gray and brown to gray, dense, finely crystalline, bituminous limestone . 15   m
3. Gray, medium crystalline dolomite with inclusions of white gypsum and grayish-green anhydrite. Beds 2 and 3 are assigned to the Zechstein Limestone. 25   m
4. Gray Lower Anhydrite with mudstone laminae 13   m
5. Brown to cherry-red mudstone and shale with anhydrite lenses and rock salt inclusions . . . . 17.4 m
6. White, coarsely crystalline rock salt. . . . . 20.5 m
7. Mudstone with halite crystals. Beds 5–7 build up the Oldest Halite. . . . .   4.5 m
8. Upper Anhydrite. . . . . 14.8 m

Farther south-east, as in all the other coastal zones of the Zechstein basin, rock salt thins out. Near Milianowa, rock salt is already absent from the section and the deposits can be divided into the lower, predominantly carbonate part and the upper, essentially anhydrite part. Dolomite is common at the base of anhydrite. Still farther south-east, in the Świętokrzyskie Góry[9], the section is entirely in carbonate facies. Basal, calcareous conglomerate can be recognized at the base followed by marly limestone, thin, platy bituminous limestone, and coarse-grained, dolomitized limestone (Szaniawski 1965).

In the eastern Central Polish Depression, salt deposits of the Lower Zechstein cycle (PZ1) have been found in the Polasie Trough (Antonowicz and Kniesner 1981, Wagner et al. 1981). They occur in a 30–40 m wide zone extending from south-east to north-west as far as the Wisla River Valley. Here, the Oldest Halite exceeds 100 m in thickness. A zone of rather thick anhydrite, which composes a sublongitudinal wall, extends west of the rock salt accumulation zone.

---

9 The Holy Cross Mountains

Another area of the Oldest Halite distribution follows the northern margins of the Central Polish Depression where it was penetrated by a number of deep boreholes near Swidwin (Poborski 1969, Marek and Znosko 1972, Antonowicz and Knieszner 1981, Wagner et al. 1981). At a depth of 3190 m, the Swidwin-2 borehole penetrated rock salt for over 50 m. Rock salt is overlain by the Upper Anhydrite, about 90 m thick.

On the north-eastern slope of the Central Polish Depression, within the Pomorsko-Kujawy Anticlinorium, the lower Zechstein cycle encompasses a 320 m anhydrite sequence forming a wall which bounds the central, subsided part of the depression to the north-east. The anhydrite wall is confined to the slope of the Kaszalin-Chojnice-Tuchola Uplift (Poborski 1969), which separates the Central Polish Depression from the Polish-Lithuanian Depression. Its apical part contains no salt either in the lower or in other Zechstein cycles.

In the interior parts of the Central Polish Depression, rock salt of the lower Zechstein cycle (PZ1) has a small thickness. A chloride-sulfate lithofacies dominates the section (Wagner et al. 1981). Figure 123 illustrates general trends of thickness variation typical of the main units of the lower Zechstein cycle in Poland.

At present, the volume of Zechstein 1 salt in the North-East German Basin and in the Central Polish Depression is difficult to evaluate. On the margins where areas of maximal salt thickness are located, the areal extent of rock salt measures probably more than 4500 km^2. The average thickness may be approximated at 100 m, thus, the volume of rock salt may exceed 4.5×10^2 km^3. In all the subsided central zones of the North-East German Basin and Central Polish Depression occupying an area of 100,000 km^2, the average thickness may be approximated at 10 m. Then the volume of salt may account for 1×10^3 km^3. The total volume of rock salt in the North-East German Basin and in the Central Polish Depression may exceed 1.45×10^3 km^3.

The Polish-Lithuanian Depression, proceeding from the lower Zechstein cycle deposits, represents an embayment of the Zechstein sea, which cuts deeply in a north-eastern direction where the accumulation of salt deposits, including potash salt, took place. The embayment was separated from the main part of the basin by the Central Pomeranian Peninsula and an anhydrite wall. In some areas of the Polish-Lithuanian Depression, salt deposits of the Upper Permian lower cycle have been studied in detail. They have been penetrated by a great number of wells near the town of Łeba in the Gdańsk area, in northern Poland, in the U.S.S.R/Poland contiguous areas, and in Lithuania.

In Wojewodztwo Olsztyńskie, the total thickness of the PZ1 deposits increases from 20 to 250—260 m from south-east to north-west, toward the Gulf of Gdańsk (Stolarczyk 1972). They vary in composition in the same direction. For example, the south-eastern areas are dominated by rocks of entirely carbonate facies, however, farther north-west, first gypsum and anhydrite, and then rock salt appear in the section. In the salt-bearing areas, the section includes, in ascending order: the Copper Slate Zechstein Limestone, Lower Anhydrite, Halite, and Upper Anhydrite. The greatest thickness for limestone (50—80 m) were recorded to the south-east, while the maximal thicknesses of the Lower Anhydrite are known farther north-west and west. The greatest thickness (up to 150 m) is confined to two longitudinal zones in the Pastek and Zaremba-Dembowec areas. Local anhydrite walls are marked by these zones. The areas where the greatest thicknesses of rock salt (up to 100–150 m) were recorded are situated either between anhydrite walls or on their slopes. In their apical parts, the thickness of rock salt decreases to 50 m.

A salt sequence can be traced farther north-west near Gdańsk and Łeba. In the latter area, it is about 100 m thick. Interbeds of polyhalite rock, as well as inclusions of carnallite, kainite, glaserite, and löweite are present north of Gdańsk.

In Lithuania, equivalents of the lower Zechstein cycle were subdivided by Suveizdis (1963), in ascending order: Kalvar, Sosnava, Novoakmyansk, and Pregol Formations. These formations can be correlated with the Werra Series horizons. The Kalvar, Sosnava, and Novoakmyansk Formations are equivalents of Zechsteinkonglomerate C1, Kupferschiefer T1, and Zechsteinkalk Ca1, respectively; and the Pregol Formation corresponds to the upper Werra Series containing the Unterer Werra-Anhydrit, Werra-Steinsalz, and Oberer Werra-Anhydrit. For composition of the above formations see Fig. 124.

Salt deposits in the Polish-Lithuanian Depression cover an area of about 35,000 km^2. The thickness of salt may reach 100–150 m and more, averaging 50–100 m. An approximate estimate of the average thickness of rock salt may be 50 m. Then the estimated total volume of halite in the Werra Series of the Polish-Lithuanian Depression is 1.7×10^3 km^3.

In the western Central European Basin, within the English Basin, the lower English Zechstein cycle (EZ1) deposits are assigned to the Don Group. It is divided, in ascending order, into the Marl Slate, Lower and Middle Magnesian Limestones, and Hartlepool (Hayton) Anhydrite (D. Smith 1972, 1981, Taylor and Colter 1975). The Marl Slate is represented by finely laminated, argillaceous, silty dolomite or limestone. Their thickness is 6 m. The Lower and Middle Magnesian Limestones reach a maximal thickness (up to 100–150 m) in the coastal areas of the basin. Eastward, the thickness decreases to a few meters or 10–15 m. Deep-water, finely laminated carbonates are developed there. Biogenic structures and reefs are common in the coastal areas. The Hartlepool Anhydrite, which continues southwards as the Hayton Anhydrite in Yorkshire, is a lenticular body about 25 km wide and up to 160 m thick. It girdles the English Zechstein Basin and is the equivalent of the Werra-Anhydrit of the German Zechstein sequences. Anhydrite builds up a sublongitudinal anhydrite wall. In the subsided, interior parts of the basin, its thickness decreases to a few meters.

In the English Basin, no thick rock salt beds have yet been found in the lower Zechstein cycle. Halite inclusions have been recorded in anhydrite, which was penetrated by the Shell 49/26-4 well (Taylor and Colter 1975).

In the Manx-Furness and Solway Firth Basins, the Basal Breccia and the Saltom Siltstone may be assigned to the lower Zechstein cycle (Arthurton and Hemingway 1972). These essentially terrigenous deposits might have accumulated in the coastal part of the basin.

The above shows that the Werra salt deposits in the Central European Basin have not been uniformly studied. They have been studied in detail in potash districts of the Werra Fulda Depression, Lower Rhine Basin, and Thüringia Depression. In many interior parts of the basin, the series has been penetrated by few wells due to the great depth of salt occurrence. Therefore, only an approximate estimate of the total volume of rock salt of the Werra Series can be given. Proceeding from the previous calculations, it may amount to no less than 6.0×10^3 km^3.

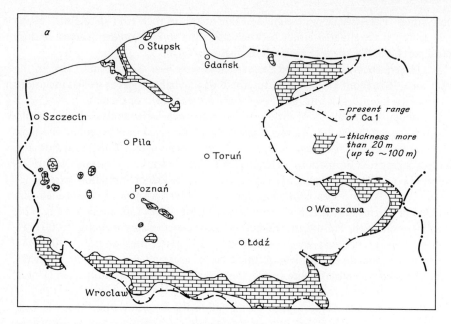

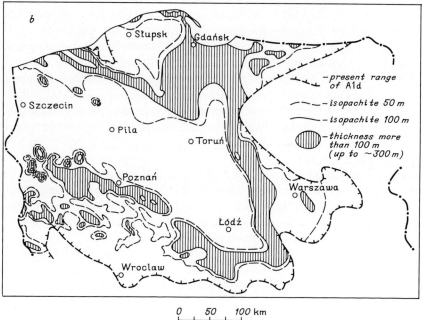

Fig. 123a, b

Fig. 123a–c. Simplified isopach maps of Zechstein Limestone (a), Lower Anhydrite (b), Oldest Halite (c), and Upper Anhydrite (d) in Central Polish Depression. After Antonowicz and Knieszner (1981)

Central European Basin

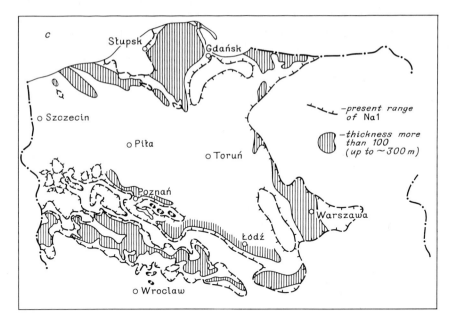

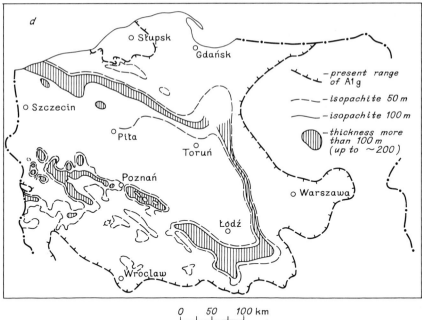

Fig. 123c, d

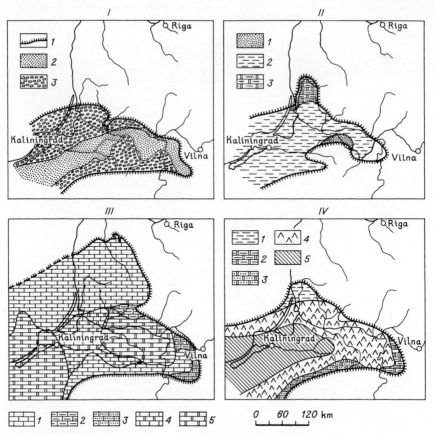

Fig. 124. Lithologic maps of Kalvar, Sosnava, Novoakmyansk, and Pregol Formations in Polish-Lithuanian Depression. After Suveizdis (1963). *I* Kalvar Formation: *1* depositional limits (for all maps), *2* sandstone, *3* conglomerate, gritstone, and coarse-grained sandstone. *II* Sosnava Formation: *1* sandy bituminous deposits, *2* clayey bituminous deposits, *3* clayey dolomite. *III* Novoakmyansk Formation: *1* limestone, *2* clayey limestone, *3* sandy limestone, *4* limestone interbedded with dolomite, *5* dolomite. *IV* Pregol Formation: *1* clay, *2* clayey dolomite, *3* sandy dolomite, *4* anhydrite and gypsum, *5* rock salt

Zechstein 2 (Stassfurt Series)

Salt deposits of the Stassfurt Series occur throughout the Central European Basin (Fig. 125). They have been found both to the west in the North Sea, in the English, Norwegian-Danish, and North-West German Basins and to the east in the North-East German Basin, and in the Central Polish and Lithuanian Depressions. Salt strata are also very common in the Weser, Subhercynian, Thüringia and Werra Fulda Depressions. In the Stassfurt Series salt deposits, potash salt is rather widespread and covers the inner Middle European Depression, central parts of the Weser, Subhercynian, and Thüringia Depressions like an almost unbroken mantle. Sulfate, sulfate-carbonate, and carbonate strata usually flank the basin. Terrigenous deposits are known from

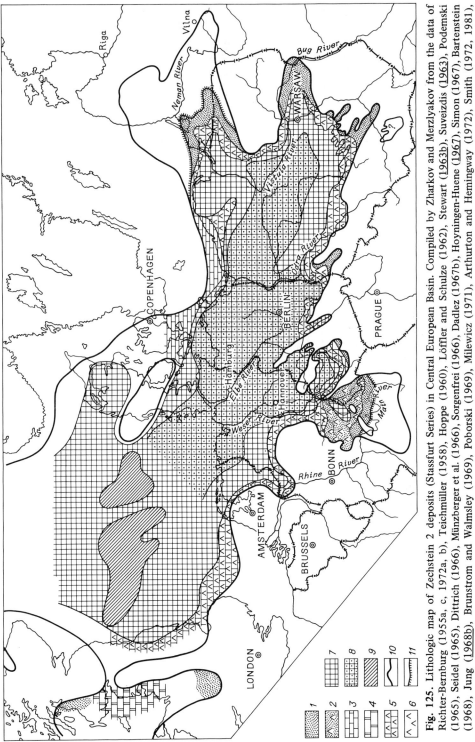

Fig. 125. Lithologic map of Zechstein 2 deposits (Stassfurt Series) in Central European Basin. Compiled by Zharkov and Merzlyakov from the data of Richter-Bernburg (1955a, c, 1972a, b), Teichmüller (1958), Hoppe (1960), Löffler and Schulze (1962), Stewart (1963b), Suveizdis (1963), Podemski (1965), Seidel (1965), Dittrich (1966), Münzberger et al. (1966), Sorgenfrei (1966), Dadlez (1967b), Hoyningen-Huene (1967), Simon (1967), Bartenstein (1968), Jung (1968b), Brunstrom and Walmsley (1969), Poborski (1969), Milewicz (1971), Arthurton and Hemingway (1972), Smith (1972, 1981), Słowarczyk (1972), Antonowicz and Kniesner (1981), Tomaszewski (1981), and Wagner et al. (1981). Areal distribution of: *1* terrigenous rock, *2* terrigenous (mainly clayey) rock with gypsum and anhydrite, *3* carbonate rock (mainly dolomite), *4* dolomite within Manx-Furness and Solway Firth Basins, *5* dolomite and anhydrite, *6* mainly anhydrite, *7* rock salt, *8* potash salt, *9* salt deposits absent on uplifts in North Sea, *10* areal extent of Upper Permian (Zechstein) deposits, *11* erosional edge of Stassfurt Series

the extreme north-eastern parts of the coastal zone of the Polish-Lithuanian Depression, to the south, in the Franconia Basin, Werra Fulda and Thüringia Depressions, in the northern Solway Firth, and southern Manx-Furness Basins.

The Stassfurt Series is best known in the Thüringia, Subhercynian, and Weser Depressions where it was penetrated by numerous deep boreholes and mined for potash salt. It is in this area that the potash districts of South Harz, Unstrut, Halle, and Mansfeld are located in the Thüringia Depression. The Subhercynian Depression encloses Aschersleben, Bernburg, Stassfurt, Schwanebeck, Schönebeck, and Allertal potash districts, while Ronnenberg-Hansa and Sarstedt-Lehrte districts are situated in the Weser Depression.

The composite section in the Thüringia Depression is composed of Stassfurt-Ton (T2), 0–2 m thick; Hauptdolomit (Ca2), 0–60 m; Basalanhydrit (A2), up to 30 m thick; Stassfurt-Steinsalz (Na2), 0–500 m; Kaliflöz Stassfurt (K2), up to 20 m; Decksteinsalz (Na2r), up to 10 m thick; and Deckanhydrit (A2r), not more than 5 m. The most subsided part of the depression, which is situated in the middle and lower Unstrut River, is composed of the thick Stassfurt-Steinsalz which wedges out near the margins. In the area with the greatest salt thickness, the underlying deposits of Hauptdolomit and Basalanhydrit do not exceed 20 m and have a peculiar composition, namely, they are represented by Stinkschiefer and mainly thin-laminated anhydrite. West, south, and east of the area, the thickness of Basalanhydrit increases and forms a belt of an anhydrite wall. Approaching the foothills, carbonate rocks become thicker and dominate the section. In the southern and south-eastern parts of the depression, a clastic mainly argillaceous facies of the Stassfurt Series is distributed (Fig. 126).

Carbonate deposits of the Stassfurt Series (Ca2) are represented mainly by two facies: (1) Stinkschiefer (Ca2st) or Unstrut and (2) dolomite (Hauptdolomit, Ca2d). The dolomite facies or Hauptdolomit in the most complete sections outside the Stinkschiefer terrain, i.e., in the western and southern Thüringia Depression, is subdivided into five stratigraphic zones, in ascending order: (1) dense, thin-platy crystalline dolomite (Ca2α), 0–5 m thick; (2) coquina dolomite (Ca2β), 0–2 m; (3) algal dolomite (Ca2γ), ranging in thickness from 0–5 m; (4) dark, dense dolomite (Ca2δ), 0–10 m; and (5) light, dense, and oolitic dolomite (Ca2ϵ), 0–20 m. When traced from west to east and from south to north, the above zones of Hauptdolomit successively wedge out in descending order until complete replacement by Stinkschiefer (Ca2st). This facies change pattern is shown in Table 5 and Fig. 127, I.

Basalanhydrit (A2) in the Thüringia Depression is 0–30 m thick. To the southeast, in the most subsided part, the thickness does not exceed 10 m. Farther west, east, and south, the thickness increases to 30 m, averaging about 20 m. It is in this area of increased thickness that an anhydrite wall forms a semicircle around the subsided zone of the depression. Still farther west, south, and east, Basalanhydrit gradually thins out; carbonate rocks appear in the section; and clayey varieties become dominant. Seidel (1965) distinguished three facies of Basalanhydrit (see Fig. 127, II) within the Thüringia Deression: (1) Unstrut, developed to the south-east and represented by thin-laminated and flaser anhydrite; (2) South Harz, composed mainly of flaser anhydrite; and (3) marginal, containing flaser anhydrite with clayey inclusions.

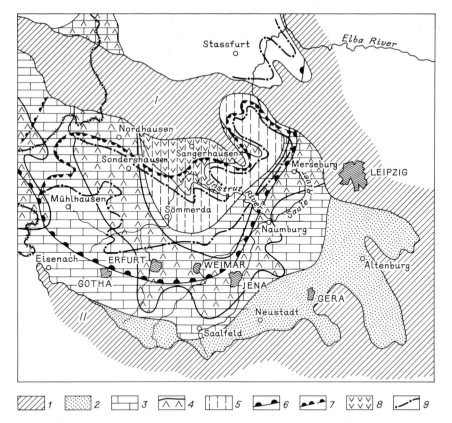

Fig. 126. Lithologic map of Stassfurt Series in Thüringia Depression. Compiled from the data of Seidel (1965), Münzberger et al. (1966) and Jung (1968b). *1* pre-Zechstein outcrops (*I* Harz, *II* Thüringer Wald), *2* clastic deposits, *3* Stassfurt carbonate deposits represented mainly by dolomite (Hauptdolomit) and over 10 m thick, *4* Basalanhydrit of Stassfurt Series over 10 m thick. *5* Stinkschiefer at the base of Stassfurt Series, *6* primary sedimentary boundary of rock salt, *7* rock salt related to erosion and salt leaching, *8* anhydrite of Sangerhausen facies, *9* Kaliflöz Stassfurt

In lithology, Basalanhydrit of the Thüringia Depression is divided into three horizons (Jung 1966, Seidel 1965, Döhner 1970b), labeled, in ascending order: A2α, A2β, and A2γ. Their description and location are given in Table 6.

The thickness of Stassfurt-Steinsalz (Na2) gradually increases from south to north where it reaches 400 m and more (see Fig. 127, III'). Farther noth-east, in the Mansfeld Syncline, rock salt is more than 500 m thick (Schulze 1958, Löffler and Schulze 1962, Simon 1967).

Richter-Bernburg (1955b) subdivided the Stassfurt Steinsalz in the Thüringia Depression into two parts: the lower one, "Älteres" Steinsalz and upper one, Übergangsschichten. "Älteres" Steinsalz can be divided into three parts: Basalsalz (Na2α), Hauptsalz (Na2β), and Hangendsalz (Na2γ). According to Bischoff, the lower part of the salt sequences is still assigned to the Anhydritregion (Jung and Lorenz 1968).

Table 5. Section types in Stassfurt carbonate deposits in the Thüringia Depression. (Adapted from Seidel 1965)

Marginal sections of Hauptdolomit	Hauptdolomit type section at Forstberg. (After Deubel 1954, Gaertner 1954)	Hauptdolomit type section at Volkenrode. (After Deubel 1954, Gaertner 1954)	Well log in Ludwigshall-Immenrode mine (Gaertner 1954)	Well log south-east of Sondershausen	South-eastern slope of the Harz. (After Jankowski and Jung 1964)
1	2	3	4	5	6
Basalanhydrit	Basalanhydrit	Basalanhydrit			
	Zone of light, dense and oolitic dolomite (zone 5)				
	Zone of dense dolomite	Zone of dense dolomite (zone 4)	Basalanhydrit	Basalanhydrit	Basalanhydrit
Undifferentiated, light, dense, and oolitic dolomite with clayey interbeds at the base	Algal zone (zone 3)	Algal zone (zone 3)	Algal zone	Stinkschiefer	$\beta_2 + \gamma$
	Coquina zone Muschelzone (zone 2)	Muschelzone (zone 2)	Muschelzone	Mueschelzone	β_1
	Zone of thin-platy crystalline dolomite	Zone of thin-platy crystalline dolomite	Light, oolitic, dark, dense dolomite	Light, oolitic, and dense dolomite	Stinkschiefer α
Werra-Anhydrit	Werra-Anhydrit	Werra-Anhydrit	Werra-Anhydrit	Alternation of anhydrite and dolomite	Werra-Anhydrit
				Werra-Anhydrit	

Seidel (1961, 1965) subdivided Stassfurt-Steinsalz into three parts: the lower, Basissalz (Na2α), the middle, Anhydritregion (Na2β), and the upper, Südharzsteinsalz (Na2γ), which is in turn divided into Südharzbasissalz (Na2γ$_1$), ungebändertes (not banded) Südharzsteinsalz (Na2γ$_2$), and gebändertes (banded) Südharzsteinsalz (Na2γ$_3$). A similar subdivision was proposed by Heynke and Zänker (1970) for the South Harz area. Different options of Stassfurt-Steinsalz division in the Thüringia Depression and correlation are shown in Table 7.

Seidel (1965) showed that proposed stratigraphic units of Stassfurt-Steinsalz are not common for the entire Thüringia Depression. So, the two lower units (Basissalz Na2α and Anhydritregion Na2β) were found only to the north-east, in the middle Unstrut River, and in the Mansfeld Syncline. As a rule, they wedge out south-, west-, and eastward near an anhydrite wall. The overlying Südharzsteinsalz (Na2γ) has a wider distribution. It is developed both in the north-eastern, subsided part of the depression and farther south, west, and east where Basissalz (Na2α) and Anhydritregion (Na2β) do not occur. In these areas, rock salt basal beds belong to Südharzbasissalz

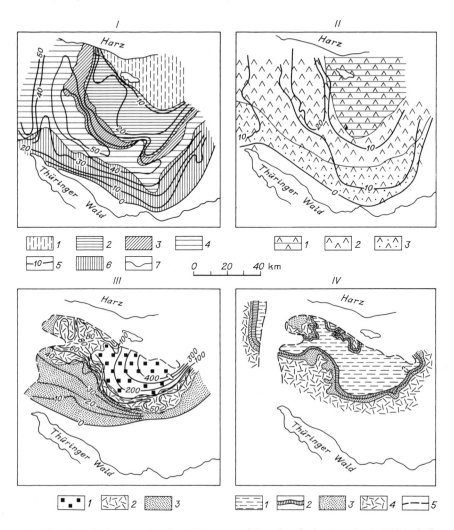

Fig. 127. Lithologic maps showing thicknesses of Stassfurt Series deposits in Thüringia Depression. After Seidel (1965). *I* Stassfurt carbonate deposits (Ca2): *1* distribution of Stinkschiefer, *2* distribution of thin-platy oolitic and dense dolomite with coquina zone, algal zone, and Stinkschiefer, *3* thin-platy, coquina algal dolomite and light dense dolomite, *4* same as *3* but upper zone of light, dense, and oolitic dolomite more common, *5* isopach contours, in meters (for all maps), *6* light, dense, and oolitic dolomite, *7* limit of different types of deposits (for all maps). *II* Basalanhydrit (A2): *1* distribution of thin-platy and flaser anhydrite, *2* mainly flaser anhydrite, *3* flaser anhydrite with clay admixture. *III* Stassfurt-Steinsalz (Na2): *1* Anhydritregion and Südharzsteinsalz; *2* ungebändertes Südharzsteinsalz und gebändertes Südharzsteinsalz; *3* gebändertes Südharzsteinsalz. Isopach contours show aggregate thickness of Stassfurt-Steinsalz (Na2), Kaliflöz Stassfurt (K2), and Decksteinsalz (Na2r). *IV* Kaliflöz Stassfurt (K2): *1* distribution of mainly carnallite rock, *2* mainly mixed potash salts, *3* mainly Hartsalz, *4* mainly barren equivalents of Kaliflöz Stassfurt, *5* limits of secondary leaching (for *III* and *IV*)

Table 6. Stratigraphic subdivision of Basalanhydrit (A2) in the Thüringia Depression (Döhner 1970b)

Marginal zones of troughs			Inner zone (Scheidtal 1 and 2)	South Harz zone
After Richter-Bernburg (1957)	After Hoyningen-Huene (1957)	After Jung (1960)	After Seidel (1965)	
1	2	3	4	5
Flaseranhydrit	III – flaser platy-fascicular anhydrite	γ (0.8–1.2 m) – anhydrite, dolomitic, uneven, thin, or thick-banded, locally in the upper part even-banded	γ (0.2 m) – anhydrite, pegmatitic with rock salt partings	γ (1–10 m) – anhydrite, pegmatitic with rock salt veinlets
	II – flaser anhydrite		β (3 m) – anhydrite, clayey and dolomitic, irregularly rimmed, thin- and thick-banded	β (3–13 m) – anhydrite, clayey and dolomitic, irregularly rimmed, thin- and thick-banded partly pegmatitic with rock salt veinlets
		β (0.7–0.9 m) – anhydrite with dolomite veinlets		
Linienanhydrit	I – platy anhydrite	α (0.4–0.7 m) – anhydrite, dolomitic, evenly-banded. Lamination often wavy; pure, pearly anhydrite at the base	α (5 m) – anhydrite, pegmatitic with rock salt veinlets, partly dolomitic, even and densely banded	α (1–3 m) – anhydrite, carbonate, usually evenly fine-banded

Table 7. Stratigraphic subdivision of Stassfurt-Steinsalz in Thüringia Depression (Döhner 1970a)

Unstrut District			South Harz District		
After Lotze (1938), and Richter-Bernburg (1955)	After Tinnes (1928)	After Jung and Lorenz (1968)		After Seidel (1961, 1965)	After Heynke and Zänker (1970)
1	2	3	4	5	6
Übergangsschichten	Übergangs-schichten	Anhydrit-region	γ	γ_3 – gebändertes Südharzsalz	γ_2 – gebändertes Südharzsalz
				γ_2 – ungebändertes Südharzsalz	γ_1 – ungebändertes Südharzsalz
				γ_1 – Südharzbasissalz	
"Älteres" Steinsalz	γ – Hangendsalz β – Hauptsalz α – Basalsalz	Anhydrit-region	β	β – Anhydrit-region	β – Anhydrit-region
			α	α – Basissalz	

($Na2\gamma_1$) and rest directly on Basalanhydrit (A2). Near the margins, ungebändertes Südharzsteinsalz ($Na2\gamma_2$) thins out. The uppermost gebändertes Südharzsteinsalz ($Na2\gamma_3$) is known only from the southern-, western-, and easternmost areas of the depression.

Kaliflöz Stassfurt (K2) is confined to the northern Thüringia Depression. It is composed of rock salt intercalated with various salts: carnallite, sylvinite, kieserite, langbeinite, löweite, kainite, and mixed salts. Potash salts occur also in the lower parts of the section within Übergangsschichten ($Na2\gamma$) where locally kieserite and polyhalite rocks are widespread. In composition, Übergangsschichten is subdivided into the lower, Polyhalitregion and the upper, Kieseritregion.

Tinnes (1928) was the first to propose a principal scheme, which is still valid, for the subdivision of Kaliflöz Stassfurt of the Unstrut potash district. He subdivided the Kaliflöz Stassfurt into the Lower and Upper Groups (Liegendgruppe and Hangendgruppe). Ten marker bands were distinguished within the Liegendgruppe, which were numbered in ascending order, e.g., 1. Unstrutbank, 2. Unstrutbank, etc. (1st UnstrutBand, etc.). There are intermediate bands (Zwischenmittel) between them composed of different potash salts. So, in the Georg-Unstrut mine, they are represented generally by langbeinite rocks. In other areas, carnallite rocks, Hartsalz, and similar rocks are found. The thicknesses of rock salt and potash salt bands for the Lower Group (Liegendgruppe) are shown in Table 8.

Table 8. Thickness of rock salt marker bands and potash salt interbeds in the Lower Group of Kaliflöz Stassfurt. Adapted from Tinnes (Löffler and Schulze 1962)

Stratigraphic unit	Thickness (m)	Remarks
10. Unstrutbank	0.45	Anhydrite, string-like laminae occur 3 cm above the base
Zwischenmittel	0.4	
9. Unstrutbank	0.5	Clayey marker bed at the top
Zwischenmittel	3.0	
8. Unstrutbank	0.32	
Zwischenmittel	0.4	
7. Unstrutbank	0.32	Locally of reddish tint
Zwischenmittel	0.4	
6. Unstrutbank	0.5	Peculiar clear-cut clayey bed at the base
Zwischenmittel	0.18	
5. Unstrutbank	0.45	Clayey bed at the base
Zwischenmittel	1.6	
4. Unstrutbank	0.3	
Zwischenmittel	2.35	
3. Unstrutbank	0.75	
Zwischenmittel	1.15	
2. Unstrutbank	0.45	
Zwischenmittel	0.75	
1. Unstrutbank	0.3	

The Upper Group (Hangendgruppe) of Kaliflöz Stassfurt (K2) in the Unstrut District consists of potash salts with distinct rock salt marking bands 10–12 cm to 1.5 m thick; the latter, locally, have proper names. Mötzung (1963), for example, recognized the following rock salt marking bands within the lower Upper Group (Hangendgruppe): Ahlbornbank-1, Ahlbornbank-2, Einfache Lage, Doppelte Lage, and above them in the upper part, bands are labeled K, L, M, N, and O. Schmale Lage (Narrow Band) can be traced below band M. The bands with letter indexes were recognized by analogy with the subdivision accepted for the potash-bearing North Harz District in the Subhercynian Depression.

The main economic potash salt horizons are associated with Hangendgruppe (Upper Group). They usually have a complex structure and contain various amounts of halite, sylvite, kieserite, langbeinite, kainite, carnallite, polyhalite, and anhydrite. Löffler and Schulze (1962) provided a detailed description for the structure of Kaliflöz Stassfurt (K2) in the Unstrut District.

In the South Harz District, Kaliflöz Stassfurt also contains various potash salts with rock salt interbeds (Stolle 1967, Döhner 1970a, b). Nineteen marking halite bands were recognized there. The lower ten bands belong to the Lower Group (Liegendgruppe) and are correlated with the corresponding marker bands of the Unstrut District; the upper nine belong to the Upper Group (Hangendgruppe).

The composition of Kaliflöz Stassfurt in the Thüringia Depression shows a regular change across the area (see Fig. 127, IV). It is noteworthy that a central zone is dominated by carnallite rocks. To the south and in places to the north and west, it is bordered by a belt containing mainly mixed potash salts. Farther south, there is another belt dominated by Hartsalz rocks. Ore-free equivalents of the potash band occur along the margins of the Thüringia Depression.

In the Halle and Mansfeld potash districts, Kaliflöz Stassfurt is subdivided into the lower, middle, and upper parts; they are, respectively, about 13 m, 8 m, and 11–12 m thick. Rock salt marker bands are easily discernible there, especially the tenth Unstrutbank (10th Unstrut Band), which lies at the top of the lower part of the section. This suggests that the lower Kaliflöz Stassfurt (K2) in the Mansfeld Syneclise and Halle District is an equivalent of the Lower Group (Liegendgruppe) of the Unstrut District and its middle and upper part is correlated with the Upper Group (Hangendgruppe) of the same region. There are areas entirely composed of carnallite rocks. In the Mansfeld Syneclise, the upper and middle part of the section locally contains kieserite and tachhydrite rocks.

In the northern Thüringia Depression, a peculiar anhydrite sequence known as the Sangerhausen Anhydrit, has developed (Langbein and Seidel 1960, Löffler and Schulze 1962, Jung 1966, 1968b). It is supposed to be an age equivalent of the Stassfurt-Steinsalz (Na2) and Kaliflöz Stassfurt (K2). The upper Sangerhausen Anhydrit may also be correlated with Decksteinsalz.

The Stassfurt Series in the Thüringia Depression is capped by thin Decksteinsalz (Na2r) and Deckanhydrit (A2r). The former is generally represented by coarse, spar halite, minor kieserite, carnallite, and anhydrite, as well as inclusions and admixture of halopelite. Deckanhydrit is 1.5–5 m thick and consists of clayey anhydrite with rare inclusions and veinlets of rock salt and is locally carnallite.

In the Subhercynian Depression, the Stassfurt Series is widespread. Its total thickness reaches 550–600 m, rock salt making up most of the section. In some areas, the thickness of the salt sequence locally increases to 800 m and even 1200 m. The following units can be recognized up the section: Stinkschiefer (Ca2st), Basalanhydrit (A2), Stassfurt-Steinsalz (Na2), Kaliflöz Stassfurt (K2), Decksteinsalz (Na2r), and Deckanhydrit (A2r). The thickness of these major stratigraphic units for different potash districts is shown in Table 9.

Table 9. Thickness of main stratigraphic units of the Stassfurt Series in the Subhercynian Depression. Adapted from Löffler (Löffler and Schulze 1962) and Meier (1969)

Stratigraphic unit	Thickness (m)					
	Aschersleben	Bernburg	Stassfurt	Schönebeck	Allertal	Schwanebeck
Deckanhydrit (A2r)	1–2	1–1.5	1.7	0.5	1.5	1.75
Decksteinsalz (Na2r)	1	1.25	0.8	–	0.3–2.5	1.75
Kaliflöz Stassfurt (K2)	45	30–40	30–40	4	6–8	17.3
Stassfurt-Steinsalz (Na2)	400–475	300–500	above 500	20–160	100–200	up to 1198.4
Basalanhydrit (A2)	2	2.3	2.3	2	2	2
Stinkschiefer (Ca2st)	4–6	5–10	5–10	5	4–7	9

Basal carbonate deposits of the Stassfurt Series throughout most of the Subhercynian Depression are represented by Stinkschiefer (Ca2st) having a small thickness. It is traced also farther west both in the Weser Depression and in subsided parts of the North-East German Basin (Münzberger et al. 1966, Jung 1968b). Dolomites similar to those of the Thüringia Depression occur only in the south-westernmost Subhercynian Depression. They are placed into Hauptdolomit and are represented by light, dense, and oolitic, locally algal varieties.

In general, Basalanhydrit (A2) in the Subhercynian Depression has a small thickness. It is represented dominantly by thin-banded and flaser anhydrite and anhydrite varvite. In two areas, namely, to the south-west and in the center west of Schwanebeck, the thickness of Basalanhydrit is more than 10 m.

Stassfurt-Steinsalz (Na2) in the Subhercynian Depression was subdivided as early as 1875 by Bischoff into three parts, namely, the lower, Anhydritregion; the middle, Polyhalitregion, and the upper, Kieseritregion. Later, minor corrections were introduced in this subdivision. Some investigators (Knak 1960) recognized an interbed (Übergangsschichten) above the kieserite zone. Others placed polyhalite and kieserite parts into Übergangsschichten (Lotze 1938, Richter-Bernburg 1955a, Löffler and Schulze 1962, Döhner 1970b).

Anhydritregion encloses most of Stassfurt-Steinsalz. Its thickness is mainly 400–500 m. Only the uppermost part of the rock salt section is built up by Übergangsschichten 20–50 m thick. The latter is composed of halite, glauberite, löweite, polyhalite, langbeinite, and kieserite in various amounts.

Kaliflöz Stassfurt (K2), as in the Thüringia Depression, is represented by the multiple alternation of rock salt forming marker bands and potash salt of carnallite, Hartsalz, or mixed composition. Kaliflöz Stassfurt is also subdivided into the Lower and Upper Group (Liegendgruppe and Hangendgruppe). It is composed of 16 marker rock salt bands labeled, in ascending order, A–P. Bands A, B, C, D, E, and F occur within Liegendgruppe; all the overlying ones belong to Hangendgruppe. Their correlation with the Unstrut District of the Thüringia Depression is as follows: F is correlated with the 10th Unstrut Band (10. Unstrutbank), E – with the 8th, C – with the 5th, and B — with the 4th. Bands 1, 2, and 3 of the Unstrut District have equivalents in the Subhercynian Depression in the section which lies between Bands A and B.

In the Aschersleben District of the Subhercynian Depression, two types of the Kaliflöz Stassfurt section, namely, the carnallite and Hartsalz sections, can be recognized. The thickness of the carnallite and Hartsalz sections is, respectively, 37 m and 27 m. It decreases at the expense of the Upper Group.

Carnallite sections are marked by the presence of carnallite rocks with various amounts of sylvinite, polyhalite, and glauberite rocks. Depending on the amount of the latter, Kaliflöz Stassfurt is often divided into the following zones, in ascending order: carnallite-anhydrite, polyhalite-sylvinite, polyhalite, which is assigned to Hangendgruppe, and glauberite, which is assigned to Liegendgruppe. Kieserite, langbeinite, löweite, sylvinite, glauberite, and anhydrite rocks are developed in a section of Hartsalz type. A zonal subdivision of Kaliflöz Stassfurt in a section of the Hartsalz type is as follows: kieserite-sylvinite, kieserite-langbeinite, and löweite zones are recognized in Hangendgruppe, while Liegendgruppe encloses the glauberite-anhydrite zone. In places, the kieserite-langbeinite zone comprises both the lower part of Hangendgruppe and the upper part of Liegendgruppe, while the löweite zone is associated with the lower part of Liegendgruppe and Übergangsschichten. In this case, the glauberite-anhydrite zone lies at the base of Übergangsschichten. Such a zonation is characteristic of Kleinschiersted I/II of the Aschersleben District.

Kaliflöz Stassfurt (K2) in the Bernburg potash district can be divided into Hangendgruppe and Liegendgruppe, which contain rock salt marker beds and interbeds (Zwischenmittel) composed of potash salts. Here the same two types of section are recognized, namely, carnallite and Hartsalz types. However, the latter is much more common. In general, potash salts are related to Hangendgruppe. In places, sections of Hartsalz type are characterized by the following succession of potash salt complexes:

Hangendgruppe

Kieserite-carnallite-halite banded complex	2.5 m
Kieserite-sylvite-halite complex (Hartsalz)	2.0 m
Sylvite-kieserite-halite complex (8% kieserite, 5% sylvite)	1.5 m
Langbeinite-halite complex	1.5 m
Vanthoffite-löweite-halite complex (6% löweite)	1.0 m
Löweite-vanthoffite-halite complex (6% vanthoffite, 4% löweite)	1.0 m
Polyhalite-halite complex (up to 9% polyhalite)	about 10.0 m
Anhydrite-halite complex (4–5% anhydrite)	about 16.0 m

Liegendgruppe

Anhydrite-halite complex (4–5% anhydrite)

Kaliflöz Stassfurt of the Subhercynian Depression was thoroughly studied. Its description and subdivision are available for many potash-bearing areas. There are also data concerning the structure and composition of some parts of the section. For example, Meier (1969) studied rock salt marker bands K and L and an interbed (Zwischenmittel) lying in between.

The Stassfurt Series in the Subhercynian Depression is capped by thin Decksteinsalz (Na2r) with Deckanhydrit (A2r) above. It is believed by some investigators that they cannot be treated as certain stratigraphic units, but that they are lithological units occupying different stratigraphic positions and marking a peculiar recessive facies of Kaliflöz Stassfurt (K2), which when traced from south to north from the South Harz to the North Harz District, show higher and higher stratigraphic levels (Döhner 1970a).

In the Weser Depression, all the major stratigraphic units of the Stassfurt Series (Fig. 128) are easily discernible. In the southern areas of the depression and in adjacent

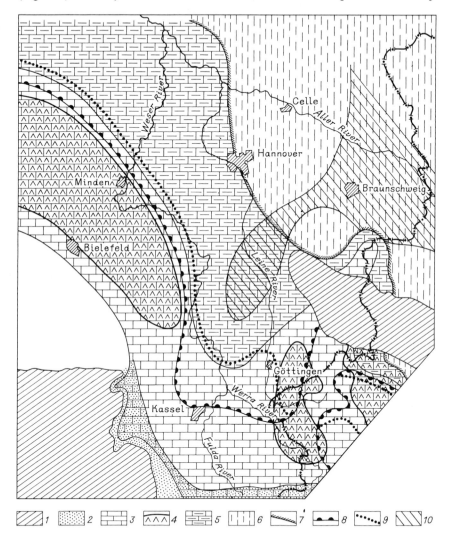

Fig. 128. Lithologic map of Stassfurt Series deposits in Weser Depression. Compiled from the data of Richter-Bernburg (1955c, 1972a, b), Münzberger et al. (1966), and Jung (1968b). *1* pre-Zechstein outcrops. Areal distribution of: *2* terrigenous, mainly pelitic deposits, *3* carbonate deposits (Ca2) represented by dolomite and placed into Hauptdolomit (Ca2d), *4* Basalanhydrit (A2) over 10 m thick and in places 20–25 m and builds up anhydrite walls, *5* carbonate (Ca2) represented by alternate dolomite and Stinkschiefer, *6* carbonate represented by Stinkschiefer (Ca2st), *7* south and south-westward extent of Stinkschiefer, *8* limit of Stassfurt-Steinsalz (Na2), *9* southward extent of Kaliflöz Stassfurt (K2), *10* Kaliflöz Stassfurt (K2) section of carnallite type

parts of the Werra Fulda Depression at the base of the series, a basal horizon of Braunroter Salzton (T2), 1–2 m or several meters thick, is found. To the north, Braunroter Salzton wedges out and in the inner Weser Depression, carbonate deposits assigned, on the basis of their composition, either to Hauptdolomit (Ca2d) or to Stinkschiefer (Ca2st), occur at the base of the series. A spatial distribution of these basal carbonate facies of the Zechstein second cycle seems to be quite regular. In the southern and south-western Weser Depression, dolomite ranging in thickness from 20–50 m and above, occurs. The greatest thickness of Hauptdolomit was recorded in the marginal south-western areas west of the Weser River and in the south-easternmost part, in the upper Leine River, where reef buildups are developed.

Starting approximately from the Weser Valley and farther east and north-east, they appear in thin-platy, bituminous varieties of dolomite and in interbeds of Stinkschiefer, which form a belt occupying the Weser-Leine Interfluve and extending from south-east to north-west. The total thickness of the Stassfurt Carbonate (Ca2) does not exceed 10–15 m in this area. Stinkschiefer (Ca2st) is developed north-east of this belt in the middle and upper Aller Basin, that is, in the boundary areas of the Weser, Subhercynian and Altmark Depressions, and North-West German Basin. It fills the most subsided parts of the salt basin. The thickness is 6–7 m, averaging 4–5 m.

Basalanhydrit (A2) conformably overlying carbonate deposits shows a considerable facies change in the Weser Depression. In the inner subsided zones, it has a small thickness, not more than 2–3 m. Anhydrite of the same thickness occurs in the areas where underlying carbonate is represented by Stinkschiefer. Most common there are varvite anhydrite, thin-banded, bituminous, and linear. Farther south-west and south, flaser, nodular, and pearly varieties appear in anhydrites; the thickness of the latter increases to 10–15 m. Finally, even farther south-west and south-east in the adjacent Thüringia Depression, anhydrites become more massive. They build up two extensive anhydrite walls (see Fig. 128). Clayey interbeds appear in Basalanhydrit on the southern and south-western margins of the Weser Depression.

Stassfurt-Steinsalz (Na2) is common throughout the central Weser Depression. The thickness of the salt sequence increases gradually to 800 m and more from the margins to the center of the depression. To the north in the Aller Basin and in the Lower Leine River, the thickness of salt probably exceeds 1000 m. In the most complete sections, the Stassfurt-Steinsalz can be divided into three parts: Basalsalz (Na2α), Hauptsalz (Na2β), and Hangendsalz (Na2γ). Furthermore, a thin kieserite-bearing interbed, Übergangsschichten (Na2k), is recognized at the very top (Lotze 1938, Richter-Bernburg 1955b). Hauptsalz and Basalsalz occur only in the subsided zones of the depression. They thin out in marginal south-western and south-eastern areas where the uppermost horizons of Hauptsalz and Hangendsalz, or only the latter, are developed. This implies that the decrease in thickness of Stassfurt-Steinsalz from the center to the periphery of the depression is due to the omission of the lower units; first, Basalsalz and then Hauptsalz. Hangendsalz occupies a larger area, including marginal parts. Such a mode of occurrence of Stassfurt-Steinsalz was studied in detail by Simon (1967) for the south-eastern Weser Depression. He showed that Stassfurt-Steinsalz was about 70 m thick near the old consedimentary Eichsfeld Swell, which separates the Weser Depression from the Thüringia Depression. It is subdivided into two parts, namely, Unteres (Lower) Stassfurt-Steinsalz (Na2a) and

Oberes (Upper) Stassfurt-Steinsalz (Na2b). The thickness of the former and the latter is, respectively, 50 m and 20 m. These horizons are correlated with a section of the central parts of the depression as follows: Unteres Stassfurt-Steinsalz is an equivalent of the upper part of Hauptsalz (Na2β); and Oberes Stassfurt-Steinsalz is correlative mainly to Hangendsalz (Na2γ) and kieserite-bearing Übergangsschichten (Na2K).

In the Weser Depression, Kaliflöz Stassfurt (K2) occupies almost the same area as that under Stassfurt-Steinsalz (Na2). Kalflöz Stassfurt is 5–10 m thick. As in the Thüringia and Subhercynian Depressions, two types of sections — carnallite and Hartsalz — were recognized. The former is developed in the south-eastern part of the depression and extends from south-west to north-east as a belt widening towards the upper Aller Basin; the latter (Hartsalz type) occupies both the periphery of the depression and its central part, the lower Leine and Aller Rivers Basin. Potash salts in sections of the Hartsalz type are represented by various kieserite-bearing sylvite-halite, langbeinite, kieserite, and kainite rocks.

The Stassfurt Series section in the Weser Depression is capped by a thin (0.5–1 m) band of Decksteinsalz (Na2r) consisting of coarse-crystalline, red-brown halite, and overlying banded Deckanhydrit (A2r) 1–1.5 m thick, represented by densely intercalated anhydrite and clay.

An approximate estimate for the rock salt volume of the Stassfurt Series in the Thüringia, Subhercynian, and Weser Depressions is inferred from the following evidence. Original thickness of rock salt in the above depressions ranges from 0–500 m and above (Fig. 129). In the central Subhercynian and Weser Depressions, the thickness of rock salt reaches 800 m and even 1000 m. The area outlined by a 500 m contour line equals about 7000 km^2. Assuming an average thickness of rock salt over this area of 700 m, then the volume in regions of greatest thickness will account for 4.9×10^3 km^3. The remaining area under Stassfurt-Steinsalz is estimated at 16,000 km^2. Here, an average thickness of rock salt is not less than 100 m and its volume amounts to 1.6×10^3 km^3. Thus, a total volume of Stassfurt-Steinsalz in the area discussed will apparently exceed 6.5×10^3 km^3.

The Werra Fulda Depression at the time of the Stassfurt Series accumulation was situated in a near-shore zone of the Zechstein Basin. Terrigenous rocks are common there, although the Stassfurt Series section differs greatly from the earlier described sections. The total thickness of the series in the Werra Fulda Depression does not exceed 30–50 m. Its lower boundary is usually drawn at the base of Braunroter Salzton (T2) (Richter-Bernburg 1955b, Roth 1955a, b). The Braunroter Salzton was determined (Dittrich 1966) to have in the south-eastern Werra Fulda Depression a larger stratigraphic extent, namely, the lower part belongs to the Zechstein lower cycle where Ton (T1r) and Oberer Werra-Anhydrit (A1r) are recognized; and only the upper part of the horizon, lying at the base and building up the basal bed of Ton (T2) 2–4 m thick, is placed into the Stassfurt Series.

Up the section, a salt unit known as Übergangsschichten is recognized. Its lower part contains anhydrite, which is an equivalent of Stassfurt Basalanhydrit (A2); and the upper part is composed of Stassfurt-Steinsalz (Na2). The thickness of anhydrite and salt is respectively, 2–8 m and 5–10 m. The remaining upper part of the series is composed of argillite, brown-red, sandy, and dolomitic with various amounts of anhydrite.

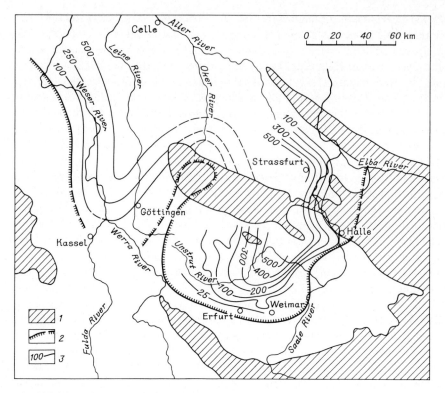

Fig. 129. Map of original thickness of Stassfurt-Steinsalz (Na2) in Weser, Subhercynian, and Thüringia Depressions. After Simon (1967). *1* pre-Zechstein outcrops, *2* areal extent of rock salt, *3* isopach contours of rock salt, in meters

The volume of Stassfurt-Steinsalz in the Werra Fulda Depression is determined as follows. Assuming an area of about 3000 km^2, and an average thickness of 10 m, the volume equals 0.3×10^2 km^3.

Farther south, in the Franconia Basin, the Stassfurt Series becomes more terrigenous. Near Hettenhausen in the apical part of the Spessart Swell, almost the entire section is composed of clay with inclusions and lenses of gypsum with an anhydrite interbed (A2); its lower part is 4 m thick. In the southernmost Franconia Basin near Kissingen, the Stassfurt Seires consists of clayey rocks with occasional gypsum lenses. The clay-marl facies is developed there (Trusheim 1964, Dittrich 1966). The northeastern Franconia Basin is dominated by argillaceous deposits where a band of Basalanhydrit (A2) can be distinguished. They are traced far south-east, in the upper Werra River, and even farther where they were penetrated at Mittelberg.

In the Lower Rhine Basin, the Stassfurt Series shows a considerable facies change. In the central most subsided parts, the following units can be recognized, in ascending order: Braunroter Salzton (T2), Basalanhydrit (A2), Stassfurt-Steinsalz (Na2), and Deckanhydrit (A2r).

In the Lower Rhine Basin, the Stassfurt Series occupies only its central areas. The thickness of salt increases northward with a drastic increase up to 500–600 m north of Oldensaal, that is, outside the Lower Rhine Basin, at the southern flank of the North-West German Basin. Carbonate rocks appear (Hauptdolomit Ca2d) in this section.

In the Lower Rhine Basin, the thickness of Stassfurt-Steinsalz, in general, does not exceed 30–40 m. South of Isselburg, the Stassfurt Series section is considerably reduced. Terrigenous, predominantly clayey material becomes abundant. Basalanhydrit and Deckanhydrit are easily discernible beyond the area of rock salt development. Farther south, they are enriched in clay material. For example, in the Kamp-4 well, most of the lower Stassfurt Series is composed of bluish-gray, banded clay with partings and an admixture of fine-grained sand, locally limy and dolomitic with thin interbeds of dolomite and calcareous marl. The upper part of the series contains, along with red and bluish clays, inclusions and partings of gypsum and anhydrite; the number of the latter increases up the section where an anhydrite interbed 6 m thick occurs, belonging to Deckanhydrit (Teichmüller 1958).

The areal extent of the Stassfurt-Steinsalz in the region is about 250 km^2. Assuming an average thickness over the entire area of 20 m, then the volume of rock salt will be 5 km^3.

Salt deposits of the Stassfurt Series are especially widespread in the North-West German and North-East German Basins and Altmark Depression. They have not been studied in detail due to a great depth of occurrence and complex tectonic structure resulting from salt tectonics. Thus, at present, the series in the area may only be described in general terms using the data for the separate parts, mainly along the periphery of the basins.

Stinkschiefer (Ca2st) 4–10 m thick is common in the central subsided zones of the North-West and North-East German Basins and Altmark Depression, in particular at the base of the second cycle of Zechstein. They are represented by dark, foliated, thin-bedded limestone often bituminous and more or less pyritized with partings and lenses of dolomite. Their occurrence in these areas was reported by Richter-Bernburg (1955c, 1972a, b), Münzberger and co-workers (1966), and Jung (1968b). In the Altmark Depression near Calvörde, the thickness of odorous shale does not exceed 2 m (Löffler and Schulze 1962). Farther west, in the Aller Basin (Philipp 1966), the thickness in boreholes is 5 m, 7 m, and 5 m. In the lower Elbe River within the Salt Dome Stade, the thickness of Stinkschiefer is about 6 m (Hofrichter 1967).

North-east and south-east of the North-East German Basin, odorous shale grades into dolomite, which is widespread in north-eastern areas of Mecklenburg and in Lausitz (Fig. 130). Münzberger and co-workers (1966) showed that in Mecklenburg odorous shale (Stinkschiefer) is replaced by dark, laminated carbonate from central areas toward the margins. The latter successively grades into limestone and dolomite, oolitic and algal dolomite, and finally clayey dolomite with anhydrite inclusions. Here, the thickness of carbonate rocks exceeds 10 m and in a narrow zone, extending from south-east to north-west near Stahlbrode, reaches 20 m and even more than 40 m. To the south-east, in Lausitz, carbonates show a similar facies pattern.

Basalanhydrit (A2) in the area discussed has a universal distribution. Its thickness is small in subsided parts (3 m in Salt Dome Stade and 1.5–2.5 m over the remaining

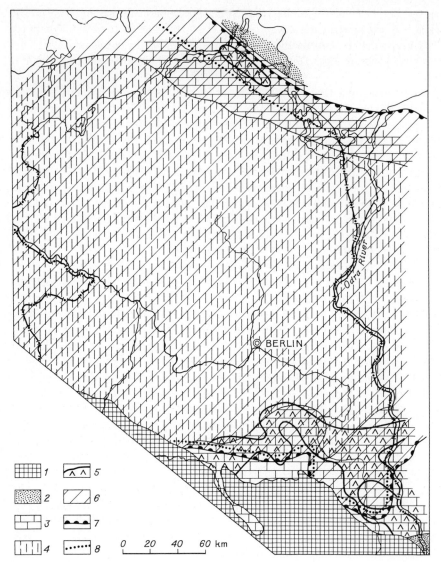

Fig. 130. Lithologic map of Stassfurt Series deposits in North-East German and Altmark Basins. Compiled from the data of Münzberger et al. (1966), Dadlez (1967b), and Jung (1968b). *1* pre-**Zechstein outcrops**, *2* mainly terrigenous (clayey) deposits, *3* carbonate deposits represented by dolomite (Hauptdolomit – Ca2d) over 10 m, *4* carbonate deposits represented by Stinkschiefer (Ca2st), *5* Basalanhydrit (A2) over 10 m thick, *6* Stassfurt-Steinsalz (Na2). Areal extent of: *7* salt, *8* Kaliflöz Stassfurt (K2)

area); anhydrite is thin-bedded, locally bituminous. In north-eastern Mecklenburg, it increases to 10 m and above and near Rügen Island, where it reaches 100 m, an anhydrite wall can be distinguished. To the south-east in Lausitz, the thickness of basal anhydrite also increases and near the mountain fringe, it builds up a separate complex east to west, where it strikes an anhydrite wall.

Most of the Zechstein section in the North-West and North-East German Basins and Altmark Depression is represented by Stassfurt-Steinsalz. In subsided parts, it is more than 600–700 m thick and in places 1000 m. A drastic increase in thickness is spasmodic near anhydrite walls and in the areas of thicker underlying carbonate (Münzberger et al. 1966). Kaliflöz Stassfurt (K2) occupies central parts of the basins. It ranges in thickness from 10–30 m. Potash salts are represented mainly by carnallite rocks. The Stassfurt Series is capped by Decksteinsalz (Na2r) 0.5–1.5 m thick and by Deckanhydrit (A2r), usually 1–2.5 m thick.

In general, the Stassfurt Series section remains the same in the south-eastern extremity of the North-East German Basin within the North Sudetic Trough and the towns of Gubin and Lubsko, Poland (Milewicz 1971). Here the series attains a thickness of 200 m. Black or gray, massive, crypto-crystalline, and finely crystalline, locally platy or porous dolomite with inclusions and small lenses of calcite and rare anhydrite lies at the base of the series. The thickness of the Main Dolomite ranges from 17–66 m. The overlying Basal Anhydrite (A2) is 1–22 m thick. In the subsided northern parts, it has a minimal thickness, while southward it increases. The Basal Anhydrite is overlain by the Older Halite. It is yellowish-gray and pink, medium- and coarsely crystalline with occasional inclusions of potash minerals and an admixture of halopelite and anhydrite. The thickness ranges widely from 27 m on the south to 201 m on the north. The greatest thickness was recorded in a narrow zone running from east to west and then from north to south near the town of Gubin. The Stassfurt Series (or the second cycle of Polish Zechstein PZ2) is crowned by a thin, clayey rock salt bed with saline anhydrite above it; the thickness of the latter does not exceed several meters. In general, the deposits of the second Zechstein cycle show considerable facies change. Farther south, where the thickness of rock salt decreases, dolomite becomes dominant in the section and the total thickness of anhydrite remains small. Still farther south within the North Sudetic Trough, the thickness of dolomite decreases, while that of anhydrite increases, thus, dominating the section. A brown mudstone bed appears at the base. Dolomite wedges out on the south-western slope of the North Sudetic Trough. Here the section becomes two-partite, namely, terrigenous-clayey and anhydrite, respectively, at the base and at the top (Kaczmarek 1970). In the piedmont areas, clayey and sandy deposits can be distinguished.

The volume of rock salt of the Stassfurt Series in the North-West and North-East German Basins and Altmark Depression can be estimated approximately at 6×10^4 km^3, assuming an areal extent and an average thickness of 150,000 km^2 and 400 m, respectively.

In the Central Polish Depression, the Polish Zechstein cycle 2 (PZ2) is divided into Main Dolomite, Basal Anhydrite, Older Halite, Older Potash, Screening Older Halite, and Screening Anhydrite. All these stratigraphic units are recognized in the central zones of the depression where deposits of the cycle are penetrated at many localities.

To the west, the complete section was penetrated by the Gorzów Wielkopolski 1 borehole (Poborski 1960). Dolomite about 20 m thick lies at the base. It is overlain by the Basal Anhydrite (A2) 20 m thick. Above is more than 110 m or rock salt. Its upper part contains anhydrite interbeds. The rock salt is overlain by a 6 m potash salt bed. There is also an interbed of anhydritic rock salt (3 m) at the top and above it lies a clayey interbed (2–2.5 m) containing anhydrite.

In the northern Central Polish Depression, where the Swidwin 2 borehole (Poborski 1960) penetrated, the thickness of the cycle PZ2 reaches 480 m. The Main Dolomite (6 m) is recognized at the base and above lies the Basal Anhydrite (2.5 m). The Older Halite, 390 m thick, is composed or rock salt with occasional partings of anhydrite and halopelite. The lower and upper parts of the Older Potash are represented respectively, by rock salt with potash salt partings and a potash salt bed 25 m, composed mainly of carnallite rocks. An interbed of the Screening Anhydrite (1–1.5 m) can be distinguished at the top and above it lies clay with inclusions of anhydrite and gypsum.

A similar thick section is known from the central Pomorsko-Kujawy Anticlinorium. Salt deposits crop out on some diapirs and are mined in many places (Werner et al. 1960). The apparent thickness of the cycle PZ2 is several hundred meters (350 m in the Inowrocław borehole and 330 m in the Kłodowa borehole). The uppermost part is represented by anhydrite only a few meters thick. The remaining lower part of the section can be divided into three parts: (1) Older Halite; (2) transition layer, consisting of rock salt with inclusions and partings of brown salt; and (3) Older Potash, composed of anhydrite-kieserite rocks about 8 m thick (within a salt field at Klodowa).

To the south-west, the cycle PZ2 was penetrated through the entire thickness by the Wschowa 1 well (Poborski 1960). Potash salts are absent there.

Dolomite about 25 m thick lies at the base, followed further up the section by anhydrite about 32 m, rock salt with occasional anhydrite partings 22 m, and Screening Anhydrite 2–3 m thick, which caps this section of the cycle.

At the south-western and south-eastern margins of the Central Polish Depression, the salt sequence wedges out. The entire section becomes sulfate-carbonate and then carbonate with terrigenous rocks occurring along the periphery. For example, in the Fore-Sudetic Monocline area between the towns of Lubin, Legnica, and Sieroszowice, the deposits of cycle PZ2 are represented by dolomite overlain by anhydrite; the thickness of the latter reaches 30–35 m. Kłapciński (1967) proposed a minute subdivision for the Basal Anhydrite of the Lubin-Sieroszowice area, in ascending order: (1) thin-bedded anhydrite (4–18.3 m); (2) lower, cavernous anhydrite (7.1–15.7 m); (3) lower anhydrite and clay breccia (0.3–2.0 m); (4) upper, cavernous anhydrite (1.5–9.5 m); and (5) upper anhydrite and clay breccia.

To the south-east in the western Nida Trough (Jurkiewicz 1970), the cycle PZ2 is represented mainly by dolomite and anhydrite. In the Pagow IG-1 well, the lower part is alternate mudstone, dolomite, and anhydrite (9.2 m), while the upper part contains an anhydrite unit 10 m thick. The Milianowa IG-1 well situated south-east of the former penetrated a carbonate section dominated by dolomite with limestone interbeds and anhydrite inclusions. A small anhydrite interbed about 1 m thick is distinguished near the top. The total thickness of the cycle does not exceed 15 m.

Still farther east in the Swiętokrzyskie Góry, the section is composed of sandy marl, bedded limestone, mudstone, and clay-marly shale with gypsum seams (Szaniawksi 1965).

In the north-eastern Central Polish Depression, the salt sequence also tapers out toward the Central Pomeranian Peninsula. According to the log data from the Chojnice 2 well (Poborski 1960), the lower part of the cycle is composed of dolomite (above 20 m). The upper part consists of anhydrite (about 45 m) with a rock salt interbed (4 m). Farther south-east, salt deposits are absent.

At present, the lithology and structure of the Main Dolomite of the Central Polish Depression are best known (Atlas . . . 1978, Depowski et al. 1981, Wagner et al. 1981). The accumulation of thin clay-carbonate deposits took place during the formation of the above deposits in the central part of the depression. This area was bounded by a shallow and barrier zone, where the thickness of the Main Dolomite reaches 40–60 m (Fig. 131). Later, thick strata of the Older Halite deposited in the inner part of the depression (Fig. 132).

The volume of rock salt of the cycle PZ2 was estimated as follows for the Central Polish Depression. The areal extent is not less than 80,000 km^2. Assuming an average thickness of 100 m, which is quite probable, then a minimal volume of rock salt will be 8×10^3 km^3.

Salt-bearing deposits of the Stassfurt Series have also been found within the Polish-Lithuanian Depression. The thickness of the series exceeds 150–200 m in its central, most subsided part situated in the western areas of Wojewodztwo Olsztyńskie. The thickness decreases to 100 m and less toward the east, south-east, and north. In marginal parts of the depression, the total thickness of the series does not exceed 10 m. The composition of deposits also changes. The Main Dolomite, Basal Anhydrite, Older Halite, and Screening Anhydrite are easily discernible in the section in the interior parts (Stołarczyk 1972).

Dolomite lying at the base of the series ranges from 9–1.5 m in thickness. It is platy and marly with inclusions and partings of clay, locally with cavities filled by salt crystals. The amount of terrigenous material in dolomite increases south-eastward. The Basal Anhydrite is 13–15 m thick. It contains dolomite and clay partings. The Older Halite reaches 72.3 m near Paslek, Wojewodztwo Olsztyńskie. The thickness of salt decreases and completely tapers out to the east, north-east, and south-east. In general, the Screening Anhydrite has a small thickness (from several centimeters to 1.5 m).

In Lithuania, Suveizdis (1963) placed the equivalents of the Stassfurt Series into the Zhalgiryay and Aistmark Formations. In south-eastern Lithuania and in the Kaliningrad Region, the former is composed mainly of dense dolomite and gray, dark gray, and yellowish-gray dolomitic limestone; the thickness ranges from 4.5–10.5 m. Marginal eastern and north-eastern areas are dominated by clayey and sandy limestone and extremely bituminous, gypsinate, yellowish or brownish-gray dolomite in some places. The Aistmark Formation is the equivalent of Basalanhydrit and Stassfurt-Steinsalz of the Stassfurt Series. In Lithuania, it consists of light gray anhydrite up to 9.2 m thick which, in coastal parts of the depression, grades into brownish-buff, marly clay with interbeds of dense, gray gypsum-anhydrite. Here the formation is 11–16.6 m thick. Salt-bearing deposits in the Aistmark Formation are known only

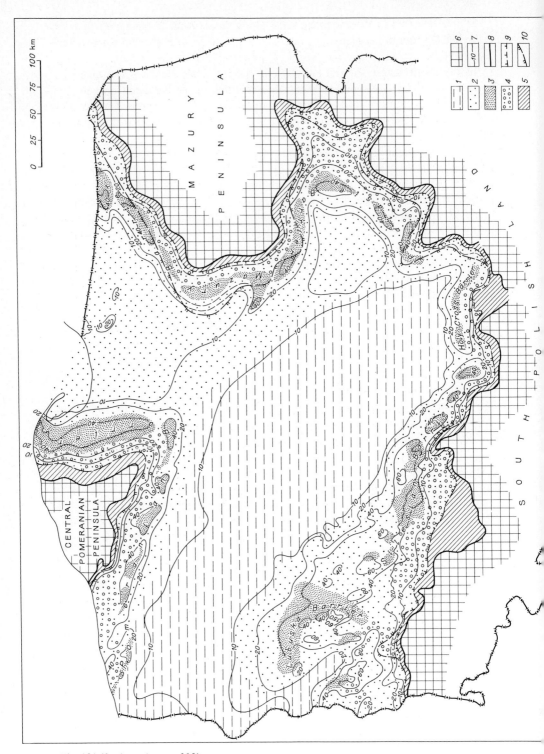

Fig. 131 (for legend see p. 299)

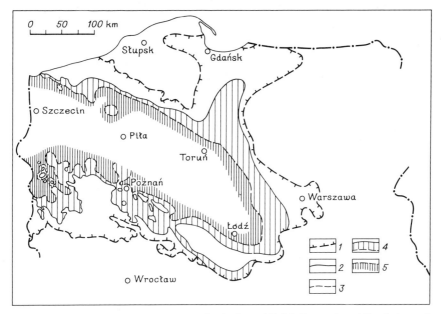

Fig. 132. Sketch isopach map of Older Halite in Central Polish Depression. After Antonowicz and Knieszner (1981). *1* present extent of Older Halite, *2* isopach line 100 m, *3* isopach line 400 m, *4* thickness 100–400 m, *5* thickness over 400 m

from the Kaliningrad Region where the South Kaliningrad (Niven) well penetrated two rock salt interbeds 5.5 m and 9 m thick (Romanov and Zotova 1962).

In the Polish-Lithuanian Depression, rock salt covers an area of about 15,000 km². The average thickness may be estimated at 30 m. Its volume then will account for 4.5×10^2 km³.

In the western Central European Basin within the English Basin, the English Zechstein cycle 2 (EZ2) encloses deposits of the Aislaby Group (D. Smith 1981). In Durham, they are subdivided, in ascending order, into: the Concretionary Limestone Formation, and Hartlepool and Roker Dolomite; in Yorkshire province, into the Kirkham Abbey Formation and the Fordon Evaporites.

The Concretionary Limestone Formation lying at the base of the cycle EZ2 in Durham reaches a thickness of about 100 m. Its lower part is dominated by finely laminated calcitic dolomite or dolomitic limestone, while the upper part contains fine-grained, unlaminated dolomite. The section also contains fine-grained or oolitic, only slightly calcitic dolomite. The central parts of the basin are dominated by thin,

◄ Fig. 131. Paleogeographical map of Main Dolomite (Ca2) in Central Polish Depression. After Depowski et al. (1981). *1* deeper part of shelf, *2* shallower part of shelf (not subdivided), *3* barrier, *4* lagoon, *5* coastal zone (evaporite-clastic sedimentation), *6* land, *7* isopach contours of Main Dolomite, in meters, *8* original extent of PZ2 deposits, *9* original extent of Main Dolomite deposits, *10* present extent of Main Dolomite deposits

dark brown, fetid, evenly, finely laminated, barren carbonates. The Hartlepool and Roker Dolomite, as well as its facies equivalent — the Kirkham Abbey Formation — consists of finely oolitic, cross-laminated dolomite. The maximal thickness of the formation (up to 150 m) was recorded in a narrow belt extending south to north along the shelf where a carbonate barrier can be distinguished. The thickness decreases to 10–15 m eastward.

The Fordon Evaporites attain a maximal thickness in the areas east of the carbonate barrier, which can be recognized by means of the underlying Kirham Abbey Formation and its equivalents (Taylor and Colter 1975, D. Smith 1981). Earlier, these deposits were assigned to the Lower Evaporites. Stewart (1949, 1951a, b) and Raymond (1953) described them in great detail. The Lower Anhydrite, Polyhalite Zone, Upper Anhydrite, and Halite-Anhydrite Zone were recognized in a section penetrated by the Eskdale No. 2 borehole drilled south-west of Whitby near Eskdale on the right side of the Esk River, Yorkshire County. The total thickness of the Lower Evaporites (Fordon Evaporites) is 146.8 m.

Raymond (1953), in a review of data obtained by prospecting for potash salts in north-eastern Yorkshire (Whitby area), subdivided the Lower Evaporites into four parts: (1) Lower Halite; (2) Anhydrite-Polyhalite; (3) Anhydrite; and (4) Upper Halite. Such a section, characteristic of the most subsided parts of the basin, was penetrated by the Sleights E-3 borehole drilled north-east of Eskdalegate. The Lower Halite, 89.3 m thick, is composed mainly of coarse-grained, usually light rock salt; the lower (16.8 m) and upper (47.2 m) parts contain interbeds of dolomitic anhydrite with dolomitic marl partings; polyhalite inclusions occur at the very top of the unit. The Anhydrite-Polyhalite is represented by alternate anhydrite and polyhalite rocks with interbeds of rock salt in the lower part. Anhydrite is generally pale buff-gray, fine-grained, and often dolomitic. Polyhalite rocks are also pale buff-gray and finely laminated. In places, there are lenses of dolomite, black shale, and an admixture of terrigenous material; the thickness is 136.8 m. The Anhydrite unit consists of pale buff-gray anhydrite with inclusions and occasional partings of rock salt; the thickness is 50.5 m. The Upper Halite is represented by crystalline, light rock salt with two anhydrite interbeds and anhydrite lenses. At the top, anhydritic marl containing abundant talc inclusions is present in rock salt. The thickness is 52.7 m. In the Sleights E-3 borehole, the Lower Evaporites have a thickness of 329.3 m.

North-west and west of Whitby in the direction of Durham, the thickness of the Lower Evaporites decreases, due primarily to wedging out of rock salt. Its thickness is 19.8 m and 18 m at Whilton (9.5 km east of Middlesbrough) and in southern areas of Durham, respectively. Rock salt occurs in the form of small interbeds mainly in the lower part. A similar decrease in thickness and wedging out of rock salt is found west and south-west of Whitby. For example, 39.5 m of anhydrite with a single rock salt band (5.8 m) was reported from the Cleveland No. 1 borehole.

In the North Sea shelf area, the Fordon Evaporites are represented by interbedded and intermixed anhydrite, polyhalite, and halite (Taylor and Colter 1975). Sulfates also predominate in a basal unit about 100 m thick on the basin floor, but halite up to several hundred meters thick is the main evaporite in all interior parts of the basin (D. Smith 1981).

On the North Sea shelf, deposits of the English Zechstein cycle 2 were penetrated by a great number of boreholes (Kent 1967a, b, 1968, Brunstrom and Walmsley 1969, Kent and Walmsley 1970, Taylor and Colter 1975). Hauptdolomit overlain by Basalanhydrit is easily discernible in the lower part of the series, while most of its upper part is represented by a salt sequence with interbeds of potash salts which are equivalents of Kaliflöz Stassfurt. In the BP 46/6-1 well, the salt sequence is about 90 m thick (Brunstrom and Walmsley 1969). In its lower part, a unit of rock salt interbedded with polyhalite rocks can be distinguished. Polyhalite interbeds occur up the section as well. Two interbeds of clayey rock salt with sylvine inclusions and sylvinite partings were found in the middle part. In the uppermost part, a potash salt bed 1.2 m was penetrated; anhydrite lies above it.

In the Norwegian-Danish Basin, salt deposits of the Stassfurt Series were penetrated by isolated wells (Sorgenfrei and Buch 1964, Sorgenfrei 1966) mainly on some salt domes. All the units of the series are present there, namely, Hauptdolomit at the base, overlying Basalanhydrit, and Stassfurt-Steinsalz. The thickness of the latter reaches several hundred meters in coastal areas of the North Sea. These deposits are placed into cycle C (Sorgenfrei and Buch 1964).

It is difficult to estimate the volume of rock salt of the Stassfurt Sieries in the English and Norwegian-Danish Basins and in adjacent areas of the North Sea. Rock salt occurs there in numerous salt domes and anticlines where salt deposits are extremely deformed and thus, subjecting their thickness to change. The approximate areal extent of rock salt in the region discussed is not less than 160,000 km^2. If it is taken into account that in marginal parts of the English Basin, the total thickness of rock salt exceeds 100 m in places, then in most subsided parts of the North Sea, the original thickness might be greater. In the western North-West German Basin, rock salt averages 400 m, thus, the average thickness over the entire area will be not less than 100 m. Assuming these minimal estimates, the volume of rock salt of the Zechstein cycle 2 in the English and Norwegian-Danish Basins and in the remaining area will then account for 1.6×10^4 km^3.

Salt-free equivalents of the English Zechstein cycle 2 were found in the Manx-Furness and Solway Firth Basins, e.g., the Saltom Dolomite and the lower Sandwith Dolomite (Arthurton and Hemingway 1972). They are represented by clayey, granular dolomite, algal dolomite, locally dolomitic breccia at the base, with limestone interbeds and inclusions as well as partings of anhydrite. Anhydrite crystal molds after gypsum are common in algal dolomite. Deposits have a thickness of about 3–4 m.

Concluding the discussion of the Stassfurt Series and its equivalents in the Central European Basin, it should be emphasized that the salt basin was the largest throughout the Zechstein period. The above estimates allow the assumption that about 9.1×10^4 km^3 of rock salt could have accumulated in the basin.

Zechstein 3 (Leine Series)

Deposits of the Zechstein cycle 3 (Leine Series) are known from the English, Norwegian-Danish, and North-West German Basins, Weser, Thüringia, Subhercynian, and Altmark Depressions, as well as the North-East German Basin and the Central Polish and Polish-Lithuanian Depression. Potash salts are generally restricted to the central Middle European Depression (Fig. 133).

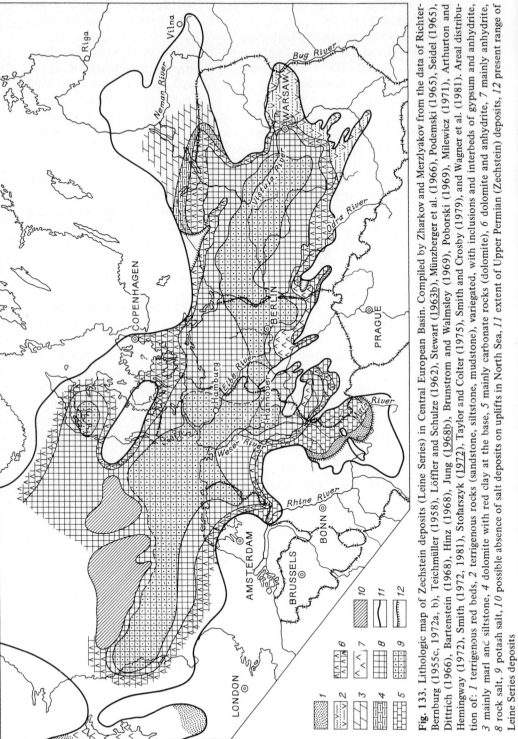

Fig. 133. Lithologic map of Zechstein deposits (Leine Series) in Central European Basin. Compiled by Zharkov and Merzlyakov from the data of Richter-Bernburg (1955c, 1972a, b), Teichmüller (1958), Löffler and Schulze (1962), Stewart (1963b), Münzberger et al. (1966), Podemski (1965), Seidel (1965), Dittrich (1966), Bartenstein (1968), Hinz (1968), Jung (1968b), Brunstrom and Walmsley (1969), Poborski (1969), Milewicz (1971), Arthurton and Hemingway (1972), Smith (1972, 1981), Stofarszyk (1972), Taylor and Colter (1975), Smith and Crosby (1979), and Wagner et al. (1981). Areal distribution of: 1 terrigenous red beds, 2 terrigenous rocks (sandstone, siltstone, mudstone), variegated, with inclusions and interbeds of gypsum and anhydrite, 3 mainly marl and siltstone, 4 dolomite with red clay at the base, 5 mainly carbonate rocks (dolomite), 6 dolomite and anhydrite, 7 mainly anhydrite, 8 rock salt, 9 potash salt, 10 possible absence of salt deposits on uplifts in North Sea, 11 extent of Upper Permian (Zechstein) deposits, 12 present range of Leine Series deposits

The type section of the series is situated in the Weser Depression. The most complete section occurs in the lower Leine and Aller Basins, in the Hannover potash district. At present, the subdivision of the Zechstein cycle 3 proposed by Richter-Bernburg (1955b) is universally accepted. Grauer Salzton (T3) lies at the base; its thickness in Bergmannssegen-Hugo and the Ronnenberg and Hansa potash districts is 3–5 m (Ahlborn and Richter-Bernburg 1955, Roth 1955b). In the northern Weser Depression, it is overlain by Hauptanhydrit (A3) up to 35 m thick. To the south and south-west between Grauer Salzton and Hauptdolomit Plattendolomit (Ca3) occurs which is developed only in the south-western Weser Depression, in the Werra and Fulda Basins; dolomite wedges out northward.

In the lower Leine and Aller Rivers, Hauptanhydrit (A3) is overlain by a thick sequence of Leine-Steinsalz (Na3). The latter is subdivided into the lower, Ronnenberg and upper, Riedel Groups. The Ronnenberg Group can be divided into three parts: Liniensalz, Kaliflöz Ronnenberg (K3Ro), and Bändersalz. Liniensalz consists of Basissalz (Na3α), up to 1.5 m thick, Graues Liniensalz (Na3β) about 10 m, and Orange Liniensalz (Na3γ) reaching 15 m in thickness. The total thickness of Liniensalz does not exceed 25–30 m. Bändersalz is composed of Banksalz (Na3δ) 2–3 m thick; Kaliflöz Bergmannssegen (K3Be) ranging in thickness from 0.5–2 m; Bändersalz (Na3ε) up to 10 m; and Buntes Salz (Na3ζ) 12 m thick. The Riedel Group consists of (in ascending order): Angydritmittelsalz (Na3η) about 40 m, Schwadensalz (Na3θ) up to 60 m, Kaliflöz Riedel (K3Ri) varying in thickness from 6–10 m, Tonmittelsalz (Na3t or Na3tm) in places with Kaliflöz Albert (K3Ab) 2–6 m, and Blauer Salzton (Na3tmI) at the base. The total thickness of Na3t does not exceed 20 m. Leine-Steinsalz (Na3) in the northern Weser Depression varies in thickness from 150–250 m.

Such a complete section of the Leine Series is known only in the interior northern part of the depression where potash salts are common (Fig. 134). Kaliflöz Riedel has a relatively small areal extent in the lower Leine and Aller Basin. Kaliflöz Ronnenberg shows a wider distribution extending far north into the North-West German Basin. Not only potash beds, but also Leine-Steinsalz wedge out south, south-west, and west of the central subsided parts of the Weser Depression. As a result, in peripheral areas, the section is first bipartite, namely, consisting of Grauer Salzton (T3) and Hauptanhydrit (A3); and with the appearance of Plattendolomit (Ca3), it becomes three-partite. Approaching the margin, Hauptanhydrit (A3) disappears and the entire series is represented by carbonate rocks (Ca3) with a unit of terrigenous, predominantly clayey deposits at the base, which is an equivalent of Grauer Salzton (T3).

In the Weser Depression, almost all the stratigraphic units of the Zechstein cycle 3 have been thoroughly studied. Kosmahl (1967) provided a detailed description of Grauer Salzton (T3) and Hauptanhydrit (A3). Grauer Salzton was subdivided into two parts, i.e., the lower, sandy-clayey part which varies in thickness from 3–4 m near Hannover to 16 m in the vicinity of Göttingen; and the upper, dolomitic part which is 0.8–2.25 m and 8 m near Hannover and Göttingen, respectively. Thirteen zones have been recognized within Hauptanhydrit (A3). The presence of numerous pits enabled the detailed study of potash salts. They are represented predominantly by sylvinite and minor Hartsalz. Kaliflöz Ronnenberg has a thickness of 4–5 m.

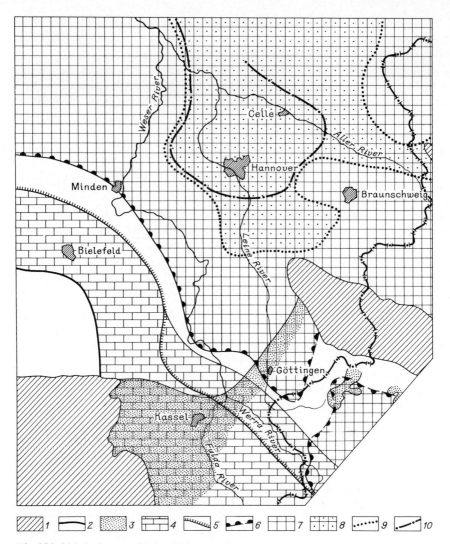

Fig. 134. Lithologic map of Leine Series deposits in Weser Depression. Compiled from the data of Richter-Bernburg (1955c, 1972a, b), Münzberger et al. (1966), and Jung (1968b). *1* pre-Zechstein outcrops, *2* depositional limit of Leine Series deposits, *3* Leine Series deposits enriched in terrigenous material (in Thüringia Depression areas where Grauer Salzton (T3) over 20 m thick), *4* mainly Plattendolomit (Ca3), *5* south-westward and southward extent of Hauptanhydrit (A3), *6* limit of Leine-Steinsalz (Na3), *7* extent of Leine-Steinsalz (Na3), *8* extent of Leine potash salt, *9* limit of Kaliflöz Ronnenberg (K3Ro), *10* limit of Kaliflöz Riedel (K3Ri)

The areal extent of rock salt in the Weser Depression exceeds 10,000 km². The average thickness is assumed to be 100 m, varying from 0–250 m. With this in mind, the estimated rock salt volume is 1×10^3 km³.

In the Werra Fulda Depression and Franconia Basin, the Leine Series is salt-free. Sandstone, siltstone, and mudstone up to 2–5 m thick lie at the base. The middle part of the series is composed of carbonate rocks assigned to Plattendolomit (Ca3) 15–20 m thick. The section is capped by a 4–6 m clayey unit. The lower and upper parts of this unit are placed into the Leine and Aller Series, respectively. The Leine Series of the Thüringia Depression has been studied most thoroughly (Schulze 1958, Löffler and Schulze 1962, Seidel 1965, Jankowski and Jung 1966, Hemmann 1970, and others). In ascending order, a composite section is as follows (Seidel 1965):

Leine-Steinsalz – Na3 (0–130 m)
Na3 + ι	Tonflockensalz and Schwadensalz	0–20 m
Na3η	Anhydritmittelsalz with minor potash salt	0–50 m
Na3γ	Kristallsalz with minor potash salt	0–40 m
Na3β	Liniensalzäquivalent	0–50 m
Na3α	Basissalz	0–7 m

Hauptanhydrit – A3 (0–60 m)
A3ζ	Schichtungslose Anhydritfolge	0–4 m
A3ϵ_3	Obere tonig-karbonatische Anhydritfolge	0–14 m
A3ϵ_2	Obere karbonatische Anhydritfolge	0–40 m
A3ϵ_1	Obere Karbonatfolge	0–2 m
A3δ	Untere tonig-karbonatische Anhydritfolge	0–20 m
A3γ	Untere karbonatische Anhydritfolge	0–8 m
A3β_2	Mittlere Karbonatfolge	0–4 m
A3β_1	Karbonatische, gemaserte Anhydritfolge	0–8 m
A3α	Untere Karbonatfolge	0–6 m

Plattendolomit – Ca3 (0–25 m)
Grauer Salzton – T3 (5–30 m)
T3δ_2	Magnesitische Wechselfolge	0–3 m
T3δ_1	Magnesitbank	0.3–2 m
T3γ_2	Graue Sandflaserlage	1–5 m
T3γ_1	Braune Sandflaserlage	0–5 m
T3β_2	Rotbrauner Salzton und Wechselfolge	0–20 m
T3β_1	Grauer anhydritischer Salzton	1–5 m
T3α	Gebänderter Anhydrit	0–2 m

The distribution of the main stratigraphic units of the Leine Series in the Thüringia Depression is shown in Fig. 135. Grauer Salzton is developed over the entire depression. In general, its thickness does not exceed 15–20 m, but in some northern and south-eastern parts, it reaches 25–30 m (Fig. 136, I). Plattendolomit (Ca3) was reported only from the southern Thüringia Depression south of the line extending from Naumburg through Erfurt to Mülhausen. It is brown-gray to dark dolomit with clayey anhydrite at the base and inclusions and lenses of anhydrite at the top. The greatest thickness was recorded in a sublatitudinal belt extending from Naumburg via Jena farther west (Fig. 136, II), where carbonate walls are developed. Hauptanhydrit (A3) is composed of gray, fine- or inequigranular anhydrite with various amounts of clay or bituminous material, often intercalated with carbonate and sulfate-carbonate rocks. Units and beds recognized in Hauptanhydrit are distinctly traceable across the area (Fig. 136, III). The thickness of Hauptanhydrit gradually increases from south to north reaching maximal value in the northern Thüringia Depression, i.e., in the

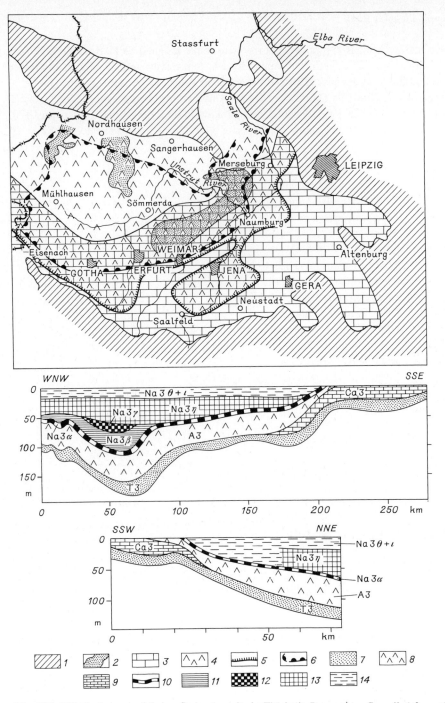

Fig. 135. Lithologic map of Leine Series deposits in Thüringia Depression. Compiled from the data of Seidel (1965), Münzberger et al. (1966), Jung (1968b), and lithologic sections of Leine Series. After Seidel (1965). Explanation used for map: *1* pre-Zechstein outcrops, *2* thickness of Grauer Salzton (T3) exceeds 20 m, *3* distribution of Plattendolomit (Ca3), *4* distribution of Hauptanhydrit (A3), *5* south- and south-eastward limit of Hauptanhydrit, *6* limit of Leine-Steinsalz (Na3). Explanation used for sections: *7* Grauer Salzton (T3), *8* Hauptanhydrit (A3), *9* Plattendolomit (Ca3), *10* Basissalz (Na3α), *11* Liniensalz (Na3β), *12* Kristallsalz (Na3γ), *13* Anhydritmittelsalz (Na3η), *14* Tonflockensalz and Schwadensalz (Na3 + ι)

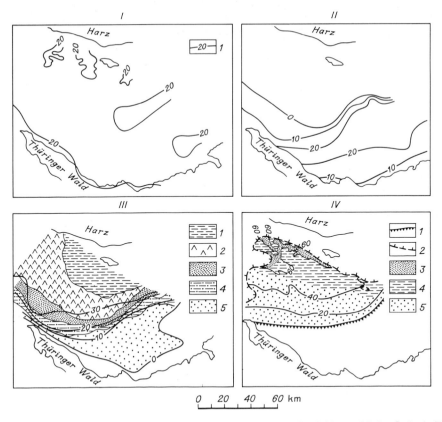

Fig. 136. Lithologic maps and thicknesses of main stratigraphic divisions of Leine Series in Thüringia Depression. After Seidel (1965). *I* Grauer Salzton (T3): *1* isopach contours, in meters (for all maps), *II* Plattendolomit (Ca3), *III* Hauptanhydrit (A3): *1* Hauptanhydrit with lower clayey carbonate unit, *2* Hauptanhydrit without above unit, *3* lower horizon of Hauptanhydrit (A3α) similar to Plattendolomit, *4* second horizon of Hauptanhydrit (A3β) similar to Plattendolomit, *5* third horizon of Hauptanhydrit rich in carbonate rocks and similar to Plattendolomit, *IV* Leine-Steinsalz (Na3): *1* consedimentary range of rock salt, *2* secondary range of rock salt related to erosion and leaching, *3* Liniensalz, Anhydritmittelsalz, Schwadensalz, and Tonflockensalz occur in one section, *4* development of upper salt horizons: Anhydritmittel-, Schwaden-, and Tonflockensalz, *5* Anhydritmittel-, Schwaden-, and Tonflockensalz absent

areas where underlying deposits of Plattendolomit (Ca3) are absent and Hauptanhydrit rests directly on Grauer Salzton (T3). Here anhydrite has a thickness of more than 30 m.

Leine-Steinsalz (Na3) is white, gray, or red. It is intercalated with numerous clayey and anhydritic partings; its middle part locally contains inclusions of potash salt. Complete sections of Leine-Steinsalz (Na3) occur generally in the subsided northwestern parts of the Thüringia Depression where the thickness of a salt sequence exceeds 60 m. Basissalz (Na3α), Liniensalz (Na3β), and Kristallsalz (Na3γ) are developed there. The latter two horizons fill the most subsided zones of the salt basin.

They wedge out toward margins where Basissalz (Na3α) is overlain by Anhydritmittelsalz (Na3η); above, Schwadensalz and Tonflockensalz (Na3 + ι) are found (see Fig. 135 and 136).

In the north-eastern Thüringia Depression within the Mansfeld Trough, the thickness of Leine-Steinsalz increases up to 100 m and above (Schulze 1958, Löffler and Schulze 1962).

Assuming an areal extent of 5000 km^2 and an average thickness of 40 m, the rock salt volume of the Leine Series in the region discussed will be not less than 2×10^2 km^3.

Salt deposits of Zechstein Cycle 3 are widespread in the Subhercynian Depression. They are composed of, in ascending order: Grauer Salzton (T3), Hauptanhydrit (A3), and Leine-Steinsalz (Na3). Thicknesses of major divisions of the Leine Series for some Potash Districts are given in Table 10.

Table 10. Thickness of major stratigraphic units of the Leine Series in the Subhercynian Depression. Adapted from Löffler and Schulze (1962)

Stratigraphic units	Thickness in potash districts (m)					
	Aschersleben	Bernburg	Stassfurt	Schwanebeck	Schönebeck	Allertal
Leine-Steinsalz (Na3)	95	100–150	100–140	up to 108	up to 80	55–85
Hauptanhydrit (A3)	25	40–90	25–90	17	30	30–40
Grauer Salzton (T3)	4	6–7	4–8	12	5–6	2.5

The greatest thicknesses of the Leine-Steinsalz (Na3) were reported from three areas: (1) the south-easternmost Aschersleben Region, (2) the eastern Bernburg Region, and (3) the central Subhercynian Depression in the Aschersleben-Stassfurt Syncline and the Atzendorf-Latdorf Graben. There, the overall thickness of salt exceeds 100 m.

Leine-Steinsalz comprises the same horizons as those of the Weser and Thüringia depressions, namely, Basissalz (Na3α), Liniensalz (Na3β), Kristallsalz (Na3γ), Anhydritmittelsalz (Na3η), Schwadensalz, and Tonflockensalz (Na3 + ι). In the Aschersleben Potash District, these horizons have the following thicknesses: Na3α, 0.5 m; Na3β, 9 m; Na3γ, 21 m; Na3η, 22 m; and Na3 + ι, 42 m.

The subdivision of Leine-Steinsalz differs slightly for the Stassfurt-Egeln District. Here, the lower part is also made up of Basissalz (Na3α), Liniensalz (Na3β), and Kristallsalz (Na3γ). Above, beds of Übergangssalz (Na3γ$_1$), Orangeaugensalz, and Bändersalz are found. Anhydritmittelsalz (Na3η) is found above Bändersalz, whereas the upper part of the section is composed of Schwadensalz (Na3) and Tonflockensalz. Apparently, Orangeaugensalz belongs to the lower part of the Na3η zone. In some areas (Berlepsch-Maybach, Neustassfurt VI/VII mines), inclusions of potash minerals are common in Liniensalz (Na3β), namely, sylvite and kieserite interbedded with polyhalite rock at the base.

The volume of Leine-Steinsalz in the depression can be approximately estimated at 1.75×10^2 km^3, assuming an area of 3500 km^2 and an average thickness of 50 m.

In the Lower Rhine Depression, Leine-Steinsalz (Na3) has been penetrated only in the northernmost area; its thickness does not exceed 50 m. Over the remaining area, the Leine Series is salt-free and is represented by Grauer Salzton (T3), Plattendolomit (Ca3), and Hauptanhydrit (A3). Anhydrites occur only in subsided interior parts. In marginal areas, the Leine Series is composed of reddish and bluish clay with lenticular inclusions and isolated limestone interbeds.

In the North-West German Depression, salt deposits of the Leine Series are very common. Due to salt tectonics these deposits are not well-known in comparison to those in the above discussed regions. Over the entire depression, the Leine Series section displays a ternary structure. At the base, there is a thin (4–6 m) bed of Grauer Salzton (T3), above is Hauptanhydrit (A3) up to 30 m thick, and the upper (up to 300 m) part of the section is composed entirely of Leine-Steinsalz (Na3), the upper part of which often contains potash salts (probably, equivalents of Kaliflöz Ronnenberg). In the southernmost part of the North-West German Depression, Leine-Steinsalz has a thickness of 150–170 m (Teichmüller 1958, Brucren 1959). In central parts, it is 300 m (Hofrichter 1967). A similar thickness was recorded at the northern margin of the depression in Denmark (Sorgenfrei and Buch 1964). Deposits of the Leine Series are distinguished here as Cycle C, which can be divided into seven zones. Two lower zones (C.1 and C.2) correspond to Grauer Salzton; zone C.3 is equivalent to Plattendolomit; zone C.4 is correlative to Hauptanhydrit; and zones C.5–C.7 encompass salt deposits.

The volume of Leine-Steinsalz in the North-West German Depression attains 1×10^4 km^3 (assuming an area of about 100,000 km^2 and an average thickness of 100 m).

In the North-East German and Altmark depressions, the deposits of the Leine Series were studied thoroughly to the north, in Mecklenburg, to the southeast within Oberlausitz, adjacent areas of Poland at Gubin and Lubsko, in the North Sudetic Trough, and at Calvörde. In subsided parts, the series begins with Grauer Salzton (T3) and followed above by Hauptanhydrit (A3) and Leine-Steinsalz (Na3). In marginal coastal areas, Plattendolomit (Ca3) appears, which pinches out in the interior parts of the depressions. In central areas, salt strata contain potash salts (Fig. 137).

In northern Mecklenburg and on Rügen Island, the variation in lithology is easily discernible from the subsided parts toward the margins. The thickness of Grauer Salzton (T3) in the interior parts is not more than 3 m. It can be divided into three parts: T3s, greenish-gray mudstone with about 0.5 m thick lenses of light gray sandstone; T3m, dark brown magnesite (0.8–1.3 m); and T3t, dark brown magnesite, alternated with greenish-gray mudstone, 0.8 m thick (Schirrmeister 1970). North- and north-eastward, the two upper beds pinch out. Only the lower bed is preserved in the section, ranging in thickness from 0.5–1 m. On Rügen Island, the thickness of the lower bed (T3s) increases to 6 m; it is dominated by sandstone. The deposits of Plattendolomit (Ca3) start to appear in areas where the upper magnesite beds of Grauer Salzton pinch out. Their lower part is composed of brownish-gray, algal, and oolitic dolomite with dark gray mudstone partings. The thickness varies between 2.5 and 5 m, increasing to 30 m at the margins.

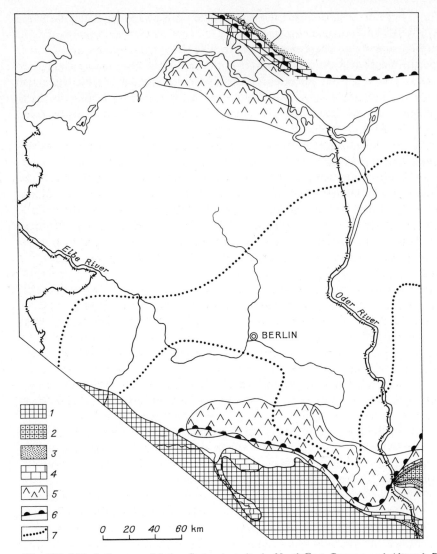

Fig. 137. Lithologic map of Leine Series deposits in North-East German and Altmark Basins. Compiled from the data of Münzberger et al. (1966), Poborski (1969), and Milewicz (1971). *1* pre-Zechstein outcrops, *2* Leine Series deposits absent, *3* terrigenous deposits, *4* mainly Plattendolomit (Ca3) deposits, *5* Hauptanhydrit deposits (A3) over 50–60 m thick, which build up anhydrite walls, *6* Leine-Steinsalz (Na3), *7* extent of Kaliflöz Ronnenberg

Hauptanhydrit (A3) in subsided parts of the northern North-East German Depression is about 25–30 m thick. North of the road, joining Rostock with the Peene Valley, the thickness of anhydrite abruptly increases to 60–80 m. This belt of increased thicknesses of Hauptanhydrit, extending from south-east to north-west, girdles a zone of an anhydrite wall. To the north, the thickness of anhydrite decreases to 20 m.

However, still farther north-east, Hauptanhydrit attains a thickness of 60—70 m, due to the presence of another narrow anhydrite wall in the south-east Rügen Island.

Leine-Steinsalz (Na3) in north-eastern Mecklenburg contains no potash salts. Southward, the thickness increases to 100 m in a sag between the two anhydrite walls and to 200 m in the Peene Basin, i.e., in subsided parts of the depression. Leine-Steinsalz can be divided here into Basissalz (Na3α), Liniensalz (Na3β), Kristallsalz (Na3γ), Anhydritmittelsalz (Na3η), Schwadensalz (Na3θ), and Tonflockensalz (Na3ι) (Jung 1968b, Hemann 1970).

In the central and southern North-East German Depression, Kaliflöz Ronnenberg (K3Ro) is 20—25 m thick.

In the south-eastern North-East German Depression, deposits of the Leine Series display a similar lithologic variation. At Oberlausitz (G.D.R.), the series has a ternary structure. From the base up the section, the series consists of Grauer Salzton (T3), Hauptanhydrit (A3), and Leine-Steinsalz (Na3). Hauptanhydrit ranges from 30—60 m; the greatest thicknesses were recorded in marginal areas, due probably to the presence of an anhydrite wall. Plattendolomit (Ca3) appears along the periphery of the Leine Series.

Leine-Steinsalz of the Gubin and Lubsko area occurs only north-west of Lubsko. Its thickness increases north-westward from 0 to 150—200 m. Grauer Salzton (T3) is dark gray, locally chocolate-brown and contains anhydrite lenses. It is 1—5 m thick. Hauptanhydrit (A3) is 3—29 m thick. Anhydrites are medium- and cryptocrystalline and dense, in places with clayey-bituminous and dolomite laminae. To the south, the Leine Series is mainly made up of anhydrite, while at the margins it is dominated by carbonates grading into terrigenous rocks in the southern North Sudetic Trough.

Within the Altmark Depression, at Calvörde, the thickness of the Leine Series attains 200 m. The series can be divided, in ascending order, into Grauer Salzton (T3), 3—4 m; Hauptanhydrit (A3) and Leine-Steinsalz (Na3), 140—150 m thick. Leine-Steinsalz is, in turn, subdivided into Basissalz (Na3α), Liniensalz (Na3β), and Kristallsalz (Na3γ), totaling about 30 m; Kaliflöz Ronnenberg II, 10 m; Steinsalzmittel, 20 m; Kaliflöz Ronnenberg I, 10 m; Anhydritmittelsalz (Na3η), not less than 30 m; and Schwaden- und Tonflockensalz (Na3θ + ι) up to 40 m thick. Kaliflöz Ronnenberg II is commonly composed of sylvinite or Hartsalz, whereas Kaliflöz Ronnenberg I is represented in some sections by Hartsalz, alternated with anhydrite and rock salt and by sylvinite with halite in the others (Löffler and Schulze 1962).

The volume of rock salt in the Leine Series in the North-East German and Altmark depressions can be estimated at 1.05×10^4 km^3, assuming an areal extent of no less than 70,000 km^2 and an average thickness above 150 m.

In the Central Polish Depression, equivalents of the Leine Series are distinguished as the Polish Zechstein Cycle 3 (PZ3) (Poborski 1960, Werner et al. 1960, Wagner et al. 1981). The deposits of the cycle can be divided into Gray Pelite, which is correlative with Grauer Salzton of the German Zechstein, Platy Dolomite, Main Anhydrite, and Younger Halite.

Gray Pelite occurs throughout the Central Polish Depression and is used as a marker horizon for correlation of the sections. It is composed of dark gray or schistose clay, locally interbedded with siltstone. It has a thickness of 5—6 m in the interior parts of the depression.

Platy Dolomite is easily recognizable in marginal parts, where its thickness ranges from 5—20 m. Extensive belts of barriers can be outlined here (Wagner et al. 1981, Dopowski et al. 1981). One belt follows the southern margin of the basin and is known as the Silesian Barrier. The Holy Cross, Podlasie, and Mazury Barriers are distinguished at the eastern margin; the Pomeranian Barrier is located to the north. In the interior parts, Platy Dolomite is dominated by argillaceous micrite; the thickness does not exceed 2—3 m.

In central parts of the depression, the Main Anhydrite is about 30 m thick. The overlying Younger Halite is sometimes subdivided into three parts. For example, at Kłodowa, the lower Younger Halite comprises linear salt, orange and light pink salt; above follows salt with sylvine inclusions and pockets and sylvinite interbeds overlain by the Younger Potash. It is made up of rock salt interbedded with carnallite and kieserite rock. The thickness ranges from 50—100 m. In the lower half, kieserite-carnallite rocks (10—15 m) occur, whereas the upper half is composed of depleted carnallite rocks. The upper part (about 100 m thick) is composed of peculiar rocks represented by clay and halite mixed in various proportions. The most common are: layered, clayey salt and Brown Zubers, containing 15—85% honey-colored halite with inclusions of dark gray and brown clay. In places, the upper part is determined as the Brown Zubers. The thickness of the unit is about 100 m.

A similar section of Cycle PZ3 is also reported from other parts of the Pomorsko-Kujawy Anticlinorium, in particular at Inowrocław. Another similar section is observed in the north-western interior part of the Central Polish Lowland (Depression), where it has been penetrated by wells at Swidwin. Only the Younger Potash displays a decrease in thickness there.

To the south-west, judging by the sections of Gorzów Wielkopolski 1 and Wschowa 1 wells, a section of Cycle PZ3 changes appreciably. The upper Brown Zubers strata are missing and potash salts disappear. In the lower part, the Gray Pelite and Main Anhydrite are preserved; the upper part of the cycle is composed of monotonous rock salt. The thickness decreases to 200—220 m at Gorzów Wielkopolski, where rock salt accounts for 150—180 m, and to 100—130 m at Wschowa, where rock salt has a thickness of 80—90 m.

More considerable changes in composition and thickness of Cycle PZ3 occur along the south-western margin of the Central Polish Depression (Podemski 1965, Kłapciński 1967a, b). On the Fore-Sudetic Monocline, near the towns of Gubin, Legnica, and Sieroszowice, the section is dominated by dolomite interbedded with mudstone in the lower part; gypsum and anhydrite inclusions also occur. The thickness does not exceed 30—40 m. Terrigenous rocks appear farther south-east in the western Nida Trough (Jurkiewicz 1970). There, the Pagow IG-1 well section consists of brownish-cherry mudstone with anhydrite interbeds and inclusions in the lower part and fine-grained, red sandstone with mudstone interbeds and white gypsum lenticules in the upper part. The thickness is 34.5 m. In the Milianowa IG-1 well, the lower part is composed entirely of sandstone interbedded with siltstone; the upper part contains conglomerate and gritstone beds.

In the south-eastern Central Polish Depression, in the Holy Cross Mountains, the equivalents of Cycle PZ3 are represented by red sandy marl, siltstone, mudstone, and layered limestone, ranging in thickness from 25 to 80—90 m (Szaniawski 1965).

The approximate rock salt volume of Cycle PZ3 in the Central Polish Depression is 8×10^3 km^3, assuming an areal extent of 80,000 km^2 and an average salt thickness of 100 m.

Salt deposits of Cycle PZ3 were also reported from the south-western Polish Lithuanian Lowland (Stołarczyk 1972), where the subsided areas are composed of dark gray clay up to 2 km thick; above, anhydrite 27.5–38.4 m thick appears. West- and south-westward, anhydrites become thinner and rock salt strata appear toward the margin, the beginning of platy dolomites is recorded in the section; in the coastal part, they are enriched in red terrigenous material. Red marly strata with gypsum, which is assigned to equivalents of the Zechstein Cycle 4, are found above. Redstones of the lower part of the Suduvsk Formation are considered as one of the equivalents of the Leine Series in the eastern Polish-Lithuanian Lowland.

The approximate areal extent of rock salt in the Polish-Lithuanian Lowland amounts to 7000 km^2; the average rock salt thickness probably does not exceed 30 m. Assuming the above estimates, rock salt volume will equal 2.1×10^2 km^3.

Within the English Basin, the English Zechstein Cycle 3 comprises deposits of the Teesside Group, which can be divided into the Seaham Formation of the Durham province or the Upper Magnesian Limestone of the Yorkshire province, the Billingham Main Anhydrite, and Boulby Halite (D. Smith 1971, 1981, Taylor and Colter 1975, D. Smith and Crosby 1979).

Carbonate rocks of the Seaham Formation and the Upper Magnesian Limestone, occurring at the base of the cycle, are easily recognizable in a belt 80–100 km wide, where their thickness attains 100 m. In the lower part, a thin, dark gray, illitic shale, equivalent to the Grauer Salzton of the German Zechstein, is distinguished.

Most of the remainder of the formation is of uniform, gray to black, fine-grained, partly pelleted, fossiliferous dolomite, diversified in marginal and shoal areas by oolites, oncolites, and thin clay beds; and toward the basin dark brown, anhydritic, laminated dolomite appears.

The Billingham Main Anhydrite varies in thickness between 15 and 55 m. It is composed of anhydritic breccia, laminated anhydrite, nodular anhydrite interbedded with dolomite, and red mudstone.

The Boulby Halite in north-east England is up to 80 m thick. It can be divided into four units: A, B, C, and D. The citation of D. Smith and Crosby (1979, pp. 398–399) is given below to describe the units.

"D. Gray, brown, and red, generally medium-grained halite with thin beds of dark gray-green or red mudstone toward the top and with patchy sylvite near the base in some coastal districts. Parts commonly contain a distended mesh of dark gray-green or red mudstone. Small amounts of anhydrite, hematite, magnesite, and quartz are generally present. The unit is from 0 to 5.2 m thick.

C. Boulby Potash. Sylvinite, colorless, pink, red, brown, or gray, commonly strikingly mottled, comprising a variable mixture of medium-grained and coarse-grained halite and generally secondary amoeboid sylvite in a weakly layered, distended and discontinuous, fine-grained mesh of dark gray-green carbonaceous mudstone and/or fine-grained, gray anhydrite and with beds of halitic or sylvinitic mudstone and anhydrite. Boracite, hematite, koenenite, magnesite, pyrite and quartz range from rare to abundant and minor carnallite is common. The unit ranges from 0 to 11 m in thickness.

B. Halite: upper part is colorless, pale gray or amber, equigranular, medium-grained, with thin laminae of fine-grained anhydrite and scattered crystals and small patches of sylvite; lower part is mainly gray and brown, coarse- to very coarse-grained, with mudstone and anhydrite forming a discontinuous coarse mesh and some thin beds. Medium-grained granular gray halite, partly pseudomorphous after gypsum, widely forms thin beds near the base. Carnallite, magnesite, sylvite, and quartz (in higher beds) are widespread minor constituents. The unit is about 30 m thick.

A. Halite with subordinate anhydrite: upper part is pale gray, amber and brown, faintly banded, generally medium-grained, commonly with a discontinuous ragged mesh and some thin beds of mudstone and fine-grained anhydrite and with upright fibro-radiate pseudomorphs after gypsum in numerous layers 0.001 to 0.003 m thick; lower part is of halite (as above) and halitic anhydrite, the latter with halite pseudo-morphing layers of fibro-radiate coarse-grained gypsum. Thin beds of halitic dolomite are present locally. The unit is about 15 m thick."

For detailed description of salt deposits of the English Zechstein Cycle 3, including that of various mineralogical changes, the reader is referred to the publications of Stewart (1951a), Raymond (1953), D. Smith and Crosby (1979), and Woods (1979).

The rock salt volume in the English Basin may exceed 3.75×10^3 km^3 (the areal extent is about 75,000 km^2 and the average thickness of salt is 50 m).

Salt deposits, including potash salts, are very common in the Norwegian-Danish Basin (Sorgenfrei and Buch 1964, Hinz 1968).

In the Manx-Furness and Solway Firth basins, the upper Sandwith Dolomite and Sandwith Anhydrite of St. Bees Evaporites may be considered as equivalents of the English Zechstein Cycle 3 (Arthurton and Hemingway 1972). The upper Sandwith Dolomite is composed of red, green, and grayish-brown, thin-bedded dolomites with laminae and lenses of anhydrite. The Sandwith Anhydrite is represented near the base by mottled anhydrite, aphanitic or granular, with anhydrite pseudomorphs after gypsum; above comes nodular anhydrite followed by anhydritic varvite; and towards the top fibro-radiate anhydrite appears. The thickness does not exceed 20 m. In marginal areas, sulfate-carbonate breccia appears.

The volume of rock salt in the Leine Series in the Central European Basin totals 3.4×10^4 km^3.

Zechstein 4 (Aller Series)

Salt deposits of Zechstein Cycle 4 are common in the Weser, North-West German, North-East German, Altmark and English Basins, and the Central Polish, Subhercynian and Thüringia Depressions (Fig. 138). They have a relatively simple structure. At the base, the Roter Salzton (T4) can be traced ubiquitously; it is overlain by the Pegmatitanhydrit (A4). Most of the section is made up of the Aller-Steinsalz (Na4). The section is crowned by the Grenzanhydrit (A4r) and the Oberste Zechsteinletten (T4r). Since these stratigraphic units are very persistent, a general characterization can be given starting with a type section in the lower Aller Basin, in the Weser Depression.

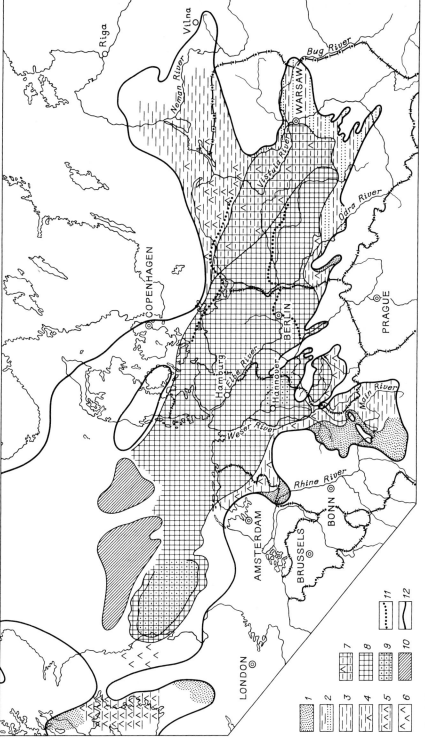

Fig. 138. Lithologic map of Zechstein 4 (Aller Series) deposits in Central European Basin. Compiled by Zharkov and Merzlyakov from the data of Richter-Bernburg (1955c, 1972a, b), Poborski (1960, 1969), Suveizdis (1962), Löffler and Schulze (1962), Seidel (1965), Podemski (1965), Bartenstein (1968), Hinz (1968), Brunstrom and Walmsley (1969), Milewicz (1971), Arthurton and Hemingway (1972), Smith (1972, 1981), Stoļarczyk (1972), Taylor and Colter (1975), Smith and Crosby (1979), and Wagner et al. (1981). Areal distribution of: *1* terrigenous red beds, *2* variegated and red sandstone, siltstone, and mudstone, *3* mainly red mudstone and marl, *4* the same as *3* with lenses and inclusions of gypsum and anhydrite, *5* the same as *3* with interbeds of gypsum and anhydrite, *6* mainly anhydrite, *7* Red Zubers, *8* salt, *9* potash, *10* probable absence of salt deposits on uplifts in North Sea, *11* extent of Zechstein Cycle 5 (Ohre Series), *12* extent of Upper Permian (Zechstein) deposits

The Roter Salzton (T4) is not the most continuous reference marker horizon of Zechstein deposits. In the Weser Depression, it is composed of thin-bedded, greenish-gray and red saline mudstone; the thickness of which near Hannover, Bergmannssegen-Hugo Potash District reaches 15 m (Roth 1955b). East- and south-eastward, in the Subhercynian Depression, reddish and chocolate-brown clay with anhydrite inclusions and lenticular partings, as well as rock salt stringers are common. The thicknesses there are as follows: at Aschersleben and Bernburg, 7 m; at Stassfurt, 4–10 m; at Schönebeck, 4–6 m; and at Schwanebeck, 19 m (Löffler and Schulze 1962). In the Allertal Graben, the thickness is 3–5 m; the lower part of the section (up to 0.5 m) encompasses greenish-light gray, layered, foliated clay with anhydrite lenses; and in the upper part, red-brown, locally greenish-gray massive clay with anhydrite lenses and inclusions is present, as well as dark and orange-red rock salt with halite and carnallite stringers, filling in fractures. In the northern Thüringia Depression, the basal horizon of Zechstein Cycle 4 is composed mainly of red, locally greenish-gray mudstone, slightly arenaceous with stringers, laminae, and inclusions of rock salt and anhydrite. There, the thickness of the Roter Salzton ranges from 1–50 m. In some areas, e.g., near Sonderhausen, three parts can be distinguished; the lower, greenish-red member 1–3 m thick; the middle, red member 10–20 m thick; and the upper, also greenish-red member up to 4 m thick (Seidel 1965). The general lithology and thicknesses of the Roter Salzton (T4) in the Thüringia Depression can be seen in Fig. 139.

In the central most subsided part of the Mid-European Trough (North-West German, North-East German, and Altmark Depressions), the thickness of the Roter Salzton ranges from 3–5 to 30 m (Jung 1968b, Hofrichter 1967, Reichenbach 1970).

In the Central Polish Depression, at the base of the Polish Zechstein Cycle 4 (PZ4), the Lower Red Pelite is easily recognized as an equivalent to the Roter Salzton of the German Zechstein. It extends into the south-west central part of the depression and varies in thickness between 2 and 35 m (Wagner et al. 1981).

The Pegmatitanhydrit (A4) displays a universal distribution. Anhydrite is commonly light gray. Its pegmatitic aspect results from characteristic inclusions of brown halite. The thickness ranges mainly from 0.4–0.5 to 1–1.5 m.

The Aller-Steinsalz (Na4) occurs in the North-West German, North-East German, Altmark, Central Polish, Weser, Subhercynian, and Thüringia Depressions. It has the following thicknesses: in the Weser Depression, 100 m; in the North-West German Depression, 100–150 m; in the North-East German Depression, 30–50 m; in the Altmark Depression, 30–45 m; in the Central Polish Depression, 80–100 m; in the Subhercynian Depression, 20–40 m; and in the Thüringia Depression, 30–35 m.

In the Weser Depression, the Aller-Steinsalz (Na4) can be divided into five parts: (1) Basissalz, Na4α, (2) Schneesalz, Na4β, (3) Rosensalz, Na4γ, (4) Tonbrockensalz, Na4δ (or Na4br), and (5) Tonbanksalz, Na4tm (Richter-Bernburg 1955b). The Tonbrockensalz (Na4br) locally contains potash salts represented by sylvite-bearing kieserite-halitic rocks, distinguished as the Kaliflöz Ottoshall (Lotze 1938, Roth 1955b). The above horizons of the Aller-Steinsalz are also reported from other regions. The most complete sections are located in subsided central parts of the depression, whereas toward the margin, some horizons pinch out. Thus, in the northwestern Subhercynian Depression within the Allertal Graben, the Aller-Steinsalz can

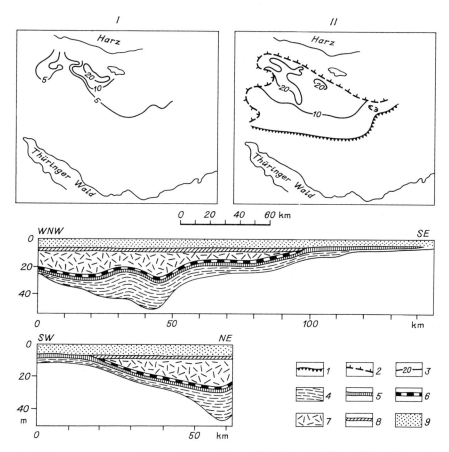

Fig. 139. Thickness maps of Roter Salzton (I) and Aller-Steinsalz (II) and lithologic sections of Aller Series in Thüringia Depression. After Seidel (1965). Explanation used for maps: *1* original extent of rock salt, *2* secondary boundary of rock salt, *3* isopach contours. Explanation used for sections: *4* Roter Salzton (T4), *5* Pegmatitanhydrit (A4), Aller-Steinsalz (Na4), *6* Basissalz (Na4α), *7* Schneesalz (Na4β), *8* Grenzanhydrit (A4r), *9* Oberste Zechsteinletten (T4r)

be divided into only three horizons: Basissalz (Na4α), Schneesalz (Na4β), and Tonbrockensalz (Na4br), which includes the Kaliflöz Ottoshall.

In the remaining Subhercynian Depression, two lower horizons, the Basissalz and Schneesalz, are easily recognizable.

In the Thüringia Depression, only two lower horizons are present, the Basissalz (Na4α) and Schneesalz (Na4β) (Seidel 1965). The Basissalz is mainly whitish-gray, rarely red with anhydrite inclusions and lenses; it is 0–3 m thick. The Schneeesalz is represented by grayish-white halite with occasional anhydrite laminae at the base; its thickness ranges from 0–37 m.

A minute subdivision of the Aller-Steinsalz in the Altmark Depression is illustrated below (Reichenbach 1970). At Calvörde, the following units are distinguished, in ascending order:

Units	Thickness
Basissalz (Na4α) – Gray and red rock salt with anhydrite inclusions and interbeds	5 m
Schneesalz (Na4β) – Light pink and gray rock salt with anhydrite inclusions below	16 m
Kristallsalz (Na4γ) – Colorless, coarse-grained rock salt interbedded with anhydrite	5 m
Flockensalz (Na4δ) – Rock salt with flake-like inclusions of red-brown clay	2 m
Anhydritmittelsalz (Na4ζ) – Colorless, coarse-grained rock salt with anhydrite laminae, one of which is 0.3 m thick at the base	5 m
Tonbrockensalz (Na4br) – Rock salt with brown-red halopelite laminae and inclusions	10 m

Within the Central Polish Depression, salt strata of Polish Zechstein Cycle 4 are assigned as the "Youngest Halite". In its middle part, there is a marker bed of the Upper Pegmatite Anhydrite 1–1.5 m thick, which divides the salt sequence into two units: the Lower Youngest Halite and the Upper Youngest Halite. Their thicknesses range between 15–40 m and 10–50 m, respectively.

The Grenzanhydrit (A4r) is easily traceable in the central Mid-European Trough. In the Weser Depression, its thickness does not exceed 1 m; in the Subhercynian Depression, 0.15–3 m (Aschersleben, 0.15 m; Schwanebeck, 0.5–0.75 m; Stassfurt, 0.5–3 m; Schönebeck, 1 m); in the Thüringia Depression, it is only 0.2 m thick (apart from the north-eastern areas, where it reaches 2.3 m at Unstrut and in the Mansfeld Syncline); and in the North-West German and North-East German Depressions, 0.5 m. In places, the Aller-Steinsalz (Na4) is directly overlain either by the Oberste Zechsteinletten or Schluffstein of Zechstein Cycle 5. The Grenzanhydrit (A4r) is unknown in most of the Central Polish Depression.

The Aller Series is crowned by the Oberste Zechsteinletten (T4r). It is common in the southern Weser, Thüringia, and south-eastern Subhercynian Depressions, south-easternmost North-East German Basin in Lausitz, and in the northern part of the same basin in north-east Mecklenburg. Within the Thüringia Depression, the Oberste Zechsteinletten (T4r) is represented by thin-bedded, red-brown interbeds and patches and greenish-gray mudstone, the thickness of which totals 5–8 m (Seidel 1965). There, the section can be divided into three members: the lower, greenish-red 0–2 m thick; the middle, reddish-brown 4–7 m thick; and the upper, greenish-gray up to 0.5 m thick.

In the Subhercynian Depression, the Oberste Zechsteinletten is locally 20–30 m thick. In the upper part, anhydrite interbeds up to 1.2 m thick appear (Löffler and Schulze 1962). In marginal areas of the North-East German Basin, the thickness varies between 5 and 8 m (Jung 1968b). The Oberste Zechsteinletten is composed of red, variegated, clayey-marly deposits with lenses and inclusions of anhydrite and gypsum.

The peripheral areas of the Central European Basin are marked by an abrupt decrease of thickness and a change of lithology of the Aller Series. Toward the margin, the thickness of the Aller-Steinsalz gradually decreases and the number of clayey red interbeds increases. As a result, the section becomes entirely clayey and is designated

Na4T (Jung 1968b). As rock salt wedges out, the thickness of the Aller Series decreases to 10–15 m. First, it has a three-partite section, since its middle part contains a 1–1.5 m anhydrite horizon, which is equivalent to the undivided Pegmatitanhydrit (A4) and Grenzanhydrit (A4r). The lower and upper parts of the series are represented by red clay, corresponding to T4 and T4r horizons. Such a section of the Aller Series is characteristic of the Werra Fulda Depression. Still nearer the margin, anhydrite also pinches out and the series becomes first entirely red and clayey-silty, then even psammitic, as in the south-western Weser Depression, in the western and south-western Werra Fulda Depression (Richter-Bernburg 1972a, b, Dittrich 1966), and eastern Thüringia Depression (Seidel 1965, Jung 1968b).

Similar transitions in lithology occurred in the interior parts of the North-West German and Lower Rhine basins toward the south (Wolburg 1958a, b, Teichmüller 1958, Richter-Bernburg 1972a, b). There, first rock salt and then anhydrite pinch out, the Aller Series then becomes entirely red clayey; in marginal areas, it contains coarse-grained, terrigenous rocks.

Similar lithologic variations are also observed in the North Sudetic Trough (Milewicz 1971). At Gubin and Lubsko, south-west Poland, the Lower Red Pelite (T4a) lies at the base of the Aller Series; its thickness in subsided northern areas is about 1 m. Above the Pegmatite Anhydrite (1–4 m) and rock salt follow with a maximal thickness of 47 m; the series is crowned by red-brown siltstone and mudstone about 9 m thick. South-eastward, toward the North Sudetic Trough, rock salt pinches out; the thickness of the Lower Red Pelite is 10 m and that of the upper red mudstone and siltstone, 16 m. In the interior parts of the North Sudetic Trough, the Aller Series becomes almost entirely red clayey, while in marginal areas of the trough, it is dominated by sandstone. Kaczmarek (1970) presented a section, in which the lower and upper Aller Series is composed of brown mudstone, 9 and 12.5 m thick, respectively; and in the middle composed of fine-grained sandstone, 5.5 m thick.

The lower part of Zechstein Cycle 4 (PZ4a) in marginal areas of the Central Polish Depression and within the Polish-Lithuanian Lowland is made up entirely of terrigenous redstones. At Lubin, Legnica, and Sieroszowice, there are common dolomitized siltstone, containing intraformational conglomerates and a bed of gypsum-anhydrite in the middle part, equivalent to the Pegmatite Anhydrite (Podemski 1965). In the western Nida Trough, the section is composed of brownish-cherry sandstone, siltstone, and mudstone about 20.5 m thick (Jurkiewicz 1970). In Świętokrzyskie Góry, Cycle PZ4a is composed mainly of mudstone, marl, and shale, in places with siltstone, sandstone, and even conglomerates. There, the thickness varies between 40 and 70–80 m (Szaniawski 1965).

In the north-eastern Central Polish Depression, there is a peculiar facies of Cycle PZ4a, represented by salt-bearing rocks, which are designated as the Red Zubers (Poborski 1960, Werner et al. 1960, Wagner et al. 1981). These deposits are 100 m in thickness.

In the south-western Polish-Lithuanian Lowland, equivalents of the Aller Series are represented by 20–25 m strata, composed of red-brown mudstone and siltstone with inclusions and stringers of gypsum and anhydrite (Stołarczyk 1972). In regard to the composition of Zechstein Cycle 4 in Poland, the upper part of the Suduvsk Formation in the eastern Polish-Lithuanian Lowland may possibly occupy the same stratigraphic level; this part is composed of red-brown marl and siltstone with gypsum inclusions (Vala 1961).

The areal extent of rock salt of the Aller Series within the North-West and North-East German Basins, and the Altmark, Central Polish, Weser, Subhercynian, and Thüringia Depressions exceeds 250,000 km^2. Over most of the area discussed, the thickness of rock salt ranges from 20–50 m, however, in some depressions it is 100–150 and even 200 m. Assuming an average salt thickness of 25 m, the volume of rock salt of the Aller Series will equal 6.25 × 10^3 km^3.

In the western Central European Basin, within the English Depression, English Zechstein Cycle 4 comprises deposits of the Staintondale Group. In the Yorkshire province, they are divided, in ascending order, into: the Carnallitic Marl, Upgang Formation, Upper Anhydrite, and Upper Halite.

The Carnallitic Marl varies in thickness between 12–17 and 20–30 m. Its lithology and structure can be exemplified by the Upgang E-6 well section west of Whitby; it has been intensely studied by Raymond (1953). In ascending order, the section consists of:

1. Green marl with abundant carnallite inclusions and crystals of flesh-pink halite . . . . 2.1 m
2. Red-brown, silty marl containing numerous, fine lenses and inclusions of halite, carnallite, and rinneite. Lenses usually occur on the bedding plane, but in places cut the rock. Inclusions of flesh-pink halite are particularly numerous in the lower 2 m. Large lenses (approximately 1 cm wide), occurring throughout the layer, contain sylvite inclusions in halite; locally they are composed entirely of sylvite. Lenses of granular carnallite are less common. . . . . 12.5 m
3. Red-brown, silty marl with lenses of granular flesh-pink halite. The upper part contains numerous rinneite grains and lenses of pink halite, locally with sylvite. . . . . 2.1 m

In other sections, inclusions and lenses of halite, rinneite, and carnallite are more or less numerous.

The Upgang Formation in coastal districts is composed of dolomite; and in the interior parts of the basin of laminated, dark gray, dolomitic shale; magnesite-rock is present as well.

The Upper Anhydrite in north-east Yorkshire is gray and more rarely, pink. It contains inclusions of gray, anhydritic carnallite marl and in places, of carbonate rocks. Lenticules of pink halite are rather common, in which grains of milky, white, and yellow sylvite occur. A 10 cm lens of coarse-grained sylvite can be recognized near the top. In some sections, halite-sylvite lenses occur at the base. The Upper Anhydrite is 5–8.5 m thick.

The Upper Halite has been thoroughly studied in north-east Yorkshire (Stewart 1951b, Raymond 1953, D. Smith and Crosby 1979, D. Smith 1981). It can be divided into the following units, in ascending order: A, B, C (Upper Potash), D, and E; their description is cited below from D. Smith and Crosby (1979, pp. 401–402):

"E. Colorless to amber, medium- to coarse-grained halite with scattered films and patches and locally a faint network of fine-grained, gray anhydrite and red silty mudstone. The unit is 1.8 to 4.9 m thick.

D. Mudstone, halitic, and argillaceous or silty halite, comprising a heterogeneous, faintly layered mixture of colorless, mainly coarsely crystalline euhedral to subhedral halite, in a distended matrix of brick-red silty mudstone or argillaceous siltstone. The relative proportions of the two main components range widely, and beds of almost halite-free mudstone occur. Anhydrite, hematite, magnesite, talc, quartz, and, locally in lower beds, sylvite are minor constituents. Unit D is 14 to 23 m thick.

C. (Upper Potash). Sylvinite, colorless, gray, brown, or mottled, comprising a variable mixture of medium- to coarse-grained halite and sylvite in a roughly concordant mesh of gray anhydrite and dark gray-green carbonaceous clay. Commonly divided into three or more parts by beds of halite or sylvinitic red or gray mudstone up to 1 m thick and by beds of thinner halite and anhydrite. . . Boracite, carnallite, hematite, magnesite, rinneite, talc, and quartz are generally minor constituents. The unit ranges from 0 to 8.5 m in thickness.

B. Halite: colorless, pink or amber, mainly medium-grained, equigranular, commonly weakly color-banded, and with a few laminae and thin beds of fine-grained gray anhydrite and gray or red-brown anhydritic mudstone. Local traces of a delicate distended mesh of anhydrite and/or mudstone in the upper part, where scattered patches of rinneite and sylvite are common. Hematite, magnesite, talc, and quartz are widely distributed minor constituents. Unit B ranges from 8 to 12 m in thickness.

A. Halite: colorless to pale pink, medium- to coarse-grained, faintly rhythmically color-banded, with delicate laminae of fine-grained gray anhydrite at intervals of up to 0.2 m and a few thin beds of halitic anhydrite. Magnesite, rinneite, sylvite, and talc are locally abundant near the base. The unit ranges from 5 to 7 m in thickness."

To the west and north-west, toward the margin of the English Basin, the Upper Halite decreases in thickness and pinches out. In marginal sections, the equivalents of carnallite marl pass into terrigenous redstones. Above comes dolomite; the whole upper part is composed of anhydrite. Such a composition is typical strata in Durham and western Yorkshire. Toward the North Sea, the thickness of the Upper Halite increases to 70–80 m. At the same time, the thickness of potash salt also increases to 50 m and above (D. Smith and Crosby 1979).

The estimated volume of the Upper Halite in the English Basin is approximately 4×10^2 km^3 assuming an areal extent of 20,000 km^2 and an average thickness of 20 m.

In the Manx-Furness and Solway Firth Basins, the Fleswick Cycle may well be an equivalent of the English Zechstein Cycle 4 (Arthurton and Hemingway 1972). In its lower part, the Fleswick Dolomite and Siltstone can be distinguished; their thickness totaling 5–6 m. Lying below is breccia of dolomite, anhydrite, and carbonate sandstone; and above gray dolomite, overlain by gray siltstone appears. The upper part of the cycle is represented by Fleswick Anhydrite; the latter contains dolomite laminae. The Fleswick Anhydrite is not more than 4 m thick.

The above data of the Aller Series and its equivalents in the Central European Basin allows an estimated volume of Zechstein 4 rock salt exceeding 6.65×10^3 km^3.

Zechstein 5 (Ohre Series)

Salt deposits of Ohre Series are reported from the North-East German Basin and the Altmark and Central Polish Depressions (see Fig. 139) (Jung 1968b, Reichenbach 1970, Wagner et al. 1981).

The thickness of the Ohre Series does not exceed 50 m. In the North-East German Basin (Jung 1968b), the Schluffstein (T5) lies at the base of the series; its thickness ranges from 0–3 m. Above follows anhydrite (A5) 0–0.3 m thick. Even higher, there is the thin (up to 1 m) Ohre-Steinsalz (Na5). The latter is overlain by the Schluffstein (T5r) 0–2.5 m thick. The Ohre-Series is crowned by anhydrite (A5r) 0–0.5 m thick.

In the Altmark Depression, the section of the series remains essentially the same. Reichenbach (1970) gave the following composite section (in ascending order) for the Calvörde Region:

Salzbrockenton (T5) – A silty, red-brown bed with fine-grained sandstone laminae and halite inclusions .... up to 5 m
Lagenanhydrit (A5) – Anhydrite with mudstone and carbonate interbeds 0.1 m
Ohre-Steinsalz (Na5) – Light gray rock salt with inclusions of anhydrite and clay.. 0.3–3.0 m
Grenzanhydrit ($A5r_1$) – Anhydrite with interbeds of brown clay...... 0.5 m
Oberer Schluffstein ($T5r_1$) – Brown, massive mudstone............. 1.5–2.0 m
Oberer Grenzanhydrit ($A5r_2$) – Almost structureless anhydrite....... 0.25 m

The greatest thickness of the Ohre Series was reported from the Central Polish Depression (Wagner et al. 1981). These deposits are assigned to Cycle PZ4b and can be divided as follows, in ascending order: the Upper Red Pelite (T4b), ranging in thickness from 3–20 m; Anhydrite (A4b), up to 1 m thick; Top Youngest Halite (Na4b), varying in thickness between 5 and 30 m; and Boundary Anhydrite (A4br), 0.5–2 m thick. Rock salt occurs in the central, most subsided part of the depression. It pinches out north-eastward and the entire section of the cycle is represented by salt-bearing, terrigenous deposits assigned to the Red Zubers.

Equivalents of the Ohre Series are known from the western Central European Basin and the English Basin (Smith 1972, 1981). English Zechstein (EZ5) comprises deposits of the lower Eskdale Group, in the Yorkshire Province, which can be divided, in ascending order, into: the Sleights Siltstone, Top Anhydrite, and Saliferous Marl, which in Durham province corresponds to the Upper Marl.

In the Manx-Furness and Solway Basins, a possible equivalent of these deposits is the lower St. Bees Shales, represented by red siltstone and marl interbedded with gypsum (Arthurton and Hemingway 1972).

The volume of rock salt in the Ohre Series is rather small. Assuming an areal extent slightly exceeding 50,000 km^2 and an average thickness of 1 m, then the volume of rock salt will be not more than 0.5×10^2 km^3.

In conclusion, it should be noted that the Central European salt basin is one of the largest basins with respect to both the areal extent of salt deposits and the volume of rock salt accumulated. The volume of the Zechstein rock salt in the basin may well exceed 1.28×10^5 km^3; in regard to the Lower Permian salt, it will amount to 1.46×10^5 km^3.

Alpine Basin

Permian salt-bearing deposits have been long known from north and north-eastern parts of the East Alps in Austria. They compose numerous salt domes or crop out along the major faults within the belt extending from Innsbruck in the west to Vienna in the east. These salt-bearing sediments were assumed to mark a separate salt basin which has not yet been outlined. Recently a considerable amount of data has become available, thus, enlightening the distribution pattern of the Permian salt-bearing deposits in the East Alps and West Carpathians (Del-Negro 1960, Zhukov 1963, 1965a, Oxburg 1968, Mahel and Vozár 1971, Zhukov and Yanev 1971, Clar 1972, Falke 1972, 1974, Frank 1972, Heissel 1972a, b, Klaus 1972, Mostler 1972a, b, c, Praehauser-Enzenberg 1972, Riehl-Herwirsch 1972, Riehl-Herwirsch and Wascher 1972, Sommer 1972, Trevisan 1972, Flügel 1975, Richter and Zinkernagel 1975, Vozárová and Vozár 1975, 1981, Schauberger et al. 1976, Zhukov et al. 1976). They have shown that the salt-bearing sediments occur not only in the East Alps, but extend eastward as far as the Spišsko-Gemerskoe Rudohoři, in the eastern part of the West Carpathians. Thus, it appears that the Alpine salt basin was a rather extensive seaway, apparently separated by the chain of uplifts from the Cental European Basin lying father north.

At present, many localities of Permian evaporite rocks are known in southern Europe. Salt-bearing deposits were also found in north-eastern Bulgaria with a separate salt basin within the Moesian Depression, as well as in the East Alps and West Carpathians. Sulfate-bearing deposits were recognized also near Rakhov in the East Carpathians and in the Dobruja Depression in Moldavia. In southernmost Europe, Permian sulfate-bearing deposits are found in northern Italy and in the Carnic Alps, in the Dinarids, in the Montenegro in Yugoslavia, in the Korab zone in north-eastern Albania, and in the Mecsek Mountains in southern Hungary. A peculiar pattern of evaporite distribution in southern Europe is shown in Fig. 140. In addition to the Alpine and Moesian salt basins, five sulfate basins are recognized: the North Italian, Dinarids, Mecsek, Rakhov, and Dobruja Basins.

Schauberger (1955) distinguished 15 regions in the northern part of the East Alps where Permian salt-bearing deposits are exposed (Fig. 141). They are studied in particular detail within the Halstatt salt stock where a sequence is exposed in salt mines. The whole sequence is composed of various salt, sulfate-salt, or sulfate-carbonate-salt rocks generally known as Haselgebirge. Several types of these rocks are recognized: gray salt, red salt, green-gray clay-salt, variegated clay-salt, bituminous-dolomite gray salt, etc. In most cases, these rocks are brecciated; primary layering is rare. The thin-bedded Haselgebirge occur in those places where rock salt alternates with anhydrite and bituminous dolomite. It should be noted that stratification of rocks within this salt sequence is obscured due to complex tectonics and that the entire sequence is strongly deformed by fluidal-tectonic processes. The mode of occurrence is also uncertain. It breaks through Triassic, Jurassic, Cretaceous, and the Neocomian sediments and is mainly assigned to the Late Permian. Schauberger (1955) assigned this sequence to the level of the Bellerophon Beds of the South Alps. The base of the sequence is not exposed. The spore-pollen analysis (Klaus 1955) showed that spores of both Triassic and Permian aspects are found in differently composed Haselgebirge

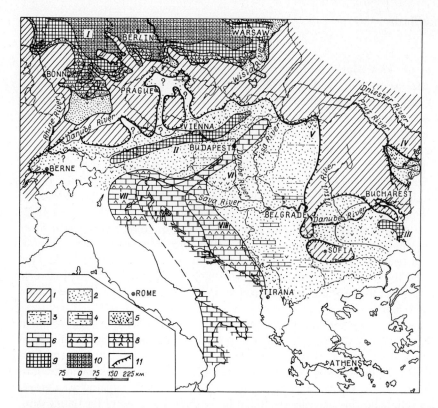

Fig. 140. Lithologic map of Upper Permian deposits in southern Europe. Compiled by Zharkov and Merzlyakov. Areal extent of: *1* deposits absent, *2* terrigenous rocks mostly sandstone, *3* terrigenous rocks, dominated by siltstone and mudstone, *4* terrigenous and carbonate rocks, *5* terrigenous rocks with lenses, inclusions, and interbeds of gypsum, *6* limestone and dolomite, *7* carbonate rocks with gypsum and anhydrite, *8* carbonate rocks with approximately equal amount of anhydrite, *9* rock salt, *10* potash salt, *11* limits of Upper Permian in southern Europe. Basins: *I* Central European, *II* Alpine, *III* Moesian, *IV* Dobruja, *V* Rakhov, *VI* Mecsek, *VII* North Italian, *VIII* Dinarids

inclusions. Thus, some investigators consider the salt-bearing sequence as Permo-Triassic and others regard it as Late Permian.

The second region where a salt sequence has been studied in some detail lies within the Bad Reichenhall Basin in the south of West Germany. Nine boreholes penetrated a considerable thickness of the salt-bearing successions. Beds of rock salt, sulfate, and carbonate are intensely folded, thus indicating the secondary occurrence of the sequence in the form of lenticular stock, traced to a depth of more than 1200 m. The upper part of the stock is represented by leached salt rocks forming a gypsum cap. Gypsification is often observed there and all the halogenic rocks grouped under the Haselgebirge are mostly secondary. Among these rocks, the clayey gypsum and anhydrite-dolomitic gypsum Haselgebirge are distinguished. These rocks are mostly composed of light gray gypsum with numerous scattered inclusions and

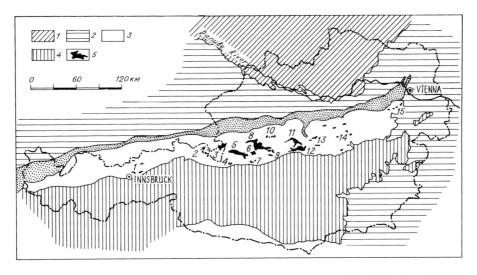

Fig. 141. Outcrops of Permian salt deposits in the northern East Alps. After Schauberger (1955).
1 Bohemian Massif, *2* Tertiary and Quaternary deposits, *3* Helvetides, *4* Central Alps, *5* North and South Limestone Alps. Outcrops of salt and sulfate deposits and their numbers: *1* Karwendel, *2* Unken-Lofer, *3* Hallein-Berchtesgaden and Bad Reichenhall, *4* Werfen, *5* Golling-Abtenau, *6* Hallstatt, *7* Rettenstein, *8* Ischl-Altaussee-Gründlsee, *9* Mitterndorf, *10* Offensee-Almsee, *11* Windischgarsten, *12* Spital-Pyrn-Bosruck, *13* Weissenbach-Gallen, *14* Rotwald-Maria-Zell, *15* Heiligenkreuz-Mödling

fragments of clay and dark to brown-gray anhydrite-dolomite. The thickness of the gypsum cap varies from 370–500 m.

The salt-bearing sequence proper, unaffected by the secondary processes, though strongly dislocated, is mostly composed of nonlayered salt rocks containing various inclusions of sulfate, carbonate, and clay. Proceeding from the correlation of the main components, the following rocks might be distinguished: more or less pure rock salt, brown to yellow-gray; massive or thin-bedded, dark-gray anhydrite; brownish-gray, dolomitic anhydrite; and variegated clay salt rocks, which are most abundant in the salt strata and are composed of clay, halite, and gray-green to brown, often saline sandstone. A detailed description of the rocks listed was given by Schauberger et al. (1976). A characteristic feature of the salt-bearing strata of the Reichenhall Region is the extensive development of anhydrite and anhydrite-dolomite rocks which are distinguished as the anhydrite-dolomite Haselgebirge and form units up to 15 m thick, which are distinct within the sequence. This led to the assumption that the Reichenhall Region was confined to a marginal zone of the salt basin. In the central part, as in Hallstatt, the red salt rocks or the variegated clay salt rocks are most widespread; minor bituminous anhydrite-dolomite and anhydrite salt Haselgebirge are also recorded, which is typical of the more deep water parts of the basin (Schauberger et al. 1976). The marginal facies is traced from the Bad Reichenhall Trough to the salt domes in Tyrol.

Structural features of a salt-bearing sequence in the Bad Reichenhall Trough are not clear. The exposed thickness varies from 737–1200 m, whereas the total normal thickness of the section studied can reach 1000 m. The sequence is overlain by the carbonate Reichenhaller Formation composed of limestone with alternate marly limestone, marly clay dolomite, and black calcareous mudstone with rare anhydrite inclusions. The carbonate rocks contain fossils of an age ranging from Early to Middle Triassic, approximately Late Scythian/Anisian. The salt sequence is non-fossiliferous. Spore-pollen analysis (Klaus 1955, Schauberger et al. 1976) suggested the Late Permian age. At the same time, the sulfur isotopic age of these sulfate rocks is Early Triassic (Scythian and Anisian). Thus, some investigators suggest Late Permian/ Early Triassic age for this salt sequence (Schauberger et al. 1976). However, most of these salt-bearing deposits probably formed at the end of the late Permian.

In addition to the above regions, salt and gypsum-anhydrite rocks in northern and north-eastern parts of the East Alps are found in numerous mines in the following areas: Karwendel (north-east of Innsbruck), Unken-Lofer, Hallein-Berchtesgaden, Werfen, Golling-Abtenau, Rettenstein, Ischl-Altaussee-Gründlsee, Mitterndorf, Offensee-Almsee, Windischgarsten, Spital-Pyrn-Bosruck, Weissenbach-Gallein, Rotwald-Maria-Zell, and Heiligenkreuz-Mödling, south-west of Vienna. Thus, the salt-bearing deposits are traced as a belt extending from west to east for more than 370 km. They are represented by the same rocks as those of the salt stocks of the Hallstatt and Bad Reichenhall regions and are also similar in age.

The easternmost salt-bearing area in the Alpine Basin lies within the Spišsko-Gemerskoe Rudohoři to the east of the West Carpathians. This region, together with the salt rock outcrops in the East Alps, is conventionally included into a single salt basin. There is a common composition of salt-bearing deposits and probably, similar age. Moreover, the probable extension of the salt basin from the East Alps to the eastern boundary of the West Carpathians is, to a certain extent, confirmed by Late Permian sulfate-bearing deposits exposed in the Low Tatra and to the west in the Tribeč Mountains (Zhukov et al. 1976).

In the eastern part of the West Carpathians, the salt-bearing deposits are penetrated by SM-1 and SM-2 boreholes near Spišska Nová Ves. Three lithological units can be distinguished in the section of the SM-1 borehole (from top to base): shale-sandstone unit — 435 m; intermediate sandstone-shale unit — 32 m; salt breccia unit — 208.9 m (Mahel and Vozár 1971). The base of the salt unit is not exposed.

There are isolated anhydrite interbeds, 10—20 cm thick confined mostly to its lower and middle parts in the upper shale-sandstone unit. A transitional unit is subdivided into two parts; the upper part, 2 m thick, is represented by shale with anhydrite bands (1—5 cm) and the lower part, 30 m thick, is composed of clayey anhydrite breccia, in which the content of anhydrite varies from 50%—70%.

The salt breccia unit is composed of rocks of the Haselgebirge type, in which the halite content varies from 40%—90%. The remaining part is represented by inclusions and clasts of shale, more rarely, dolomite and anhydrite. There are isolated interbeds of shale, 1—8 m thick, anhydrite, 1—2 m, clay-anhydrite breccia, and dolomite in the succession. Cores taken at depths of 457.5 m and 494.0 m contain sylvite; they have a potassium content of 6.8%—8.55%.

The halogenic unit is overlain by red and variegated arenaceous-argillaceous deposits containing anhydrite and interbeds of calcareous shale. The underlying strata were penetrated by the SM-2 borehole. They are represented mostly by the same red shale, often with an admixture and intercalation of volcanic rocks.

Evaporite deposits found in the Spišsko-Gemerskoe Rudohoři are confined to the interiors of the North Hemeride Syncline. Toward the margins, they grade into terrigenous red beds, in places with anhydrite bands. Most investigators suggest the Late Permian age for these salt and sulfate sequences (Zhukov 1965a, Mahel and Vozár 1971, Zhukov et al. 1976).

The Alpine Basin extends for more than 650 km from west to east, 40–50 km in width. The present salt-bearing area is not less than 26,000 km^2. Though one salt sequence seemed to be developed over the entire area, the data confirming this are not available at present. At the present time, two isolated regions can be outlined more or less precisely where separate salt sequences are recorded: one in the East Alps and the other in the east of the West Carpathians within the North Hemeride Syncline. The rock salt sequences amount to about 5.2×10^3 km^3.

Midcontinent Basin

Within the Midcontinent Basin, salt deposits are known to occur throughout the entire Permian section over a vast area of USA, i.e., from southern to northern boundaries embracing Texas, New Mexico, Oklahoma, Kansas, Colorado, Nebraska, Wyoming, Montana, and South and North Dakota (Fig. 142). The description of salt deposits covering the Midcontinent is based on data from the following publications: Darton (1921), Wilder (1923), Udden (1924), Hoots (1925), Baker (1929), Cartwright (1930), Spooner (1932), Muir (1934), Lang (1935, 1937, 1939, 1941, 1942, 1950), Kroenlein (1939), Dickey (1940), Fritz and Fitzgerald (1940), Page and Adams (1940), Bates (1942), P.B. King (1942, 1948, 1967), Roth (1942, 1945), Adams (1944, 1963, 1965), Taft (1946), R.H. King (1947), Pepper (1947), McGregor (1952), Maher and Collins (1952), Moore (1953, 1959), Jones (1959, 1965, 1968, 1972), McGinnis (1954), Maher (1954), Burk and Thomas (1956), Kulstard et al. (1956), McCauley (1956), Anderson and Hansen (1957), Galley (1958), Ham (1958, 1960, 1962, 1963), Ham and Curtis (1958), Vertrees et al. (1959), Weber and Kottowski (1959), Hall (1960), Sloss et al. (1960), Mear and Yarbrough (1961), Baars (1962), Bissel (1962, 1964a, b, 1967, 1970), Fay (1962), Hoyt (1962, 1963), Pierce and Rich (1962), Campbell (1963), Johnson (1963a, b), Jordan and Vosburg (1963), Kerr and Thomson (1963), MacLachlan and Bieber (1963), Momper (1963), Sheldon (1963), Ward (1963), S.B. Anderson (1964), Jordan (1964), Vosburg (1964), Haun and Kent (1965), Kottlowski (1965), Martin (1965), Gerrard (1966), Maughan (1966, 1967), Peirce and Gerrard (1966), Snider (1966), R.J. Anderson (1967, 1968), R.J. Anderson et al. (1972), Crosby (1967), Dixon (1967), Hallgarth (1967a, b), Ketner (1967), MacLachlan (1967), McKee (1967), McKee et al. (1967), Mudge (1967a, b), Oriel (1967a, b), Oriel et al. (1967), Sheldon et al. (1967a, b, c), Hills (1968), Mear (1968), Lefond (1969), Silver and Todd (1969), Jones and Madsen (1972), and others.

The thickness distribution pattern and structure of Permian deposits allow the division of the Midcontinent Basin into the southern part, known as the West Texas

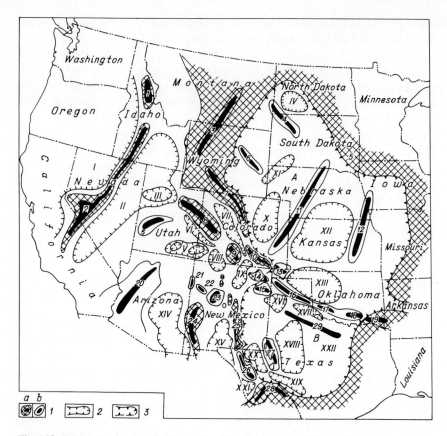

Fig. 142. Paleotectonic map of the western USA for the end of Wolfcampian. After McKee et al. (1967). *1* positive tectonic elements where Wolfcampian deposits are absent *(a)* or are present *(b)*, and their numbers: (*1* Western Idaho positive element, *2* Central Nevada Ridge (Antler orogenic belt), *3* Northern Rocky Mountains Uplift, *4* Cedar Creek, *5* Siouxia Landmass, *6* Ancestral Casper Arch, *7* Ancestral Front Range Uplift, *8* Uncompahgre Uplift, *9* San Luis Uplift, *10* Apishapa Uplift, *11* Las Animas Arch, *12* Nemaha Arch, *13* Cimarron Uplift, *14* Sierra Grande Arch, *15* Bravo Dome, *16* Amarillo Uplift, *17* Wichita Uplift, *18* Arbuckle Uplift, *19* Ouachita structural belt, *20* Western Arizona Platform, *21* Defiance positive element, *22* Zuni positive element, *23* Joyita positive element, *24* Southwest New Mexico Uplift, *25* Pedernal positive element, *26* Diablo Platform, *27* Central Basin Platform, *28* Marathon Uplift, *29* Matador Arch), *2* sedimentary depressions and their numbers: (*I* Cordilleran Eugeosyncline, *II* Cordilleran Miogeosyncline, *III* Oquirrh Basin, *IV* Williston Basin, *V* Southeastern Utah Basin, *VI* Uncompahgre Trough, *VII* Colorado Trough, *VIII* San Juan Basin, *IX* Rowe-Mora Basin, *X* Denver Basin, *XI* Julesburg Basin, *XII* Kansas Basin, *XIII* Anadarko Basin, *XIV* Sonoran Geosyncline, *XV* Orogrande Basin, *XVI* Palo Duro Basin, *XVII* Hollis and Hardeman Basins, *XVIII* Midland Basin, *XIX* Val Verde Trough, *XX* Delaware Basin, *XXI* Marfa Basin, *XXII* Eastern Shelf), *3* boundary of Midcontinent Basin. Letters refer to: *A* northern Midcontinent, *B* southern Midcontinent (West Texas Permian Basin)

Permian Basin and the northern part. They are separated by a zone formed by the Ouachita, Arbuckle, Wichita, Amarillo, and Cimarron Uplifts. Within the West Texas Permian Basin, the Palo Duro, Hollis, Hardeman, Midland, and Delaware Basins, Val Verde Trough, the Central Basin Platform, Matador Arch, and Bravo Dome are recognized. The northern Midcontinent Basin embraces the Anadarko, Kansas, Denver, Julesburg, and Williston Basins. Among major positive structures developed in the northern Midcontinent of particular interest are: the Las Animas and Nemaha Arches, Cedar Creek Anticline, and Ancestral Casper Uplift. The southern Midcontinent Basin differs from its northern part due to thicker Permian deposits. Their total thickness reaches 2700 m and 4500 m in the Midland and Delaware Basins, respectively, and exceeds 5100 m in the Val Verde Trough. In most of the northern Midcontinent Basin, the thickness of Permian deposits ranges from 150–500 m. Only to the south, in the Kansas and Anadarko Basins, it increases to 1000 m and even 1800–2000 m.

Permian deposits have been divided into four series: Wolfcampian, Leonardian, Guadalupian, and Ochoan (Fig. 143). In general, the Wolfcampian is correlated with the Asselian and a larger, lower part of the Sakmarian (Lower Permian); the Leonardian is the equivalent of the Upper Sakmarian, the whole Artinskian and Lower Kungurian (Lower Permian); the Guadalupian is correlated with the Upper Kungurian (Lower Permian) and with the Ufimian, Kazanian, and Lower Tatarian (Upper Permian); and the Ochoan corresponds to the Upper Tatarian (McKee et al. 1967, Oriel 1967a, b).

The definition of the age position of transitional salt sequences between the Leonardian and Guadalupian is the most difficult problem concerning the stratigraphy of Permian deposits. In some areas, these sequences cap the Leonardian sedimentary cycles, although their assignment implies the Guadalupian. On lithofacies and paleotectonic maps (McKee et al. 1967), the sequences of the northern Midcontinent Basin were assigned to the Leonardian. The same uncertainty holds for salt sequences of the San Andres Limestone, Blaine Formation, and their equivalents.

As a whole, salt rocks have been recognized in Permian deposits of the Midcontinent at five stratigraphic levels: (1) in the Upper Wolfcampian; (2) in the Leonardian beneath the San Andres Limestone, Blaine Formation, and their equivalents; (3) at the transition from the Leonardian to Guadalupian within the San Andres Limestone, Blaine Formation, and their equivalents; and (5) in the Ochoan. Areal distribution of rock salts at these stratigraphic levels is shown in Figs. 144 and 145.

The Wolfcampian evaporites occupy mainly the northern Midcontinent Basin (Fig. 146). Rock salt beds are common in the Denver, Julesburg, and Williston Basins. In Colorado, within the Denver Basin, thin rock salt bands occur in terrigenous red beds and variegated rocks interbedded with dolomite and anhydrite. Two rock salt units have been recognized in the middle and upper Ingleside Formation. Rock salt lenses are believed to have formed when the Denver Basin was isolated from the surrounding basins (Mudge 1967b). It was separated from the Kansas Basin by an uplift within the Las Animas Arch and from an evaporite basin situated in the Julesburg and Williston Basins by a carbonate reef zone. Wolfacampian deposits filling in the interior part of the Denver Basin are represented mainly by sandstone, siltstone, and mudstone. The amount of siltstone and mudstone increases toward the east and north-east where carbonate interbeds, chiefly dolomite, appear. The thickness of Wolfcampian salt

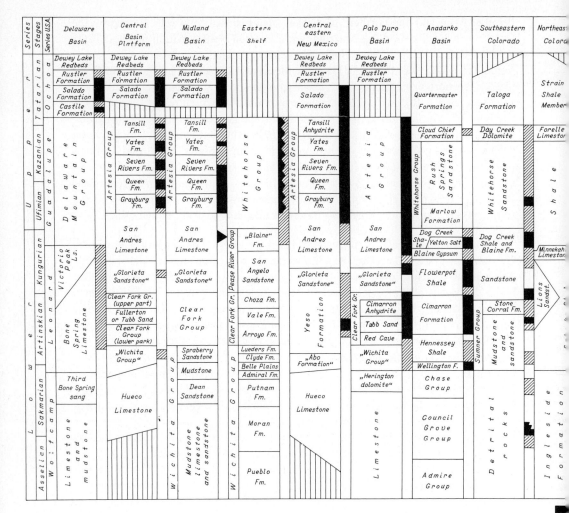

Fig. 143. Correlation chart of Permian deposits in the western USA and stratigraphic position of evaporites. Compiled from the data of McKee et al. (1967). *1* stratigraphic range of salt, *2* stratigraphic range of anhydrite and/or gypsum

rocks in the Denver Basin is still unknown. Their areal extent measures approximately 4500–5000 km².

In the Julesburg and Williston Basins in Nebraska, Wyoming, and South and North Dakota, Wolfcampian evaporites are fairly common (Maughan 1966, 1967). Rock salt is known to occur only in the Julesburg Basin, in north-western and south-western Nebraska. It is associated with the upper Minnelusa Formation or with the Broom Creek Group. The thickness of rock salt does not exceed 25–30 m. The total areal extent of rock salt in the Julesburg Basin exceeds 11,000 km² and its volume accounts for 200 km³. In addition to rock salt, the upper Minnelusa Formation also contains dolomite interbedded with anhydrite. These rocks underlie and overlie the salt sequence and replace it along the strike. Maughan (1966) believes that in the area under discussion, the Minnelusa Formation deposited in the southern part of a large evaporite basin with the center located near the Black Hills Uplift. The number of anhydrite and gypsum interbeds gradually decreases from the salt-bearing zone toward the margins. In the same direction, toward the mountain fringe, the Wolfcampian deposits become enriched in terrigenous rocks, namely, mudstone, siltstone, and then sandstone. To the north-east and east, terrigenous rocks are mainly gray, while to the north-west and west, red and variegated rocks prevail.

Over most of the central and southern Midcontinent, the Wolfcampian deposits consist of limestone, argillaceous limestone, and limy marl, which are divided in the

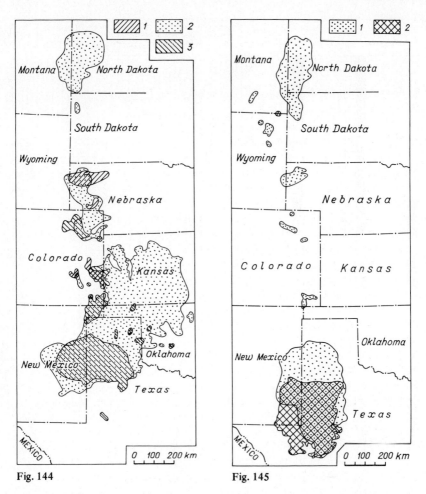

Fig. 144

Fig. 145

Fig. 144. Distribution of rock salt in Wolfcampian and Leonardian deposits in Midcontinent. After McKee et al. (1967). *1* area of development of rock salt in Wolfcampian deposits, *2* area of development of rock salt in Leonardian deposits occurring below San Andres Limestone, Blaine Formation, and their equivalents, *3* area of distribution of rock salt in San Andres Limestone, Blaine Formation, and their equivalents

Fig. 145. Distribution of rock salt in Guadalupian and Ochoan deposits in Midcontinent. After McKee et al. (1967). *1* area of development of rock salt in Guadalupian deposits occurring above San Andres Limestone, Blaine Formation, and their equivalents, *2* area of development of rock salt in Ochoan deposits

Kansas and Anadarko basins (in ascending order) into the Admire, Council Grove, and Chase Groups; in Texas, they are assigned to the Wichita Group. In the eastern Midcontinent, carbonate deposits are replaced first by gray, marl-clayey and then by clay-sandy sediments.

Structural features and distribution of Leonardian salt deposits have received the most study in the Midcontinent. Figure 147 shows that most of the Midcontinent

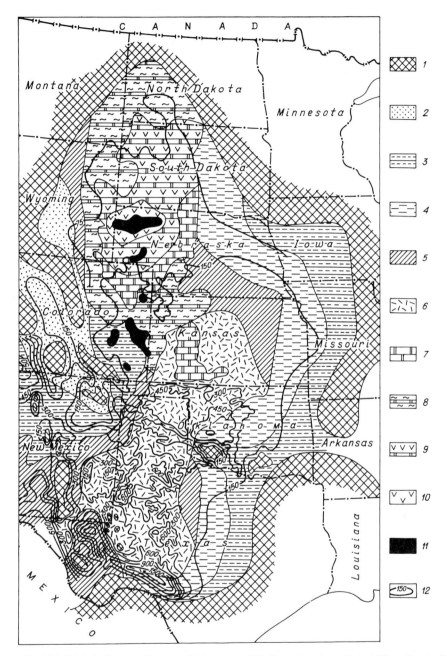

Fig. 146. Lithologic map showing thicknesses of Wolfcampian deposits in Midcontinent. Compiled from the data of McKee et al. (1967). *1* Wolfcampian absent. Limits of: *2* sandstone, *3* sandstone, siltstone, and mudstone, *4* siltstone and mudstone, *5* terrigenous rocks and limestone, *6* mainly limestone, *7* dolomite, *8* terrigenous rocks with dolomite interbeds, *9* dolomite and anhydrite (gypsum), *10* mainly anhydrite (gypsum), *11* rock salt; *12* isopach contours, in meters

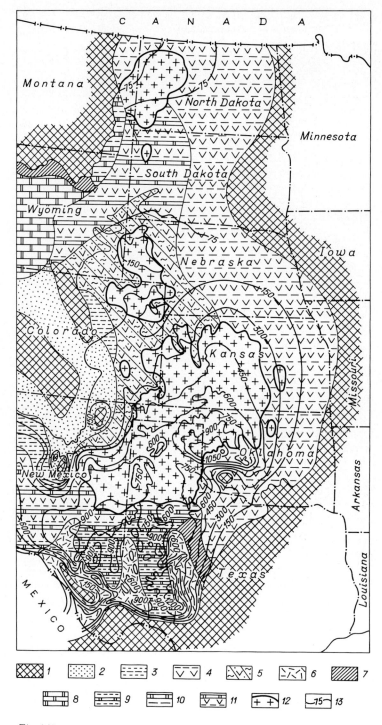

Fig. 147

was the site of evaporite sedimentation. Salt series deposited in the Anadarko, Kansas, Julesburg, Williston, and northern West Texas Permian Basins. Variable amounts of gypsum accumulated over the remaining area, except for some southern parts of the Delaware Basin, Central Basin Platform, Eastern Shelf, and small districts around the fringe.

Leonardian salt deposits are well-known in the Anadarko Basin in Oklahoma and north Texas. Four salt sequences, namely, the Wellington, Cimarron, Flowerpot, and Yelton have been recognized (Jordan and Vosburg 1963). Initially, the two upper sequences — the Flowerpot and Yelton — were assigned to the Guadalupian; the two lower ones, to the Leonardian; the boundary between them was drawn at the base of the Flowerpot Shale (Jordan and Vosburg 1963). At present, the boundary is drawn much higher in the section. The Flowerpot Shale and part of the Blaine Formation are tentatively assigned to the Leonardian (Dixon 1967, MacLachlan 1967, McKee et al. 1967). On lithofacies and paleotectonic maps of the USA (McKee et al. 1967), as well as on the lithological map compiled on the basis of the above data (Fig. 147), the Yelton Salt and Dog Creek Shale are also assigned to the Leonardian, although they are Guadalupian in age.

On the basis of their structural features, salt deposits of the Anadarko Basin are placed into three evaporite series. The lower and middle series include the Wellington and Cimarron Evaporites; the upper Beckham Series incorporates the Flowerpot and Yelton Salts (Figs. 148 and 149).

In most of the Anadarko Basin, the Wellington Evaporite overlies carbonate deposits and only near the Amarillo and Ouachita Uplifts rests on terrigenous rocks of the Pontotoc Group and its equivalents. The Wellington Evaporite is overlain by clastic rocks known as the Garber Sandstone to the east and placed into the Hennessey Shale in the central areas. The Wellington Evaporite is developed in southern Kansas, throughout northern Oklahoma, and northern Texas. The thickness varies over a wide range; in northern Oklahoma, it ranges from 250–340 m and farther east, it measures 200–206 m. On the southern flank of the Anadarko Basin, near the Ouachita Mountains, the thickness is 150 m. From north to south, the following thicknesses occur: 290 m in north Harper County, Oklahoma; 400 m farther south, in north Beckham County, Oklahoma; 225 m still farther south, in central Beckham County. In northern Texas, the thickness ranges from 300–360 m.

In the Anadarko Basin, the Wellington Evaporite can be divided into three parts. The lower salt-anhydrite member lies at the base, the middle part consists of a clayey member, and the section is capped by the upper anhydrite member.

The lower member of Wellington Evaporite is the most common. It is made up of rock salt and anhydrite locally interbedded with mudstone and dolomite. The thickness varies from 175–280 m. Its lower half is represented by anhydrite and mud-

---

◀ **Fig. 147.** Lithologic map showing thicknesses of Leonardian deposits in Midcontinent. Compiled from the data of McKee et al. (1967). *1* Leonardian absent. Limits of: *2* sandstone, *3* sandstone, siltstone, and mudstone, *4* clay-gypsiferous rock, *5* sandstone, siltstone, and mudstone interbedded with anhydrite and gypsum, *6* mainly limestone, *7* terrigenous rocks (mainly sandstone) and limestone, *8* mainly dolomite, *9* sandstone and dolomite, *10* dolomite, siltstone and mudstone, *11* terrigenous rocks, dolomite, and anhydrite, *12* rock salt; *13* isopach contours, in meters

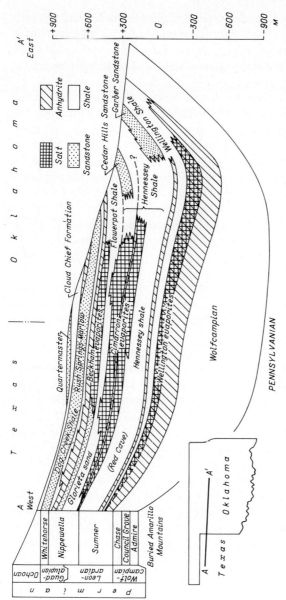

Fig. 148. Generalized geologic structure section in Permian strata in western Oklahoma and Panhandle of Texas. After Jordan and Vosburg (1963)

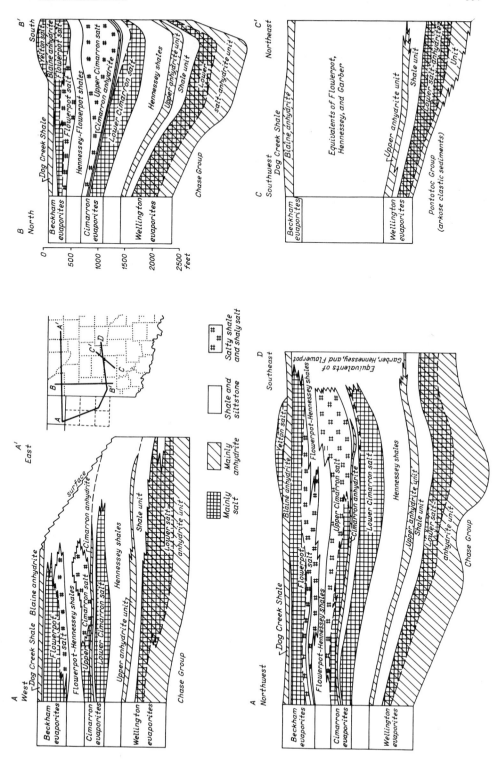

Fig. 149. Structure of Leonardian salt deposits in Anadarko Basin. After Jordan and Vosburg (1963)

stone, while the upper half consists predominantly of rock salt with the thickness attaining 75 m. In the lower part of the member, anhydrite totals about 30 m and accounts for 50–55% of the section. In the upper salt part, the thickness of rock salt beds rarely exceeds 10 m and on the average ranges from 1–2 m to 3–4 m. They are repeatedly interbedded with anhydrite, mudstone, and locally with dolomite. Rock salt amounts on the average to 40–45% of the section. Eastward, the amount of anhydrite decreases and that of mudstone increases.

The clayey member, which is the marker horizon in the Anadarko Basin, is hardly distinguishable in the adjacent area. The member consists of gray, greenish-gray, and chestnut-green mudstone and shale with small thin interbeds and inclusions of anhydrite and dolomite. The thickness ranges from 60–75 m to 100 m. Westward, the amount of anhydrite increases and thus, in some areas, joins the upper anhydrite member.

The upper anhydrite member can be traced in most of Oklahoma and Texas. It is represented by interbeds of anhydrite, dolomite, and in places by gray mudstone and shale. Anhydrite and dolomite account for no more than 20% and only locally in western districts, for 30% of the section. The thickness of the upper anhydrite member varies from 30–54 m and reaches 75 m in the subsided parts.

As a whole, the Wellington Evaporite in Oklahoma covers an area of about 40,000 km^2; the average thickness is 65 m and the rock salt volume totals 2.6×10^3 km^3. Halite beds lie at a depth of 240–1170 m (Jordan and Vosburg 1963).

The Hennessey Shale is easily recognizable in the area of the Anadarko Basin where the underlying and overlying deposits are salt-bearing. Conversely in the areas where salt deposits thin out, the Hennessey Shale joins the Flowerpot Shale. This is illustrated by the sections located in eastern Oklahoma (see Fig. 149). The Hennessey Shale is represented by red and reddish-green mudstone and shale with thin gray mudstone laminae at the base; inclusions and interbeds of anhydrite or dolomite occur locally. In some districts of northern Texas, clay is interbedded with sandstone and siltstone. The thickness of Hennessey Shale varies from 80–200 m.

The Cimarron Evaporite is fairly common in Oklahoma and in the adjacent areas of Texas (Fig. 150). Its thickness varies over a range of 0–350 m. The maximal values were recorded in the areas located in Beckham and Roger Mills Counties, Oklahoma. The top of rock salt lies at a depth of 60–730 m. The Cimarron Evaporite is divided into the Lower Cimarron Salt, Cimarron Anhydrite, and Upper Cimarron Salt. The Lower Cimarron Salt consists of light crystalline rock salt beds 2–5 m, locally up to 10 m thick; they are interbedded with tawny and greenish-gray mudstone and shale. On the average, rock salt accounts for 75% of the section, the rest being mudstone. A total thickness of the member ranges from 62–135 m. The depth to the salt top ranges from 450–500 m. Above the Lower Cimarron Salt, a thin unit of fulvous, tawny, and gray mudstone with ribs of siltstone and sandstone occurs, as well as dolomite and anhydrite at the top; its thickness does not exceed 15 m.

The Cimarron Anhydrite, 12–25 m thick, is represented by alternate anhydrite, dolomite, and mudstone with some rock salt bands.

The Upper Cimarron Salt is developed in north-western Oklahoma and northern Texas (see Fig. 150). It is made up of clayey rock salt interbedded with fulvous mudstone; the thickness does not exceed 60 m. Some beds are 6–7 m thick.

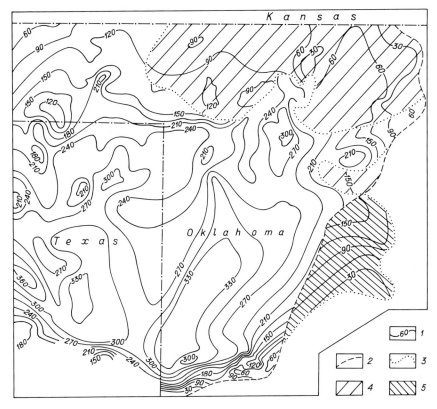

Fig. 150. Thickness of Cimarron Evaporites in Oklahoma and Texas. After Jordan and Vosburg (1963). *1* isopach contours, in meters, *2* limit of Lower Cimarron Salt, *3* limit of Upper Cimarron Salt, *4* area where only Lower Cimarron Salt occurs, *5* area where only Upper Cimarron Salt occurs

According to Jordan and Vosburg (1963), the areal distribution of the Cimarron Salt in Oklahoma is about 34,000 km^2. On the average, the total thickness measures 150 m; the volume of salt can exceed 5.0×10^3 km^3. Rock salt beds lie at a depth of 100–650 m.

The Cimarron Evaporite is overlain by the Flowerpot Shale or Flowerpot-Hennessey Shale represented by thin-bedded, schistose, buff, red, greenish-gray, and gray mudstone with tawny sandstone interbeds, as well as anhydrite and dolomite lenses. The thickness ranges from 60–100 m.

The maximal thickness of the Beckham Evaporite is above 200 m (Jordan and Vosburg 1963). It is subdivided (in ascending order) into the Flowerpot Salt, Blaine Anhydrite, and Yelton Salt.

The Flowerpot Salt has a more limited distribution than the Cimarron Salt (Fig. 151). Its thickness ranges mainly from 60–100 m and only locally reaches 120–150 m. Rock salt is usually interbedded with red or buff, thin-laminated, schis-

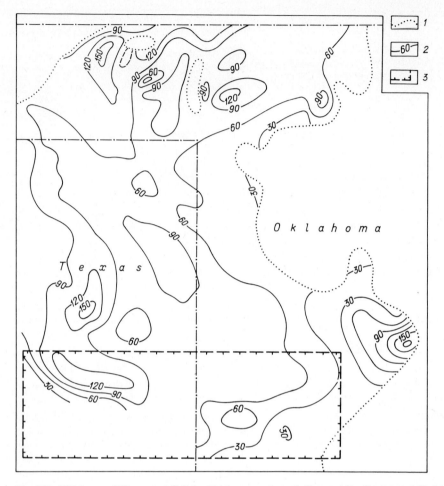

Fig. 151. Thickness of Flowerpot Salt in Oklahoma and north Texas. After Jordan and Vosburg (1963). *1* limit of rock salt, *2* isopach contours, in meters, *3* limit of area where Yelton Salt occurs

tose mudstone. Pink or red prevails. The depth of occurrence varies from 10–500 m. The total thickness of halite averages 60 m and the areal extent in Oklahoma measures about 19,000 km^2. The volume of rock salt in this case can exceed 1.1×10^3 km^3. The lower Flowerpot Salt is represented mainly by mudstone, more or less saline, with interbeds and lenses of highly argillaceous rock salt. The upper part contains a mudstone bed with delicate, individual laminae of siltstone and sandstone. Anhydrite and gypsum in the form of thin interbeds and lenticular inclusions occur throughout the entire section.

The Blaine Anhydrite is represented by alternate anhydrite (or gypsum), dolomite, and mudstone. Beds of pink and red rock salt, 2–3 m thick, can be encountered in some parts of the Anadarko Basin. The thickness of the Blaine Anhydrite ranges from 25–75 m.

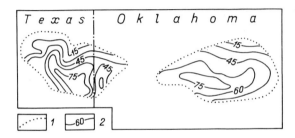

Fig. 152. Thickness of Yelton Salt in Oklahoma and Texas. After Jordan and Vosburg (1963). *1* limit of salt deposits, *2* isopach contours, in meters. Location of the district shown in Fig. 151

In Oklahoma and northern Texas, the Leonardian salt section is crowned with the Yelton Salt, which has a local distribution and was reported only from two isolated areas: Beckham County, Oklahoma, and Wheeler County, Texas (Fig. 152). Rock salt accounts for 75–80% of the section, the remainder being brownish and greenish-gray mudstone and shale (Jordan and Vosburg 1963).

The thickness of the Yelton Salt varies from 0 to 75–80 m. The top of rock salt in Oklahoma lies at depths of 160 to 350–400 m. The average thickness of halite totals 50 m; the areal extent of the rock salt in Oklahoma and Texas exceeds 2000 km^2, and the volume of the Yelton Salt is 1.0×10^2 km^3. In the part of the Anadarko Basin, in Oklahoma, the total volume of rock salt in all the Leonardian evaporite series is about 0.9×10^4 km^3.

In the northern West Texas Permian Basin, in New Mexico and Texas, evaporites are very common among the Leonardian deposits. They are widely distributed in the Palo Duro Basin, where the upper two-thirds of the Leonardian section forms, in essence, a single salt sequence (Dixon 1967, McKee et al. 1967, Oriel et al. 1967). However, in the northern West Texas Permian Basin, salt rocks have not been found in the Lower Leonardian at the level where the Wellington Evaporite occurs in the Anadarko Basin. To the north-west, the deposits are represented by dolomite with small interbeds of anhydrite and green marl assigned to the "Wichita Group", which is equivalent to the upper part of an eponymous group located in the more southerly areas of the Midcontinent. To the west and south-west toward the Pedernal positive element, carbonate deposits pinch out passing into terrigenous, mainly clay-marley red beds. It is only at a single locality in New Mexico, in the west Orogrande Basin, i.e., outside the West Texas Permian Basin, where thin rock salt beds alternated with anhydrite have been found in the Lower Leonardian. They apparently lie higher in the section than the Wellington Evaporite of the Anadarko Basin.

The Upper Leonardian deposits in most of the Palo Duro Basin are placed into the Clear Fork Group, which can be divided, in ascending order, into: the Red Cave, Tubb Sand, Cimarron Anhydrite, and unnamed upper unit (Dixon 1967). The Red Cave consists mainly of red mudstone, and to the west, of mudstone, siltstone, and sandstone. In north-western Texas, an anhydrite band appears in the section with rock salt in its upper part. The Leonardian salt section in the Palo Duro Basin starts with the Tubb Sand. It is represented by alternate anhydrite, rock salt, red marl, siltstone and sandstone. The thickness ranges from 75–200 m. The sequence consists mainly of anhydrite, which accounts for 70–80% of the section. Rock salt is confined to the lower part. The uppermost Tubb Sand consists of red mudstone and sandstone up to 20—25 m thick. In regard to the stratigraphic position of the Tubb Sand, it may well be an equivalent of the Lower Cimarron Salt of the Anadarko Basin.

The Cimarron Anhydrite of the Palo Duro Basin consists chiefly of anhydrite. Rock salt interbeds are small in number, accounting for 10—15% of the section. The Cimarron Anhydrite is 25—60 m thick.

The upper Clear Fork Group is represented in the northern West Texas Permian Basin by red mudstone, siltstone, and sandstone interbedded with anhydrite and rock salt. The thickness is 125—360 m. The lower part of the section is salt-bearing, while the upper part is composed of mudstone interbedded with anhydrite, 50—60 m thick. The upper Clear Fork Group may be considered an equivalent of the Upper Cimarron Salt and lower Backham Evaporite, including the Flowerpot Salt and Blaine Anhydrite taken together.

The data available indicate that in the northern West Texas Permian Basin in Texas and New Mexico, the Leonardian salt deposits occupy an area of more than 100,000 km^2. The thickness of rock salt there may total 100 m, thus the volume of rock salt is estimated at 1.0×10^4 km^3.

North of Oklahoma, the Leonardian salt deposits occur in much of Kansas, namely, in the Kansas Basin, Las Animas Arch, Nemaha Anticline, and in some districts of south-east Colorado. In the Kansas Basin, rock salt lies at four stratigraphic levels: (1) in the lowermost Leonardian within the Wellington Formation; (2) in the upper Ninnescah Shale underlying the Stone Corral Formation; (3) above the Stone Corral Formation within the Salt Plain Formation; and (4) in the Upper Leonardian within the Flowerpot Shale and Blaine Formation. Salt deposits are very common in the Wellington Formation, Ninnescah, and Flowerpot Shales (Fig. 153). Rock salt of the Salt Plain Formation has been found in a small area near the Kansas-Oklahoma boundary.

The Wellington Formation in the Kansas Basin is divided into the lower, middle, and upper members. The lower member is composed of red anhydritic mudstone

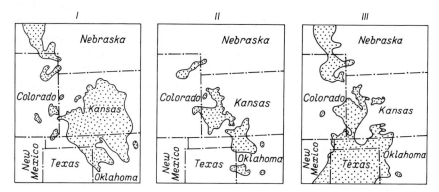

Fig. 153. Distribution of Leonardian salt deposits in Kansas and adjacent areas. After Mudge (1967a). *I* distribution of salt of Wellington age, *II* distribution of salt of Ninnescah age, *III* distribution of salt of Nippewalla age

interbedded with anhydrite. The middle member consists of red mudstone, anhydrite, and rock salt; it is often known as the Hutchinson Salt Member. This member is an equivalent of the upper salt part of the Wellington Evaporite in the Anadarko Basin, extending southward into Oklahoma and Texas (Fig. 153, I). The upper Wellington Formation in Kansas consists mainly of gray anhydritic mudstone interbedded with red mudstone. The Hutchinson Salt Member attains a thickness of 150–180 m. Salt occurs at a depth of 100–600 m. The approximate areal extent of rock salt in Kansas and in south-east Colorado is 80,000 km^2. Assuming an average thickness of 100 m, then the volume will be 8.0×10^3 km^3.

The Ninnescah Shale in the Kansas Basin and in the adjacent area is composed of red anhydritic mudstone, in places, particularly in north-west Kansas, it contains interbeds of sandstone and arenaceous mudstone. In south and west Kansas, the upper part locally encompasses rock salt interbedded with anhydrite and red mudstone. The areal distribution of rock salt within the upper Ninnescah Shale is shown in Fig. 153, II. The stratigraphic position of the salt-bearing unit may be considered an equivalent of some part of the Lower Cimarron Salt in Oklahoma. The Ninnescah Shale in Kansas has a thickness of 80–85 m (Mudge 1967a). The thickness of salt-bearing deposits, based on the data of Carlstadt (Lefond 1969), is 20–30 m and reaches a maximal value of 75 m in the most subsided parts of the Kansas Basin. The average thickness of salt in the Ninnescah Shale does not exceed 30 m. The areal distribution of rock salt in Kansas and in the adjacent areas of Colorado is about 13,000 km^2. The volume of rock salt is estimated at 4.0×10^2 km^3.

Above the Ninnescah Shale in Kansas and in south-east Colorado, dolomite and anhydrite of the Stone Corral Formation occur. On the surface, the formation is composed mainly of dolomite interbedded with mudstone; however, it is also rich in anhydrite, which dominates the deeper section. In southern Kansas, the Stone Corral Formation consists chiefly of anhydrite interbedded with dolomitic mudstone. To the south-west, two anhydrite units separated by red mudstone are easily recognizable. The thickness of the Stone Corral Formation ranges from 2–30 m. The majority of investigators (Jordan and Vosburg 1963, Maughan 1966, 1967, Mudge 1967a) correlate the Stone Corral Formation with the Cimarron Anhydrite of Oklahoma and Texas.

All the Leonardian sedimentary strata lying above the Corral Stone Formation in Kansas and in east Colorado are placed into the Nippewalla Group. Its lower part contains mainly red siltstone and mudstone and locally sandstone assigned to the Harper Siltstone and Salt Plain Formation, which is salt-bearing. Rock salt here shows a local distribution. The upper Nippewalla Group encompassing the Cedar Hills Sandstone, Flowerpot Shale, and Blaine Formation is composed of red mudstone and minor sandstone and siltstone; their amount increasing northward. The uppermost Flowerpot Shale just below the overlying Blaine Formation in some western and south-western districts of Kansas and south-east Colorado contains a salt unit composed of rock salt interbedded with anhydrite and marl. It is recognized under the name of the Nippewalla Salt (Mudge 1967a). This unit is an equivalent of the Flowerpot Salt of the Beckham Series in the Anadarko Basin. The Flowerpot Salt extends from Texas and Oklahoma far northward, where it is common in west Kansas, east Colorado, and Nebraska (Fig. 153, III). The thickness of the Flowerpot Salt in Kansas and Colorado ranges from 0–150 m. The depth to the top varies from 140–685 m. The average thickness does not exceed 80 m. The areal extent of the Flowerpot Salt is approximately estimated at 12000 km^2. The volume may equal 9.6×10^2 km^3.

The Blaine Formation in Kansas is represetned by gypsum and anhydrite with rare interbeds of dolomite and red marl.

The Upper Leonardian, confined to the Nippewalla Group west of Kansas toward eastern Colorado, is rich in red mudstone, sandstone, and siltstone with minor anhydrite and rock salt. Rock salt bands occur there at the level of the Flowerpot Shale, upper Ninnescah Shale, and locally of Wellington Formation. Farther west, evaporites pinch out passing into red beds of the Lyons Sandstone.

In north-eastern Colorado, in the Denver Basin, salt deposits occur at the same levels of the Leonardian as in the Kansas Basin. They are common in the following sections: the lower Satanka Shale, where interbeds of rock salt and anhydrite are considered equivalents of the Wellington sequence of Kansas; the middle Satanka Shale, where salt rocks lying at the level of the Ninnescah salt member occur below anhydrite of the Stone Corral Formation; and the upper Satanka Shale, just below the Minnekahta Limestone correlative to the Blaine Formation. The areal extent of the lower salt member (equivalent to the Wellington) is over 3000 km^2, that of the middle salt member (equivalent to the Ninnescah or Lower Cimarron Salt) exceeds 2000 km^2, and that of the upper salt member (equivalent to the Flowerpot Salt) is more than 3500 km^2. In the Denver Basin, anhydrite is developed within the Stone Corral Formation and Minnekahta Limestone, which contains salt bands as in the Blaine Formation. Assuming that the average total thickness of the Leonardian salt in the Denver Basin is not less than that of the Kansas Basin, then the volume of salts will amount to 6.4×10^2 km^3.

The total volume of the Leonardian salt in Kansas and Colorado may be approximately estimated at 1.0×10^4 km^3.

The Leonardian salt series north of Kansas and Colorado have been found in two other regions of the Midcontinent, namely, in the Julesburg and Williston Basins.

In the Julesburg Basin, in east Wyoming and west Nebraska, the Leonardian deposits are divided into three formations, in ascending order: the Owl Canyon Formation, Opeche Shale, and Minnekahta Limestone, which shows a transitional age between the Leonardian and Guadalupian.

The Owl Canyon Formation in the Julesburg Basin is salt-bearing. In the most subsided parts, it is subdivided into the lower, anhydrite-salt and the upper, red clay parts. The lower part consists of three members: the lower, anhydritic member, where anhydrite is interbedded with red mudstone 20–25 m thick; the middle, salt-bearing member represented by alternate rock salt, anhydrite, and mudstone, 15–20 m thick; and the upper, anhydritic member, 10–15 m. The upper clayey part has a thickness of 35–50 m (Maughan 1966).

The salt member pinches out along the strike and in the area where it is absent, the entire lower part is represented by anhydrite interbedded with red and variegated mudstone. In these districts, the Owl Canyon Formation is divided into three parts: the lower and upper, red terrigenous and middle, clayey-anhydritic. Salt deposits of the Owl Canyon Formation are in all probability equivalent to the Wellington Formation and Ninnescah Shale of Kansas and Colorado.

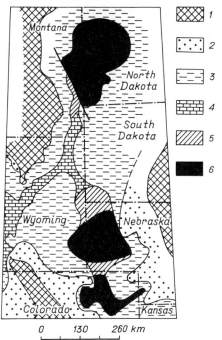

Fig. 154. Lithologic map of Opeche Shale deposits and equivalent rocks in the northern Midcontinent. After Maughan (1966). *1* Opeche Shale absent, *2* mainly sandstone, *3* mainly red mudstone and siltstone, *4* carbonate rocks (dolomite), *5* anhydrite and gypsum, *6* rock salt

Northward, the thickness of the Owl Canyon Formation decreases; it becomes salt-free and in most of the Williston Basin is eroded.

An overlying salt sequence associated with the Opeche Shale is common in the Julesburg and Williston Basins (Fig. 154). In the most subsided parts of the Julesburg Basin, thin anhydrite beds can be recognized at the base. All of the overlying part is represented by rock salt with occasional laminae of anhydrite and mudstone. The thickness of salt deposits reaches 25–30 m. Along the strike, the rock salt grades first into anhydrite and then the latter passes into red mudstone, which contains siltstone and sandstone near the source areas.

In the Williston Basin, in Montana and North Dakota, the Opeche Shale is composed of anhydrite (or gypsum), rock salt, and mudstone. In a salt-bearing area, rock salt is developed in the lower part of the formation, while its upper part generally consists of anhydrite (gypsum) alternated with mudstone. A maximal thickness of the Opeche Salt in the Williston Basin reaches 55 m, but ranges mainly from 15–30 m (Fig. 155). The depth to the top of salt deposits varies from 1730–2285 m.

In general, the upper Leonardian salt deposits in the Julesburg and Williston Basins pass into red, mainly clayey rocks at the margins. Only north-west of the Julesburg Basin and south-west of the Williston Basin, the salt series grade into anhydritic and then dolomitic strata. Carbonate deposits of the same age are developed mainly in central Wyoming. The similarity in lithological variations implies that marine water entered the Leonardian salt zones in the northern Midcontinent from the west via an embayment located in central Wyoming (Maughan 1966, 1967).

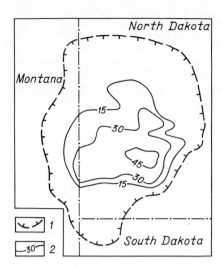

Fig. 155. Map of aggregate salt thicknesses in Leonardian deposits of Williston Basin. After Anderson and Hansen (1957). *1* distribution of rock salt, *2* isopach contours of rock salt, in meters

The section of the Leonardian evaporite series in the Julesburg and Williston Basins is capped by the Minnekahta Limestone, in which the upper part is assigned to the Guadalupian. This formation in most of the Midcontinent is represented by dolomite and to the west by limestone. In some districts, it contains interbeds of gypsum and locally of rock salt. The latter has been found in the Julesburg Basin, while gypsum is common in the Williston Basin. In general, the thickness of the Minnekahta Limestone does not exceed 10—15 m.

The approximate volume of rock salt in the Opech Shale and equivalent rocks both in Julesburg and Williston Basins can be estimated from the following data. Assuming an areal extent of rock salt of above 50,000 km² and an average total thickness of halite of about 20 m, then the volume will be not less than 1.0×10^3 km³.

The above data indicate that the total volume of the Leonardian rock salt within the Midcontinent exceeds 3.0×10^4 km³. The volume does not include salt of the Blaine Formation, Minnekahta Limestone, Owl Canyon, and Salt Plain Formations. Taking these areas into account, then the volume of the Leonardian rock salt will amount to 3.5×10^4 km³.

The Guadalupian salt series in the Midcontinent is known mainly from the widely spaced districts, namely, from the southernmost West Texas Permian Basin in Texas and New Mexico and from the northern Williston Basin, in North Dakota, Montana, and South Dakota. Furthermore, rock salt strata of Guadalupian age occur in small areas of the Denver Basin in Colorado, in the Julesburg Basin, north-west Nebraska,

Fig. 156. Lithologic map showing thicknesses of Guadalupian deposits of Midcontinent. Compiled ▶ from the data of McKee et al. (1967). *1* extent of Guadalupian deposits. Distribution of: *2* sandstone, *3* sandstone, siltstone, and mudstone, *4* mudstone, *5* limestone and dolomite, *6* limestone and dolomite interbedded with mudstone, *7* terrigenous and carbonate rocks, *8* dolomite interbedded with anhydrite (gypsum), *9* terrigenous-carbonate rocks and anhydrite, *10* anhydrite (gypsum), *11* sandstone, siltstone, and mudstone interbedded with anhydrite (gypsum), *12* mudstone and anhydrite, *13* rock salt, *14* isopach contours, in meters

Midcontinent Basin

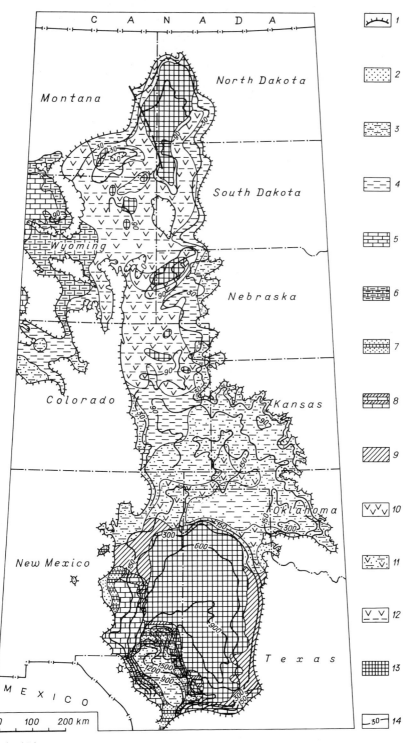

Fig. 156

in north-east Wyoming, and south-east Montana. Lithology and thickness of the Guadalupian deposits and the areal distribution of rock salt are given in Fig. 156. The accumulation of Guadalupian salt deposits in the southern and northern Midcontinent occurred in two separate basins. The northern basin extended longitudinally from central East Colorado up to the Canadian boundary. Marine water entered the basin from the west through an embayment situated in central Wyoming. Carbonate Guadalupian deposits are common and pass eastward first, into sulfate-terrigenous and then, into salt formations. The influx of sea water into the southern basin was from the south judging from lithological variations of carbonate to sulfate-carbonate, sulfate, and salt deposits. The southern and northern basins were separated by a wide, shallow zone, where terrigenous rocks deposited.

In the West Texas Permian Basin, the Guadalupian salt deposits are fairly well-known. The most comprehensive studies are: King (1942), Mear and Yarbrough (1961), Tait et al. (1962), Dixon (1967), Oriel et al. (1967), Hills (1968), and Mear (1968). Over most of the area within the Midland Basin, Central Basin Platform, and Palo Duro Basin, the Guadalupian contains the San Andres Limestone at the base; in the upper part, the Artesia Group, consisting of the Grayburg, Queen, Seven Rivers, Yates, and Tansill Formations can be recognized.

Salt deposits associated with the San Andres Limestone are developed only in the northern West Texas Permian Basin in central-east New Mexico and north-west Texas, in the Palo Duro Basin. The San Andres Limestone is represented by a sulfate-carbonate unit at the base and is the alternation of anhydrite and dolomite, 40—70 m thick. Above, a salt unit follows composed in the Palo Duro Basin of rock salt interbedded with anhydrite and dolomite. The thickness of the salt unit ranges from 100—200 m. In the most subsided parts, the salt unit consists of rock salt and anhydrite. Southward, dolomite and then limestone begin to dominate the section, starting from the lower member to the upper one, e.g., on the Central Basin Platform, the San Andres Limestone contains mainly dolomite and limestone. In general, the thickness exceeds 400 m.

Evaporites of the Artesia Group in the West Texas Permian Basin are widespread. They are common in the Palo Duro and Midland Basins, on the Central Basin Platform, and Eastern Shelf. In the south-eastern Midland Basin, the Artesia Group consists of formations showing variations in lithology. Thus, the Grayburg Formation is represented by brownish dolomite, anhydritic dolomite, and anhydrite. Farther east and south, red sandstone and red gypsinate mudstone appear in the section. In outcrops, the Queen Formation is composed of evenly bedded red sandstone 45 m thick. To the west, its thickness increases to 180 m and in addition to red sandstone, red mudstone, anhydrite, and rock salt appear. The thickness and amount of rock salt increase westward and north-westward. In the central Midland Basin, anhydrite in the Queen Formation accounts for about 44% and halite for 29%; the remainder is sandstone and mudstone. The maximal content of rock salt (68%) was reporte from some districts of Glasscock County, Texas.

The Seven Rivers Formation in the south-east Midland Basin is represented by yellow and red sandstone alternated with anhydrite and rock salt. In "salt-free" sections, the formation has a thickness of 70 m; and in the salt section, above 200 m. The amount of anhydrite and halite increases westward. In Glasscock County, Texas, sandstone and mudstone average 63%, while anhydrite and rock salt constitute 41% and 36%, respectively. The maximal total salt thickness reaches here 75 m, which accounts for about 58% of the section in subsided parts. The Yates Formation in the south-east Midland Basin is composed mainly of red and gray, evenly bedded sandstone with thin limestone interbeds; anhydrite and locally rock salt occur in subsided parts. The maximal thickness of the formation is 45–46 m. The Tansill Formation consists of anhydrite and rock salt; it varies in thickness from 5–15 m.

The above data show that in the south-east Midland Basin salt deposits of the Artesia Group occur at four levels, i.e., in the Queen, Seven Rivers, Yates, and Tansill Formations. The amount of salt rock in the section increases westward and northwestward. Evaporite strata have a rhythmic structure and are represented by multiple alternations of rock salt, anhydrite, red terrigenous, and minor carbonate rocks.

The deposits of the Artesia Group have a similar composition in the remainder of the Midland Basin and in the more northerly districts of the West Texas Permian Basin. The salt saturation of the section increases here and along with rock salt and anhydrite, mudstone interbeds appear. East of the Palo Duro Basin, salt deposits are rich in mudstone and sandstone. They are assigned to the Whitehorse Group.

In contrast, south and south-west of the salt accumulation area, lithological variations are quite different. On most of the Central Basin Platform, the Artesia Group is represented by sulfate-carbonate strata and upon the wedging out of rock salt, anhydrite dominates the section, which then passes into anhydrite interbedded with dolomite. There, the Artesia Group has a three-partite structure. Dolomite interbedded with anhydrite (Grayburg Formation) is recognized at the base; above, a part of the section is represented by an even intercalation of dolomite and anhydrite equivalent to the Queen, Seven Rivers, and Yates Formations taken together; and even higher in the section, the Tansill Anhydrite is easily recognizable. West and south-west of the Central Basin Platform at the Delaware Basin margins, the equivalents of the Artesia Group become entirely carbonate. A carbonate reef belt extends here. Carbonate deposits of the area are placed into the Delaware Mountain Group, the upper part of which contains the Altuda Formation and Capitan and Gilliam Limestones (King 1942, 1967). The Capitan Formation builds up an extensive barrier reef which runs along the Delaware Basin margins and is about 6.5 km wide; the thickness reaches 600 m (King 1967). Reef structures almost entirely bound the area where terrigenous rocks are developed in the Delaware Basin.

On the whole, salt deposits of the Artesia and Whitehorse Group in the West Texas Permian Basin cover an area of about 130,000 km^2. The salt saturation of the section generally constitutes more than 30%. An approximate estimate for the average total thickness is 100 m. Proceeding from the above estimates, the volume of rock salt of the Upper Guadalupian will be 1.3×10^4 km^3. The volume of salt in the underlying San Andres Limestone (areal extent is 60,000 km^2; average total thickness is at least 30 m) may account for 1.8×10^3 km^3. The total volume of the Guadalupian rock salt may well exceed 1.5×10^4 km^3.

In the Williston Basin, the lower Spearfish Formation is placed into the Guadalupian, but in South Dakota, it is divided, in ascending order, into the Glendo Shale, Forelle Limestone, and Freezeout Shale, in which only the lower part is assigned to the Guadalupian (Maughan 1966, 1967). In north-west South Dakota, east Montana, and west North Dakota, the upper Spearfish Formation contains the Pine Salt. Thin salt deposits have been found in north-east Wyoming as well. The Glendo Shale is composed of red mudstone and siltstone with numerous gypsum interbeds. The lower shale member of the Spearfish Formation in the Williston Basin has a similar composition. The thickness does not exceed 25–30 m. In southern South Dakota and in Wyoming, in the salt-free sections above, the easily recognizable Forelle Limestone 15 m thick appears, which then wedges out north and north-westward.

The Pine Salt is marked by a different stratigraphic range in the Williston Basin. In some districts, it comprises not only the lower Spearfish Formation, but the middle, Triassic part as well. The Pine Salt is represented by alternate rock salt, anhydrite, red mudstone, and siltstone. In the most subsided parts, it exceeds 100 m in thickness. The depth to the salt top ranges from 1200–2285 m (Sandberg 1962). The total thickness of rock salt varies mainly between 30 and 60 m, only locally increasing to 90 m (Fig. 157). The areal distribution of the Pine Salt in South and North Dakotas and in Montana is about 60,000 km^2; the appropriate volume equals 1.8×10^3 km^3.

Salt deposits of small areal extent and thickness have been reported from the Guadalupian sedimentary strata in north-west Nebraska, in the Julesburg Basin, and in some eastern areas of Colorado, within the Denver Basin. In Nebraska (Maughan 1966, Mudge 1967a, b), the lower Glendo Shale is salt-bearing; rock salt interbedded with anhydrite occur among red mudstone and siltstone.

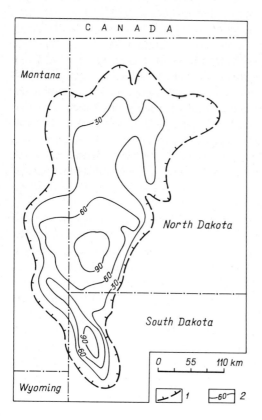

Fig. 157. Map of aggregate thickness of Pine Salt. After Ziegler (Lefond 1969). *1* limit of Pine Salt, *2* isopach contours of rock salt, in meters

In Kansas and Colorado, the Whitehorse Sandstone and Day Creek Dolomite are assigned to the Guadalupian (Mudge 1967a, b). The former is represented by well-sorted red sandstone, siltstone, and mudstone, which in Colorado, locally contain rock salt and anhydrite interbeds. In Kansas and Colorado, the Day Creek Dolomite forms a marker horizon. It is composed of dolomite and anhydrite interbedded with red mudstone. The section is capped by the Taloga Formation represented by red, arenaceous mudstone and clayey sandstone alternated with delicate laminae of anhydrite and dolomite. The Taloga Formation was placed into the Ochoan.

The total volume of the Guadalupian salt in the Midcontinent Basin may well exceed 1.5×10^4 km^3. This estimate seems to be rather low because the Guadalupian salt developed in Wyoming, Nebraska, and Colorado has not been taken into account.

The Ochoan salt series in the south Midcontinent have been studied intensively. Their distribution is restricted to the West Texas Permian Basin, where they fill the Delaware and Midland Basins, and are also common in the Central Basin Platform and adjoining areas of the Eastern Shelf in Texas and the Northwestern Shelf in New Mexico.

The Ochoan sedimentary strata in the southern Midcontinent covers an area of almost 150,000 km^2, extending for 420 km from south to north and for 350 km from west to east (Jones 1972). The areal extent of salt rocks there was assumed to be 36,000 km^2 (Pierce and Rich 1962). By now, it is known that it almost equals 95,000 km^2, i.e., about two-thirds of the entire area under the Ochoan deposits (Jones 1972). The total thickness of the Ochoan deposits ranges from 60–1500 m, attaining its maximum in the Delaware Basin (Fig. 158). On the remainder of the Midland Basin and the adjacent shelf, it does not exceed 300–400 m.

Lithologic composition of the Ochoan deposits is shown in Fig. 159. Salt strata occur on most of the central part of the area. South-westwards, they grade into anhydrite and gypsum, while in the south-western end, anhydrite and dolomite prevail. Along the eastern boundary, the deposits are enriched in terrigenous material; and in the coastal district, the entire section is composed of red sandstone and siltstone. To the north, salt strata pass into clayey deposits interbedded with gypsum and anhydrite. Limestone is found only on a small area to the north-west. The Ochoan salt series are characterized by the presence of potash salts, confined mainly to the Salado Formation and developed to the south-east.

A composite section of the Ochoan can be divided, in ascending order, into four formations: Castile, Salado, Rustler Formations, and Dewey Lake Red Beds. Originally, all the Ochoan salt deposits were placed into a single Castile Formation. Cartwright (1930) was the first to show that the lower part of this formation differed appreciably from the upper one in composition and structure. He divided it into two parts, namely, the Lower und Upper Castile. The Lower Castile Formation was confirmed to occur only in the Delaware Basin, whereas the Upper Castile Formation has a wider distribution. Later, Lang (1935, 1939) proposed to use the term "Castile" only for the lower Ochoan developed in the Delaware Basin; and the "Upper Castile" denoted as the Salado Formation. He found that the Castile Formation (Lower Castile) to the north-east and east was bounded by the Capitan Carbonate Reef; and thus, the top of the formation was placed at the base of beds overlying the reef, i.e., at the Fletcher Anhydrite Member. The upper boundary of the Salado

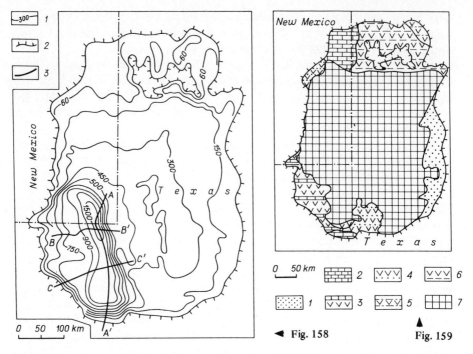

Fig. 158. Isopach map of Ochoan deposits in West Texas Permian Basin. After McKee et al. (1967). *1* isopach contours, in meters, *2* limit of Ochoan deposits, *3* section lines shown in Fig. 160

Fig. 159. Lithologic map of Ochoan deposits in West Texas Permian Basin. Compiled from the data of McKee et al. (1967). Areal distribution of: *1* sandstone, *2* carbonate rocks, *3* anhydrite alternated with dolomite, *4* anhydrite, sandstone, and siltstone, *5* anhydrite, sandstone, siltstone, and mudstone, *6* mudstone interbedded with anhydrite, *7* rock salt

Formation was drawn on the omission of potassium-bearing deposits from the section and the appearance of numerous anhydrite interbeds. A salt unit, overlying the Salado Formation and represented by alternate anhydrite and rock salt interbedded with dolomite, was placed into the Rustler Formation overlain by the Dewey Lake Red Beds.

A further subdivision of the Ochoan deposits was porposed by Kroenlein (1939) and Adams (1944). Divisions of the Salado Formation, containing all potash salt horizons were especially detailed. Kroenlein divided the formation into the lower and upper parts containing potash salts, which he named pre-potash and potash-bearing. He distinguished a salt unit above the basal Fletcher Anhydrite and above the unit, 44 salt and anhydrite bands. The lowermost anhydrite interbed was named the "Gowden Anhydrite" (Kroenlein 1939). In the middle part of the Salado Formation, the McNutt potash zone has been distinguished, named after McNutt, who discovered the first economic potash salt deposit in the USA.

In the Salado Formation, sandstone, siltstone and mudstone interbeds occur regionally at two stratigraphic levels. A siltstone interbed lies near the base and is known as La Huerta Siltstone Member; another one occurs in the upper part of the

Formation, below Kroenlein's anhydrite bed 12 and is known as the Vaca Triste Sandstone Member. Lang (1935) refined the nomenclature of the Rustler Formation and distinguished two marker beds, the Gulebra and Magenta dolomite members.

The present division of the Ochoan deposits in the West Texas Permian Basin was proposed with regard to the above data. It is rather well-illustrated on diagrammatic sections, which show the structure of the Ochoan salt series in the Delaware Basin (Fig. 160).

The Castile Formation is composed of massive and layered anhydrite, rock salt, minor limestone, and terrigenous rocks. No potash rocks have been found as yet within the formation. It is dominated by white and banded anhydrite and rock salt. Gypsum occurs in the section near the surface and at depths to 150 m. At a greater depth, only anhydrite was found. The most characteristic rock of the formation is layered or banded anhydrite, composed of thinly intercalated calcite and anhydrite beds. Calcite beds are brown and more or less bituminous. They contrast sharply with white anhydrite beds. Three units composed of laminated calcite-anhydrite and anhydrite rocks are distinguished. The lower, middle, and upper units are 70 m, 25–30 m and 140–150 m thick, respectively. Along with anhydrite and gypsum, the formation also contains glauberite and minor polyhalite. Rock salt beds are grouped into three units. The lower and the middle salt units are of an almost universal distribution. The upper unit is made up of heterogeneous rock salt, alternated with anhydrite. Salt content of the formation rarely attains 60%. The composition of the formation is as follows: 30% rock salt, 11% carbonate rocks (limestone, dolomite,

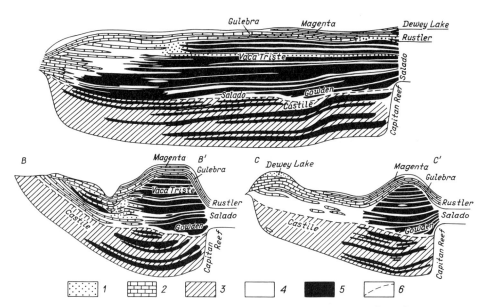

Fig. 160. Diagrammatic sections of Ochoa rocks in Delaware Basin across A-A^1, B-B^1, and C-C^1 profiles (for section lines see Fig. 158). After Adams (1944). *1* sandstone, *2* dolomite and limestone, *3* banded anhydrite, *4* anhydrite and gypsum, *5* salt, *6* boundary between Castile and Salado Formations

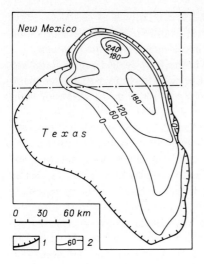

Fig. 161. Isopach map of Castile Formation salt in New Mexico and Texas. After Hayes (Lefond 1969). *1* extent of Delaware Basin in Castile period, *2* isopach contours, in meters

and magnesite), and 59% anhydrite and gypsum (Jones 1972). The thickness of the Castile Formation varies between 450 and 560 m, attaining a maximal thickness of 640 m in the north-west Delaware Basin. The total rock salt thickness in the Castile Formation ranges from 0 to 180–200 m, locally reaching 250–300 m in most subsided parts (Fig. 161). The areal extent of rock salt is about 13,000 km^2. Assuming an average thickness of rock salt for the entire area of 100 m, then the salt volume in the Castile Formation will equal 1.3×10^3 km^3.

The Salado Formation has a wider distribution and is common within the Midland Basin and adjacent shelf. The Salado Formation has a rather complex composition. Along with rock salt, anhydrite, and gypsum, it also contains polyhalite, magnesite, glauberite, thenardite, sylvite, carnallite, langbeinite, kieserite, kainite, leonite, and other rocks. The section is dominated by rock salt, sylvite, and carnallite rocks, which account for 85% of the section. The remaining section consists of 12% sulfate rocks (anhydrite, gypsum, glauberite, and polyhalite rocks); 1% carbonate rocks and 2% terrigenous rocks (siltstone and mudstone).

As mentioned earlier, Kroenlein (1939) distinguished 43 salt and anhydrite beds above the Gowden Anhydrite in the section of the Salado Formation. Jones (1972) reported a cyclic structure of the formation. Thus, some cyclical sedimentary units are composed of anhydrite at the base; above, rock salt – clayey rock salt –, and a thin, clayey band follow. The thickness of cycles varies between 1 and 10 m. The anhydrite member of the cycle is genearlly represented by a rhythmic alternation of anhydrite, magnesite, and minor halite. The halite member is either massive, structureless, or locally layered and contains anhydrite and polyhalite inclusions. Transitions from the halite member to the overlying argillaceous halite are gradual in some cycles. Furthermore, there are more complete cycles of potash salts; potash cycles, possibly, amount to over 66.

Cyclical sedimentation units with potash salts lie in a fairly compact group near the middle of the Salado Formation; they form an important potassium-rich zone, the McNutt potash zone. Its thickness in the Carlsbad District ranges from 50—140 m; in the north-western shelf, it is about 40 m; and in the eastern Delaware Basin, often exceeds 180 m.

The lower sub-potash part of the Salado Formation has the following structure. The basal Fletcher Anhydrite is overlain by a salt member almost 45 m thick. Above, the Gowden Anhydrite, No. 44 in Kroenlein's (1939) division, follows. In some regions, particularly on the north-western shelf, it contains polyhalite and anhydrite. Between the McNutt potash zone and the Gowden Anhydrite, Kroenlein distinguished 19 beds (Nos. 25—43), four of which are anhydrite with different amounts of polyhalite; the remainder is halite. The maximal thickness of this part of the section reaches 150 m.

The upper part of the Salado section above the potash-bearing section, is made up of rock salt, alternated with anhydrite (polyhalite) interbeds. At its base, above the McNutt potash zone, the variegated, terrigenous Vaca Triste interbed occurs. Kroenlein distinguished three anhydrite beds in this part of the formation (Nos. 12, 7, and 4), which are marker beds. Their thickness varies between 6 and 9 m. These beds are also rich in polyhalite. It is only in the uppermost part, above anhydrite bed No. 4 that the carnallite rock interbed containing red clay is locally developed. The McNutt potash zone is distinguished, primarily, on the basis of the presence of potash salts, namely, sylvinite, carnallite, kainite, langbeinite, leonite, and other rocks. A section of this zone comprises 11 potash horizons, known also as ore zones (Jones 1972). In the lower half of the McNutt zone, the white marker polyhalite 1.2 m thick, bed No. 24 is easily recognizable (Kroenlein 1939). Four potash horizons lie below; and seven lie above the bed.

Potash horizons in the McNutt zone occupy a small area in the West Texas Permian Basin. They are developed within the north-western shelf and in places in the Delaware and Midland Basins (Fig. 162). Potash salts are, in fact, concentrated within the Carlsbad area. The horizons are generally polymineral in composition. They contain mainly sylvite, carnallite, kainite, and langbeinite; the remaining minerals have a local distribution. Jones (1972) distinguished two major types of sylvite deposits — massive and disseminated— and two subordinate types — vein deposits and lens deposits. Two major types of potash salts differ mainly in their position in parts of various thickness. Disseminated potash salts occur in salt units 75—100 m thick. White massive sylvite deposits are confined to interbeds 0.75—1.5 m, and rarely, 7 m thick.

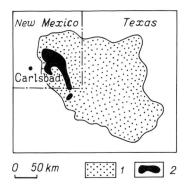

Fig. 162. Distribution of massive potash deposits in evaporites of the Ochoan Series. After Jones (1972). *1* extent of polyhalite deposits in anhydrite beds, *2* extent of sylvite- and langbeinite-bearing deposits in halite beds

As a rule, massive sylvite builds up the middle and upper parts of a cyclical sedimentation unit. These rocks form ore zones.

Polyhalite deposits occur almost universally in the Salado Formation. In the Delaware Basin, their lower boundary is recognized first below the Gowden bed, and then can be traced up the section. Polyhalite deposits are either massive or disseminated, or they fill in fractures and lenses in rock salt, anhydrite, and clayey rock. The known area of their distribution is 325 km long and 220 km wide. Massive polyhalite deposits are, in general, dense, fine-grained rocks, differing in color and structure. They are concentrated mainly in the anhydrite beds. In some parts of the Carlsbad area, many massive polyhalite deposits grade along the strike into anhydritic Hartsalz.

Salt deposits of the Salado Formation cover an area of about 95,000 km^2 (Fig. 163). The total thickness of rock salt varies between 60 and 300 m. It is only in the most downwarped part of the Delaware Basin that their thickness increases to 360 and even to 480 m. The minimal average thickness for the entire area may be estimated at 150 m. In this case, the volume of salt in the Salado Formation will attain 1.43×10^4 km^3.

The Ochoan salt section in the West Texas Permian Basin is crowned by the Rustler Formation. Salt beds within this formation, although rather thin (generally not more than 10–15 m), constitute a considerable part of the formation. The approximate composition of the Rustler Formation, according to Jones (1972), is as follows: 10% carbonate rocks, 30% sulfate rocks (anhydrite, polyhalite), 43% rock salt, and 17% terrigenous rocks. The above synopsis implies that the Rustler Formation, as compared to the underlying Ochoan salt series, contains a larger amount of terrigenous

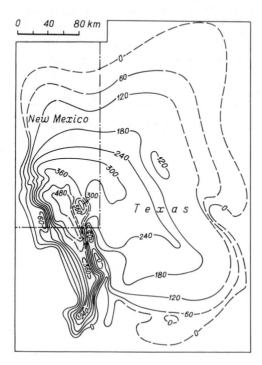

Fig. 163. Isopach map of rock and potash salts in Salado Formation in New Mexico and Texas. After Hayes (Lefond 1969). Isopach contours of rock salt, in meters

rocks. They are represented by gray and red sandstone, siltstone, and mudstone. Two dolomite marker beds are distinguished in the Rustler section, the Gulebra and Magenta, each about 10 m thick. In general, the formation is often divided into two parts: the upper, 45–55 m thick, composed of anhydrite or gypsum with rare interbeds of red sandstone, siltstone, and rock salt; and the lower, made up of dolomite, anhydrite, sandstone, mudstone, and rock salt up to 110–120 m thick. The rock salt volume in the Rustler Formation is not large and probably does not exceed 9×10^2 km^3.

The Dewey Lake Red Beds are composed of red mudstone, siltstone, and minor sandstone with interbeds, stringers, and lenses of gypsum. The maximal thickness is slightly above 100 m.

In concluding the characterization of the Midcontinent Basin, it can be assumed that the total volume of rock salt accumulated exceeds 6.3×10^4 km^3.

Supai Basin

Permian salt deposits in the western United States are also known from an area outside the Midcontinent, namely, at the southern margin of the Colorado Plateau, in the Supai Basin (also called Holbrook Basin).

Supai Basin is situated in central-east Arizona. It is bounded to the south-west by the escarpment of the Colorado Plateau — the Mogollon Rim — which is reflected in topography, and to the north-east, by the Defiance Positive Element.

Evaporites in the Supai Basin are confined to the Supai Formation, which comprises both Late Carboniferous and Early Permian deposits. The formation is composed of red and orange-brown sandstone, siltstone, and clay. Locally, there are carbonate interbeds in its middle and upper parts, while at Fort Apache and north of the Mogollon Rim, anhydrite and rock salt appear in the section. There, the Supai Formation is divided into the lower, middle, and upper parts. The lower part is Carboniferous, whereas the middle and the upper ones are Lower Permian. The middle part seems to correspond to Wolfcampian and the upper one is placed at the level of the lower Leonardian. All the evaporite deposits in the Supai Formation are restricted to the upper part of the middle and upper parts proper.

In the area underlain by evaporites, the Permian part of the Supai Formation is generally divided into the Middle Supai unit and the Upper Supai unit; the boundary between them is drawn at the base of the Fort Apache Limestone Member. The base of the same carbonates also defines the Wolfcampian/Leonardian boundary (McKee 1967).

The Wolfcampian deposits in Arizona and adjacent areas of New Mexico, Colorado, and Utah are characterized by a considerable facies change (Fig. 164). The Middle Supai unit and its equivalents are composed of terrigenous red beds, namely, siltstone, mudstone, and sandstone. Only in north-west Arizona and in Nevada, the section contains numerous carbonate interbeds of dolomite and limestone. Evaporrites of the Middle Supai unit were reported in a small area in central-east Arizona. They occur in the upper part and are represented by alternations of rock salt with

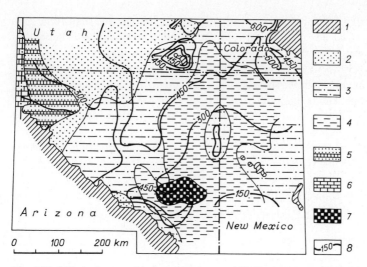

Fig. 164. Lithologic map showing thicknesses of Wolfcampian deposits (Middle Supai and equivalent rocks) in Arizona and adjacent areas. Compiled from the data of McKee et al. (1967). *1* Wolfcampian deposits absent. Distribution of: *2* terrigenous rocks, mainly sandstone, *3* siltstone, sandstone, and mudstone, *4* mainly mudstone and siltstone, *5* terrigenous rocks interbedded with carbonates, *6* carbonate rocks, *7* rock salt, *8* isopach contours, in meters

red siltstone, mudstone, minor sandstone, and anhydrite. The salt series passes along the strike first into sulfate-terrigenous and then, terrigenous rocks. The thickness of Wolfacampian ranges from 200–450 m. The highest salt saturation of 75% was recorded in a belt, extending from north-east to south-west, south of Holbrook (McKee 1967). In the remaining area of rock salt distribution, salt content does not exceed 25–50%, averaging 10–15%. At present, rock salt covers an area of about 5000 km^2. Its volume is probably slightly over 50 km^3 (with an average rock salt thickness of 10 m).

In the Upper Supai unit, salt deposits occur in red and orange-brown terrigenous rocks represented by siltstone and mudstone (Fig. 165). Along the strike, salt deposits pass into sulfate-terrigenous rocks. To the south-west and north-east, red sandstone with rare siltstone and mudstone interbeds prevail. In north-western Arizona, the Leonardian deposits are enriched by carbonate rocks. In some districts in south Utah, the section becomes entirely terrigenous-carbonate in places. The Leonardian sedimentary strata in Arizona vary in thickness between 150 and 450 m; the greatest thicknesses are confined to the south-western margin of the Colorado Plateau, at the junction with the Mogollon Rim.

Within the Supai Basin, the thickness of the Leonardian deposits ranges between 150 and 300 m, attaining a maximal thickness of 340 m. Some marker carbonate beds also occur in the upper Supai section, above the Fort Apache Limestone Member.

Rock salt beds rapidly pinch out. They are intercalated with anhydrite, mudstone, and siltstone. Gerrard (1966) distinguished four carbonate rock marker horizons in this section; the lower, Fort Apache, which is the thickest and most persistent; the other three were designated Nos. 1, 2, and 3. The maximal amount of anhydrite was

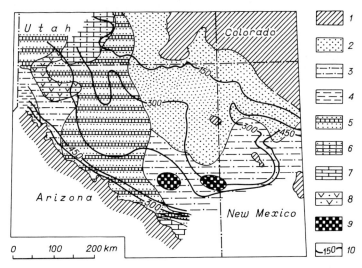

Fig. 165. Lithologic map showing thicknesses of Leonardian deposits (Upper Supai and equivalent rocks) in Arizona and adjacent areas. Compiled from the data of McKee et al. (1967). *1* Leonardian absent. Distribution of: *2* terrigenous rocks, mainly sandstone, *3* sandstone, siltstone, and mudstone, *4* mainly mudstone, *5* terrigenous rocks interbedded with carbonate, *6* terrigenous-carbonate rocks, *7* carbonate deposits, *8* terrigenous-carbonate rocks interbedded with anhydrite and gypsum, *9* rock salt, *10* isopach contours, in meters

reported from the north-east and east. With regard to the distribution pattern of rock salt in the section, Peirce and Gerrard (1966) divided the Upper Supai unit into four zones. The first zone lies between the Fort Apache carbonate beds and bed 1; the second zone occurs between carbonate beds 1 and 2; the third zone, below carbonate bed 3; and the fourth zone, above them. The thickest salt beds were reported from the third and fourth zones. The third salt zone is also characterized by the presence of sylvinite interbeds in rock salt (Peirce and Gerrard 1966).

The areal extent of rock salt in the upper Supai Formation, in Arizona, does not exceed 6000 km². The greatest total thicknesses of salt are confined to areas south of Holbrook, the total rock salt thickness varies between 60 and 140 m. Despite a lenticular mode of occurrence of rock salt and its rapid pinching out, the average salt thickness for the whole area may be estimated at 50 m. Thus, the rock salt volume in the upper Supai Formation will equal 3×10^2 km³; and in the entire Supai Basin, it may exceed 3.5×10^2 km³.

North Mexican Basin

In Chihuahua State in north Mexico, near the border of the USA (Fig. 166), salt strata of uncertain age have been found. They were penetrated by deep wells below Jurassic formations and some investigators assign to them the Permian age. The stratigraphic position of these salt deposits is debatable, but the North Mexican evaporite basin can be tentatively outlined.

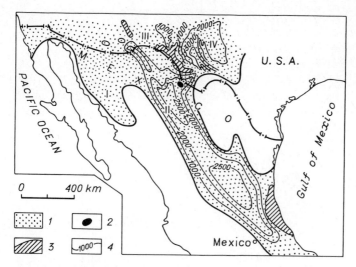

Fig. 166. Distribution of Permian deposits in Mexico. After Lopez-Ramos (1969). *1* distribution of Permian deposits. *2* Cuchillo Prado area with probable Permian salt strata, *3* plateau-like uplifts, *4* isopach contours, in meters. Basins: *I* Sonoran, *II* Chihuahua, *III* Orogrande, *IV* Midland

Permian deposits in Mexico show relatively limited distribution. They were reported from the Sonoran Basin in north-west Mexico, Sonoran State, and Chihuahua Basin in the south-eastern, eastern, and north-eastern part of the eponymous State, and from some other central districts of Mexico as well. In general, Permian deposits fill in a narrow trough, 200–400 km wide, extending from south-east to north-west, and situated in central Mexico. From south-west and north-east, the trough is bounded by uplifts. The thickness of the Permian sedimentary strata in the Chihuahua Basin attains 2500 m. As the Sonoran Basin is approached, the thickness decreases to 1000 m and even 500–300 m. For more detailed information on Permian deposits in Mexico, the reader is referred to the publications of Lopez-Ramos (1969), Salas (1968a, b), and Lefond (1969). The following description is based on their works.

Within the Chihuahua and Sonora basins, the Permian deposits are represented mainly by carbonate rocks. The section is dominated by limestone. Dolomite, flint, sandstone, and mudstone are minor constituents. Locally, carbonate structures and flysch-like, terrigenous-carbonate strata occur. In most districts, Permian deposits are of Wolfcampian and Leonardian age. The Upper Permian strata were recorded only in the northern Chihuahua Basin, in the Rio Grande Basin, where they are distinguished as the Concha Limestone at Palomas and as the Plamosos Formation near Placer De Guadalupe.

Probable Permian evaporites were penetrated by deep boreholes in the northern Chihuahua Basin near Cuchillo Prado. The Cuchillo Prado No. 1 well, about 3.2 km north-west of Cuchillo Prado, penetrated rock salt at a depth of 270 m; the well penetrated 2585 m of salt, but did not reach the base of the salt series. The Cuchillo Prado No. 2 well, 10 km north-west of the town, penetrated salt strata over a depth interval of 1350.0–1737.4 m. It is assumed that this well penetrated a rock salt lens, thrusted on the overlying deposits on an overturned fold limb. Four other wells, south of the Cuchillo Prado No. 1 well, penetrated salt strata at a depth of about 180 m. Only the above data are available on salt series of the North Mexican Basin. At present, their areal extent and structural features are unknown. In the vicinity of the above described areas, within the Chihuahua Basin, salt deposits of probable Permian age have also been penetrated by two wells in the Rio Grande Basin, in south-western Texas (USA) (Oriel et al. 1967). However, at present the rock salt volume in the North Mexican Basin cannot be estimated.

Peru-Bolivian Basin

The Peru-Bolivian Basin seems to be rather large. It extends for about 2500 km along the Central and East Andes (South American Cordillera) from the Frontier regions of Columbia, Ecuador, and Peru on the north, to Argentina in the south. The width of the basin locally exceeds 800 km. Its total area can attain 1,250,000 km^2.

In the Peruvian part of the basin, the Permian is subdivided into two formations: Capacabana and Mitu. They belong to the Wolfcampian and Leonardian, respectively. Sulfate rocks are found in both formations. The Capacabana Formation is composed mostly of limestone and shale with dolomite, siltstone, and sandstone interbeds. In its upper part, gypsum and gypsified rock interbeds were recorded near Cuzco (Newell et al. 1953). The Mitu Formation lying above is represented by terrigenous red beds among which gypsinate rocks and gypsum layers up to 8 m thick occur. They were reported from northern Peru, in the Marañon River Basin and in the south, near Cuzco (Newell et al. 1953, Benavides 1968). The Mitu Formation comprieses, presumably, the salt-bearing deposits, which compose most of the diapiric and salt structures in Peru. The salt domes are most abundant in the Huallaga River Basin and its left tributary, the Mayo River. Gypsum, anhydrite, and rock salt alternating with pink, red, and purple mudstone and siltstone crop out along the crest of the dome. The age of the salt-bearing sediments of the domes is uncertain, probably Permian, although some authors assign them to the Triassic and even Jurassic (Newell et al. 1953, Benavides 1968).

In Bolivia, the Permian succession is often placed into the Capacabana Group (Helwig 1972), but in some regions all the Permian sedimentary series are included either in the Yarichambi Formation (Schlatter and Nederlof 1969) or in the Titicaca Group. In the Titicaca Lake area, they are subdivided into two parts: the lower Capacabana Group and the upper Chuquichambi (Lohmann 1970). Helwig (1972) tentatively assigned all the evaporites occurring in this region to Permian red beds.

In general, the Permian succession in Bolivia is quite similar to that of Peru. Two levels of the evaporite rocks can also be distinguished: the lower one, confined to the Capacabana Group and the upper one, enclosing the salt-bearing Chuquichambi formations, which might possibly be correlated with the presumably salt-bearing Mitu Series. The Chuquichambi salts in Bolivia make up several domes in the area between Titicaca and Poopo Lakes. In fact, salt domes of Peru and Bolivia fall into a single tectonic zone within which a thick salt sequence might be developed.

Another viewpoint expressed by Lohmann (1970) seems quite probable; it presupposed that in Bolivia and in adjacent regions of Chile there might be three salt basins of different age. The Peruvian Basin might also be a separate basin. In view of the uncertain age and spatial distribution of salt-bearing units, the Peru-Bolivian Basin is conventionally distinguished as a Permian salt basin. Two salt-bearing sequences (Mitu in the north, in Chile, and Chuquichambi in the south, in Bolivia) can be tentatively outlined.

An approximate estimate of rock salt volume of these strata can be given, proceeding from the following tentative assessments. Assuming an areal extent of the Mitu Evaporite of 3×10^5 km^2, and an average total thickness of salt of 100 m, then the rock salt volume will be 3×10^4 km^3. The areal extent of the Chuquichambi Evaporite is 2×10^5 km^2, the salt thickness is 100 m, and the rock salt volume is 2×10^4 km^3. The total salt volume may well exceed 5×10^4 km^3.

Permian Evaporite Deposits in Other Basins

In addition to the salt basins discussed earlier, Permian evaporites have also been reported from the following basins: Darvaza, Karasu-Ishsay, Dobruja, Rakhov, Mecsek, Dinarids, North Italian, Spitsbergen, East Greenland, Sverdrup, Rio Blanco, Parnaiba (Maranhão), Arabian, and Browse.

The Darvaza Basin is outlined on the basis of the distribution of gypsum-bearing deposits; those of probable Early/Late Permian age are distributed within the Darvaza Range in Tadzhikistan. They are distinguished either as a gypsum-bearing sequence or as the Shakarsen Formation (Dutkevich 1937, Vinogradov 1959), which is composed of purple clay, fine-grained sandstone, gypsum, and limestone. The thickness of gypsum layers often attains 10–12 m; the total thickness is 400 m.

The Karasu-Ishsay Basin lies in Uzbekistan, in the north-eastern Fergana Valley. Data on the occurrence of gypsum and anhydrite in terrigenous-carbonate deposits of the Karasu Formation were published by Biske and Kushnar (1976). The sulfate Karasu Formation is upper Early Permian and lower Late Permian in age.

The Dobruja Basin is confined to a depression in South Moldavia. Sulfate rocks in the Permian deposits were penetrated by several deep boreholes. They compose two sulfate-bearing sequences: the first and the second anhydrite-bearing strata, separated by a bit of terrigenous red beds. They are 120 m and 110 m thick, respectively (Kaptsan and Safarov 1969b, Kaptsan et al. 1963, Bobrinksy et al. 1964).

The Rakhov Basin is conventionally distinguished as a separate basin. Zhukov (1965b, Zhukov et al. 1964, 1976) recorded abundant sulfate rocks in the Rakhov Zone of the East Carpathians. There a gypsum-bearing sequence, about 100 m thick, is distinguished in the upper part of the Permian section; it is composed of variegated mudstone and siltstone, among which interbeds of gypsinate shale and small gypsum lenses are recorded. The age of this sequence is probably Late Permian.

The Mecsek Basin is tentatively distinguished on the basis of Late Permian gypsum-bearing clay deposits developed in the Mecsek Mountains, in southern Hungary. However, the data on the thickness and areal extent of gypsum-bearing strata are very scarce. It can be noted that gypsum layers are thin and occur among variegated terrigenous formations (Vadas 1964, Trunkó 1969, Zhukov et al. 1976).

In the Dinarids Basin, sulfate rocks are abundant among the Late Permian deposits. They occur in the Durmitor Massif where gypsum beds, 20—30 m thick, crop out (Cadet 1966, 1970a, b), as well as in the High Karst Zone around the Čersnica Mountains (Charvet 1970); they occur in a carbonate succession. Additionally, gypsum-bearing, terrigenous red beds are recorded in the Korab Zone of the Dinarids in north-eastern Albania (Papa 1970). Miljush (1973) outlined an extensive zone of Permian evaporite deposits in the Dinarids, embracing all the above regions.

The North Italian Basin is outlined on the basis of extensive gypsum and anhydrite deposits in northern Italy. They are confined mostly to the Upper Permian Bellerophon Formation, where they form two sulfate-bearing sequences: the lower one, up to 150 m thick, represented mostly by laminite gypsum; and the upper one, 200—300 m thick, composed of a cyclic alternation of gypsum and dolomite. These sequences are separated by a unit of dense, fine-grained dolomite (Bosellini and Hardie 1973). Sulfate-bearing deposits are traced there from the northern regions of Italy into the Dolomitic and Carnic Alps and farther into the Dravt Ridge, where layers of gypsum and anhydrite occur in the Late Permian Bellerophon Series (Kahler 1972, Mostler 1972a, b, Rutten 1972, Buggisch et al. 1976). They probably continue eastward into the Karawanke Range. Sulfate rocks are less abundant among the Lower Permian terrigenous red beds distinguished in the South Alps as Groden Formation and in North Italy as Gardena Sandstone (Oxburg 1968, Falke 1972, Rutten 1972, Bosellini and Hardie 1973, Flügel 1975, 1981, Buggisch et al. 1976). These sulfate-bearing red beds can be regarded as a separate unit. They probably occur along the northern marginal zone of the basin. Besides the above sulfate-bearing deposits, the Lower Permian (Autunian) deposits are tentatively included into the North Italian Basin; these deposits are developed in the Bergamasche Alps in Lombaridia, Italy, and are designated as the Collio sequence. There gypsum lenses and layers are recorded among red siltstone, marl, and siltstone (Sitter and Sitter-Koomans 1949, Falke 1972).

Within the Spitsbergen Basin, gypsum and anhydrite are confined to the Upper Gypsum-Bearing Formation of Artinskian age (Harland 1964, Sosipatrova 1967). In the East Greenland Basin, a gypsum-bearing sequence occurs at the base of the Zechstein; this sequence is represented by massive gypsum alternated with limestone, dolomite, and calc-dolomite rocks. Gypsum lenses are about 50 m thick. Gypsum-bearing beds are sometimes recorded throughout the entire Zechstein section, which is about 300 m thick (Maync 1964).

The Sverdrup Basin, as discussed earlier, was an evaporite basin not only during the Carboniferous epoch, but also at the beginning of the Early Permian, when the Mount Bayley sulfates were being formed.

The Rio Blanco Basin is located in north-western Argentina. It is highly probable that this basin was a part of the Peru-Bolivian Basin situated farther north. Gypsum and gypsinate interbeds are recorded from the upper Patquia Formation, composed mostly of red siltstone and sandstone. Gypsum is most abundant in the La Rioja and San Juan Provinces (Teruggi et al. 1969, Volkheimer 1969).

The Parnaiba Basin lies in western Brazil. Gypsum and anhydrite are found among the Pedra-do-Fogo red beds; the thickness of the formation is about 300 m. Gypsum layers are sometimes up to 18 m thick.

During Permian time, in the Arabian Basin reef limestone, sandstone, variegated siltstone, and clay with gypsum interbeds, lenses, and inclusions were accumulated (North East . . . 1973). In northern Iraq, gypsum-bearing deposits are of Upper Permian age, whereas their lower part is known as the Chia-Sapri Formation. In southeastern Turkey, variegated gypsum-bearing rock units belong to the Upper Permian Inbirik Formation. The sulfate-bearing deposits of the Arabian Basin have not as yet been thoroughly studied.

Browse Basin is situated in the north-western coastal districts of Australia. There, anhydrite interbeds, inclusions, and lenses occur among the Sakmarian sandstone, which was penetrated by the Rob Roy No. 1 well at a depth of 1895.8–1897.4 m (Wells 1980).

References

Adams JE (1944) Upper Permian Ochoa Series of Delaware basin, West Texas and Southeastern New Mexico. Am Assoc Pet Geol Bull 28:1596–1625
Adams JE (1963) Permian salt deposits of West Texas and New Mexico. In: Symposium on salt. Cleveland, Ohio. North Ohio Geol Soc, pp 124–130
Adams JE (1965) Stratigraphic-tectonic development of Delaware basin. Am Assoc Pet Geol Bull 49:2140–2148
Ageev BD (1982) On the deposition history of Middle Paleozoic evaporites in the Chu-Sarysu Depression related to exploration for oil and gas in southern Kazakhstan (in Russian). In: Oil-and-gas potential of old salt accumulation regions. Nauka Novosibirsk, pp 89–97
Ahlborn O (1934) Die Ausbildungsformen des Stassfurt-Lagers am Rosslebener and Bernburger Sattel in ihrer Beziehung zur Tektonik. Z Dtsch Geol Ges 86:663–699
Ahlborn O (1955) Die Flöze "Thüringen" und "Hessen" der Werraserie und ihre wechselseitigen Beziehungen. Z Dtsch Geol Ges 105:664–673
Ahlborn O, Richter-Bernburg G (1955) Exkursion zum Salzstock von Benthe (Hannover) mit Befahrung der Kaliwerke Ronnenberg and Hansa. Z Dtsch Geol Ges 105:855–865
Aitken JD (1981) Stratigraphy and sedimentology of the Upper Proterozoic Little Dal Group, Mackenzie Mountains, Northwest Territories. In: Campbell FHA (ed) Proterozoic Basins of Canada. Geol Surv Can Pap 81-10:47–71
Aizenshtadt GEA, Nevolin NV, Eventov YaS (1960) Geological structure and oil-and-gas potential of the Caspian salt dome area (in Russian). Proc of the XXI IGS Session. Izd Akad Nauk SSSR, Moscow, pp 30–39
Aizenshtadt GEA, Gershtein EI (1963) On initial thickness of the Kungurian salt-bearing complex in the Caspian Depression (in Russian). Dokl Akad Nauk SSSR 151:1156–1158
Aizenshtadt GEA, Gorfunkel MV (1965) Tectonics and oil potential of the Caspian and North German Derpessions (in Russian). Tr VNIGRI 246, 256 pp
Aizenshtadt GEA, Pinchuk IA (1961) South Emba 2 and Tugarakchan 5 key wells (in Russian). In: Key wells of the USSR. Gostoptekhizdat, Leningrad, 294 pp
Aliev MM, Vysotsky VI, Golenkova NP, Timonin LS (1979) Geological structure and oil-and-gas potential of North Africa and Middle East (in Russian). ELM, Baku, 547 pp
Aliev MM, Zabanbarg A (1974) Geological structure and oil-and-gas potential of Iran (in Russian). ELM, Baku, 104 pp
Anderle JP, Crosby KS, Waugh DCE (1979) Potash at Salt Springs, New Brunswick. Econ Geol 74:389–396
Anderson RJ (1967) Carbonate deposition in standing bodies of water (abs). In: Guidebook, Defiance-Zuni-Mt. Taylor region, Arizona and New Mexico. New Mexico Geol Soc 18th Ann Field Conf Socorro, New Mexico dept Mines and Mineral Resources, pp 225
Anderson RJ, Egleson CC (1970) Discovery of potash in the A-1 Salina Salt in Michigan. In: Sixth Forum on the geology of industrial minerals. Mich Geol Surv, Lansing, Michigan
Anderson RJ, Dean WE Jr, Kirkland DW, Snider HI (1972) Permian Castile Varved Evaporite Sequence, West Texas and New Mexico. Geol Soc Am Bull 83:59–86
Anderson SB (1964) Salt deposits in North Dakota. In: The mineral resources of North Dakota. Grand Forks, North Dakota Gen Extension Div, pp 60–79

Anderson SB, Hansen DE (1957) Halite deposits in the Dakota. North Dakota Geol Surv Inv 28: 3 sheets

Anderson SB, Swinehart RP (1979) Potash salts in the Williston Basin, USA. Econ Geol 74:358–376

Andreas D, Enderlein F, Michael J (1966) Zur Entwicklung des Rotliegenden im Thüringer Wald. Ber Dtsch Ges Geol Wiss A, Geol Paläont 11:119–130

Andrichuk JM (1959) Ordovician and Silurian stratigraphy and sedimentation in southern Manitoba, Canada. Am Assoc Pet Geol Bull 43:2333–2398

Antonowicz L, Knieszner L (1981) Reef zones of the Main Dolomite, set out on the basis of paleogeomorphologic analysis and the results of modern seismic techniques. In: International Symposium Central European Permian. Proc Wydawnictwa Geologiczne Warszawa, pp 356–368

Antsupov P, Akaeva V, Bakuń N, Snegireva O, Korab L, Bojarska J, Głowacki E (1981) The paleogeography of the southern part of Permian Basin in the Polish Lowland during the Saxonian age. Symp Central European Permian. Proc Wydawnictwa Geologiczne Warszawa, pp 253–261

Arabadzhi MS, Skvortsov II, Charygin MM (1966) On the thickness of hydrochemical formations in the Caspian Depression (in Russian). In: Geology of the Russian Platform, Caspian and Orenburg areas. Nedra, Moscow, pp 3–9

Arkhipov AYa (1964) Permian system (in Russian). In: Geology and oil-and-gas potential of the southern USSR. Kara-Bogaz area. Nedra, Leningrad, pp 22–24

Aronova SM, Gassanova IG, Krems AY, Lotsman OA, Lyashenko AI, Maksimov SP, Hechitailo SK, Pistrak RM, Radionova KF, Sikolova LI (1967) The Devonian of Russian Platform. In: Int Symp on the Devonian System, vol I. Calgary, Alberta, Canada, pp 376–396

Arthurton R, Hemingway J (1972) The St. Bees Evaporites – a carbonate-evaporite formation of Upper Permian age in West Cumberland, England. Proc Yorkshire Geol Soc 38:565–592

Ashirov KB, Efremov PE (1969a) Environmental conditions of Permian salt basins in the southeastern Russian Platform (in Russian). Tr Giprovostokneft 12:245–253

Ashirov KB, Efremov PE (1969b) Paleogeographic and sedimentation conditions of Permian sulfate-halogenic strata in the Transvolga Kuibyshev-Orenburg area (in Russian). Tr Giprovostokneft 12:254–265

Atlas of lithofacies-paleogeographical maps of the Permian of platform areas of Poland, 1:500 000 (1978) Inst Geologiczny, Warszawa

Auden JB (1974) Afghanistan – West Pakistan. In: Spencer AM (ed) Mesozoic-Cenozoic Orogenic Belts. Academic Press, London, pp 235–253

Avrov PYa, Kosmacheva LG (1963) Geological structure and oil-and-gas potential of the Cis-Urals at Aktyubinsk and western Mugodzhary area (in Russian). Izd Akad Nauk Kaz SSSR, Alma-Ata, 163 pp

Azizov AI, Tikhvinsky IN (1978) Distribution and accumulation regularity pattern of potassium sulfate salts in the Cis-Uralian–Caspian Basin (in Russian). Litol I Polezn Iskop 2:124–136

Baar A (1966) Der Bromgehalt im Steinsalz als stratigraphischer und genetischer Indikator im norddeutschen Zechstein. Z Dtsch Geol Ges 115:572–608

Baars DL (1962) Permian System of Colorado plateau. Am Assoc Pet Geol Bull 46:149–218

Badham JPN, Stanworth CW (1977) Evaporites from the Lower Proterozoic of the East Arm, Great Slave Lake. Nature 268:516–518

Baker CL (1929) Depositional history of the red beds and saline residues of the Texas Permian. Texas Univ Publ 2901, pp 9–72

Bakirov SB, Belyashov DN (1966) Lithofacies and paleogeographical conditions of accumulation of the Upper Paleozoic deposits in the southwestern Chu Depression (in Russian). In: Temat sb Ministerstva vysshego i srednego spetsialnogo obrazovaniya Kaz SSR Geologiya 2:38–45

Banera NI (1964) Some problems of salt tectonics in the Caspian Depression from geophysical data (in Russian). Tr VNIIG 45:55–61

Baranov KA (1946) A secondary mirabilite deposit in the Arctic (in Russian). Priroda 6:28–41

Baranov KA (1947) On mirabilite geology in the Nordvik salt dome deposits (in Russian). Interiors of the Arctic 2:177–189

Barkhatova VP (1963) Permian system. Timan (in Russian). In: Geology of the USSR, vol II (Arkhangelsk, Vologda regions and Komi ASSR), pt 1. Gosgeoltek hizdat, Moscow, pp 532–554

Bartenstein H (1968) Present status of the Paleozoic Paleogeography of Northern Germany and adjacent parts of North-West Europe. In: Geology of shelf seas. Oliver and Boyd, Edinburgh, London, pp 73–86

Bates RL (1942). Occurrence and origin of Permian evaporites (abs). Am Assoc Pet Geol Bull 26: 907

Bazhenova SN (1972) On stratigraphy of Tatarian deposits in the Mezen-Vychegda Basin (in Russian). Izv Vyssh Uchebn Zaved Geol I Razved 11:19–26

Bell RT, Jackson GD (1974) Aphebian halite and sulphate indications in the Belcher Group, Northwest Territories. Can J Earth Sci 11:722–728

Belyea HR, Grayston LD, Grey FF, Moyer GL (1964) Upper Devonian. Chap 6, part I. In: Geological history of western Canada. Calgary, Alberta. Alberta Soc Pet Geol, pp 60–66

Berberian M, King GCP (1981) Towards a paleogeography and tectonic evolution of Iran. Can J Earth Sci 18:210–265

Beydoun LR (1966) Geology of the Arabian Peninsula. East Aden Protectorate and part of Dhufar. US Geol Surv Prof Pap 560-H, 49 pp

Bigarella JJ (1975) Brazil. In: Fairbridge RW (ed) The encyclopedia of World regional geology, part 1. Distributed holsted press. Wiley, pp 427–433

Birina GM, Sorskaya LS, Rozhdestvenskaya KK, Fomina EV (1971) The Carboniferous system. Lower section (in Russian). In: Geology of the USSR, vol IV (Centre of the European part of the USSR). Nedra, Moscow, pp 194–258

Bissell HJ (1962) Pennsylvanian and Permian rocks of Cordilleran area. In: Pennsylvanian system in the United States. Am Assoc Pet Geol Bull 46:188–263

Bissell HJ (1964a) Ely, Arcturus and Park City groups (Pennsylvanian-Permian) in Eastern Nevada and Western Utah. Am Assoc Pet Geol Bull 48:565–636

Bissell HJ (1964b) Patterns of sedimentation in Pennsylvanian and Permian strata of part of the Eastern Great Basin. Kans Geol Surv Bull 169:43–56

Bissell HJ (1967) Pennsylvanian and Permian basins in northeastern Nevada and south-central Idaho: Discussion. Am Assoc Pet Geol Bull 51:791–802

Bissell HJ (1970) Realms of Permian tectonism and sedimentation in western Utah and eastern Nevada. Am Assoc Pet Geol Bull 54:285–312

Blizeev BI (1970) Structure, environmental conditions of the Kungurian halogenic strata in the southeastern Russian Platform and adjacent part of the Cis-Uralian Trough and their potassium potential (in Russian). Avtoref Kand Diss Kazan, 20 pp

Blizeev BI (1971a) Rhythmical pattern of Kungurian salt strata in the southeastern Russian Platform and adjacent part of the Cis-Uralian Trough (in Russian). In: Geology and genesis of mining chemical raw material. Kazan, pp 142–146

Blizeev BI (1971b) Structure of Kungurian salt strata in the South Cis-Urals (in Russian). In: Geology and genesis of mining chemical raw material. Kazan, pp 165–171

Blizeev BI, Vishnyakov AK (1969) Rhythmicity of Kungurian halogenic strata in platform areas of Bashkiria and stratigraphic position of their potash salt deposits. Regional Geol 15, ONTI VIEMS, pp 3–8

Blom GI, Vavilov GA (1970) On Kazanian deposits in northern Tokmovo Arch (in Russian). In: Materials on geology of the eastern Russian Platform. Kazan, pp 73–75

Bobrov AK (1964) Geology of the Baikal Foredeep. Structure and oil-and-gas potential (in Russian). Nauka, Moscow, 228 pp

Bobrov VP, Korenevsky SM (1965) Lithology, rhythmicity and geochemistry of the Lower Permian halogenic deposits in the northwestern Donets Basin (in Russian). Sov Geol 10: 110–126

Bobrov VP, Korenevsky SM, Rybichenko OP (1968) Lithology and stratigraphy of the Kramatorsk Formation in the Donets Basin: mineralogy and petrology of rocks in potash-bearing horizons (in Russian). In: Geology of salt and potash deposits. Leningrad, pp 80–116

Bobukh VA, Dvurechensky VA, Shcherbakov VP (1969) New data on geological structure and oil-and-gas perspectiveness of the southwestern Caspian Depression and adjacent areas (Astrakhan Region and the Kalmyk ASSR). Tr Nizhne-Volzhsk Nauch-Issl Geol Inst Geofis 13: 28–39

Bocharov VM, Khalturina II (1981) On the generation of various potash-magnesium salts in the Caspian halogenic formation (in Russian). In: Structure and formation environments of potash salt deposits. Nauka, Novosibirsk, pp 138–144

Bondarenko SP (1963) New data on the Kramatorsk and Dronovo Formations of the Donets Permian (in Russian). Geol Zhurnal 2:32–40

Borchert H (1959) Ozeane Salzlagerstätten. Borntraeger, Berlin, S 129–142

Borchert H (1972) Secondary replacement processes in salt and potash deposits of oceanic origin. In: Geology of saline deposits. Proc Hannover Symp 1968 (Earth Sciences 7). Unesco, Paris, pp 61–68

Borozdina ZI (1959) Stratigraphy and paleogeography of the Permian deposits of the northern Volga-Ural area (in Russian). Tr VNIGNI 25:118–200

Brailovsky GS (1961a) On the distribution of the Lower Permian chemogenic strata in the Shebelino-Alekseevo district and in adjacent areas (in Russian). In: Materials on geology and gas potential of the Lower Permian deposits of the southern Russian Platform. Izd Kharkov Univ, Kharkov, pp 275–279

Brailovsky GS (1961b) Distribution of the Lower Permian salt strata and its bearing on the formation and structures of the Shebelino-Alekseevo district (in Russian). Geol Nefti I Gaza 12:18–22

Braitsch O (1962) Mineralogie und Petrographie in Einzeldarstellungen, Bd III. Entstehung und Stoffbestand der Salzlagerstätten. Springer, Berlin Göttingen Heidelberg, 232 pp

Braitsch O, Herrman A (1963) Zur Geochemie des Broms in salinären Sedimenten. Geochim Cosmochim Acta 27:361–392

Brandt RT, Burton CCJ, Maree SC, Woakes ME (1961) Mufulira. In: The geology of the Northern Rhodesian Copperbelt. MacDonald, London, pp 411–461

Briggs LI, Briggs DL (1974) Niagaran salina relationships in the Michigan Basin. In: Silurian reef evaporite relations: Michigan Basin. Geol Soc Field Conf, pp 1–23

Brinkmann R (1954) Abriß der Geologie, Bd 2. Enke, Stuttgart, 330 pp

Brucren JWR (1959) The stratigraphy of the Upper Permian Zechstein Formation in the eastern Netherlands. In: I giacimenti gassiferi dell'Europa occidentale, vol 1: Rome, Italy. Accad Naz Lincei, pp 243–274

Brunstrom RGW, Walmsley PJ (1969) Permian evaporites in North Sea Basin. Am Assoc Pet Geol Bull 53:870–883

Budanov GF, Molin VA (1968) Types of sections of the Lower Kazanian deposits in the northern Russian Platform (in Russian). Dokl Akad Nauk SSSR 183:1147–1150

Bulekbaev ZE, Muldakulov GG, Torekhanov I (1969) Upper Permian deposits in the Kenkiyak-Shengelshy district of the eastern flank of the Caspian Depression (in Russian). In: Permo-Triassic of the Russian Platform with regard to oil-and-gas potential. Nedra, Moscow, pp 99–102

Burk CA, Thomas HD (1956) The Goose Egg Formation (Permo-Triassic) of eastern Wyoming. Wyoming Geol Surv Rep Inv 6, 11 pp

Burkovskaya EG (1956a) On geology of the Zhylyany potash salt deposit in the Cis-Urals, Aktyubinsk area (in Russian). In: Problems of geology of agricultural ores. Izd Akad Nauk SSSR, Moscow, pp 218–221

Burkovskaya EG (1956b) On discovery of potash salts in the Cis-Urals, Aktyubinsk area (in Russian). In: Materials on Geol and Miner Res of South Urals, vol 1. Gosgeoltekhizdat, Moscow, pp 92–95

Butkovsky YuM, Ginodman AG, Kozhevnikov II (1965) New data on the structure of the northern flank zone of the Caspian Depression of Uralsk (in Russian). In: Geology and oil-and-gas potential of some regions of the USSR. Nedra, Moscow, pp 138–162

Butkovsky YuM, Omelchenko DA (1968) Geology and oil-and-gas potential of the northern flank zone of the Caspian Depression and its central part. In: Geology and oil-and-gas potential of some regions of the USSR and problems dealing with underground gas storage. Nedra, Moscow, pp 27–58

Campbell JA (1963) Permo-Triassic red beds, northern Denver basin. In: Guidebook to the geology of the northern Denver basin and adjacent uplift, Rocky Mountains. Assoc Geol 14-th Field Conf Denver, Colorado, Rocky Mountains. Assoc Geol, pp 105–110
Campbell FHA (1978) Geology of the Helikian rocks of the Bathurst Inlet area. Current Research, Part A. Geol Surv Can Pap 78-1A:97–106
Campbell FHA, Cecile MP (1981) Evolution of the Early Proterozoic Kilohigok Basin, Bathurst Inlet-Victoria Island, Northwest Territories. In: Campbell FHA (ed) Proterozoic Basins of Canada. Geol Surv Can Pap 81-10:103–131
Cartwright LD Jr (1930) Transverse section of the Permian basin, West Texas and southeastern New Mexico. Am Assoc Pet Geol Bull 14:969–981
Chandler FW (1978) Geology of part of the Wollaston Lake Fold Belt, North Wollaston Lake, Saskatchewan. Geol Surv Can Bull 277, 25 pp
Chechel EI (1969a) Relations of salt-bearing and carbonate formations in the southeastern Irkutsk Amphitheater (in Russian). In: Comparative analysis of sedimentary formations. Nauka, Moscow, pp 113–122
Chechel EI, Mashovich YaG, Chauzova GI (1981) Nepa potassium basin of East Siberia and deposition environments (in Russian). In: Main problems of salt accumulation. Nauka, Novosibirsk, pp 113–122
Chechel EI, Mashovich YaG, Gilev YuG (1977) Structural regularities of salt-bearing Cambrian deposits in the southern Siberian Platform (in Russian). Nedra, Moscow, 144 pp
Chechel EI, Mashovich YaG, Gilev YuG (1980) New data on potassium content of the Angara Formation in the Nepa deformation zone (in Russian). Sov Geol 5:100–109
Chirvinskaya MV, Zabello GD (1969) Maps of salinity and thicknesses of Devonian deposits in the Dnieper-Donets Depression (in Russian). Atlas of lithological-paleogeographical maps of the USSR, vol II, Moscow
Chmyrev VM, Azimi NA, Dronov VI, Slavin VI, Kafarsky AKh, Stazhilo-Alekseev KF (1977) Main features of geological structure of Afghanistan (stratigraphy) (in Russian). Izv Akad Nauk SSSR, Ser Geol 2:29–48
Chochia NG (1955) Geological structure of the Kolva-Vishera area (in Russian). Gostoptekhizdat, Leningrad, 404 pp
Churpin NE, Kovtunov LP, Larchenkov AYa, Alekseeva LP (1968) Recent ideas on stratigraphy and oil-and-gas potential of Devonian deposits in the Dnieper-Donets Depression. In: Materials on geology and oil-and-gas potential of Ukraine. Nedra, Moscow, pp 127–134
Cohee GV (1979) Evaporites in the Michigan Basin. In: Paleotectonic investigations of the Mississippian system in the United States, part II. US Geol Surv Prof Pap 1010-R:437–439
Colman-Sadd SP (1978) Fold development in Zagros simply folded belt, southwest Iran. Am Assoc Pet Geol Bull 62:984–1003
Colton GW (1970) The Appalachian basin – its depositional sequences and their geologic relationships. In: Studies of Appalachian geology: central and southern. Interscience, New York, pp 5–47
Craig LC, Varnes KL (1979) History of the Mississippian system. An interpretative summary. In: Paleotectonic investigations of the Mississippian system in the United States, part II: Interpretative summary and special features of the Mississippian system. US Geol Surv Prof Pap 1010-R:371–406
Crick IH, Muir MD (1979) Evaporites and uranium mineralisation in the Pine Creek Geosyncline. Int Uranium Symp of the Pine Creek Geosyncline, NT Australia. Extended Abstracts, pp 30–33
Crist RP, Hobday M (1973) Diapiric features of the offshore Bonaparte Gulf Basin, Northwest Australian Shelf. Aust Soc Expl Geophys Bull 4:43–66
Crosby EJ (1967) Gulf coast region, chap B; Paleotectonic investigations of the Permian system in United States. US Geol Surv Prof Pap 515:13–15
Dadlez R (1967a) New sections of the Zechstein substratum in north-west Poland (in Polish). Kwart Geol 11:572–586
Dadlez R (1967b) Zum Permomesozoikum und seinem Untergrund in Nordwestpolen. Ber Dtsch Ges Geol Wiss, A Geol Paläont 12:353–368
Dadlez R, Dembowska J (1965) Geological structure of the Pomeranian Paranticlinorium (in Polish). Prace Inst Geol, vol 40. Warszawa, 263 pp

Dadlez R, Marek S (1969) Structural style of the Zechstein-Mesozoic complex in some areas of the Polish Lowland (in Polish). Kwart Geol 13:543–565

D'Ans J (1967) Bemerkungen zu Problemen der Kalisalzlagerstätten. 1. Kali und Steinsalz, vol 4, pp 369–386

Darton NH (1921) Permian salt deposits of the south-central United States. US Geol Suv Bull 715:205–223

Davies WE, Krinsley DB, Nicol AH (1963) Geology of the North Star Bugt area, northwest Greenland. Medd Gron, vol 162, 68 pp

Dawes PR (1976) Precambrian to Tertiary of northern Greenland. In: Escher AE, Watt WS (eds) Geology of Greenland. Geol Surv Greenland, pp 248–303

Dennison JM, Head JW (1975) Sea level variations interpreted from the Appalachian Basin Silurian and Devonian. Am J Sci 275:1080–1120

Depowski S, Peryt TM, Piątkowski TS, Wagner R (1981) Paleogeography versus oil-and-gas potential of the Zechstein Main Dolomite in the Polish Lowland. In: Int Symp Central Eur Permian. Proceedings. Wydawnictwa Geologizcne, Warszawa, pp 587–595

Derevyagin VS (1981) Paleotectonic and paleogeographical conditions of salt accumulation in the North Caspian area during the Lower Permian (in Russian). In: Structure and deposition conditions of salt formations. Nauka, Novosibirsk, pp 36–44

Derevyagin VS, Makarov AS, Kovalsky FI (1981a) Geology and potassium content in the Lower Permian halogenic formation of the North Caspian area (in Russian). In: Structure and deposition conditions of salt formations. Nauka, Novosibirsk, pp 3–15

Derevyagin VS, Morozov GN, Svidzinsky SA (1979) Structure and depositional features of halogenic strata at the Elton salt-dome potash salt deposits (in Russian). Litol I Polezn Iskop 1:122–138

Derevyagin VS, Svidzinsky SA, Sedletsky VI (1981b) Lower Permian halogenic formation of the North Caspian area (in Russian). Izd Rostov GosUniv, 400 pp

Destombes J (1952) Gypse, sel et autres substances des dépots d'évaporation. XIX Congrès Géol Int Monogr région, 3. Ser: Maroc N 1 (Géologie des Gites minéraux Marocains). Rabat, pp 359–370

De Witt W Jr (1979) Evaporites in the Appalachian Basin. In: Paleotectonic investigations of the Mississippian System in the United States, part II. US Geol Surv Prof Pap 1010R:431–432

Diarov MD (1971a) Correlation of anhydrite of salt domes in the central Caspian area on the basis of Ba^{2+}/Sr^{2+} (in Russian). Geokhimia 9:1131–1137

Diarov MD (1971b) Sedimentary features of potash salts in the Caspian Depression (in Russian). In: Problems of geology of Western Kazakhstan. Nauka, Alma-Ata, pp 205–212

Diarov MD (1972) On prospecting for potash salts in the Caspian Depression (in Russian). Sov Geol 5:125–134

Diarov MD, Dogalov AB (1971) On Kungurian bishofite rocks in the North Caspian Depression (in Russian). Izv Akad Nauk Kaz SSSR, Ser Geol 2:72–76

Diarov M, Utarbaev GS, Tukhfatov K (1981) Types and some distribution peculiarities of potash salt deposits in salt-dome structures of the Caspian Depression (in Russian). In: Structure and deposition conditions of potash salt deposits. Nauka, Novosibirsk, pp 124–126

Dickey RI (1940) Geological section from Fisher County through Andrews County, Texas, to Eddy County, New Mexico. Am Assoc Pet Geol Bull 24:37–51

Dietz C (1955) Exkursion zu den Tagesaufschlüssen im Bereich des Sarstedt-Lehrter Salzstockes. Z Dtsch Geol Ges 105:872–875

Dimroth E, Kimberley MM (1976) Precambrian atmospheric oxygen: Evidence in the sedimentary distributions of carbon, sulfur, uranium and iron. Can J Earth Sci 133:1–22

Dittrich E (1962) Zur Gliederung der Werra-Serie (Zechstein 1) im Werra-Kaligebiet. Ber Geol Ges 2:296–301

Dittrich E (1966) Einige Bemerkungen über Rand- und Schwellenausbildungen im Zechstein Südwest-Thüringens. Ber Dtsch Ges Geol Wiss, A Geol Paläont II:185–198

Dixon GH (1967) Northeastern New Mexico and Texas – Oklahoma Panhandles, Chap D. In: Paleotectonic investigations of the Permian System in the United States. US Geol Surv Prof Pap 515:61–80

Dneprov VS, Koltypin SN (1971) Section penetrated by Biikzhal SG-2 well (in Russian). In: Biikzhal superdeep SG-2 well. Sb No 1, Leningrad, pp 16–22

Döhner C (1967) Neue Ergebnisse feinstratigraphischer Untersuchungen im Grenzbereich Kalilager Stassfurt – Grauer Salzton des Südharzreviers. III Int Kali-Symp, part II, Leipzig 1967, pp 99–107

Döhner C (1970a) Die Feinstratigraphie des Stassfurtflözes im Südharz and ihre Parallelisierung mit dem Unstrutgebiet. Z Angew Geol 17:517–538

Döhner C (1970b) Stratigraphie des Stassfurtsalinars und Probleme der Liegend- und Hangendabgrenzung. Ber Dtsch Ges Geol Wiss, A Geol Paläont 15:517–538

Döhner C, Flert K-H, Koch K, Mötzung R (1968) Gegenüberstellung unterschiedlicher Umbildungsprozesse im Kaliflöz „Stassfurt" in der Deutschen Demokratischen Republik. Rep of the 23-rd Sess Int Geol Congr 8:269–282

Donovan DT, Dingle RU (1965) Geology of part of the southern North Sea. Nature 207:1186–1187

Douglas RJW, Gabrielse H, Wheeler JO, Stott DF, Belyea HR (1970) Geology of Western Canada. In: Douglas RJW (ed) Geology and economic minerals of Canada. Geol Surv Can Econ Geol Rep 1:367–488

Dubinina VN (1954) On mineralogy and petrology of Verkhnekamsk deposit (in Russian). In: Materials on petrology of salt accumulation areas. Gostoptekhizdat, Leningrad, pp 3–128

Dunham KC (1948) A contribution to the petrology of the Permian evaporite deposits of northeastern England. Proc Yorkshire Geol Soc 27:217

Dunham KC (1967) England potash beds offer challenge for deep mining. Min World 3:22–24

Eastwood T (1966) Stanfords geological Atlas of Great Britain. Stanford, London, 288 pp

Edgerley DW, Crist RP (1974) Salt and diapiric anomalies in the southeast Bonaparte Gulf Basin. APEA J 14:85–94

Efremov PE, Egorov VN, Kologreeva IA (1982) Structural features of anhydritic strata of the Oka superhorizon with relation to oil-and-gas potential of chemogenic dolomite in the Kuibyshev-Orenburg Transvolga area (in Russian). In: Oil-and-gas potential of old salt accumulation areas. Nauka, Novosibirsk, pp 71–74

Ekiert F (1960) Neue Anschauungen über die Herkunft des in den Sedimenten des unteren Zechsteins auftretenden Kupfers. Freiburger Forsch 79:190–201

Ermakov VA, Isaev AYa, Getmonova EI (1968) Hydrochemical strata on the western fringe of the Caspian Syneclise (in Russian). Geol Nefti I Gaza 5:33–38

Eroshina DM (1981) Late Frasnian stage of salt accumulation in the Pripyat Depression (in Russian). In: Structure and deposition environments of salt formations. Nauka, Novosibirsk, pp 82–91

Eroshina DM, Kislik VZ (1980) On deposition environments of potash salts of the Upper Frasnian salt formation in the Pripyat Depression (in Russian). Sov Geol 10:43–50

Eroshina DM, Vysotsky EA (1972) Structural and compositional features of the Lower Salt Unit in the northern Pripyat Depression (in Russian). In: Regional geology and mineral resources of BSSR. Nauka I Tekhnika, Minsk, pp 45–58

Eroshina DM, Vysotsky EA (1975) Structural and compositional features of the Lower Salt Unit in the Pripyat Depression (in Russian). In: Mineral resources of BSSR. BelNIGRI, Minsk, pp 3–12

Eventov YaS (1957) Paleozoic deposits in the western Caspian Depression: perspectiveness in exploration for oil-and-gas in this area (in Russian). Sov Geol 57:130–153

Eventov YaS (1962) Formation history and features of tectonics of the western Caspian Depression in connection with evaluating its oil-and-gas potential (in Russian). In: Materials on tectonics of the Lower Volga area. Gostoptekhizdat, Moscow, pp 62–81

Fabian H-J (1957) Die Zechsteinprofile im Gebiete Barnsdorf – Diepholz – Rehden – Wagenfeld. Erdöl Kohle (Hamburg) 10:741–747

Fabian H-J (1958) Die Faziesentwicklung des Zechsteins zwischen Bielefeld und Hameln. Geol Jb 73:127–134

Falcon NL (1974) Southern Iran: Zagros Mountains. In: Spencer AM (ed) Mesozoic-Cenozoic orogenic belts. Academic Press, London, pp 199–211

Fay RO (1962) Stratigraphy and general geology of Blaine County. In: Geology and mineral resources of Blaine County, Oklahoma. Okla Geol Surv Bull 89:10–99

Fergusson WB, Prather BA (1968) Salt deposits in the Saline Group in Pennsylvania. Pennsylvania Topog Geol Surv, 4th Ser Bull 58 (Miner Res Rep), 41 pp

Fiveg MP (1948) On the annual cycle of rock salt sedimentation in the Verkhnekamsk deposit (in Russian). Dokl Akad Nauk SSSR 61:1087–1090

Fiveg MP (1955a) Deposition environments of potash salts (in Russian). Byull MOIP, Otd Geol 30:3–15

Fiveg MP (1955b) On deposition environments of the Verkhnekamsk salt series (in Russian). Tr VNIIG 30:182–194

Fiveg MP (1959) On the structures of lower rock salt of the Verkhnekamsk deposit (in Russian). Tr VNIIG 35:244–250

Fleck A (1950) Deposits of potassium salt in North-East Yorkshire. Chem Ind, 17 October, F1–F15

Flügel E (1981) The Permian of the Carnic Alps – depositional environments and microfacies. In: Int Symp Central Eur Permian. Proceedings. Wydawnictwa Geologiczne, Warszawa, pp 84–94

Forman DJ, Wales DW (1981) Geological evolution of the Canning Basin, Western Australia. Bur Min Res Aust Bull 210: 91 pp

Forsh NN (1951) Stratigraphy and facies of the Kazanian in the Middle Volga area (in Russian). Tr VNIGRI, Nov Ser 45:34–80

Forsh NN (1955) Volga-Uralian oil-bearing area. Permian deposits. The Ufimian Formation and the Kazanian Stage (in Russian). Tr VNIGRI, Nov Ser 92: 156 pp

Forsh NN (1979) Structure of Permian halogenic formation in the Caspian Depression (in Russian). In: Structural features and oil-and-gas potential of sedimentary rocks in the Caspian Depression. VNIGRI, Leningrad, pp 31–40

Forsh NN, Goryacheva LP (1979) On oil-and-gas potential of the Podsolevye Permian deposits at the eastern flank of the Caspian Depression (in Russian). In: Structural features and oil-and-gas potential of sedimentary rocks in the Caspian Depression. VNIGRI, Leningrad, pp 47–58

Frisch TO, Moegan WC, Dunning GR (1978) Reconnaissance geology of the Precambrian Shield on Ellesmere and Coburg Islands, Canadian Arctic Archipelago. Current Research, part A. Geol Surv Can Pap 78-1A:135–138

Fritz WC, Fitzgerald J (1940) South-north cross section from Pecos County through Ector County, Texas, to Roosevelt County, New Mexico. Am Assoc Pet Geol Bull 24:15–28

Fuchtbauer H (1964) Fazies, Porosität und Gasinhalt der Karbonatgesteine des norddeutschen Zechsteins. Z Dtsch Geol Ges 114:484–531

Fuchtbauer H (1972) Influence of salinity on carbonate rocks in the Zechstein Formation, North-Western Germany. In: Geology of saline deposits. Proc Hannover Symp 1968 (Earth Science 7). Unesco, Paris, pp 23–31

Fuchtbauer H, Goldschmidt H (1965) Beziehungen zwischen Calciumgehalt und Bildungsbedingungen der Dolomite. Geol Rundsch 55:29–40

Fuller JGCM (1961) Ordovician and contiguous formations in North Dakota, South Dakota, Montana and adjoining areas of Canada and United States. Am Assoc Pet Geol Bull 45:1334–1363

Gailite LK (1963) Distribution of gypsum in the Frasnian deposits of the Latvian SSR (in Russian). In: Frasnian deposits of the Latvian SSR. Izd Akad Nauk Latv SSR, Riga, pp 201–211

Galabuda NI (1978) Geological structure of Frasnian salt formation of the Dnieper-Donets Depression (in Russian). Geol I Geochem Gor Iskop 50:59–68

Galley JE (1958) Oil and geology of the Permian basin of Texas and New Mexico. In: Habitat of oil – a symposium. Am Assoc Pet Geol, pp 395–446

Galloway MC (1970) Adavale, Queensland – 1:250,000 Geological Series. Bur Miner Res Aust, Explanatory Notes SG/55-5

Gansser A (1964) Geology of the Himalayas. Wiley, New York, 289 pp

Garlick WG (1981) Sabkhas, slumping, and compaction of Mufulira, Zambia. Econ Geol 76: 1817–1847

Gavrish VK (1963) On Permian halogenic deposits of the Dnieper-Donets Depression (in Russian). Dokl Akad Nauk SSSR 152:175–178
Gavrish VK (1964) Lower Permian stratigraphy of the Dnieper-Donets Depression (in Russian). Sov Geol 2:124–129
Gavva TI, Ryb'yakov BL (1981) Potash shows at the Bratsk rock salt deposit (in Russian). In: Structure and deposition environment of potash salt deposits. Nauka, Novosibirsk, pp 47–50
Gelfand MS, Karpenko YuA (1968) New data on Kungurian age of halogenic sediments in the central Ural-Volga Basin (in Russian). In: Geology and oil-and-gas potential of some regions of the USSR related to problems of underground gas storage. Nedra, Moscow, pp 178–180
Geology of oil-and-gas of the Siberian Platform (1981) (in Russian). Kontorovich AE, Trofimuk AA (eds). Nedra, Moscow, 552 pp
Geology and mineral resources of Afghanistan (1980) (in Russian). Book 1 – Geology, 535 pp, Book 2 – Mineral resources, 336 pp. Nedra, Moscow
Gerhard LC, Anderson SB, Le Fever JA, Carlson CG (1982) Geological development, origin, and energy mineral resources of Williston Basin, North Dakota. AAPG Bull 66:989–1020
Gerrard TA (1966) Environmental studies of Fort Apache Member, Supai Formation, east-central Arizona. Am Assoc Pet Geol Bull 50:2434–2463
Gill D (1977) Salina A-1 sabkha cycles and the Late Silurian paleogeography of the Michigan basin. J Sed Pet 47:979–1017
Gill D (1979) Differential entrapment of oil and gas in Niagaran Pinnacle – Reef Belt of Northern Michigan. AAPG Bull 63:608–620
Gill D, Briggs LI, Briggs DL (1978) The Cain Formation; a transitional succession from open marine carbonates to evaporites in a deep water basin, Silurian, Michigan basin (abs): 10th IAS Int Congr Sed Abs, vol 1, pp 244–245
Glennie KW, Beuf MGA, Hughes CMW, Moody-Stuart M, Pilaar WFH, Reinhardt BM (1974) Geology of the Oman Mountains. Ned Geol Mijnbouwkd Genoot, Verh 31:423 pp
Glover JE (1973) Petrology of the halite-bearing Carribuddy Formation, Canning Basin, Western Australia. J Geol Soc Aust 20:343–359
Glushko VV, Dickenstein GKh, Kalinko MK, Maksimov SP, Takaev YuG, Fedorov DL (1976) Features of oil-and-gas-bearing areas in North-Western Europe (in Russian). Sov Geol 12:127–134
Golding LY, Walter MR (1979) Evidence of evaporite minerals in the Archaean Black Flag Beds, Kalgoorlie, Western Australia. BMR J Aust Geol Geophys 4:67–72
Golubev BM, Brovchenko OF (1981) Variability pattern of structural and textural features of sylvinite and rock salt beds in the Verkhnekamsk potash salt deposit area based on underground mapping (in Russian). In: Structure and deposition environments of potash salts. Nauka, Novosibirsk, pp 100–106
Golubtsov VK (1961) Lower Permian deposits in the Pripyat Trough (in Russian). In: Materials on geology and gas potential of Lower Permian deposits of the southern Russian Platform. Izd Kharkov Univ, Kharkov, pp 299–317
Golubtsov VK (1966) Brest Depression (in Russian). In: Stratigraphy of the USSR . Permian system. Nedra, Moscow, pp 160–161
Golubtsov VK, Poznyakevich ZA (1963) On stratigraphy of Devonian sub-salt (podsolevye) deposits of the Pripyat Trough (in Russian). In: Geology and hydrogeology of the Pripyat Trough. Izd Akad Nauk BSSR, Minsk, pp 3–19
Gordievich VA, Sanarov IV (1961). Lower Permian deposits and oil-and-gas potential of the southeastern Dnieper-Donets Depression (in Russian). In: Materials on geology and gas potential of the Lower Permian deposits in the southern Russian Platform. Izd Kharkov Univ, Kharkov, pp 281–290
Gordon-Yanovsky FA, Bespalov IM (1964) On deposition environments of Lower Permian sediments in the Donets Basin (in Russian). Dokl Akad Nauk SSSR 159:109–110
Gottesmann W (1964) Zur Mineralogie des Werraanhydrits in Südbrandenburg. Wiss Z Humboldt Univ Berlin, Mathem Naturwiss R, Jg XIII:167–169

Graifer BI, Romanov PI, Zalkind IE (1962) Stratigraphy and lithofacies of the Kungurain in the Kama area at Perm (in Russian). In: Stratigraphic schemes of Paleozoic deposits. Permian. Gostoptekhizdat, Moscow, pp 98–105

Grebennikov NP, Ermakov VA (1980) On some distribution pattern of potash and magnesium salts in the western Caspian Depression (in Russian). In: Peculiar structure of bishofite and potash salt deposits. Nauka, Novosibirsk, pp 66–70

Greenwood JEGW, Bleackley D (1967) Geology of the Arabian Peninsula. Aden Protectorate. US Geol Surv Prof Pap 560-C: 96 pp

Grigoriev VN, Repina LN (1956) Stratigraphy of Cambrian deposits at the western margin of the Siberian Platform. Izv Akad Nauk SSSR, Geol Ser 7:17–24

Haase G (1966) Die Bedeutung neuer geologischer Erkenntnisse für die Einschätzung der Entwicklungsperspektive des Kalibergbaues im thüringischen Werragebiet. Ber Dtsch Ges Geol Wiss, RA 11:425–437

Hajash GM (1967) The Abu sheikhdom – the onshore oil-fields history of exploration and development. Proc 7th World Pet Congr, Mexico, vol 2

Hall WE (1960) Upper Permian correlations in southeastern New Mexico and adjacent part of West Texas. In: Geology of the Delaware basin and field trip guidebook. Midland, Texas, West Texas Geol Soc, pp 85–87

Hallagarth WE (1967a) Environment of western Colorado and southern Utah, intervals A and B (Pl 10, Maps B–C). In: Paleotectonic maps of the Permian System. US Geol Surv Misc Geol Inv Map 1–450, pp 54–56

Hallagarth WE (1967b) Western Colorado, southern Utah, and northwestern New Mexico, chapter I. In: Paleotectonic investigations of the Permian System in the United States. US Geol Surv Prof Pap 515:171–197

Ham WE (1958) Stratigraphy of the Blaine Formation in Beckham County, Oklahoma. Okla Acad Sci Proc 38:88–93

Ham WE (1960) Middle Permian evaporites in south-western Oklahoma. 21-st Int Geol Congr Rep, part 12, pp 138–151

Ham WE (1962) Economic geology and petrology of gypsum and anhydrite in Blaine County. In: Geology and mineral resources of Blaine County, Oklahoma. Okla Geol Surv Bull 89:100–151

Ham WE (1963) Cyclicity in Permian evaporites of western Oklahoma (abs). Am Assoc Pet Geol Bull 47:359

Ham WE, Curtis J (1958) Gypsum in Weatherford-Clinton district, Oklahoma. Okla Geol Min Rep 35: 32 pp

Hartwig G (1958) Zur Feinstratigraphie des Stassfurt-Lagers in der Thüringer Mulde. Kali und Steinsalz 2:206

Harwood G (1981) Barytes mineralisation related to hydraulic fracturing in English Permian Z1 carbonates. Int Symp Central Eur Permian. Proceedings. Wydawnictwa Geologiczne, Warszawa, pp 379–387

Havlena V (1966a) Upper Carboniferous and Permian Basins. The Upper Silesian Basin. In: Regional geology of Czechoslovakia, part 1. The Bohemian Massif. Czechoslovak Ac Sci Prague, pp 414–433

Havlena V (1966b) The Carboniferous of Central Bohemia and the Permo-Carboniferous of the Krušne Hory Mountains; the Blanice and Boskovice Furrows; the Central Sub-Cretaceous Basin. In: Regional geology of Czechoslovakia, part 1. The Bohemian Massif. Czechoslovak Ac Sci, Prague, pp 452–467

Haynes SJ, McQuillan H (1974) Evolution of the Zagros suture zone, southern Iran. Geol Soc Am Bull 85:739–744

Hecht G (1960) Über Kalkalgen aus dem Zechstein Thüringens. Freiberger Forsch H, C 89, Berlin, pp 125–176

Helmuth H-J (1968a) Zur Gliederung des Zechsteinkalkes (Cal) in NE-Mecklenburg. Geologie (Berlin) 17:164–175

Helmuth H-J (1968b) Zur Fazies der Oberrotliegenden in NE-Mecklenburg. Geologie (Berlin) 17: 52–59

Hemmann M (1970) Zur Feinstratigraphie und stratigraphischen Korrelation im Hauptanhydrit und Leinesteinsalz. Ber Dtsch Ges Geol Wiss, A Geol Paläont 15:543–554

Herrmann A (1956) Der Zechstein am südwestlichen Harzrand. Geol Jb 72:1–72

Herrmann A, Hinze C, Stein V (1967) Die halotektonische Deutung der Elfas-Überschiebung im südniedersächsischen Bergland. Geol Jb 84:407–462

Herrmann A, Richter-Bernburg G (1955) Frühdiagenetische Störungen der Schichtung und Lagerung im Werra-Anhydrit (Zechstein 1) am Südwestharz. Z Dtsch Geol Ges 105:669–702

Heybroek L, Haanstra U, Erdman DA (1967) Observations on the geology of the North Sea area. In: 7-th World Pet Congr, vol 2, Paneb discussion 9 (11). Mexico, pp 905–916

Heynke A, Zänker G (1970) Zur Ausbildung und Leitbankgliederung des Stassfurtsteinsalzes im Südharz-Kalirevier. Z Angew Geologie 16:344–356

Hills J (1968) Permian Basin, Field area, West Texas and Southeastern New Mexico. In: Saline deposits. Geol Soc Am Spec Pap 88:17–27

Hinz K (1968) A contribution to the geology of the North Sea according to geophysical investigations by the Geological Survey of German Federal Republic. In: Geology of shelf seas. Oliver and Boyd, Edinburgh London, pp 55–71

Hodenberg R, Kuhn R (1967) Zur Kenntnis der Magnesiumsulfathydrate und der Effloreszenzen des Keiserifs von Hartsalzen. Kali und Steinsalz 4:326–430

Hoffman PF (1968) Stratigraphy of the Lower Proterozoic (Aphebian) Great Slave Supergroup, East Arm of Great Slave Lake, District of Mackenzie. Geol Surv Can Pap 68: 93 pp

Hofrichter E (1967) Suberosion und Bodensenkung am Salzstock von Stade. Geol Jb 84:327–340

Hoots WH (1925) Geology of a part of western Texas and southeastern New Mexico with special reference to salt and potash. US Geol Surv Bull 780 B:33–126

Hoppe W (1958) Die Bedeutung der geologischen Vorgänge bei der Metamorphose der Werra-Kalisalzlagerstätte. Freiberger Forsch A 123:41–60

Hoppe W (1960) Die Kali- und Steinsalzlagerstätte des Zechstein in der DDR. Teil 1: Das Werragebiet. Freiberger Forsch 97/I: 32 pp

Hoyt JH (1962) Pennsylvanian and Lower Permian of northern Denver basin, Colorado, Wyoming, and Nebraska. Am Assoc Pet Geol Bull 46:46–59

Hoyt JH (1963) Permo-Pennsylvanian correlation and isopach studies in the northern Denver basin, Wyoming. In: Guidebook to the geology of the northern Denver basin and adjacent uplifts. Rocky M Assoc Geol 14-th Field Conf, Denver, Colorado. Rocky M Assoc Geol, pp 68–83

Huh JM (1973) Geology and diagenesis of the Salina-Niagaran pinnacle reefs in the northern shelf of the Michigan basin. PR D thesis, Univ Michigan, 253 pp

Ignatiev VI (1967) Permian system. Tatarian (in Russian). In: Geology of the USSR, vol XI (Volga and Kama areas). Nedra, Moscow, pp 395–427

Ivanov AA (1932) Verkhnekamsk potash salt deposits (in Russian). Tr Vses Geol Razved Ob'edin 232: 154 pp

Ivanov AA (1965) Permian salt basins of the Pechora-Kama Cis-Ural area (in Russian). Red Izd Otdel Sib Otdelen Akad Nauk SSSR, 98 pp

Ivanov AA, Voronova ML (1968) Geology of the Upper Pechora salt basin and its potassium content (in Russian). In: Geology of salt and potash deposits. Nedra, Leningrad, pp 3–79

Ivanov AV, Fotieva NN, Osipova RP, Konovalova MV (1959) Stratigraphy and oil-and-gas potential of Permian deposits in the southeastern Pechora Depression and the Upper Pechora Basin (in Russian). In: Geology and oil-and-gas potential of the Timan-Pechora area. Gostoptekhizdat, Leningrad, pp 204–232

Ivanov YuA, Gusev ON, Efremova GD (1969) Upper Permian deposits in the eastern Caspian Depression and marginal areas (in Russian). In: Permo-Triassic of the Russian Platform in relation to oil-and-gas potential. Nedra, Moscow, pp 106–112

Jackson GD, Ianelli TR (1981) Rift-related cyclic sedimentation on the Nechelikian Borden Basin, northern Baffin Island. In: Campbell FHA (ed) Geol Surv Can Pap 81-10:269–302

Jackson MJ, Van de Graaff WJE (1981) Geology of the Officer Basin in Western Australia. Bur Min Res Aust Bull 206: 102 pp

Jackson PR (1966) Well completion report, No. 2 Yowalga, Officer Basin, Western Australia. Bur Min Res Aust, Pet Search Subside Acts File 66:4191

Jahne H (1966) Der Aufschluß des Zechsteins im Schacht "Marx-Engels" II (Werrakaligebiet). Ber Dtsch Ges Geol Wiss, A Geol Paläont 11:439–460

Jahne H, Oettel S, Voitel R (1970) Die feinstratigraphische Gliederung des Salinars im Zechstein 1 des Werra-Kaligebietes. Ber Dtsch Ges Geol Wiss, A Geol Paläont 15:505–515

Jahne H, Pielert P (1966) Zur stratigraphisch-petrographischen Gliederung des Kaliflözes "Hessen" im Bereich der Grube "Marx-Engels" Unterbreizbach (Werrakaligebiet). Ber Dtsch Ges Geol Wiss, A Geol Paläont 11:461–474

Jankowski G, Jung W (1962) Zum Zechsteinkalk (Ca1) im Bereich der Sangerhäuser und Mansfelder Mulde. Geologie 11:943–953

Jankowski G, Jung W (1964a) Die Ausbildung des Stinkschiefers (Ca2st) im südöstlichen Harzvorland. Geologie 13:929–941

Jankowski G, Jung W (1964b) Erweiterung der Feinstratigraphie der Werraanhydrite (Z1) im südöstlichen Harzvorland. Geologie 13:1091–1098

Jankowski G, Jung W (1966) Deszendenzen im Leinesteinsalz des südöstlichen Harzvorlandes und ihre Beziehungen zu den "Auslangungsresten". Ber Dtsch Ges Geol Wiss, A Geol Paläont 11: 199–208

Johnson KS (1963a) Brine-well production of Permian salt at Sayre, Beckham County, Oklahoma. Okla Geol Notes 23:83–93

Johnson KS (1963b) Salt in the El Reno Group (Permian), Elk City area, Beckham and Washita Counties, Oklahoma. In: Western Oklahoma and adjacent Texas. Univ Oklahoma Geol Symp, 8-th Bien Proc. Norman, Oklahoma, pp 79–92

Jones CL (1959) Potash deposits in the Carlsbad region, southeastern New Mexico (abs). Econ Geol 54:1349; Geol Soc Am Bull 70:1625

Jones CL (1965) Petrography of evaporites from the Wellington Formation near Hutchinson, Kansas. US Geol Surv Bull 1201-A:1–70

Jones CL (1968) Some geologic and petrographic features of Upper Permian evaporites in southeastern New Mexico (abs). In: Saline deposits. Geol Soc Am Spec Pap 88:538–539

Jones CL (1972) Permian basin potash deposits, southwestern United States. In: Geology of saline deposits. Proc Hannover Symp 1968 (Earth Sciences 7). Unesco, Paris, pp 191–201

Jones CL, Madsen BM (1972) Evaporite geology of fifth ore zone Carlsbad district southeastern New Mexico. US Geol Surv Bull 1252-B:1–21

Jong KA (1967) Paläogeographie des ostalpinen oberen Perms, Paläomagnetismus und Seitenverschiebungen. Geol Rundsch 56:103–115

Jordan L (1964) The geology and economic future of the salt deposits of Oklahoma (abs). Tulsa Geol Soc Dig 32:170–171

Jordan L, Vosburg DL (1963) Permian salt and associated evaporites in the Anadarko basin of the western Oklahoma-Texas Panhandle. Okla Geol Surv Bull 102: 76 pp

Jung W (1958a) Zur Feinstratigraphie der Werraanhydrite (Zechstein 1) im Bereich der Sangerhäuser und Mansfelder Mulde. Geologie Beih 24:88

Jung W (1958b) Zur Feingliederung des Basalanhydrits (Z2) und des Hauptanhydrits (Z3) im SE-Harzvorland. Geologie 9:526–555

Jung W (1966) Nochmals zum Sangerhäuser Anhydrit (Zechstein 2). Geologie 15:443–460

Jung W, Lerenz S (1968) Die Ausbildung des Stassfurtsteinsalzes (Na2) im südöstlichen Harzvorland. Bergakademie 20:509–515

Jungwith J, Sellert J (1966) Zur Stratigraphie und Facies des Zechsteins in Südwest-Thüringen. Geologie 15:421–433

Jurkiewicz H (1970) Lithologic development of Zechstein deposits in the western part of Nida Trough. Kwart Geol 14:79–87

Jux U, Schultz G (1971) Zur Altersfrage der Salzlagerstätte von Rukh, Afghanistan (The age of the salt deposit of Rukh, Afghanistan). N Jb Geol Paläont Mh 3:157–170

Kaczmarek A (1970) Gypsum-anhydrite deposits at Gierałtow near Lubań Slaşki (in Polish). Kwart Geol 14:65–78

Kahle CF, Floyd JC (1972) Geology of Silurian rocks, north-western Ohio. Ohio Geol Soc Guidebook, 91 pp

Kamen-Kaye M (1970) Geology and productivity of Persian Gulf Synclinorium. Am Assoc Pet Geol Bull 54:2371–2394
Kaptsan VKh, Safarov EI (1969) Devonian system (in Russian). In: Geology of the USSR, vol XLV (Moldavian SSR). Nedra, Moscow, pp 88–91
Karpov PA (1970) Devonian system (in Russian). In: Geology of the USSR, vol XLVI. Nedra, Moscow, pp 65–111
Katz HR (1961) Late Precambrian to Cambrian stratigraphy in East Greenland. In: Raash GO (ed). Geology of the Arctic. Toronto Univ Press, pp 299–328
Katzung G (1966) Sedimentation und Paläogeographie des tieferen Unterrotliegenden im SE-Teil des Thüringer Waldes. Ber Dtsch Ges Geol Wiss, A Geol Paläont 11:131–136
Katzung G (1968) Rotliegendes. In: Grundriß der Geologie der Deutschen Demokratischen Republik, Bd I, S 201–218
Katzung G (1970) Das Permosiles im Südteil der DDR. Ber Dtsch Ges Geol Wiss, A Geol Paläont 15:7–27
Katzung G (1972) Stratigraphie und Paläogeographie des Unterperms in Mitteleuropa. Geologie 21:570–584
Kayak KF (1962) On geology of Southeastern Estonia from deep drilling data (in Russian). In: Paleozoic Geology. Tr Akad Nauk Eston SSR, Tallinn, pp 33–40
Kel'bas BI (1963) Lithologic composition and thicknesses of the Lower Permian salt-bearing formation in the area of the Shebelino, Balakleya and Chervony Donets Uplifts (in Russian). In: Geology and petroleum province of Ukraine. Gostoptekhizdat, Moscow, pp 43–48
Kelley DG (1970) Geology of Southeastern Canada. Carboniferous and Permian. In: Douglas RJW (ed) Geology and economic minerals of Canada. Geol Surv Can Econ Geol Rep, pp 284–296
Kent PE (1967a) Progress of exploration in North Sea. Am Assoc Pet Geol Bull 51:731–741
Kent PE (1967b) Outline geology of the southern North Sea Basin. Proc Yorks Geol Soc 36:1–22
Kent PE (1968) Geological problems in North Sea exploration. In: Geology of Shelf Seas. Edinburgh London, pp 75–86
Kent PE (1980) The structural framework and history of subsidence of the North Sea Basin. In: Geology of Europe (from Precambrian to post-Hercynian sedimentary basins). Mem BRGM 108:281–288
Kent PE, Walmsley PJ (1970) North Sea progress. Am Assoc Pet Geol Bull 54:168–181
Kerkmann K (1966) Zur Kenntnis der Riffbildungen in der Werraserie des thüringischen Zechsteins. Freib Forsch H 213:123–143
Kerr SD Jr, Thomson A (1963) Origin of nudular and bedded anhydrite in Permian shelf sediments, Texas and New Mexico. Am Assoc Pet Geol Bull 47:1726–1732
Ketner KB (1967) West Coast Region, Chapter K. In: Paleotectonic investigations of the Permian system in the United States. US Geol Surv Prof Pap 515:225–238
Khatyanov FI (1968a) Geological and geophysical features of buried reef masses related to petroleum exploration (in Russian). In: Fossil reefs and methods used in their study. Sverdlovsk, pp 226–247
Khatyanov FI (1968b) On relation of reef masses and anticlinal folds with Kungurian salt structures in the Bashkiria-Orenburg Cis-Urals area (in Russian). Tr VNIGNI 59:87–94
Khavin EI (1971) Permian system (in Russian). In: Geology of the USSR, vol I (Leningrad, Pskov and Novgorod regions). Nedra, Moscow, pp 295–297
Khomenko VA (1977) Lithology of Devonian deposits in the Dnieper-Donets Depression (in Russian). Naukova Dumka, Kiev, 146 pp
Khu Bin, Wang Jing-bin, Gao Zhen-jca, Fang Xiao-di (1965) Some geological problems of the Paleozoic of the Tarim Platform (in China). Geol J China 45:131–141
Khursik VZ, Shirinkina AP, Zolotova VP (1970) Lower boundary of Kungurian in the Cis-Urals area at Perm (in Russian). Sov Geol 6:118–124
Kijewski P, Salski W, Tomaszewski J (1979) Das Auftreten von Steinsalzen in Zechsteinablagerungen im Südwestteil der Vorsudetischen Monoklinale. Z Geol Wiss 7:879–889
King PB (1942) Permian of West Texas and southeastern New Mexico. Am Assoc Pet Geol Bull 26:535–763

King PB (1948) Geology of the southern Guadalupe Mountains, Texas. US Geol Surv Prof Pap 215: 183 pp
King PB (1967) Reefs and associated deposits in the Permian of West Texas. In: Paleotectonic maps of the Permian system in the United States. US Geol Surv Misc Geol Inv Map 1–450, pp 36–44
King RH (1947) Sedimentation in Permian Castile Sea. Am Assoc Pet Geol Bull 31:470–477
Kirkinskaya VN, Vasilevsky AF, Datsenko VA et al (1975) Paleogeography of the Siberian Platform in Early Cambrian-Amga time (Middle Cambrian) (in Russian). In: Paleogeography of Late Proterozoic and Paleozoic basins of the Siberian Platform. VNIGRI, Leningrad, pp 62–83
Kityk VI, Galabuda NI (1975) Devonian salt-bearing formations in the Dnieper-Donets Depression and deposition environments (in Russian). Geol I Geochem Gor Iskop 45:14–23
Kityk VI, Galabuda NI (1981) Paleogeography of Devonian salt accumulation periods in the Dnieper-Donets Depression (in Russian). In: Structure and deposition environments of salt-bearing formations. Nauka, Novosibirsk, pp 74–82
Kityk VI, Galabuda NI, Petrichenko OI, Shaidetskaya VS (1980) On the study of Lower Devonian salt accumulation in the Pripyat Trough and Dnieper-Donets Depression (in Russian). In: Lithology and geochemistry of salt strata. Naukova Dumka, Kiev, pp 77–95
Kłapciński J (1964) The paleogeography of the Fore-Sudetic Monocline during the Zechstein age (in Polish). Roczn Polskiego Towarz Geol 34:551–573
Kłapciński J (1967) Stratigraphy of the Stassfurt Anhydrites in the vicinities of Lubin and Sieroszowice (Fore-Sudetic Monocline) (in Polish). Kwart Geol 11:303–312
Klaus W (1955) Über die Sporendiagnose des deutschen Zechsteinsalzes und des alpinen Salzgebirges. Z Dtsch Geol Ges 105:756–788
Knak I (1960) Feinstratigraphische Aufnahme des Kalilagers Stassfurt auf den Schachtanlagen Neustassfurt VI/VII, Berlepsch-Maybach und Ludwig II. Freib Forsch H 90:7–51
Knoth W (1970) Zur Lithologie und Paläogeographie des höheren Rotliegenden im Thüringer Wald. Ber Dtsch Ges Geol Wiss, A Geol Paläont 15:47–65
Kölbel F (1961) Die Entwicklung des Zechsteins in Südbrandenburg. Z Angew Geol 7:58–63
Kolchanov VP, Kulakov VV, Mikhailov KYa, Pashkov BP (1971) New data on stratigraphy of Precambrian and Paleozoic formations in northern piedmont areas of Western Hindu Kush (in Russian). Sov Geol 3:130–136
Komissarova IN (1981) Recognition of potassium-bearing zones within the Caspian Depression from paleotectonic evidence (in Russian). In: Structure and deposition environments of potash salt deposits. Nauka, Novosibirsk, pp 126–133
Konishchev VS (1980) Tectonics of halokinesis areas on ancient platforms (South American, African-Arabian and Australian) (in Russian). Nauka I Tekhnika, Minsk, 240 pp
Konitz O (1966) Zur Feinstratigraphie des Kalisalzflözes „Thüringen". Ber Dtsch Ges Geol Wiss RA Geol Paläont 11:475–487
Konstantynowicz E (1965) Signs of mineralization in the Zechstein of the north Sudetic Syncline (in Polish). Prace Geol 28, Warszawa, 100 pp
Kontorovich AE, Evtushenko VM, Ivlev NB, Larichev AI (1981) Regularities in accumulation of organic matter on the Siberian Platform during Precambrian and Cambrian (in Russian). In: Lithology and geochemistry of oil-and-gas-bearing strata on the Siberian Platform. Nauka, Moscow, pp 19–42
Kopnin VI (1962) On salt content of Upper Kungurian rocks near the Verkhnekamsk salt deposit (in Russian). Dokl Akad Nauk SSSR 144:1123–1125
Kopnin VI (1963) On areal distribution of potassium chloride in sylvinite beds of the Verkhnekamsk deposit (in Russian). Dokl Akad Nauk SSSR 149:416–419
Kopnin VI (1965) Regular features of the Verkhnekamsk salt deposit formation. Avtoref Kand Diss, Moscow, 26 pp
Kopnin VI, Korotaev MA (1981) Stratification of salt strata in the Verkhnekamsk potash salt deposit (in Russian). In: Structure and deposition environments of potash salt deposits. Nauka, Novosibirsk, pp 79–94
Kopnin VI, Kukharchuk AK (1980) Detailed study of Krasnyi II sylvinite bed structure in the southern part of the Verkhnekamsk deposit (in Russian). In: Structural features of bishofite and potash salt deposits. Nauka, Novosibirsk, pp 33–44

Kopnin VI, Maloshtanova NE (1980) On mineral composition of sylvinite ores, the Verkhnekamsk deposit (in Russian). In: Structural features of bishofite and potash salt deposits. Nauka, Novosibirsk, pp 44–47

Korenevsky SM (1970) Distribution pattern of lithologo-facies complexes of halogenic formations (in Russian). In: State of the art and tasks of Soviet lithology, vol III. Nauka, Moscow, pp 32–39

Korenevsky SM (1981) Halogenic formation of the Russian Platform: typization, distribution pattern, and the associated mineralization (in Russian). In: Main problems of salt accumulation. Nauka, Novosibirsk, pp 122–131

Korenevsky SM, Poborsky YuV (1981) Zechstein halogenic formation in the Gdansk-Kaliningrad side Basin and its potassium content (in Russian). In: Structure and deposition environments of potash salt deposits. Nauka, Novosibirsk, pp 161–171

Korenevsky SM, Voronova ML (1966) Geology and deposition environments of potash deposits in the Caspian Syneclise and South Cis-Uralian Trough (in Russian). Nedra, Moscow, 280 pp

Korenevsky SM, Bobrov VP, Supronyuk KS, Khrushchov DP (1968) Halogenic formations of the northwestern Donets Basin and Dnieper-Donets Depression and their potassium content (in Russian). Nedra, Moscow, 239 pp

Korenevsky SM, Kazanov YuV, Protopopov AL (1980) Potassium content of Zechstein halogenic deposits in the Baltic area (in Russian). Lithol I Polezn Iskop 6:116–121

Korenevsky SM, Protopopov AL, Shaporev AA (1983) Kainite deposits in Zechstein of the Baltic area (in Russian). Lithol I Polezn Iskop 1:71–80

Korich D (1967) Eruptivgesteine im Rotliegenden des Nordteils der DDR. Ber Dtsch Ges Geol Wiss, A Geol Paläont 12:231–242

Kosmahl W (1967) Grauer und Hauptanhydrit des Zechsteins in Nordwestdeutschland. Geol Jb 84:367–406

Kotlyarov AM, Tsaregradsky VA, Kolik AL (1965) Deposition history of halogenic formations in the Dzhezkazgan-Sarysu Depression related to oil-and-gas potential (in Russian). Vestn Akad Nauk Kaz SSR 6:53–59

Kottlowski FE (1965) Southwestern New Mexico Basins. In: Rocky Mountains sedimentary basins. Am Assoc Pet Geol Bull 49:2120–2139

Kovalsky FI (1981) Halogenic formation of salt domes in North Caspian area (in Russian). In: Main problems of salt accumulation. Nauka, Novosibirsk, pp 131–134

Kovanko ND (1962) New data on correlation of Permian sections in the Orenburg Region (in Russian). In: Stratigraphic scales of Paleozoic deposits. Permian System. Gostoptekhizdat, Moscow, pp 181–183

Kovanko ND, Larionova EN, Sofronitsky PA (1939) Kungurian and Kazanian deposits of the Kama area at Perm (in Russian). Izv Akad Nauk SSSR, Ser Geol 5:111–127

Krason J (1962) Sedimentation cycles of the Lower Zechstein (in Polish). Przegl Geol 6:284–288

Krason J (1964) Stratigraphic division of North-Sudetic Zechstein in the light of facial investigations (in Polish). Geol Sudetica 1:221–256

Kroenlein GA (1939) Salt, potash, and anhydrite in Castile Formation of southeast New Mexico. Am Assoc Pet Geol Bull 23:1682–1693

Kröll D, Nachsel G (1967) Zur Ausbildung des Stassfurt-Steinsalzes im Südharz-Kalirevier. Geologie 16:269–279

Kühn R (1955) Tiefenberechnung des Zechsteinmeeres nach dem Bromgehalt der Salze. Z Dtsch Geol Ges 105:646–663

Kühn R, Dellwig LF (1971) Salt deposits of Permian, Triassic and Tertiary age in West Germany. In: Sedimentology of parts of Central Europe (Guidebook to excursions held during the VIII Int Sediment Congr 1971 in Heidelberg, Germany), pp 303–326

Kulstad RO, Fairchild PW, McGregor DJ (1956) Gypsum in Kansas. Kans State Geol Surv Bull 113: 110 pp

Kumpan AS, Dobretsov GL, Mitrofanova KV (1963). Upper Paleozoic formations of East Kazakhstan (in Russian). Nedra, Leningrad, 212 pp

Lambert IB, Donnolly TH, Dunlop JSR, Groves DI (1980) Stable isotopic compositions of early Archean sulphate deposits of probable evaporitic and volcanogenic origins. Nature 276:808–811

Lang WTB (1935) Upper Permian formation of Delaware Basin of Texas and New Mexico. Am Assoc Pet Geol Bull 19:262–270

Lang WTB (1937) The Permian formations of the Pecos Valley of New Mexico and Texas. Am Assoc Pet Geol Bull 21:833–898

Lang WTB (1939) Salado Formation of the Permian Basin. Am Assoc Pet Geol Bull 23:1569–1572

Lang WTB (1941) New source for sodium sulfate in New Mexico. Am Assoc Pet Geol Bull 25:152–160

Lang WTB (1942) Basal beds of Salado Formation in Fletcher core test, near Carlsbad, New Mexico. Am Assoc Pet Geol Bull 26:63–79

Lang WTB (1950) Comparison of the cyclic deposits of the Castile and the Salado Formations of the Permian of the southwest (abs). Geol Soc Am Bull 61:1479–1480

Langbein R (1965) Zur Petrographie eines Hauptanhydrit-Plattendolomit-Übergangsprofils. Geologie 14:47–57

Langbein R, Seidel G (1960) Zur Frage des „Sangerhäuser Anhydrits". Z Geol 9:23–31

Lapchik FE (1965) On lithology and facies of the Upper Permian deposits on the western continuation of the Greater Donets Basin (in Russian). Dokl Akad Nauk Ukr SSR 1:81–83

Lapkin IYu (1961a) On Lower Permian stratigraphy of the southern Russian Platform (in Russian). Dokl Akad Nauk SSSR, 137:143–145

Lapkin IYu (1961b) Lower Permian of the southern Russian Platform (in Russian). In: Materials on geology and potential Lower Permian gas deposits in the southern Russian Platform. Izd Kharkov Univ, Kharkov, pp 19–49

Lapkin IYu (1964) Correlation of the Lower Permian of the southern Russian Platform, Western Europe and Caucasus (in Russian). In: Upper Paleozoic and Mesozoic stratigraphy of southern biogeographic provinces. Nedra, Moscow, pp 41–62

Lapkin IYu (1966) Donets Trough, Dnieper-Donets and Pripyat depressions (in Russian). In: Stratigraphy of the USSR. Permian system. Nedra, Moscow, pp 161–176

Lapparent A, Blaise I (1966) Un intinéraire géologique en Afghanistan central: de Tirin à Penjao, par Ghizao. Rev Geogr Phisic Geol Dyn 8:343–349

Lappo VI (1946) Nordvik (Yurung-Tumus) oil deposit (in Russian). Interiors of the Arctic 1:74–129

Lappo VI, Kusov NI (1947) Nordvik rock salt deposit (in Russian). Interiors of the Arctic 2:147–176

Latskova VE (1981) New data on structural pattern of Kungurian deposits in the Caspian Depression and adjacent areas (in Russian). In: Structure and deposition environments of salt-bearing formations. Nauka, Novosibirsk, pp 16–21

Laws RA, Brown RS (1976) Bonaparte Gulf Basin – southeastern part. In: Leslie RB, Evans HJ, Knight CL (eds) Economic geology of Australia and Papua New Guinea, 3. Petroleum. Aust Inst Min Metallurgy, Monogr 7:200–208

Laws RA, Kraus GP (1974) The regional geology of the Bonaparte Gulf – Timor Sea area. APEA 14:77–84

Lees GM (1927) Salzgletscher in Persien. Mitt Geol Ges Wien 22:29–34

Lees GM (1931) Salt – some depositional and deformational problems. J Inst Pet Tech 17:259–280

Levenshtein ML (1961) New data on composition and structure of Lower Permian deposits in the Donets Basin (in Russian). In: Materials on geology and potential Lower Permian gas deposits of the southern Russian Platform. Izd Kharkov Univ, Kharkov, pp 105–112

Levina VS, Stepanov VV, Bondarenko LS, Chernova NI (1981) Structure and relations of Kungurian salt strata in the inner and outer flank areas of the Caspian Syneclise (in Russian). In: Structure and deposition environments of salt-bearing formations. Nauka, Novosibirsk, pp 21–24

Link WK (1959) The sedimentary framework of Brazil. 5th World Pet Congr NY Pr Sec 1:901–923

Lipatova VV (1962) Upper Permian stratigraphy of the Cis-Urals, Aktyubinsk area (in Russian). In: Stratigraphic scales of Paleozoic deposits. Permian system. Gostoptekhizdat, Moscow, pp 132–138
Lithofacies-Paleogeographical Atlas of the Permian of Platform area of Poland 1:500,000, Inst Geol, Warszawa (1978) 30 pp
Lobanova VV (1968) A petrographic-mineralogical characteristic of potash horizons at the Zhilyany deposits (in Russian). Tr VNIIG 40:137–156
Loczy L (1963) Palaeogeography and history of the geological development of the Amazonas Basin. Jahrb Geol B Aust 106:449–502
Loczy L (1964) Stratigraphic and paleogeographic problems of the Gondwanic Parana Basin, South America. Rep of the XXII Session Int Geol Congr, part IX – Gondwanes. New Dehli, pp 87–110
Löffler J, Schulze G (1962) Die Kali- und Steinsalzlagerstätten des Zechsteins in der DDR. Teil III, Sachsen-Anhalt. Freib Forsch H 97: 348 pp
Lohmann HH (1972) Salt dissolution in sub-surface of British North Sea as interpreted from seismograms. AAPG Bull 56:472–479
Lopez-Ramos E (1969) Marine Paleozoic rocks of Mexico. Am Assoc Pet Geol Bull 53:2399–2417
Lorenc S (1971) Preliminary petrographic description of the Werra Anhydrites from the Fore-Sudetic Monocline. Kwart Geol 15:56–66
Lotze F (1958a) Geschichte der Zechsteinforschung am Niederrhein und einige heutige Probleme. Geol Jb 73:1–6
Lotze F (1958b) Der englische Zechstein in seiner Beziehung zum deutschen. Geol Jb 73:135–140
Lupinovich YuI, Kislik VZ, Vysotsky EA (1969a) Potash horizons of Davydov area (in Russian). Dôkl Akad Nauk BSSR 13:1100–1103
Lupinovich YuI, Kislik VZ, Zelentsov II (1969b) Geological structure and deposition environment of the halogenic formation of the Pripyat Depression (in Russian). In: Geology and petrography of potash salts of Byelorussia. NaukaI Tekn, Minsk, .pp 7–28
Lützner H (1966) Fazies und Transportrichtung im Oberrotliegenden von Elgersburg (Thüringer Wald). Ber Dtsch Ges Geol Wiss, A Geol Paläont 11:137–160
Lützner H, Ellenberg J, Falk F (1981) Sedimentology and basin development of Rotliegendes deposits in the Saale Trough. In: International Symposium Central European Permian. Proceedings. Wydawnictwa Geologiczne, Warszawa, pp 239–252
Lyashenko AI, Aronova SM, Gassanova IG, Lotsman OA, Sokolova LI, Lyashenko GP, Arkhangelskaya AD, Batrukova LS, Karpov PA, Pan'shina LN (1970) Oil-and-gas-bearing and prospective rocks in the central and eastern Russian Platform. V. II. Devonian deposits in the Volga-Ural oil-and-gas-bearing area (in Russian). Nedra, Leningrad, 272 pp
MacLachlan ME (1967) Oklahoma, Chapter E. In: Paleotectonic investigations of the Permian system in the United States. US Geol Surv Prof Pap 515:81–92
MacLachlan J, Bieber A (1963) Permian and Pennsylvanian geology of the Hartville uplift – Alliance basin. Chadren arch area. In: Guidebook to geology of the northern Denver Basin and adjacent uplift. Rocky M Assoc Geol 14th Field Conf Denver, Colorado. Rocky M Assoc Geol, pp 84–94
Maher JC (1954) Lithofacies and suggested depositional environment of Lyons Sandstone and Lyons Formation in southeastern Colorado. Am Assoc Pet Geol Bull 38:2233–2239
Maher JC, Collins J (1952) Correlation of Permian and Pennsylvanian rocks from western Kansas to the Front Range of Colorado. US Geol Surv Oil Gas Inv Chart OC 46, 3 sheets
Makarov AS (1981) Assessment of potash-magnesium Kungurian salt strata of northern Caspian area (in Russian). In: Structure and deposition environments of potash salt deposits. Nauka, Novosibirsk, pp 133–137
Makarova TV (1957) Permian deposits of the central Russian Platform (in Russian). Gostoptekhizdat, Leningrad, 123 pp
Makarova TV (1959) Stratigraphy and facies of Permian deposits of the eastern Russian Platform (in Russian). Tr VNIGRI 22:147–154

Mallory WW (1975) Middle and Southern Rocky Mountains, northern Colorado Plateau, and eastern Great Basin Region. In: McKee ED, Crosby EJ (eds) Paleotectonic investigations of the Pennsylvanian System in United States, part I: Introduction and regional analyses of the Pennsylvanian System. Geol Surv Prof Pap 853:265–278

Maree SC (1960) Lithology of the Mufulira copper deposits. In: Lombard J, Nicolina P (eds) Stratiform copper deposits of Africa. Part 1. Lithology, sedimentation. Paris Assoc Ser Geol Afr 159–172

Marek S, Znosko J (1972) Tectonics of the Kujaw. Kwart Geol 16:1–18

Marr U (1962) Zur Faziesdifferenzierung im Unstrut-Kali-Revier. Ber Geol Ges 6:246–255

Martin CA (1965) Denver Basin. Am Assoc Pet Geol Bull 49:1908–1925

Martynov AA, Khnykin VI (1963) Areal distribution of Lower Permian salt deposits in the Dnieper-Donets Depression (in Russian). In: Geology and oil-and-gas potential of Ukraine. Gostoptekhizdat, Moscow, pp 30–33

Maughan EK (1966) Environment of deposition of Permian salt in the Williston and Alliance basins. In: Second Symposium on salt. Cleveland, Ohio. North Ohio Geol Soc 1:35–47

Maughan EK (1967) Eastern Wyoming, Eastern Montana, and the Dakotas, In: Paleotectonic investigations of the Permian System in the United States. US Geol Surv Prof Pap 515:125–152

Mayr U (1978) Stratigraphy and correlation of Lower Paleozoic formations, subsurface of Carnwallis, Devon, Somerset, and Pussell Islands, Canadian Arctic Archipelago. Geol Surv Can Bull 276: 28 pp

Mayr U (1980) Stratigraphy and correlation of Lower Paleozoic formations, subsurface of Bathurst Island and adjacent smaller islands, Canadian Arctic Archipelago. Geol Surv Can Bull 306: 52 pp

McCauley VT (1956) Pennsylvanian and Lower Permian of the Williston basin. In: Williston basin symposium. Bismarck, North Dakota. North Dakota Geol Soc, pp 150–164

McGinnis CJ (1954) Habitat of oil in the Denver Basin. In: Habitat of oil – a symposium. Am Assoc Pet Geol, pp 328–343

McGregor DJ (1952) Geology of the gypsum deposits near Sun City, Barber Country, Kansas (abs). Geol Soc Am Bull 63:1277–1278

McKee ED (1975) Arizona. In: McKee ED, Crosby EJ (eds) Paleotectonic investigations of the Pennsylvanian system in the United States. Part I: Introduction and regional analyses of the Pennsylvanian system. Geol Surv Prof Pap 853:293–309

Mear CE (1968) Upper Permian sediments in southern Permian Basin. In: Salina deposits. Geol Soc Am Spec Pap 88:349–358

Mear CE, Yarbrough DV (1961) Yates Formation in Southern Permian Basin of West Texas. Am Assoc Pet Geol Bull 45:1545–1556

Meier R (1969) Beitrag zur Geologie des Kaliflözes Stassfurt (Zechstein 2). Geologie 65:1–99

Meier R (1981) Clastic resedimentation phenomena of the Werra Sulphate (Zechstein 1) at the eastern slope of the Eichsfeld Swell (Middle European Basin) – an information. In: International Symposium Central European Permian. Proceedings. Wydawnictwa Geologiczne, Warszawa, pp 369–373

Meinhold R, Reinhardt H-G (1967) Halokinese im Nordostdeutschen Tiefland. Ber Dtsch Ges Geol Wiss, A Geol Paläont 12:329–353

Melnikov NV, Efimov AO, Kilina LI et al (1978) Detailed correlation of Vendian and Cambrian sections of the southern Siberian Platform (in Russian). In: Geology and oil-and-gas potential of East Siberia. SNIIGGiMS, Novosibirsk, pp 3–26

Melnikov NV, Kilina LI (1981) Lithology and deposition environments of Vendian and Cambrian deposits in the southern half of the Lena-Tunguska petroleum province (in Russian). In: Lithology and geochemistry of oil-and-gas-bearing strata of the Siberian Platform. Nauka, Moscow, pp 51–66

Merz G (1966) Vergleich eines Zechsteinprofils von Südthüringen mit Zechsteinprofilen des Thüringer Beckens. Ber Dtsch Ges Geol Wiss, RA 11:175–183

Merzlyakov GA (1979) Permian salt basins of Eurasia (in Russian). Nauka, Novosibirsk, 144 pp

Mesolella KJ (1978) Paleogeography of some Silurian and Devonian reef trends, Central Appalachian Basin. Am Assoc Pet Geol Bull 62:1607–1644

Mesolella KJ, Robinson JD, McCormick LM, Ormistan AR (1974) Cyclic deposition of Silurian carbonates and evaporites in Michigan basin. Am Assoc Pet Geol Bull 58:34–62

Meyer HG (1965a) Stratigraphy of the Permian evaporites, NW England. Proc Yorks Geol Soc 35:71

Meyer HG (1965b) Revision of the stratigraphy of the Permian evaporites and associated strata in north-western England. Proc Yorks Geol Soc 35:71–89

Miall AD, Kerr JW (1977) Phanerozoic stratigraphy and sedimentology of Somerset Island and northeastern Boothia Peninsula. Geol Surv Can Pap 77-1A:99–106

Miall AD, Kerr JW (1980) Cambrian to Upper Siluriian stratigraphy, Somerset Island and northeastern Boothia Peninsula, District of Franklin, N.W.T. Geol Surv Can Bull 315: 43 pp

Milewicz J (1971) Zechstein in the region of Gubin (in Polish). Kwart Geol 15:605–623

Milewicz J (1981) The influence of the tectonics on sedimentation of the Rotliegendes in south-western Poland. In: Int Symp Central Eur Permian. Proceedings. Wydawnictwa geologiczne, Warszawa, pp 273–280

Miller RL (1976) Silurian nomenclature and correlations in Southwest Virginia and Northeast Tennessee. US Geol Surv Bull 1405-H: 25

Momper JA (1963) Nomenclature, lithofacies, and genesis of Permo-Pennsylvanian rocks – Northern Denver Basin. In: Guidebook to the geology of the northern Denver Basin and adjacent uplift, Rocky M Assoc Geol 14th Field Conf Denver, Colorado. Rocky M Assoc Geol, pp 41–67

Moore DF (1953) Occurrence of the Permian salt – western Kansas (abs). Oil and Gas 52:195. Am Assoc Pet Geol Bull 37:2609

Moore GW (1959) Alteration of gypsum from the Capitan Limestone of New Mexico (abs). Geol Soc Bull 70:1647

Morozov LN, Sedletskaya NM, Anoshin LV, Svidzinsky SA (1980) Structure of potassium-bearing beds of the Elton deposit (in Russian). In: Structural features of bishofite and potash salt deposit. Nauka, Novosibirsk, pp 47–59

Mossop GD (1979) The evaporites of the Ordovician Baumann Fiord Formation, Ellesmere Island, Arctic Canada. Geol Surv Can Bull 298: 52

Mostofi B, Frei E (1961) Major sedimentary basins of Iran and their oil-and-gas potential (in Russian). In: V World Pet Congr, vol I. Gostoptekhizdat, Moscow, pp 77–83

Mötzung R (1963) Neue Ergebnisse bei der Parallelisierung der Steinsalzleitbänke des Unstrutkaligebietes mit denen des Nordharzkaligebietes. Bergakademie 11:759–763

Movshovich EV (1968) On the thickness of salt formations in the North Caspian Syneclise (in Russian). In: Geological structure of mineral resources of the Volga-Don and adjacent areas. Nizhne-Volzhsk, Volgograd, pp 244–251

Mosvshovich EV (1970) Permian system (in Russian). In: Geology of the USSR, vol 46 (Rostov, Volgograd, Astrakhan regions and Kalmyk ASSR). Nedra, Moscow, pp 183–228

Movshovich EV (1977) Paleogeography and paleotectonics of the Lower Volga area during the Permian and Triassic (in Russian). Izd Saratov Univ, Saratov, 241 pp

Mudge MR (1967a) Central Midcontinent Region, Chapter F. In: Paleotectonic investigations of the Permian System in the United States. US Geol Surv Prof Pap 515:93–125

Mudge MR (1967b) Environment of Central Midcontinent Region: Interval B (pl. 10, map A). In: Paleotectonic maps of the Permian System in the United States. US Geol Surv Misc Geol Inv Map 1–450, pp 47–48

Muir JL (1934) Anhydrite-gypsum problem of Blaine Formation, Oklahoma. Am Assoc Pet Geol Bull 18:1297–1312

Münzberger E, Rost U, Wirth J (1966) Vergleichende Darstellung der Sedimentationsverhältnisse des Zechsteins von Thüringen mit denen des nordostdeutschen Flachlandes. Ber Dtsch Ges Geol Wiss RA 11:161–174

Murris RJ (1980) Middle East: stratigraphic evolution and oil habitat. Am Assoc Pet Geol Bull 64:597–618

Nalivkin VD (1950) Facies and geological history of the Ufimian Plateau and the Yuryuzan-Sylva Depression (in Russian). Gostoptekhizdat, Leningrad Moscow, 126 pp

Nemec W (1981) Tectonically controlled alluvial sedimentation in the Skupiec Formation (Lower Permian) of Intrasudetic Basin. In: Int Symp Central Eur Permian. Proceedings. Wydawnictwa Geologiczne, Warszawa, pp 294–311

Nemec W, Porebski SJ (1981) Sedimentary environment of the Weissliegendes sandstones in Fore-Sudetic monocline. In: Int Symp Central Eur Permian. Proceedings. Wydawnictwa Geologiczne, Warszawa, pp 281–293

Nesterenko LP (1956) Permian stratigraphy of the Donets Basin (in Russian). Izv Akad Nauk SSSR Ser Geol 7:33–48

Nesterenko LP (1958) On correlation technique of Lower Permian deposits of the Donets Basin, its northwestern margins and the eastern Dnieper-Donets Depression (in Russian). Izv Akad Nauk SSSR Ser Geol 2:118–122

Nesterenko LP (1970) Lower Permian stratigraphy of the Donets Basin (in Russian). In: Essays on geology of the Kuznetsk and Donets basins. Nedra, Leningrad, pp 85–93

Nesterenko LP (1978) Lower Permian deposits in the Kalmius-Torez Trough of the Donets Basin (in Russian). Vishcha Shkola, Kiev, 148 pp

Nikolaev YuD, Sivkov SN (1982) Relationship between evaporite deposits of the Timan-Pechora province and oil-and-gas deposits (in Russian). In: Oil-and-gas potential of old salt accumulation areas. Nauka, Novosibirsk, pp 79–84

Oborin AA (1968) Kungurian lagoonal and lagoonal marine formations in the Sylva Depression of the Cis-Uralian Trough and deposition environments (in Russian). Tr VNIGRI 65:100–130

Odoleev OG (1972) Generation pattern of the western Caspian Depression and Volga Monocline (in Russian). In: Problems of geology and oil-and-gas potential of the Lower Volga areas. Nizhne-Volzhsk, Volgograd, pp 53–61

Oettel S, Voitel R (1966) Erste Untersuchungsergebnisse einer feinstratigraphischen und petrographischen Gliederung im Kaliflöz „Thüringen" der Gruben „Springen" und „Alexandershall". Ber Dtsch Ges Geol Wiss, RA 11:489–499

Oliveira AI (1956) Brazil. Geol Soc Am Mem 65:1–62

Oriel SS, Myers DA, Crosby EJ (1967) West Texas Permian Basin Region, Chapter C. In: Paleotectonic investigations of the Permian system in the United States. US Geol Surv Prof Pap 515:21–60

Oshakpaev TA (1974) Chelkar gigantic salt dome (in Russian). Nauka, Alma-Ata, 184 pp

Paech H-J (1970) Über von Porphyriten überlagerte permosilesische Deckgebirgssedimente der Flechtinger Scholle (Bezirk Magdeburg). Ber Dtsch Ges Geol Wiss, A Geol Paläont 15:75–86

Page LR, Adams JE (1940) Stratigraphy, eastern Midland Basin, Texas. Am Assoc Pet Geol Bull 24:52–64

Paleotectonic investigations of the Mississippian system in the United States (1979) Part I. Introduction and regional analysis of the Mississippian System. US Geol Surv Prof Pap 1010

Pashkevich EI, Pistrak RM (1964) On Devonian stratigraphy of the Dnieper-Donets Depression (in Russian). Tr VNIIGaz 22/30a:5–16

Patchen DG, Smosna RA (1975) Stratigraphy and petrology of Middle Silurian McKenzie Formation in West Virgnia. Am Assoc Pet Geol Bull 59:2266–2287

Patterson JK (1961) Ordovician stratigraphy and correlations in North America. Am Assoc Pet Geol Bull 45:1364–1378

Paul J (1981) Textural analysis of Permian algal stromatolite reefs (Harz Mountains) (FRG). In: Int Symp Central Eur Permian. Proceedings. Wydawnictwa Geologiczne, Warszawa, pp 374–378

Pawłowska K (1963) New sites of the Rotliegendes deposits in Poland (in Polish). Prace Inst Geol 30:209–214

Pawłowska K (1964) On the subdivision of the Zechstein in the Góry Swietokrżyskie into four sedimentation cycles (in Polish). Przegl Geol 9:367–371

Pawłowska K, Poborski J (1963) Main stratigraphic and paleogeographic problems of the Western Permian margin of the East-European Precambrian Platform (in Polish). Prace Inst Geol 30:215–220

Peterson JA, Loleit AJ, Spencer CW, Ullrich RA (1965) Sedimentary history and economic geology of San Juan Basin. Am Assoc Pet Geol Bull 49:2076–2119

Petri S (1958) Sobre o facies de evaporites do Carbonifero do Amazônia. Geol Soc Brazil Bull 7:35–47

Petrova NS, Sedun EV (1981) Structural-textural and geochemical features of rocks of potash horizon II at the Starobino deposit (in Russian). In Structure and deposition environments of potash salt deposits, Nauka, Novosibirsk, pp 57–67

References

Philipp W (1966) Zechstein und Buntsandstein in Tiefbohrungen zwischen Harz und Lüneburger Heide. Geol Jb (Hannover) 77:711–740

Pinchuk IA (1971) A lithopetrographic characteristic of the Lower Permian deposits in the Biikzhal SG-3 well over the interval 4,535–4,979.5 m (in Russian). In: Biikzhal superdeep SG-SG-2 well, 1:23–37

Pinneker EV (1968) Mineral waters of Tuva (in Russian). Kyzyl, 112 pp

Pisarchik YaK (1963) Lithology and facies of Cambrian deposits in the Irkutsk Amphitheater related to oil-and-gas potential and salinity (in Russian). Gostoptekhizdat, Leningrad, 346 pp

Pistrak RM (1963) Devonian and Tournaisian deposits in the Dnieper-Donets Depression and their oil-and-gas potential (in Russian). Gostoptekhizdat, Moscow, 138 pp

Pistrak RM, Lyashenko AI, Pashkevich EI, Voloshina ZG, Galitsky IV (1972) Devonian deposits of the southern marginal zone of the Dnieper-Donets Depression. Dokl Akad Nauk SSSR 203:175–178

Plumhoff F (1966) Marines Ober-Rotliegendes (Perm) im Zentrum des nordwestdeutschen Rotliegend-Beckens. Erdöl und Kohle 19:713–720

Poborski J (1960) Central-European Zechstein Salt Basin in Poland (in Polish). Prace IG 30:355–366

Poborski J (1964) Facial relations in the Zechstein Basin of Poland (in Polish). Kwart Geol 8:111–117

Poborski J (1969) New outline of lithological relations in the Zechstein Basin of Poland (in Polish). Kwart Geol 13:93–99

Podemski M (1965) Development of sedimentation of the Zechstein deposits in the regions Lubin Legnicki-Sieroszowice (in Polish). Kwart Geol 9:115–130

Podemski M (1968) Some remarks on sedimentological bases of the Zechstein stratigraphy (in Polish). Kwart Geol 12:875–883

Podemski M (1972) Some remarks on sedimentological bases of Zechstein stratigraphy, In: Geology of saline deposits. Proc Hannover Symp 1968 (Earth Sci 7). Unesco, Paris, pp 219–223

Podemski M, Wagner R, Pawłowska K (1968) Permian. Geological atlas of Poland. Warszawa

Pokorski J (1971) Lower Permian in the Podlasie Depression (in Polish). Kwart Geol 15:589–604

Pokorski J (1981) Paleogeography of the Upper Rotliegendes in the Polish Lowland. In: Int Symp Central Eur Permian. Proceedings (in Polish). Wydawnictwa Geologiczne, Warszawa, pp 56–68

Polivko IA (1967) Facies, thicknesses and mode of occurrence of the Narva horizon rocks in Latvia (in Russian). In: Problems of Middle and Upper Paleozoic geology of the Baltic area. Znanie, Riga, pp 16–25

Poole WH, Sanford BV, Williams H, Kelley DG (1970) Geology of southeastern Canada. In: Douglas RJW (ed) Geology and economic minerals of Canada. Geol Surv Can Econ Geol Rep 1:229–304

Porfiriev GS (1963) Volga-Ural petroliferous area. Lower Permian deposits (in Russian). Gostoptekhizdat, Leningrad, 287 pp

Porter JW, Fuller JG (1959) Lower Paleozoic rocks of northern Williston Basin and adjacent areas. Am Assoc Pet Geol Bull 43:124–189

Predtechensky NN (1980) Lithogenetic typization of Silurian deposits of the Siberian and Russian Platforms (in Russian). In: Silurian of the Siberian Platform. Key sections of the northwest Siberian Platform. Nauka, Novosibirsk, pp 150–162

Predtechensky NN, Tesakov YuI (1979) Regular pattern of sedimentation and typization of Silurian facies complexes of the Siberian Platform (in Russian). In: Silurian of Siberian Platform, new regional and local stratigraphic units. Nauka, Novosibirsk, pp 28–43

Protasevich BA (1976a) Potassium rocks in the central zone of the Pripyat Trough (in Russian). In: Geology of non-metallic minerals of BSSR. BelNIGRI, Minsk, pp 21–29

Protasevich BA (1976b) On potassium rocks in the southern zone of the Pripyat Trough (in Russian). In: Geology of non-metallic minerals of BSSR. BelNIGRI, Minsk, pp 30–35

Quester H (1964) Petrographie des erdgashöffigen Algendolomits im Zechstein 2 zwischen Weser und Ems. Z Dtsch Geol Ges (Hannover) 114:461–483

Ramsay CR, Davidson LR (1970) The origin of scapolite in the regionally metamorphosed rocks of Mary Kathleen, Queensland, Australia. Contr Mineral Pet 25:45–51

Raymond LR (1953) Some geological results from the exploration for potash in North-East Yorkshire. Geol Soc Ld Quart J 108:283–310
Reichel W (1970) Abriß des Rotliegenden im Döhlener Becken. Ber Dtsch Ges Geol Wiss, A Geol Paläont 15:67–74
Reichenbach W (1970) Die lithologische Gliederung der rezessiven Folgen von Zechstein 2–5 in ihrer Beckenausbildung – Probleme der Grenzziehung und Parallelisierung. Ber Dtsch Ges Geol Wiss, A Geol Paläont 15:555–564
Reinson GF (1978) Carbonate-evaporite cycles in the Silurian rocks of Somerset Island, Arctic Canada. Geol Surv Can Pap 76-10: 13 pp
Reinson CE, Kerr JW, Stewart WD (1976) Stratigraphic field studies, Somerset Island, District of Franklin (58B to F). Geol Surv Can Pap 76-AA:497–499
Rentzsch J (1965) Die feinstratigraphisch-lithologische Flözlagenparallelisierung im Kupferschiefer am Südrand des norddeutschen Zechsteinbeckens. Z Angew Geol 11:11–14
Repina LN (1966) Lower Cambrian trilobites of southern Siberia (Superfamily *Redlichioidea*) (in Russian). Nauka, Moscow, 204 pp
Richter-Bernburg G (1955a) Über salinare Sedimentation. Z Dtsch Geol Ges 105:593–645
Richter-Bernburg G (1955b) Stratigraphische Gliederung des deutschen Zechsteins. Z Dtsch Geol Ges 105:843–845
Richter-Bernburg G (1955c) Der Zechstein zwischen Harz und Rheinischem Schiefergebirge. Z Dtsch Geol Ges 105:876–899
Richter-Bernburg G (1959a) Zur Paläogeographie des Zechsteins. In: Accad Nazionale dei Lincei, 1 Giacimenti gassiferi dell Europa occidentale (Rome), vol 1, pp 87–99
Richter-Bernburg G (1959b) Die Korrelierung isochroner Warven im Anhydrit des Zechstein 2 (zweiter Beitrag). Geol Jb 75:629–646
Richter-Bernburg G (1960) Zeitmessung geologischer Vorgänge nach Warven-Korrelationen im Zechstein. Geol Rdsch 49:215–222
Richter-Bernburg G (1972a) Saline deposits in Germany: a review and general introduction to the excursion. In: Geology of saline deposits. Proc Hannover Symp 1968 (Earth Sci 7). Unesco, Paris, pp 275–287
Richter-Bernburg G (1972b) Sedimentological problems of saline deposits. In: Geology of saline deposits. Proc Hannover Symp 1968 (Earth Sci 7). Unesco, Paris, pp 33–39
Richter-Bernburg G (1981) Some particularities in the sedimentation of the German Permian. In: Int Symp Central Eur Permian. Proceedings. Wydawnictwa Geologiczne, Warszawa, pp 48–55
Rickard LV (1969) Stratigraphy of the Upper Silurian Salina Group, New York, Pennsylvania, Ohio, Ontario. NY State Mus Sci Serv Map Chart Ser 12, 57 pp
Ricketts BD, Donaldson JA (1981) Sedimentary history of the Belcher Group of Hudson Bay. In: Campbell FHA (ed) Proterozoic basins of Canada. Geol Surv Pap 81-10:235–254
Romanov FI, Zotova AI (1962) Key wells of the USSR. South Kaliningrad (Nivenskoe) key well (in Russian). Gostoptekhizdat, Leningrad, 128 pp
Roniewicz P, Czapowski G, Gizejewski J, Karnkowski PH (1981) Variability in depositional environment of the Rotliegendes of the Poznan area. In: Int Symp Central Eur Permian. Proceedings. Wydawnictwa Geologiczne, Warszawa, pp 262–272
Rose ER, Sanford BV, Hacquebard PA (1970) Economic minerals of Southeastern Canada. In: Douglas RJW (ed) Geology and economic minerals of Canada. Geol Surv Can Econ Geol Rep 1:305–362
Rost W, Schimanski W (1967) Übersicht über das Oberkarbon und das Rotliegende im Nordteil der DDR. Ber Dtsch Ges Geol Wiss, A Geol Paläont 12:201–221
Rotai AP (1963) Permian system. Pechora coal basin, Urals and Pai-Khoi. Lower Permian (in Russian). In: Geology of the USSR, vol II (Arkhangelsk, Vologda regions and Komi ASSR). Gosgeoltekhizdat, Moscow, pp 554–593
Roth H (1955a) Ausbildung und Lagerungsformen des Kaliflözes „Hessen" im Fuldagebiet. Z Dtsch Geol Ges 105:674–684
Roth H (1955b) Befahrung des Kaliwerkes „Bergmannssegen-Hugo" der Wintershall AG. Z Dtsch Geol Ges 105:868–871
Roth H (1972) Deformations in subhorizontal salt deposits of German Zechstein 1. In: Geology of saline deposits. Proc Hannover Symp 1968 (Earth Sci 7). Unesco, Paris, pp 225–233

Roth RI (1942) West Texas Barred Basin. Geol Soc Am Bull 53:1659-1674
Roth RI (1945) Permian Pease River Group of Texas. Geol Soc Am Bull 56:893-907
Rozen OM (1982) Geochemical variations in sulfate-bearing deposits and their metamorphic derivatives: in relation to Precambrian evaporites (in Russian). Lithol I Miner Res 2:94-103
Rutten MG (1969) The geology of western Europe. Elsevier, Amsterdam London New York, 520 pp
Ruzhentsev VE (1948) Major facies zones of Sakmarian-Artinskian complex in the South Urals (in Russian). Izv Akad Nauk SSSR Ser Geol 1:101-120
Rybakov FF (1958) Stratigraphy of Lower Permian deposits of the Volga area at Kuibyshev (in Russian). In: Geology and exploration of oil deposits. Gostoptekhizdat, Moscow, pp 53-67
Rybakov FF (1960) Stratigraphy, facies and paleogeography of Permian deposits in the Kuibyshev and Orenburg areas and subdivision of the Permian system into three series (in Russian). In: Geology, geochemistry, geophysics. Kuibyshev, pp 65-110
Rybakov FF (1962) Stratigraphy and correlation of Permian sections in Kuibyshev and Orenburg areas (in Russian). In: Stratigraphy scales of Paleozoic deposits. Permian system. Gostoptekhizdat, Moscow, pp 61-72
Rykovskov AE (1932) Problems relevant to sulphate-free Solikamsk potassium deposit (in Russian). Tr Glavn Geol Razved Upr 43:4-27
Sable EG (1979) Evaporites in the Eastern Interior Basin. In: Paleotectonic investigations of the Mississippian system in the United States, part II. US Geol Surv Prof Pap 1010-R:432-437
Safarov EI, Kaptsan VKh (1967) On stratigraphy of Devonian and Carboniferous flank base deposits of the Dobruja Trough (in Russian). In: Paleontology, geology and mineral resources of Moldavia. Izd Akad Nauk Mold SSR 2:61-74
Salski W (1967) Clay deposits of the Upper Zechstein in the Lubin region (in Polish). Kwart Geol 11:587-598
Sanford BV (1967) Devonian of Ontario and Michigan. In: Int Symp on Devonian System, vol I. Calgary, Alberta, Canada, pp 975-1000
Sanford BV (1968) Oil and gas in southwestern Ontario. In: Natural gases of North America, vol II. Am Assoc Pet Geol Mem 9:1798-1818
Sannemann D (1963) Über Salzstock-Familien in NW Deutschland. Erdoel Z 11:3-10
Sapegin BI, Yanin VN (1981) Main tectonic features of the Verkhnekamsk potassium deposit (in Russian). In: Structure and deposition environments of potash salt deposits. Nauka, Novosibirsk, pp 118-124
Savitzky VE, Astashkin VA (1978) Cambrian reef systems of West Yakutia (in Russian). Sov Geol 8:27-37
Savitzky VE, Evtushenko VM, Egorova LI et al (1972) Cambrian of the Siberian Platform (the Yudoma-Olenek type of section, Kuonam complex) (in Russian). Nedra, Moscow, 198 pp
Saxby DB, Lamar JE (1957a) Gypsum and anhydrite in Illinois. Geol Surv Prof Pap 260E:293-300
Saxby DB, Lamar JE (1957b) Gypsum and anhydrite in Illinois. Ill State Geol Surv Circ 226: 26 pp
Schettler H (1965) Untersuchungsmethoden und stratigraphische Ergebnisse von Trias und Zechstein-Bohrungen im westlichen Emsgebiet. Z Dtsch Geol Ges 115:216-227
Schettler H (1972) The stratigraphical significance of idiomorphic quartz crystals in the saline formations of the Weser-Ems area, north-western Germany. In: Geology of saline deposits. Proc Hannover Symp 1968 (Earth Sci 7). Unesco, Paris, pp 111-127
Schindewolf OM, Seilacher A (1955) Beiträge zur Kenntnis des Kambriums in der Salt Range (Pakistan). Abh Akad Wiss Lit, Mainz 10:342-446
Schirrmeister U (1970) Zur Feinstratigraphie des Grauen Salztones (T3) und seiner Übergänge sowie stratigraphischen Korrelation mit T3-Profilen aus verschiedenen Gebieten der DDR, Niedersachsens und der nördlichen VR Polen. Ber Dtsch Ges Geol Wiss, A Geol Paläont 15: 517-538
Schreiber A Van (1960) Das Rotliegende des Riechtunger Höhenzuges. Akademie-Verlag, Berlin, p 132

Schulze G (1958) Beitrag zur Stratigraphie und Genese der Steinsalzserien 1–4 des mitteldeutschen Zechsteins, unter besonderer Berücksichtigung der Bromverteilung. Freiberger Forsch H A 123, 175 pp
Schulze G (1960) Stratigraphische und genetische Deutung der Bromverteilung in den mitteldeutschen Steinsalzlagern des Zechstein. Freiberger Forsch H C 83, 114 pp
Schulze G (1962) Zur Faziesdifferenzierung im Nordharz-Kali-Gebiet. Ber Geol Ges 6:256–257
Schwab M (1970) Tektonik, Sedimentation und Vulkanismus im Permosiles Mitteleuropas. Ber Dtsch Ges Geol Wiss, A Geol Paläont 15:29–45
Schwandt A (1962) Zur Faziesdifferenzierung im Werra-Kaligebiet. Ber Geol Ges 2:288–295
Seidel G (1961) Zur Stratigraphie des Zechsteins und Buntsandsteins im Südharzkalirevier (Thüringer Becken). Geologie 10:952–972
Seidel G (1965) Zur geologischen Entwicklungsgeschichte des Thüringer Beckens. Geologie 50:1–115
Seidel G (1966) Zechstein-Randausbildung in Süd- sowie Südostthüringen und Westsachsen (Tagungsbericht). Ber Dtsch Ges Geol Wiss, A Geol Paläont 11:501–502
Seifert J (1970) Zur stratigraphischen Einordnung der Zechsteinrandausbildung in Südostthüringen und Sachsen. Ber Dtsch Ges Geol Wiss, A Geol Paläont 15:565–567
Seiful-Mulyukov RB, Sheremetiev GA (1969) Permian and Triassic deposits in the northern flank zone and adjacent areas of the Caspian Syneclise: assessment of oil-and-gas potential (in Russian). In: Permo-Triassic of the Russian Platform: assessment of oil-and-gas potential. Nedra, Moscow, pp 38–46
Sementovsky YuV (1969) Geological history of Late Permian in the eastern Russian Platform (in Russian). Tr Geol Inst (Kazan) 24:126–132
Serdyuchenko DP (1972) Salt-bearing sedimentary rocks in Precambrian strata of the Earth and their scapolite-bearing metamorphic derivatives (in Russian). In: Precambrian geology. Nauka, Leningrad, pp 31–41
Shablovskaya RK, Avkhimovich VI, Vysotsky EA (1976) Lithogeochemical and stratigraphic subdivision of the supersalt Devonian at the Petrikov potash salt deposit (in Russian). In: Geology of non-metallic minerals of BSSR. BelNIGRI, Minsk, pp 36–48
Shcherbakov VP, Volgina MG (1969) On the problem of geological structure of the Kalmyk-Astrakhan part of the Caspian Depression flank zone (in Russian). In: Geological structure and mineral-raw material potential of the Volga-Don Region. Izd Rostov Univ, Rostov-on-Don, pp 11–17
Shcherbina VN (1959) Pripyat Salt Basin (in Russian). Izv Akad Nauk BSSR, Ser Phys Techn Sci 2:76–80
Shcherbina VN (1960a) Mineralogical-petrographic zones of potassium-bearing horizons in the Pripyat salt basin (in Russian). Dokl Akad Nauk BSSR 4:66–69
Shcherbina VN (1960b) Structure of salt strata at the Starobino potash salt deposit (in Russian). Izd Vyssh Uchebn Zaved Geol Inst Razv 11:82–89
Shcherbina VN 1961) General characteristic of halite rocks in the Pripyat salt basin (in Russian). Tr Inst Geol Nauk Akad Nauk BSSR 3:259–269
Shcherbina VN (1962) Layered sylvinite rocks in the Pripyat salt basin (in Russian). Dokl Akad Nauk BSSR 6:376–378
Sheldon RP (1963) Physical stratigraphy and mineral resources of Permian rocks in Western Wyoming. US Geol Surv Prof Pap 313-B: 269 pp
Sheldon RP, Cressman ER, Cheney TM, McKelvey VE (1967a) Middle Rocky Mountains and Northeastern Great Basin, Chapter H. In: Paleotectonic investigations of the Permian system in the United States. US Geol Surv Prof Pap 515:153–170
Sheldon RP, Maughan EK, Cressman ER (1967b) Environment of Wyoming and adjacent states. Interval B (pl 11). In: Paleotectonic maps of the Permian System in the United States. US Geol Surv Misc Geol Inv Map 1–450, pp 48–54
Sheldon RP, Maughan EK, Cressman ER (1967c) Paleogeography and sedimentary environments at time of maximum transgression during Leonard time in Wyoming and adjacent areas (pl 11). In: Paleotectonic maps of the Permian System in the United States. US Geol Surv Misc Geol Inv Map 1–450

Shevchenko TA (1971) Lithology and correlation of sections of the Pyarnu-Narva series of Devonian in the Vishany, Davydovo, Ostashkovich and Tishkovo areas (in Russian). Dokl Akad Nauk SSSR 15:254–257

Shmelev NV (1963) Permian system. Pechora coal basin. Urals and Pai-Khoi. Upper series (in Russian). In: Geology of the USSR, vol II (Arkhangelsk, Vologda regions and Komi ASSR). Gosgeoltekhizdat, Moscow, pp 593–605

Shurawlew WS (1967) Ähnlichkeiten der Salztektonik in der exagonalen Kaspi- und der Norddeutsch-Polnischen Senke. Ber Dtsch Ges Geol Wiss, A Geol Paläont 12:421–436

Sidorenko AV, Rozen OM (1981) Carbonate and the associated chemogenic deposits of Precambrian and processes leading to continental crust formation (in Russian). In: Problems of sedimentary geology of Precambrian. Nauka, Moscow 6:9–19

Silver BA, Todd RG (1969) Permian cyclic strata Northern Midland and Delaware Basins, West Texas and Southeastern New Mexico. Am Assoc Pet Geol Bull 53:2223–2252

Simon P (1967) Feinstratigraphische und paläogeographisch-fazielle Untersuchungen des Stassfurt-Steinsalzes (Zechstein 2) im Kalisalzbergwerk „Königshall Hindenburg". Geol Jb 84:341–366

Sims PK, Card KD, Lumbers SB (1981) Evolution of early Proterozoic basins of the Great Lakes region. In: Campbell FHA (ed) Proterozoic basins of Canada. Geol Surv Pap 81-10:379–398

Singleton AE (1966) Sahara No. 1 well. Bur Min Res Aust Pet Search Subsidy Acts 80: 19 pp

Slivkova RP, Konovalova MV, Bogatsky VI, Ioffe GA (1972) Permian deposits in the Upper Pechora Depression (in Russian). In: Geology and oil-and-gas potential of the north-east European part of the USSR. Komi, Syktyvkar, pp 132–144

Smith DB (1970) Permian and Trias. In: Geology of Durham County. Trans Nat Mist Soc Northumb 41:66–91

Smith DB (1971) Possible displasive halite in the Permian Upper Evaporite Group of north-east Yorkshire. Sedimentology 17:221–232

Smith DB (1972) Foundered strata, collapse-breccias and subsidence features of the English Zechstein. In: Geology of saline deposits. Proc Hannover Symp 1968 (Earth Sci 7). Unesco, Paris, pp 255–269

Smith DB (1981) The evolution of the English Zechstein Basin. In: Int Symp Central Eur Permian. Proceedings. Wydawnictwa Geologiczne, Warszawa, pp 9–47

Smith DB, Crosby A (1979) The regional and stratigraphic context of Zechstein 3 and 4 potash deposits in the British Sector of the Southern North Sea and adjoining land areas. Econ Geol 74:397–408

Smith HJ (1938) Potash in the Permian salt basin. Ind Eng Chem 30:854–860

Smith JP (1950) Geology of the potash deposits of Germany, France and Spain. Mining Eng 187: 117–121

Smosna R, Patchen D (1978) Silurian evolution of Central Appalachian Basin. Am Assoc Pet Geol Bull 62:2308–2328

Snider HI (1966) Stratigraphy and associated tectonics of the Upper Permian Castile-Salado-Rustler Evaporite Complex, Delaware Basin, West Texas and Southeastern New Mexico (abs). Diss Abs, Sec B 27:1992

Sokolov BA, Movshovich EB (1968) Geological history of the Suleiman-Kirtar mountain folded structure (West Pakistan) (in Russian). Izv Akad Nauk SSSR Ser Geol 5:75–94

Sokolov BS, Khomentovsky VV (1981) Age of oil-and-gas-bearing strata in the southwestern Siberian Platform (in Russian). Sov Geol 5:45–53

Sokolova EI (1966) Caspian Depression (in Russian). In Stratigraphy of the USSR. Permian system. Nedra, Moscow, pp 176–186

Sokolova EI, Ivanova EN, Egorov IP (1961) Permian and Triassic of the South Emba and their oil potential (in Russian). Gostoptekhizdat, Leningrad, 195 pp

Sokolowski J (1966) The role of halokinesis in the development of Mesozoic and Cenozoic deposits of the Mogilno Structure and of the Mogilno-Lodź Synclinorium (in Polish). Prace Inst Geol, vol 50. Warszawa, 112 pp

Solak M, Zołnierczuk T (1981). Some regularities of the distribution of oil-and-gas deposits in the Permian (Polish Lowland). In: Int Symp Central Eur Permian. Proceedings. Wydawnictwa Geologiszne, Warszawa, pp 596–601

Soloviev UK (1956) On Lower Permian in the Volga area at Gorki (in Russian). Uchen Zapiski Kazan Univ 115:59 68

Sorgenfrei T (1966) Strukturgeologischer Bau von Dänemark. Geologie 15:641–660

Spooner W (1932) Salt in Smackover field, Union County, Arkansas. Am Assoc Pet Geol Bull 16: 601–608

Steiner W (1966) Das Rotliegende des Ilfelder Beckens und seine Beziehungen zu benachbarten Rotliegend-Vorkommen. Ber Dtsch Ges Geol Wiss, A Geol Paläont 11:67–118

Stepanov DL (1969) Paleozoic stratigraphy of Iran (in Russian). Byull MOIP 44:5–16

Stepanov DL, Forsh NN (1966) Central and eastern Russian Platform (in Russian). In: Stratigraphy of the USSR. Permian system. Nedra, Moscow, pp 53–117

Stewart FH (1949) The petrology of the evaporites of the Eskdale No. 2 boring, East Yorkshire. Pt 1, The Lower evaporite bed. Min Mag 28:621–675

Stewart FH (1951a) The petrology of the evaporites of the Eskdale No. 2 boring, East Yorkshire. Pt II, The middle evaporite bed. Min Mag 29:445–475

Stewart FH (1951b) The petrology of the evaporites of the Eskdale No. 2 boring, East Yorkshire. Part III, The upper evaporite bed. Min Mag 29:557–572

Stewart FH (1953) Early gypsum in the Permian evaporites of North-Eastern England. Proc Geol Assoc Univ Durham 64:33–39

Stewart FH (1954) Permian evaporites and associated rocks in Texas and New Mexico compared with those of Northern England. Prof Yorkshire Geol Soc 29:185

Stöcklin J (1961) Lagunäre Formationen und Salzdome in Ostizan. Eclogae Geol Helv 54:1–27

Stöcklin J (1966) Tectonics of Iran (in Russian). Geotektonika 1:3–21

Stöcklin J (1974a) Northern Iran. Alborz Mountains. In: Spencer AM (ed) Mesozoic-Cenozoic orogenic belts. Scott Acad Press, pp 213–234

Stöcklin J (1974b) Possible ancient continental margins in Iran. In: Burk CA, Drake CL (eds) The geology of continental margins. Springer, Berlin Heidelberg New York, pp 873–888

Stöcklin J, Nabavi M, Samimi M (1965) Geology and mineral resources of the Soltanich Mountains (Northwest Iran). Geol Surv Iran Rep 2: 44 pp

Stolle E (1962) Zur Faziesdifferenzierung im Südharz-Kaligebiet. Ber Dtsch Geol Ges 2:266–287

Stolle E (1967) Die Kali- und Steinsalzlagerstätten des Zechsteins in der DDR. Teil II, Das Südharzgebiet. Freib Forsch H C97/II

Stoneley R (1974) Evolution of the continental margins bounding a former Southern Tethys. In: Burk CA, Drake CL (eds) The geology of continental margins. Springer, Berlin Heidelberg New York, pp 889–906

Summet EYu (1971) Devonian system (in Russian). In: Geology of the USSR, vol I (Leningrad, Pskov and Novgorod areas). Nedra, Moscow, pp 174–245

Supronyuk KS, Tsypko YaI, Shevchenko GD (1968) Stratigraphy of Permian deposits in the northwestern Dnieper-Donets Depression (in Russian). In: Materials on geology and oil-and-gas potential of Ukraine. Nedra, Moscow, pp 69–78

Suveizdis PI (1963) Upper Permian deposits of the Polish-Lithuanian Syneclise (in Russian). In: Problems of geology of the Lithuania. Mintis, Vilnius, pp 225–372

Suveizdis PI (1968) Die Zechstein-Sedimentation in der Polnisch-Litauischen Syneklise. Z Angew Geol 14:25–28

Suveizdis PI (1975) Stratigraphy (in Russian). In: Permian system of the Baltic area. Mintis, Vilnius, pp 10–25

Suveizdis PI (1976) Stratigraphy of Permian deposits in the Baltic area (in Russian). In: Materials on stratigraphy of the Baltic area. Mintis, Vilnius, pp 86–87

Svidzinsky SA (1971a) Geological structure and study of the Elton salt dome potash salt deposit (in Russian). Avtoref Kand Diss, Rostov-on-Don, 26 pp

Svidzinsky SA (1971b) A composite normal section of the northeastern Elton salt dome (in Russian). In: Geology and genesis of mining-chemical raw material deposits. Tr Geol Inst 31, Kazan, pp 132–141

Svishchev MF (1961) Geological structure and oil-and-gas potential of the Orenburg area (in Russian). Gostoptekhizdat, Moscow, 228 pp

Szaniawski H (1965) New stratigraphical subdivision of the Zechstein in the Gałezice-Kowala Syncline, Swiety Krzyź Mountains (in Polish). Kwart Geol 9:575–590

Szaniawski H (1966) Facial development and paleogeography of the Zechstein within the elevation of Łeba (in Polish). Acta Geol Polonica 16:229–247

Szatmari P, Carvalho RS, Simoes IA (1979) A comparison of evaporite facies in the Later Paleozoic Amazon and the Middle Cretaceous South Atlantic salt basins. Econ Geol 74:432–447

Taft R (1946) Kansas and the nation's salt. Kansas Acad Sci Trans 49:223–272

Tait DB, Ahlen JL, Gordon A, Scott GL, Motts WS, Spitler ME (1962) Artesia Group of New Mexico and West Texas. Am Assoc Pet Geol Bull 46:504–517

Tasler R (1966) Upper Carboniferous and Permian Basins. The Infra-Sudetic Basin (pp 434–444). The Krkonose Piedmont Basin (pp 445–451). In: Regional geology of Czechoslovakia. Part I. The Bohemian Massif. Czechoslovak Acad Sci, Prague

Taylor JCM, Colter VS (1975) Zechstein of the English section of the Southern North Sea Basin. In: Petroleum and the continental shelf of north-west Europe, vol I. Appl Sci: Barking, England, Appl Sci, pp 249–263

Teichmüller R (1958) Ein Querschnitt durch den Südteil des Niederrheinischen Zechsteinbeckens. Geol Jb 73:39–50

Tesakov YuI (1981) Evolution of ecosystems of old platform sedimentary basins (in Russian). In: Problems of evolution of geological processes. Nauka, Novosibirsk, pp 186–199

The encyclopedia of World regional geology, part I (1975) Fairbridge RW (ed) Distributed Holsted Press. Wiley

Thorsteinsson R, Tozer ET (1970) Geology of the Arctic Archipelago. In: Douglas RJW (ed) Geology and economic minerals of Canada. Geol Surv Can Econ Geol Rep 1:549–590

Tikhvinskaya EI (1967) Permian system. Kazanian stage (in Russian). In: Geology of the USSR, vol XI (Volga area and Kama area). Nedra, Moscow, pp 369–395

Tikhvinskaya EI, Chepikov KR (1967) Permian system. Upper series. Ufimian stage (in Russian). In: Geology of the USSR, vol XI. Nedra, Moscow, pp 357–369

Tikhvinsky IN (1965) Principles of Lower Permian stratigraphy of the Volga-Vyatka Interfluve (in Russian). Dokl Akad Nauk SSSR 164:654–657

Tikhvinsky IN (1967) Permian system. Lower series (in Russian). In: Geology of the USSR, vol XI (Volga and Kama areas). Nedra, Moscow, pp 339–357

Tikhvinsky IN (1970a) Lithofacies characteristic and sedimentary environment of Kazanian deposits in the Western Transkama area, southwestern Tataria and northern Ulyanovsk region and paleogeography of the area during Kazanian age (in Russian). In: Stratigraphy and lithology of the Volga-Ural area. Tr Geol Inst 26, Kazan, pp 205–225

Tikhvinsky IN (1970b) Sedimentary environment in the Asselian-Sakmarian Sea in Tataria and adjacent areas (in Russian). In: Paleozoic stratigraphy and lithology of the Volga-Ural area. Kazan, pp 170–181

Tikhvinsky IN, Tikhvinskaya EI, Igonin VM, Zolotova VP, Khursik VZ (1967) Permian system. Kungurian, its extent and range (in Russian). In: Materials on geology of the eastern Russian Platform, vol 2. Nedra, Moscow, pp 3–53

Tinnes A (1928) Die ältere Salzfolge Mitteldeutschlands unter besonderer Berücksichtigung des Unstrutgebietes. Arch Lagerstättenforsch (Berlin) 38

Tokarski A (1959) The profile of Zechstein at Chojnice (in Polish). Roczn PT Geol (Kracow) 29:129–150

Tomaszewski JB (1962) Stratigraphic problems of Fore-Sudetic Monocline (in Polish). Rudy Met Niézel (Katowice) 7:547–551

Tomaszewski JB (1981) Development of Zechstein deposits in the vicinity of Lubin and Sieroszowice. In: Int Symp Central Eur Permian. Proceedings. Wydawnictwa Geologiczne, Warszawa, pp 341–355

Tonndorf H (1965) Beiträge zur Geochemie des randnahen Zechsteins in den Mulden von Zeitz bis Schmölln und Borna unter besonderer Berücksichtigung der Stratigraphie, Fazies und Paläogeographie. Freib Forsch H C 187:140

Treesch MI (1973) Depositional environments of the Salina Group (upper Silurian) in New York State. Ph D thesis, Rensselaer Polytechnic Inst, 127 pp

Tretyakov YuA, Sapegin BI (1981) Stratification of salt-marl strata in the Verkhnekamsk potash salt deposit area (in Russian). In: Structure and sedimentary environment of salt-bearing formations. Nauka, Novosibirsk, pp 52–59

Trofimuk AA (1950) Oil potential of Paleozoic in Bashkiria (in Russian). Gostoptekhizdat, Moscow Leningrad, 246 pp

Trusheim F (1957) Über Halokinese und ihre Bedeutung für die strukturelle Entwicklung Norddeutschlands. Z Geol Ges 109:111–158

Trusheim F (1960) Mechanism of salt migration of northern Germany. Am Assoc Pet Geol Bull 44:1519–1540

Trusheim F (1964) Über den Untergrund Frankens. Ergebnisse von Tiefbohrungen in Franken und Nachbargebieten. Geol Bav 54:1953–1960

Tumanov RR (1981) Sedimentary environment of the Sakmarian salt strata of the Moscow Syneclise (in Russian). In: Structure and sedimentary environment of salt-bearing formations. Nauka, Novosibirsk, pp 69–74

Turkov OS, Shudavaev K (1981) Structural features of the Permian salt-bearing formation in the southern Caspian Depression (in Russian). In: Structure and sedimentary environment of salt-bearing formations. Nauka, Novosibirsk, pp 29–32

Tyshchenko LF, Faizulina ZKh (1982) New data on stratigraphy of the boundary Cambrian/Precambrian deposits in the Irkutsk Amphitheater (in Russian). Sov Geol 6:52–62

Udden JA (1924) Laminated anhydrite in Texas. Geol Soc Am Bull 35:347–354

Ullrich H (1964) Zur Stratigraphie und Paläontologie der marin beeinflußten Randfazies des Zechsteinbeckens in Ostthüringen und Sachsen. Freib Forsch H C169:229–230

Urusov AV, Ketat OB, Koltsova VV (1967) Correlation of Lower Permian deposits in the Volga-Don Interfluve and their comparison with some areas of the Russian Platform (in Russian). In: Problems of geology and oil-and-gas potential of the Lower Volga area. Nizhne-Volzhsk, Volgograd, pp 86–95

Urusov AV, Koltsova VV, Ketat OB (1962) Stratigraphic scheme for the Permian and Triassic in the Volga area at Volgograd (in Russian). Tr Volgograd NII Neft I Gaz Promyshl, vol 12. Gostoptekhizdat, Moscow, pp 91–110

Vaag OV, Matukhin RG (1982) Environment of carbonate accumulation in Early Carboniferous in the northern Siberian Platform (in Russian). In: Carbonate formations of Siberia and the associated mineral resources. Nauka, Novosibirsk, pp 62–71

Vainblat AB (1969) Stratigraphy and lithology of the Upper Permian deposits in the eastern Caspian Depression (in Russian). In: Permo-Trias of the Russian Platform in relation to its oil-and-gas potential. Nedra, Moscow, pp 79–86

Vakhrameeva VA (1956) On stratigraphy and tectonics of the Verkhnekamsk deposit (in Russian). Tr VNIIG 32:277–313

Vakhrameeva VA, Gorkin OP (1960) Petrographic description of rocks in the Sub-Salt (Podsolevaya) strata and Lower Salt at the Verkhnekamsk deposit (in Russian). In: Petrology of halogenic rocks. Gostoptekhizdat, Leningrad, pp 371–391

Vala AI (1961) Permian system (in Russian). In: Geology of the USSR, vol 36 (Lithuanian SSR). Gostoptekhizdat, Moscow, pp 85–93

Varentsov MI, Ditmar VI, Li AB, Shmakova EI (1964) On the age of rock salt in diapiric structures of the Chu-Sarysu Depression (in Russian). Dokl Akad Nauk SSSR 159:327–329

Varlamov NP (1965) Lithofacies features of the Kungurian halogenic deposits in Bashkiria (in Russian). In: Materials on geology and mineral resources of the South Urals. Nedra, Moscow, pp 67–76

Vary JA, Elenbaas JR, Johnson MA (1968) Gas in Michigan Basin. In: Natural gases of North America, vol II. Am Assoc Pet Geol Mem 9:1761–1797

Vasiliev YuM (1968) Geological structure of the Caspian Depression and oil-and-gas distribution pattern of the interior part. Nedra, Moscow, 278 pp

Vertrees C, Atchison CH, Evans GL (1959) Paleozoic geology of the Delaware and Val Verde Basins. In: Geology of the Val Verde Basin and field trip guidebook. Midland, Texas, West Texas Geol Soc, pp 64–73

Vinogradov VI, Reimech TO, Leites AM, Smelov SB (1976) Oldest sulfates in Archean formations of the South African and Aldan Shields and evolution of oxygen atmosphere of the Earth (in Russian). Lithol I Min Res 4:12–27

Vinogradova NG, Oshchepkov KF (1969) On geological structure of the Permian sulfate-halogenic deposits in the northeastern Volga Monocline (in Russian). In: Problems of geology and oil-and-gas potential of the Lower Volga area. Nihzne-Volzhsk, Volgograd, pp 143–147

Volchegursky LF, Demidov VA, Kopaevich LP (1969) Stratigraphy of the Upper Permian and Triassic deposits in the Inder-Chelkar Region (in Russian). In: Permo-Trias of the Russian Platform in relation to its oil-and-gas potential. Nedra, Moscow, pp 113–119

Vosburg DL (1964) Permian subsurface evaporites in the Anadarko Basin of the western Oklahoma-Texas Panhandle (abs). Diss Abs 25:1852–1853

Voskresensky IA, Kravchenko KN, Movshovich EB, Sokolov BA (1971) An assay on Pakistan geology (in Russian). Nedra, Moscow, 168 pp

Vozárová A, Vozár J (1981) Division of the Permian in the West Carpathians. In: Int Symp Central Eur Permian. Proceedings. Wydawnictwa Geologiczne, Warszawa, pp 95–102

Vysotsky EA, Kislik VZ, Protasevich BA (1981) On paleotectonic setting of potash salt accumulation in the Famennian in the Pripyat Trough (in Russian). In: Structure and formation conditions of potash salt deposits. Nauka, Novosibirsk, pp 67–73

Wagner R, Peryt TM, Piatkowski TS (1981) The evolution of the Zechstein sedimentary basin in Poland. In: Int Symp Central Eur Permian. Proceedings. Wydawnictwa Geologiczne, Warszawa, pp 69–83

Walger G (1962) Vorläufige Mitteilungen zum Vorkommen sulfatischer Kaliminerale in der Schachtanlage Pöthen im Bereich der Störungszone des Schlotheimer Grabens. Ber Geol Ges 6:266–287

Ward PE (1963) Shallow halite deposits in the Flowerpot Shale in southwestern Oklahoma. US Geol Surv Prof Pap 450-E:40–42

Wardel RI, Bailey DG (1981) Early Proterozoic sequences in Labrador. In: Campbell FHA (ed) Proterozoic Basins of Canada. Geol Surv Pap 81-10:331–360

Weber RH, Kottowski FE (1959) Gypsum resources of New Mexico. New Mexico Bur Min Miner Res Bull 68: 69 pp

Weippert D, Wittekindt H (1964) Ein Vorkommen von paläozoischem Salz im westlichen Zentralafghanistan. Geol Jb 82:99–102

Wells AT (1980) Evaporites in Australia. BMR Bull 198. Aust Govern Publ Serv Canberra, 104 pp

Wells AT (1981) Evaporites in Australia. Bur Miner Res Bull 198: 104 pp

Wengerd SA (1962) Pennsylvanian sedimentation in Paradox Basin, Four Corners region. In: Pennsylvanian system in the United States – a symposium. Am Assoc Pet Geol, pp 264–330

Wengerd SA, Matheny ML (1958) Pennsylvanian system of Four Corners region. Am Assoc Pet Geol Bull 42:2048–2106

Wengerd SA, Strickland JW (1954) Pennsylvanian stratigraphy of Paradox salt basin, Four Corners region, Colorado and Utah. Am Assoc Pet Geol Bull 38:2157–2199

Werner Z, Poborski J, Orska J, Bakowski J (1960) A geological and mining outline of the Kladowa salt deposits. Prace Inst Geol 30:467–505

White AJR (1959) Scopolite-bearing marbles and calc-silicate rocks from Tungkillo and Milendella, South Australia. Geol Mag 96:285–306

Wienholz R, Wirth J (1965) Über die Erdöl- und Erdgasführung des Zechsteins im Nordostdeutschen Flachland. Erdöl Erdgas Inform. VVB Erdöl Erdgas, Gommern 7:5–17

Wijhe DH van (1981) The Zechstein 2 carbonate exploration in the eastern Netherlands. In: Int Symp Central Eur Permian. Proceedings. Wydawnictwa Geologiczne, Warszawa, pp 574–586

Wilder FA (1923) Gypsum, its occurrence, origin, technology and uses, with special chapters devoted to gypsum in Iowa. Iowa Geol Surv 28:47–537

Wilkening W (1959) Die Kalisalzlagerstätten Niedersachsens. Mitt Markscheidewesen 66:66–79

Wilson RF (1975) Eastern Colorado. In: McKee ED, Crosby EJ (eds) Paleotectonic investigations of the Pennsylvanian system in the United States, part I: Introduction and regional analyses of the Pennsylvanian system. Geol Surv Prof Pap 853:243–264

Wolburg J (1958a) Ein Querschnitt durch den Nordteil des niederrheinischen Zechsteinbeckens. Geol Jb 73:7–38

Wolburg J (1958b) Zur Frage der basalen Fazies des 2. Zechsteinzyklis im Innern des Niederrheinbeckens. Geol Jb 73:165–170

Wold RJ, Paull RA, Wolosin CA, Friedel RJ (1981) Geology of Central Lake Michigan. Am Assoc Pet Geol Bull 65:1621–1632

Wood J (1973) Stratigraphy and depositional environments of Upper Huronian rocks of the Rawhide Lake-Flack Lake Area, Ontario. In: Young GM (ed) Huronian stratigraphy and sedimentation. Geol Assoc Can Spec Pap 12:73–96

Woods PJE (1979) The geology of the Boubly Mine. Econ Geol 74:409–418

Worsley N, Fuzesy A (1979) The potash-bearing members of the Devonian Prairie Evaporite of southeastern Saskatchewan, south of the mining area. Econ Geol 74:377–388

Yanev S (1981) The Permian in Bulgaria. In: Int Symp Central Eur Permian. Proceedings. Wydawnictwa Geologiczne, Warszawa, pp 103–126

Young GM (1981) The Amundsen embayment. Northwest Territories; relevance to the Upper Proterozoic evolution of North America. In: Campbell FHA (ed) Proterozoic basins of Canada. Geol Surv Can Pap 81-10:203–218

Zaitsev NS (1940) On the structure of the Sarysu domes (in Russian). Izv Akad Nauk SSSR Ser Geol 5:88–113

Zamaraev SM, Malykh AV, Geletiy NK (1981) Tectonic setting of the halogenic formation deposition and postsedimentary tectonics of potassium-bearing deposits in the Nepa-Gazhenka Region (in Russian). In: Structure and depositional environments of potash salts. Nauka, Novosibirsk, pp 40–47

Zamarenov AK (1962) Upper Paleozoic of the Cis-Urals, Aktyubinsk area. Izd Akad Nauk Kaz SSR, Alma-Ata, 90 pp

Zamarenov AK (1970) Middle and Upper Paleozoic on the eastern and southeastern flank of the Caspian Depression (in Russian). Nedra, Leningrad, 170 pp

Zamel A (1975) Recent exploratory development in the Middle East. Proc 9th World Pet Congr, vol 3

Zharkov MA (1981) History of Paleozoic salt accumulation. Springer, Berlin Heidelberg New York, 310 pp

Zharkov MA, Chechel EI (1964) Late Precambrian and Cambrian deposits in the Chaya River Basin (in Russian). Dokl Akad Nauk SSSR 159:85–88

Zharkov MA, Chechel EI, Knyazev IM (1964) Cambrian deposits of the Middle and Lower Kirenga River (in Russian). Dokl Akad Nauk SSSR 149:822–824

Zharkov MA, Zharkova TM, Merzlyakov GA et al (1980) Bishofite deposits of the Volga monocline (in Russian). In: Structural features of bishofite and potash salt deposits. Nauka, Novosibirsk, pp 4–32

Zharkov MA, Merzlyakov GA, Yanshin AL (1982) Discovery of potash salts in Siberia (in Russian). Priroda 7:3–10

Zharkov MA, Sovetov YuK (1969) Irkut horizon, its range and stratigraphic position (in Russian). In: Lower Cambrian and Upper Precambrian in the southern Siberian Platform. Nauka, Moscow, pp 34–55

Zholtaev GZh (1971) Geological history of the eastern margin of the Caspian syneclise in the Kungurian age (in Russian). In: Geology and prospecting of the Earth's interiors, vol 2. Alma-Ata, pp 65–77

Zhuravlev VS (1972) Comparative tectonics of the Pechora, Caspian and Severomorsk exogonal depressions on the European Platform (in Russian). Nauka, Moscow, 400 pp

Zhuravlev VS, Dal'ian IB, Soloviev BA, Fomin GV (1972) Kazanian salt deposits in the eastern Caspian Depression and its northern periphery (in Russian). Byull MOIP Otd Geol Nov Ser 47:40–54

Zhuravlev VS, Svitoch AA (1971) On original thickness of Permian salt deposits in the Caspian Depression (in Russian). In: Problems of geology of western Kazakhstan. Nauka, Alma-Ata, pp 197–204

Ziegler PA (1977) Geology and hydrocarbon provinces of the North Sea. Geol J 1:7–32

Ziegler PA (1978a) North Sea rift and basin development. In: Ramberg IB, Neumann ER (eds) Tectonics and geophysics of continental rifts. NATO Adv Study Inst Ser Math Phys Sci Ser 2:249–277

Ziegler PA (1978b) North-western Europe: Tectonics and basin development. Geol Mijnbouw 57:589–626

References

Ziegler PA (1980) Northwestern Europe: subsidence patterns of Post-Variscan basins. In: Geology of Europe (from Precambrian to post-Hercynian sedimentary basins). Mem Bur Rec Geol Min 108:249–280

Zoricheva AI (1963) Permian system. Northern Russian Platform and southwestern Timan area. Upper Division. Kazanian and Tatarian (in Russian). In: Geology of USSR, vol II. Gosgeoltekhizdat, Moscow, pp 488–532

Zubov IP, Kunin NYa, Volozh YuA et al (1972) New data on the structure of Caspian Depression in relation to assessment of its oil-and-gas potential (in Russian). Sov Geol 11:25–37

Subject Index

A, A-1, A-2 Carbonate, Evaporite, Unit of Michigan Basin 70–74, 76
Abakan Formation 158
Abava River 124
Abo Formation 330, 331
Abu Dhaby (Abu Zaby) Group of Salt Plugs, Domes 46
Achikkul Lake, Basin 161, 196, 200
Adak horizon 90
Adamovo field 120, 121
Adavale Basin 91, 155–157
Adelaide Geosyncline 4, 13
Aden 52
Adirondack Mountains 4, 12
Admiral Formation 330
Admire Group 330–332, 336
Adze District 105
Afghanistan 53, 160, 163
Africa 4, 13, 161, 201
Agakukan Formation 95
Agaleva Formation 18
Aghagrania Formation 199
Agra town 50
Ahlbornbank 285
Ainslie Formation 176
Ainslie Lake 180
Aislaby Group 254, 257, 299
Aistmark Formation 255, 257, 297
Akan cycle 187, 188, 191
Akarshura District 126
Akka town 160
Akkol Anticline 162
Akmene area 124
Akniste town 90
Akron Formation 71
Akshat Formation 210
Aksu Basin 161, 196, 200
Aksu River 163
Aktastin horizon 210, 211
Aktaylyak Formation 164
Aktyuba section type 235
Aktyubinsk area, town 214, 235, 239

Aktyubinsk Formation 210
Akul District 126
Al Buza Salt Plug 48
Alanda area 124
Alay Range 163
Albania 323, 363
Albany River 151
Albert Formation, rock salt 175–177, 181
Alberta 64, 66, 128–135, 140–145, 170
Alberta Basin 177
Alcoota area 59
Aldan Anteclise, Shield 11, 14, 16, 19, 28, 36, 41, 42
Alebastrovyi Island 196
Aleksandrovsk section type 235
Alekseevo Village 239
Alexandershall Mine 258
Algalzone 282
Algerian Sahara 201
Algonquin Arch 69, 145
Alice No. 1 borehole 7, 56, 57
Alida Member 170, 171
Alkali Gulch cycle 187–189
Alkali Trough 192, 193
Allegheny Basin, county 71, 149
Allen Bay Formation 64
Aller-Anhydrit 254
Aller River Basin 289–293, 303, 304, 314
Aller Series 256–258, 305, 317–321
Aller-Steinsalz (Na3) 252, 254–256, 314, 316–318
Allertal Graben 269, 316
Allertal town, Potash District 280, 287, 308
Alma-Ata 163, 240
Alpine Basin 203, 323–327
Altenburg town 266, 281, 306, 309
Altmark Basin, Depression 249, 251, 255, 256, 258, 269, 271, 290, 293–295, 301, 310, 311, 314, 316, 317, 320, 322
Altuda Formation 349
Aluksne town 90
Alva No. 1 well 155, 156

Amadeus Basin 4–9, 11, 15, 56–58, 87
Amakan Formation 210
Amarillo Mountains, Uplift 328, 329, 335, 336
Amarkan Formation 67
Amazon River Basin, Depression 161, 194–196
Amelia Dolomite 13
Amga stage 18, 24, 25, 44, 45
Amherstberg Formation 145, 146, 148
Amsterdam 246, 248, 250, 251, 253, 259, 279, 302, 315
Amul Formation 123, 124
Amund Ringnes Island 165, 166
Amundsen Basin 4
Amundsen Embayment 12
Anabar Anteclise 16, 31, 41
Anadarko Basin 60, 68, 328–330, 332, 335–338, 340–343
Ancestral Casper Arch, Uplift 328, 329
Ancestral Front Range Uplift 328
Anderson Formation 201
Andes 58
Andizhan town 163
Andylakh Formation 99
Aneth area 188
Angara fold zone, folds 16, 18
Angara River 16, 18, 36, 38
Angara salt sequence 20, 44
Angara-Lena Trough 16–20, 25, 31, 36, 42, 43
Angir Formation 67
Anhydrite Member 234
Anhydrite of West Canadian Basin 130
Anhydritfolge 305
Anhydritknotenschiefer (CaA1) 254, 256, 260, 271
Anhydritmittelsalz (Na3η) 254, 255, 303, 305–308, 311
Anhydritmittelsalz (Na4ξ) 318
Anhydritregion (Na2β) 282–284, 287
Anisian 326
Anisimovka Village 105, 109
Annet Basin 161, 201
Anthraconite horizon 211, 232, 233
Anti-Atlas 15, 58, 160
Anticosti Island 174
Antigonish town 174, 180
Antigonish-Mabou sub-basin, Upland 174–176, 180
Antipovka rhythmic unit 231, 232
Antler orogenic belt 328
Antoinette Formation 167, 169
Antonovo field 120
Antrim Formation 146–148
Anyba well 122

Apishapa Uplift 328
Appalachian Basin, Depression 69–74, 76–78, 145–147, 181, 183
Appalachians 59
Arabian Basin, salt area 47, 203, 362, 363
Arabian Platform 46
Arabian Peninsula 5, 46, 51, 53
Arabian-Iran Hindustan Platform 53
Arbuckle Uplift 328, 329
Archean 4, 9, 11, 14
Arckaringa Basin 9, 15, 58, 91, 158, 160
Arctic Bay Formation 12
Arctic Coastal Plain 61
Arctic Ocean 166
Arctic Platform 61, 62
Aregala area, borehole, District 105, 124, 125
Arenigian 84
Areyonga Formation 5, 6, 8, 57
Argentina 361, 363
Arizona 184, 191, 328, 331, 357–359
Arkansas 328, 333, 334
Arktag Formation, sequence 200
Armorican Massif 247
Arrithrunga Formation 59
Arroya Formation 330
Artesia Group 330, 348, 349
Artinskian 123, 167–169, 208–211, 218, 219, 241, 329, 330, 363
Artygan Formation 95
Arumbura Sandstone 56–58
Aschersleben town, Potash District 269, 280, 287, 288, 308, 316, 318
Aschersleben-Stassfurt Syncline 308
Asgan-Bulag-Taga area 200
Ashchi Region 239
Ashern Formation or Third Red Beds 130, 133, 135
Asia 15, 45, 64, 161
Askyz Formation 158, 159
Asselian 167, 169, 208–215, 218, 241, 329, 330
Assistance Formation 167, 169
Astrakhan town 236
Atbashi Range 163, 165
Athens 324
Atlantic Continental Shelf 174, 175
Atlantic Ocean 194
Atoka Series 186
Atovo borehole, field 17, 27
Atyrdakh Formation 95
Atyrkan Formation 95, 96
Atzendorf-Latdorf Graben 308
Audhild Formation 167
Australia 2, 4, 5, 12, 14, 15, 59, 60, 64, 68, 83, 87, 155, 157, 161, 201, 364
Austria 323

Subject Index 399

Austrian Alps 59
Autunian 249, 363
Axel Heiberg Island 165, 166
Ayagkumkul Lake 200
Ayana borehole, field 17

B Evaporite, Unit of Michigan Basin 70, 71, 75, 76
Babbagoola Beds 10, 11
Bache Peninsula 61
Bad Reichenhall Basin, Trough 324–326
Baden Trough 245, 247
Baffin Island 12
Bagaryak District 200
Baggy complex of seams, Member 51
Bagovitst Formation 79, 80
Baigendzhin Formation 210
Baikal Lake, area 19, 27–29, 31, 36, 37, 41, 42, 100
Baikal-Patom Highland 16, 17, 27–29, 31, 36, 41, 42
Baikit Anteclise 16–18
Baitugan town 127, 197
Bakhmut Trough 205, 209, 211, 213, 217
Bakhta Arch 16, 17
Bakken Formation 170
Baku 47
Balakhna town, District 126
Balakhonikha town 126
Balakovo town 127
Balkhash Lake 240
Baltic area 102, 105
Baltic Basin, Depression 60, 68–70, 89
Baltic Sea 249
Baltic Shield 206
Baltic Syneclise 106, 123–125, 128
Balykhta Trough, field 27, 164
Balykla town 127
Balyklei rhythmic unit 231
Bamble Trough 247
Bancannia Trough, Basin 91, 158, 160
Bändersalz (Na3ϵ) 254, 255, 303
Bangemall Basin 9
Banksalz (Na3δ) 254, 303
Barents Sea 206
Barker Creek cycle 187, 188, 191
Barrier Reef 134
Barrow Dome 166
Barut Formation 52
Baryatino town 104
Basal Anhydrite (Basalanhydrit) A2 254–256, 272, 280–297, 301
Basal Conglomerate 273
Basal Halite Member 176, 178
Basal Red Beds of West Canadian Basin 130, 131, 133

Basal Zechsteinkonglomerat 256
Basalsalz 290
Bashkir Arch 205
Bashkiria 228, 229
Bashkirian 163–165, 167–169, 197, 200
Basissalz (Na2a) 254, 255, 282, 303, 305
Basissalz (Na3a) 284, 306–308, 311
Basissalz (Na4a) 316–318
Bass Island Formation, Group 70, 71, 78
Bastak area 46
Bastakh Formation 95
Bathurst Caledonian River J-34 borehole 61–63
Bathurst Inlet District 11
Bathurst Island 61–64, 166
Batten Trough 13
Baumann Fiord Formation 61, 62
Baurchin Anticline, area 81, 106
Baus Formation 123, 124
Bay Fiord Formation 62, 63
Bayandor Formation 52
Bazun Pir area 48, 53
Bear Member 130, 136
Beaufort Sea 56
Beaverhill Lake Formation, Group 130, 140
Beckham County 335, 338, 339, 341
Beckham Series, Evaporite 335–337, 342, 343
Bedford Formation 146, 148
Begleitlager 263, 264
Beisk (Beysk) Formation 98, 158, 159
Belaya Trough 229
Belbasol horizon 211
Belbasov Motley Bed 215
Belbazh rock salt deposit 217
Belcher Channel Formation 167, 169
Belcher Group 11
Belden Formation 185, 186, 191, 192
Belebeevo Formation 210
Belfast 246, 248, 250, 251
Belgrade 80, 324
Bell Arch 149
Belle Plaine Member 130, 135, 137, 139, 140
Belle Plains 330
Bellerophon Beds, Formation, Series 323, 363
Belokatai Formation 210
Belsk Beds, Formation, horizon, sequence 5, 18, 20, 44, 53, 210
Belsk 1-0 borehole, field 17, 27
Berdov borehole 159
Berea Formation 146, 148
Berezniki Formation, sequence 210, 221, 225
Berezniki town, Region 224–227, 235
Bereznyansk field 120
Berezovaya Basin, Depression 16, 19, 21, 23, 28–33, 36, 41, 42, 67, 81
Berezovka Village 127

Bergamasche Alps 363
Bergmannssegen Beds 250
Bergmannssegen-Hugo Potash District 303, 316
Berlepsch-Maybach Mine 308
Berlin 246, 248, 250, 251, 253, 259, 279, 294, 302, 310, 315, 324
Bernburg town, Potash District 269, 280, 287, 288, 308, 316
Berne 324
Bertie Formation 70–72, 78
Bestube Dome, District 100, 101, 162
Betpak-Dala Salt Domes, area 100, 101
Bezvoditsa Village, area 242
Bezvoditsa 9–75 well 243
Bielefeld 289, 304
Big Snowy Formation, Group
Big Valley Formation 130, 131
Bighorn Group 64, 65
Billingham Main Anhydrite 254
Bilovo town 126
Bird Fiord Formation 159
Birdbear Formation 130, 144
Birsk Cap, Saddle 205, 228, 229
Birzhai town 124
Bistcho Member 136
Bitter Springs Formation, sequence, series 5, 6, 8, 9, 11, 56, 57, 87
Black Creek Formation 131, 134
Black Flag Beds 11
Black Hills Uplift 129, 331
Black Mesa Basin 186
Black Sea 78, 80, 243
Blacksburg town 181, 182
Blackwater Trough 155, 157
Blagodarnen Formation 210
Blaine Anhydrite, Formation, Gypsum 329–332, 335, 337, 339, 340, 342–344, 346
Blasenkalk 254, 271
Blauer Salzton (Na3tmI) 303
Blidene District 105
Block Inlier area 12
Bloods Range 8
Bloomsburg Formation 71, 72, 75, 76
BMR Alice Springs No. 3 borehole 8
BMR Lake Amadeus No. 3, No. 3A, No. 3B boreholes 7, 8
BMR Madley No. 1 borehole 7, 10
BMR Mount Liebig No. 1 borehole 6, 7
BMR Warri No. 20 borehole 7, 10
Bobr River 272
Bochakta borehole, field 17, 27, 33
Bocholt town 270
Bohemian Massif 245, 325
Bois Blanc Formation 145, 146, 148
Bokaly River 164

Bokhan borehole, field, Village 17, 27
Bolivia 361, 362
Bolivian Andes 58
Bolshekinelsk Formation 210
Bolsherazvodnaya borehole 27
Bolshesognino Basin 204
Bolshoi Kinel River 239
Bolshoi Patom River 18
Bolshoi Yllymakh River Basin 11
Bonaparte Gulf Basin 91, 157, 158
Bone Spring Limestone 330
Bonn 246, 248, 250, 251, 253, 259, 279, 302, 315, 324
Bonnie No. 1 well 155, 156
Bootia Uplift 61
Borden Basin, Group, Formation 4, 12, 157, 201
Borden Island 166
Boree Member 155, 156
Boree Nos 1, 3 wells 155, 156
Borodulino town 125
Borovka Village 127
Borovsk Anticline 159
Borsul-Varde town 81
Borthot town 270
Borup Fiord Formation 167, 168
Boskovice Trough 245, 247, 249
Boss Point Formation 176
Boulby Halite, Potash 254, 258, 313
Boulder Knoll Anticline 190
Boundary Anhydrite (A4br) 255, 322
Bozhekhan No. 1sp borehole 17
BP 46/6-1 well 301
Bragin (Bragin-Chernigov) Uplift 106, 107
Bratsk Bulge, town 16
Bratsk Formation, field 17, 23, 67
Bratsk No. 1-r, 3-r boreholes 17
Braunroter Salzton 254, 260, 264, 290–292
Braunschweig town 289, 304
Bravo Dome 328, 329
Brazil 194, 363
Brazos State Foster No. 1 well 59
Brinzeny Village 79
Britain 245
British Columbia 129, 133, 134
Broom Creek Group 330, 331
Broome Platform 83, 87
Brown Beds, diapir 7–11
Brown Sound Formation 11
Brown Zubers 257, 312
Browne Nos 1, 2 boreholes 7
Browse Basin 203, 362, 364
Bruss Formation 95
Brussels 246, 248, 250, 251, 253, 259, 279, 302, 315
Bryantsevo Bed 211, 213

Bucharest 80, 324
Buckabie Formation 156
Budapest 324
Bug River 246, 248, 250, 251, 253, 259, 279, 302, 315
Bugrovka Village, District 127
Buguruslan town 239
Buinovich Depression 117
Bulai Formation 18, 20
Bulgaria 159, 242, 323
Bullara Syncline 88
Buntes Salz (Na3ξ) 254, 303
Burakovo town 120
Bureinak Dome 101
Burgundy Trough 245, 247
Burkhala Formation 96
Bury Limestone Member 156
Bury No. 1 well 155
Butte area 188
Buzbash town 127
Buzuluk Depression, area 205, 210, 229, 230, 239, 240
Buzuluk town 239
Byelorussian Massif, Shield 105–107, 245
Byro sub-basin 88
Bysyuryakh Formation 67

C Unit of Michigan Basin 70, 71, 75, 76
Cache Valley Anticline 190
Cahill Formation 12
Cain Formation 71, 72
Călărasi town 81
Caledonian 245
Caledonian Upland 174
Calgary 129, 133–135, 142
California 328
Callanna Beds 13
Callville Limestone 331
Calmar 130
Calumet 130
Calvörde District, region 269, 271, 293, 309, 311, 317, 322
Cambell Settlement Member 176
Cambrian 2, 5–11, 15–61, 68, 69, 157, 160, 167, 181, 182, 184
Camillus Formation 70–72, 77
Camrose 130
Canada 12, 54, 64, 68, 69, 128, 142, 170–173, 347, 350
Canadian Arctic Archipelago, Basin 60–64, 69, 70, 89–91, 158, 159, 165
Canadian Shield 14, 60, 128, 145, 149
Cane Creek Anticline 187, 189, 190
Canning Basin 9, 60, 68–70, 83–87, 91, 158, 160
Canol 130

Canon Fiord 169
Canso Group 176, 181
Canyon Fiord Formation 167
Capacabana Formation, Group 361
Cape Breton Island, Upland 174, 180
Cape Colquhoun Dome 166
Cape Phillips Formation 64
Capitan Carbonate Reef, Limestone 349, 351, 353
Carbonate horizon, sequence of East European Basin 211, 213
Carbonate-terrigenous sequence of East European Basin 211
Carboniferous 8, 83, 95, 97, 99, 100, 101, 122, 123, 157, 159, 161–202, 357, 363
Carlsbad District 355, 356
Carmichall diapir (structural) 7, 8
Carnallite-Halite Member of East European Basin 211
Carnallitic Marl 320
Carnarvon 88
Carnarvon Basin 69, 70, 87–89, 91, 158, 160
Carnic Alps 323, 363
Carribuddy Formation 84–87
Caspian Depression 102, 199, 205, 206, 209, 210, 214, 218, 221, 228–231, 234, 235, 237, 239, 240
Caspian Sea 47, 206
Caspian Syneclise 2
Cassiar Platform 54
Cassidy Lake Formation 176, 178
Castile Formation 330, 351, 353, 354
Castle Valley Anticline 190
Cattle Creek Anticline, town 192, 193
Caucasus Range 205
Cayugan 71, 75
Cedar Creek Anticline 67, 328, 329
Cedar Hills Sandstone 331, 336, 343
Celia Dolomite 12
Celle town 289, 292, 304
Cenozoic 87, 157
Central Afghanistan 160
Central Alberta sub-basin 128, 129, 131–136, 140
Central Andes 361
Central Appalachians 59
Central Asia 200, 201
Central Basin Platform 328–330, 335, 349, 351
Central Bohemian Trough 245, 247, 249
Central England Basin 161, 196, 199
Central European Basin 2, 203, 244, 245, 252, 256–259, 272, 275, 278, 279, 299, 301, 302, 314, 315, 318, 320–324
Central Fars (Group of Salt Plugs) 46
Central Iowa Basin 91, 158, 159

Central Iran 5
Central Kazakhstan 99
Central Montana Uplift 129
Central Mount Stuart Beds 59
Central Nevada Ridge 328
Central North Sea Basin 247
Central Polish Depression 249, 251, 252, 255, 257, 258, 272–274, 276, 278, 295–297, 299, 301, 311–314, 316, 318–322
Central Pomeranian Peninsula 274, 297, 298
Central Urals 200
Central West Europe 253, 254
Čersnica Mountains 363
Český-Budějovice Trough 245, 247
Chaarkuduk Formation 200
Chan Benu Salt Plug 48
Chandler Evaporite, Limestone 56–58
Changala River 95
Channel Basin 247
Chara Formation, horizon 18–20, 36, 40–44
Charles Formation 170, 171
Chase Group 330–332, 336, 337
Chasov gravity low 122
Chasov-Yar horizon 215, 217
Chastin No. 2-ch borehole 17, 19, 20, 23
Chatkal Range 163, 165
Chattanooga Shale 181
Chaya River 18
Chebotarikha No. 1-sp borehole 17
Cheepash River 152
Chelkar Dome, Uplift, Beds 211, 230, 231, 233, 234
Chellokkoin No. 6 borehole 164
Chełmo Lowland 247
Chemanda Formation 164
Chemkis-Iol 127
Cherdyn Region 198
Chereshovo 242
Chernigov Uplift 205
Chernukha Anticline 118
Chervonaya Sloboda 116
Chervonaya Sloboda field 105, 112
Cheverie Formation 176
Cheverie town 180
Chia-Sapri Formation 364
Chichino area 126
Chihuahua Basin 360, 361
Chihuahua State 359
Chile 362
Chimkent Basin, Depression 161–163, 165, 196, 200
Chimkent town 163
China 59, 163, 200
Chinarevo field 230
Chinchaga Formation 130, 132, 133
Chipewyan Member 130, 135, 136

Chojnice 2 well 297
Choza Formation 330
Christina 130
Chu Depression 241, 242
Chu Ili (Chu Iliyskiy) Mountains (Range) 99, 100
Chu River 100, 163, 240
Chubovka 127
Chudovo Beds 125
Chugwater Formation 331
Chukhloma town 126, 198
Chulym-Yenisei Basin, Depression, deposit 91, 98, 158, 159
Chuquichambi Evaporite, Formation, Salts 361, 362
Chu-Sarysu Basin, Depression 2, 91, 99–101, 161–163, 203, 241, 242
Chusovaya Basin, Depression, Salt Anticline, Dome 121, 204, 209, 210, 221, 227
Chuya River, Depression 18
Cimarron Anhydrite, Evaporite, Formation, Salt 330, 335–343
Cimarron Uplift 328, 329
Circle Cliffs Uplift 184
Cis-Andean Basin, Depression, Trough 15, 194
Cis-Carpathian Depression 80
Cis-Sayans 19, 22, 28, 31, 33, 36, 41–43
Cis-Sayans-Yenisei Syneclise, Trough 16–19, 25, 29, 31, 36, 41, 42
Cis-Sulaiman Foredeep 50
Cis-Uralian Trough, area 2, 122, 204–206, 209, 210, 214, 218, 221, 228–230, 235
Cis-Urals 125, 198, 235, 239
Cis-Verkhoyansk Trough 16
Clay-Carbonate Sulfate-Salt Member of Solikamsk Depression 221, 222
Clear Fork Group 330, 341, 342
Clelend Anticline 8
Cleveland 1 borehole 300
Clinton Formation 70, 71
Cloud Chief Formation 330, 336
Clover Hill Formation 178
Clyde-Ouse Trough 247
Cobalt Group 11
Cobequid Upland 174
Cochabamba-Villa Tunari Road 58
Cold Lake Formation 130–133
Collio sequence 363
Colorado 184, 191, 327, 328, 330–334, 342–347, 350, 351, 357–359
Colorado Plateau, Trough 184, 328, 357, 358
Columbia 361
Concha Limestone 360
Concretionary Formation, Limestone 299
Cononino Sandstone 331
Constanta town 81

Subject Index 403

Contact Rapids Formation 130, 132, 133
Cooladdi Dolomite, Dolomite Member, Trough 155, 157
Coolcalalaya sub-basin 88
Coomalie Dolomite 12
Cootanoorina well 160
Copenhagen 246, 248, 250, 251, 253, 259, 279, 302, 315
Copes Bay Formation 61
Copper Shale (Kupferschiefer T1) 255, 257, 272, 273
Coquina zone 282
Cordilleran Eugeosyncline, Geosyncline, Miogeosyncline 129, 328
Cornish Platform 247
Cornwallis Formation, Group 62–64
Cornwallis Island 61, 62, 166
Corry Member 199
Council Grove Group 330–332, 336
Cove Creek Limestone 183
Cover Salt of East European Basin 211
Craignish Formation 176
Cretaceous 8, 9, 83, 89, 134, 157, 323
Crimea 206
Cuchillo Prado 360
Cuchillo Prado Nos 1, 2 wells 360, 361
Cumberland area, Group, sub-basin 174–176, 178, 179, 181
Curtin Springs 7, 8
Cutler Group 186, 191
Cuzco 361
Cyclops Member 6
Czechoslovakia 80

D Evaporite, Unit of Michigan Basin 70, 71, 75, 76
Dadonkov deposit 98
Dakota Shelf 129, 133
Daldyn Uplift 196
Daly River, Basin 15, 59
Danilovo No. 124 borehole, field 17
Dankov horizon 126
Dankov-Lebedyan Beds, horizon 126–128
Danube River 246, 248, 250, 324, 325
Danynie River 96
Darakhov town 78–80
Darvaza Basin 362
Darvaza Range 203, 362
Darwin block 157, 158
Dashti (Group of Salt Plugs) 46
Daugava Formation, River 123, 124
Davernay Majeau Lake 130
Davidson Evaporite, Formation 130, 131, 140, 142, 143
Davydovka area, Village 105, 114
Dawson Bay Formation 129, 130, 133, 137, 139

Dawson-Settlement Member 175, 176
Day Creek Dolomite 330, 331, 351
De Chelly Dandstone 331
Deadwood Formation 65
Dean Sandstone 330
Deckanhydrit (Screening Anhydrite A2r) 254–256, 280, 286–288, 291–293, 295
Decksteinsalz (Screening Halite Na2r) 254–256, 280, 283, 286–288, 291, 295
Defence Plant Corporation Reeder 1 borehole 189
Defiance Uplift 184, 328, 357
Deforest Lake Member 176
Degerbols Formation 167, 169
Delaware Basin 328–330, 335, 349, 351, 353–356
Delaware Mountain Group 330, 349
Delgei field, No. 2-r borehole 17, 21
Delhi 50
Demidkovo Member 210, 221, 222, 229
Denault Formation 11
Denmark 244, 309
Denver Basin 328–330, 344, 346, 350
Des Moines Series 186
Desert Creek cycle 187, 188, 191
Detroit River Group 145, 146
Devine Corner Member 176, 178
Devon Island 61, 63, 64, 166
Devonian 2, 8, 60, 62, 69, 71, 83–85, 87, 90–160, 162, 196
Devyatino Formation 198
Dewey Lake Red Beds 330, 351–353, 357
Dezhnev Formation 96
Diablo Platform 328
Dibba transform fault 46, 47
Dichów 272
Dietlas Potash District 261
Dinarids 323, 363
Dinarids Basin 203, 323, 324, 362, 363
Dinsmore Formation, Member 130, 131, 144
Dirin-Yuryakh borehole, field 17
Dirk Hartog Formation 87–89
Divino Formation 210
Dnieper River 204, 207, 208, 212, 216, 219, 220, 238
Dnieper-Donets Depression, Trough 2, 102, 106, 107, 118–121, 127, 128, 205, 206, 209, 213–215, 217
Dniester River, Basin, Region, Valley 78–80
Dniester-Prut Basin, interfluve 69, 70, 78–81
Dobruja 78, 81, 106, 161
Dobruja Basin, Depression, Trough 196, 199, 203, 323, 324, 362
Dog Creek Shale 330, 331, 335–337
Döhlen Trough 247
Dolina rhythmic unit 231, 232

Dolomitic Alps 363
Dolores Anticline 190
Domnin District 197
Domol diapir, District 7, 8, 58
Don Group 254
Don River 204, 207, 208, 212, 216, 219, 220, 238
Don-Medveditsa Swell 105
Donets Basin 121
Doppelte Lage 285
Dorsten town 270
Downtonian 93
Dra'a Hammada 160
Dravt Ridge 363
Dronov Group 211
Druvas town 90
Dublin 246, 248, 250
Dublyany Formation 79
Duisburg town 270
Dundas Formation 13
Dundee Formation 145–148
Duperow Formation 130, 143
Duperwon Formation 130
Durham Province 254, 299, 300, 313, 321, 322
Durmitor Massif 363
Dvina-Sukhona Basin, Depression, interflive, Region 209, 211
Dyudyunbel Mount 164
Dzhamandavan Range 162, 164
Dzhamatan 163
Dzhambul town 240
Dzhardzan Uplift 95
Dzhentula 95
Dzhezkazgan Depression, sub-basin 99–101, 162, 241, 242
Dzhezkazgan town 240
Dzhukste borehole 124

E. Thälmann I-III mine 258, 261
E Unit of Michigan Depression 70, 71, 75, 76
Eagle Basin, River, Valley, town 161, 184–186, 191–193
Eagle Valley Evaporite 185, 186, 192, 193
East Alps 323, 325–327
East Andes (South American Cordillera) 361
East Carpathians 323, 362
East Cis-Balkan area 243
East Colorado 348
East Elsk field 112
East Europe 204
East European Basin 161, 196–198, 203, 205–221, 235–240
East European Platform 78, 80, 90, 102, 106, 107, 128, 214, 235
East European Upper Devonian Basin 91, 106, 128

East Franconian Trough 247
East Greenland Basin 203, 362, 363
East Iran 5, 46, 49
East Karakol Trough 162
East Podolia 78
East Sayan 16, 17, 19
East Siberia 5, 16, 92, 94
East Siberian Basin 2, 4, 5, 15–19, 25, 32, 36, 43–45, 54, 59
East Thüringia Uplift 247
East Uralian (East Urals) Basin 161, 196, 200
East Wyoming Basin 161, 196, 202
Eastern Shelf 328, 330, 335, 348, 351
Ebelyak Formation 95
Ecuador 361
Edmonton town 129, 133–135, 140, 142
Eichsfeld Swell 290
Eids Formation 159
Eifelian 93–95, 97, 102, 105–107, 119, 130, 145, 146, 152, 159, 160
Einfache Lage 285
Eisenach town 266, 281, 306
El Adeb Larach Formation, sequence 201
El Paso Region 202
Elanka Formation 18
Elbe River 246, 248–250, 253, 259, 266, 271, 279, 281, 292, 293, 302, 306, 310, 315
Elborz (Elburz) Mountains 52, 53
Eleanor River Formation 62
Elegest River 99
Elenovo 242
Elets horizon 114, 120, 126
Elets (Upper) salt sequence of Dnieper-Donets Depression 119, 121
Elgyan Formation, horizon 5, 18, 32–35, 44
Elk Point Basin, Group 129, 130, 133, 139, 140, 142
Elkino Member 210, 221, 228, 229
Ellef Ringness Island 165, 166, 168
Ellesmere Island 61, 64, 90, 159, 165–167
Ellsworth Formation 146, 148
Elovka borehole 27
Elsk area, Depression, field 111, 112, 117
Elstow 130
Elton Dome, Uplift, Beds 211, 230–234
Emba River Basin 235
Emma Fiord Formation 167
Emsian 95, 152
Emyaksa Formation 95
England 244, 313
English Basin, Depression 245, 247, 249, 251, 252, 254, 257, 258, 275, 278, 299, 313, 314, 320–322
English Zechstein 257, 275, 299, 301
English Zechstein Cycle 3 313, 314
English Zechstein Cycle 4 320, 321

English Zechstein (EZ5) 322
Enrage Formation 176
Eocene 9
Erfurt 266, 281, 292, 305, 306
Erie Lake 73–75, 77, 147
Erldunda No. 1 borehole 6, 7
Ermakov borehole 159
Ernestina Lake Formation 130, 132, 133
Eruslan rhythmic unit 231, 232
Erzgebirge Trough 245, 247
Esayoo 167
Esk River 300
Eskdale Group 254, 257, 322
Eskdale No. 2 borehole 300
Eskdale town 300
Esterharzy Member 130, 135, 137, 138
Estonia 90
Etonvale Formation 156, 157
Eucla Basin 9
Euphrates River 47
Eurasia 4
Europe 64, 161, 245, 249, 323, 324
Evaporite Member of West Canadian Basin 130
Evlanovo horizon 108, 109, 120, 125–127
Evlanovo-Liven age, horizon 109, 110, 120
Evlanovo-Liven (Lower) salt sequence of Dnieper-Donets Depression 119–121, 127
Ezere borehole 124

F, F-1, F-2 Evaporite, Unit of Michigan Basin 70, 71, 75–77
4-G well 100
Fahud Salt area 46, 47
Famennian 95, 107, 114, 120, 125–130, 144, 146, 147, 153, 157, 159
Fars 48
Fedorovo Formation 11
Fehmarn Island 249
Fehmarn Z-1 well 249
Fenno-Scandian High 245, 247
Fergana Range, Valley 163, 165, 362
Fido Sandstone 183
Filippielsk horizon, sequence 90
Filippovka Village 127
Filippovo Formation, horizon 210, 211, 221, 225–228
Findlay Arch, Uplift 69, 145
Firebag 130
First Clastic Formation 51
First Dolomite Formation 51, 52
First Red Beds of West Canadian Basin 130, 133
First Salt of Appalachian Basin 77
Fisher Valley Anticline 190
Fisset Brook Formation 176
Fitzroy Basin 83, 161, 196, 201

Fitzroy Graben (Trough) 87
Flaseranhydrit 284
Flat Pebble Bay 61
Flechtingen Uplift 247, 249, 268
Fleswick Anhydrite, cycle, Dolomite, Siltstone 254, 258, 321
Fletcher Anhydrite (Member) 351, 352, 355
Flockensalz (Na4δ) 263, 264
Flockensalzpartie 262
Flowerpot Salt, Shale 330, 331, 335–340, 342–344
Fokin Formation 92, 93, 95–97
Fordon Evaporite 254, 258, 299, 300
Forelle Limestone 330, 331, 349
Fore-Sudetic Monocline 273, 312
Forstberg town 282
Fort Apache 357
Fort Apache Limestone Member, carbonate beds 357–359
Fort Nelson Basin 129
Fort Rayne Formation 201
Fort Simpson Formation 130
Four Corners Platform 184
Fowler Brook Member 176
Franconia Depression, Basin 252, 258, 264, 265, 280, 292, 304
Franklin Mountains 56
Franklinian Eugeosyncline 61, 62
Franklinian Miogeosyncline 61–64
Frankonish facies 265
Fraserdale Arch 149
Frasnian 84, 94, 95, 107, 110, 118, 123–127, 130, 131, 140, 143, 146, 147, 153
Frederick Brook Member 175, 176
Freezeout Shale, equivalent 331, 349
Frewena No. 1 borehole 59
Fridrih-Henrih 57 borehole 271
Frobisher Member, Anhydrite 170, 171
Frome Rocks No. 1 well 84
Frome Rocks Salt Dome 87
Front Range Uplift 184, 185, 192, 193
Frunze town 100, 163, 240
Fulda River Basin, Potash District 258, 260, 264, 289, 292, 303, 304
Fullerton Sand 330
Fundy Bay 174

G Unit of Michigan Basin 70, 71, 75, 77
Galindas Formation 255, 257
Garber Sandstone 335–337
Garden Island Formation 145, 146
Gardena Sandstone 363
Gardiner Range 6
Gascoyne sub-basin 87, 88
Gasper Formation 331
Gatar-Mund Thrust 46

Gautreau Member 175–177
Gavrilchitsa Village 105
Gavrilov Yam town 198
Gazhenka Member 37
Gdańsk area, Gulf, town 274–277, 299
Geauya County 71
Gedinnian 71, 84, 93, 95, 151
Gelsenkirchen town 270
Georgina Basin 15, 59, 60, 68
Georg-Unstrut Mine 285
Gera town 266, 281, 306
German Democratic Republic 244, 258, 272
German Zechstein 256, 257, 275, 311, 313, 316
Ghabar Group 51, 52
Gibraltar Formation 12
Gibson Dome 190
Giles Creek Dolomite 56, 57
Gillen Member 6
Gilliam Limestone 349
Girkin Limestone 183
Givetian 84, 93–95, 97, 107, 129, 130, 146, 153, 159
Glasscock County 348
Glazov key borehole, town 125, 126, 197
Glendo Shale 331, 349, 350
Glinyan Formation 81
Glorieta Sandstone 330, 331, 336
Gneudna Formation 160
Gnidif District 58
Godomichy Village 79
Goldwyer Formation 68, 84
Golling-Abtenau area 325, 326
Golyushurma town 127, 197
Gondwanaland (Gondwana) 53
Goose Fiord 159
Gordon Lake Formation 11
Görgeyite-Halite Member 211, 234
Gorky District 217
Gorky town 126, 204, 207, 208, 216, 219, 220
Gorky-Volga Region 197
Gorlovo town 103, 104
Gorodische horizon 232
Gorodok-Hatets tectonic step 109
Gorokhov area, field 112, 113
Gorzów Wielkopolski No. 1 well 296, 312
GOS-4 borehole 92
Gosses Bluff diapir 7, 8
Gotha 266, 306
Göttingen 289, 292, 303, 304
Goulburn Group 11
Gowden Anhydrite, Bed 352–356
Goyder Formation 56–58
Goyder Pass structure (diapir) 7, 8
Gramina 130

Grampian High 247
Grand Range No. 1 borehole 84
Grand Traverse County 72
Grauer Salzton (T3) 254–257, 303–309, 311, 313
Grauers Liniensalz (Na3β) 254, 303
Gray Pelite 255, 257, 311, 312
Grayburg Formation 330, 348, 349
Great Bear Lake 56
Great Lakes area 11
Great Sandy Desert 83
Great Slave Lake, area 12, 56
Greendale Syncline 181–183
Greenland 13, 166
Gregory sub-basin 83
Grenville Series, epoch 12
Grenzanhydrit (A4r) 254–256, 314, 317–319
Grenzanhydrit (A5r$_1$) 322
Gröden Formation 363
Grosmont Cooking Lake 130
Grossland Platform 83, 87
Guadalupe Series, Guadalupian 329–332, 335, 346, 348–351
Guapore Shield 195
Gubin town 272, 295, 309, 311, 312, 319
Guelph Formation 70, 71
Gulebra Dolomite Member 353, 357
Gumbardo Formation 156
Gunton Formation 65, 67
Guragir Formation 67
Gurev town 236
Gurupa Swell 194, 195
Guyana Shield 195
Gypsum Valley Anticline 190
Gzhelian 169, 198, 199

H Unit of Michigan Basin 70
Habbard Evaporite 130
Hadramaut 51
Halite-Anhydrite Zone 300
Halle town, Potash District 280, 286, 292
Halle-Wittenberge Uplift 247
Hallein-Berchtesgaden 325, 326
Hallstatt 325, 326
Hallstatt salt stock 323
Halogenic Formation 210
Hamayrah Dome 48
Hamburg 246, 248, 250, 251, 253, 259, 279, 302, 315
Hamelin Pool Nos 1, 2 wells 87–89
Hamilton Formation 147
Hangendgruppe 285–288
Hangensalz (Na2γ) 254, 284, 290, 291
Hannover town, Potash District 246, 248, 250, 251, 253, 259, 279, 289, 302–304, 315, 316

Hansa Potash District 303
Hardeman Basin 328, 329
Hare Fiord Formation 167, 168
Hare Indian 130
Harirud River 160
Harper County 335
Harper Siltstone 331, 343
Harris Formation, Member 130, 143
Harrodsburg Limestone 201
Hartlepool (Hayton) Anhydrite, Dolomite 254, 275, 299, 300
Hartsalz 260–263, 285, 286, 288, 291, 303, 311, 356
Hartsalzlager 263
Harut Formation 52
Harz 266, 281–283, 307, 317
Haselgebirge 59, 249, 257, 323–326
Hatfield Formation 130
Hathern town 199
Hattorf Mine 258
Haughton Dome 62, 63
Hauptanhydrit (Main Anhydrite A3) 254–256, 303–311
Hauptdolomit 255–257, 280–282, 287, 289, 290, 293, 294, 301, 303
Hauptsalz (Na2β) 254, 284, 290, 291
Hay River 130
Hayton Anhydrite 254, 275
HB Fina Northumberland Strait F-26 well 180
Heath Formation 173
Heavitree Formation, Quartzite 5, 6
Hecla Beds 64
Heilinkreuz-Mödling 325, 326
Helderberg Group 71
Helikian 12
Helmand-Archandab High 53
Helvetides 325
Hennessey Shale 330, 335–339
Henriette-Maria Arch, arched uplift 149
Hercynian 245
Herington Dolomite 330
Hermosa Group 185, 186, 188, 189, 191
Hessen Beds 256
Hettenhausen town 292
High Karst Zone 363
High Zagros 5, 46, 48, 53
Hillsborough 177
Hillsborough Syncline, Member 176, 177
Hillsdale Limestone 183
Himalayas 50
Himalayas Foreland 50
Hindustan Peninsula, Platform, Shield 50, 51, 53
Hiram Brook Member 175, 176
Hirlău town 81
Holbrook Basin 357

Holbrook town 358, 359
Holland 244
Hollis Basin 328, 329
Holy Cross Barrier, Mountains 273, 298, 312
Honaker Trail Formation 185, 186, 188, 189, 191
Hondo 130
Hoodoo Dome, diapir 166, 168
Hoodoo L-41 borehole 165, 168
Hopewell Group 176, 181
Hormoz (Hormuz) area 46, 47
Hormoz Evaporite, Formation, Island 5, 46–53
Horton Bluff Formation 176
Horton Group 175–177
Horton-Winsor area, District 179, 180
Huallaga River 361
Hubbard Evaporite 131, 133, 139, 141
Hudson Basin 91, 149–154
Hudson Bay 11, 68, 149, 153, 154
Hudson Platform 149
Hudson Walrus A-71 borehole 149, 152
Hueco Limestone 330
Hugh River Shale 56
Hume 130
Hungary 80, 323, 362
Hunsrück Uplift 247
Hunt Oil Brown borehole No. 1 9
Huqf Group 49, 51
Huron Lake 73–75, 77, 145, 147, 148
Huronian Supergroup, Biostrome 11, 146
Hutchinson Salt Member 343
Hycrochemical Formation 210, 239

Ichera Formation 18
Ida District 58
Idaho 129, 134, 328, 331
Iengra Series 11, 14
Ievsk superhorizon 68
Ikhedushiingol Formation 97–99
Ilga Depression 16, 41, 43
Ili River, Depression, Basin 91, 158, 159, 240
Ilimsk No. 1-r borehole 17
Illinois 73, 201
Illinois Basin 91, 145, 158, 159, 161, 196, 201
Illisie Basin 161, 196, 201
Ilma Beds 9
Iltyk Formation 67
Ilyich River 90, 196
Imgantau Range 163
Imperial Formation 130
Imperial Port Hood No. 1 well 180
Imperial Vermilion Ridge No. 1 well 54, 55
Inbirik Formation 364
Inder Dome 211, 230, 234

Inderbor Bed 231
Indian Ocean 87
Indiana 73–75, 77, 201
Infracambrian 5, 48, 52
Ingleside Formation 329, 330
Innsbruck town 323, 325, 326
Inowroclaw 296, 312
Interior Platform 56
Interlake Formation 65, 67, 90
Inverness Formation 176
Ioachim Formation 68
Ionishkis town 125
Iowa 328, 333, 334
Iquitos Arch 194
Iran 45, 46, 49, 50, 53
Iran-Arabian salt zone, salt basin 46, 47, 53
Iran-Pakistan Basin 4, 5, 15, 45–54, 59
Iraq 363
Irbukla Formation 67
Iren Formation, horizon 210, 211, 221, 222, 225–227
Irene Bay Formation 62, 63
Ireton 130
Irginsk Formation 210
Irish Massif, High 247
Irkineevo Formation 18
Irkut horizon, sequence 5, 17–24, 44, 53
Irkutsk Amphiteater 5, 18, 19, 21, 25, 27–29, 32, 33, 36, 37, 41–43, 67
Irkutsk period 5
Irkutsk town 16, 32
Irkutsk-Tunguska facies zone 17
Isachkov Anticline, Formation 118, 120
Isachsen Dome 166
Isakly Beds 239
Isakovtsy Beds 79
Ischl-Altaussee–Gründlsee area 325, 326
Ismay cycle 187–189, 191
Issa town 126
Isselburg town 293
Issyk-Kul Lake 163, 200, 240
Istra Formation 123, 124
Itaituba Formation, town 195
Italy 323, 363
Itfer horizon 68
Ivanovka Village 78
Ivanovo town 126
Izhma Limestone 127

Jacque Mountain Limestone, Member 192
James Bay 154
Jarkent River 163
Jay Creek Dolomite, Limestone 56–58
Jean Marie Formation 130
Jefferson ville Formation 159
Jena town 266, 281, 305, 306

Johnstone Hill diapir 7, 8
Jones Arch 83
Julesburg Basin 328–330, 335, 344–346, 350
Julie Member 6
Jurassic 8, 46, 83, 157, 323, 359, 361

Kaa-Bulak Village 164
Kadzham Nura Dome 101
Kaibab Limestone 331
Kakisa 130
Kalabagh area, field 51
Kalaida field 120
Kalargon Formation, sequence 95–97
Kalgoorlie Beds 11
Kaliakra 242
Kaliflöz Albert (K3Ab) 254, 256, 303
Kaliflöz Bergmannssegen 254, 256
Kaliflöz Hessen (K1H) 252, 254, 256, 260–264
Kaliflöz Ottoshall (K4ot) 254–256, 316, 317
Kaliflöz Riedel (K3Ri) 252, 254, 256, 303, 304
Kaliflöz Ronnenberg (K3Ro) 252, 254–256, 303, 304, 309–311
Kaliflöz Ronnenberg I, II 311
Kaliflöz Stassfurt (K2) 254–256, 280, 281, 283, 285–289, 291, 294, 295, 301
Kaliflöz Thüringen (K1Th) 252, 254, 256, 260, 261
Kalinin District 127
Kaliningrad District 278, 297, 299
Kalinkovichi Depression 116, 117
Kalinovo field 112
Kalinovsk Depression, Formation 210, 239
Kalizone 254, 270, 271
Kalmius-Torets Trough 205, 213, 217
Kalvar Formation 255, 275, 278
Kalpintag 163
Kaluga town 103, 126
Kama River, Basin, area 204, 207, 208, 212, 216, 219–221, 224–227, 235, 238
Kamenka borehole, field 17
Kamensk Anticline, Beds 128, 211, 213
Kamp 1, 2, 4 boreholes 270, 271, 293
Kamp-Lintfort town 270
Kan-Taseeva Depression, 54 and 56 wells 16, 19, 22, 28, 29, 32, 33, 36, 41, 42
Kankakee Arch 145
Kansas 327–329, 331–335, 339, 342–345, 347
Kansas Basin 328, 335, 342–344
Kapseda town 124
Kara Depression 204
Karachauli River, Formation 164
Karaimishkyai town 124

Karakyr Formation 241, 242
Karalarga River 164
Karampur 50
Karampur well 50
Karamurun Formation 210
Karasu Formation 362
Karasu-Ishsay Basin 203, 362
Karatau Range 99, 100, 163
Karateke Range 163
Karavanke Range 363
Karbonatfolge 305
Karelino No. 3-p borehole, field 17, 20, 28, 33
Karfagen Beds 211, 213
Karpensk rhythmic unit 231
Kartamysh Formation 209, 211
Karwendel area 325, 326
Kascattama No. 1 borehole 149–152
Kasegalik Formation 11
Kashgar River 163
Kashin town 197
Kashira horizon 198
Kashira-Myachkov sequence 198
Kasimirow town 273
Kasimov town, superhorizon 198
Kassel town 289, 292, 304
Kasyanka field 27, 33
Kaszalin-Chojnice-Tuchola Uplift 274
Katanga Supergroup 13
Kauno-Vone 105
Kavak-Tau Mountains 164
Kaverino town 105
Kavir Fault 47
Kavyuk-Su River 164
Kazachinsk 1-r borehole, field 17, 27
Kazakhstan 200
Kazan town, district 126, 204, 207, 208, 212, 216, 219, 220, 238
Kazan-Kazhim Trough 126, 127
Kazanian 167, 169, 208–211, 235, 238–241, 329, 330
Kazarka borehole, field 17, 27
Kechika Trough 54
Kedainyai town 124
Keefer Sandstone 71, 72
Keele River Basin 55
Keg River Formation, Limestone 130, 132–136
Kemeri Village 124
Kempendyai River, Trough, Anticline, Depression 92–97, 196
Kendyktas Range 99
Kenogami River Formation 90, 149–151
Kentaral Dome 100
Kentucky 201
Kenwood sequence 159

Kerman 5, 46, 47, 49, 50, 160
Keyser Formation 71, 72
Khaastyr horizon 52, 81
Khaiyrkan Mount 99
Khamzas deposit 98, 159
Khanelbir Formation 95
Khantai-Rybinsk Swell 16
Kharkov 119, 204, 207, 208, 212, 216, 219, 220, 238
Kharyalakh Formation, map area, Range 67, 93, 95
Khatanga Trough, Depression, River 16, 92–95, 196
Khewra area 51
Khishchnikov River 97
Khobochalo River 96
Kholmsk area, field 120
Kholyukhan horizon 82
Khoper Monocline 105
Khristoforovo borehole, field 17
Khurat Formation 95
Kibbey Formation 170, 173
Kidson No. 1 borehole 84, 85, 160
Kidson sub-basin 83, 84, 87
Kieseritische Übergangsschichten Hartsalzbank (Na2K) 254, 261
Kieseritregion 285
Kiev 80, 204, 207, 208, 212, 216, 219, 220, 238
Kikino town 126
Kilian Formation 12
Kilohigok Basin 11
Kimberley Block 83, 157, 158
Kingir Formation 241, 242
Kirdonis 124
Kirenga River 18
Kirenga-Lena watershed 42
Kirensk No. 1-o borehole, field 17
Kirey Formation 200
Kirghiz Range 99, 100, 162, 163
Kirghizia 200
Kirkham Abbey Formation 254, 299, 300
Kirov-Kazhim Trough 122, 123
Kisbey Member 170, 171
Kishiburul Mountains 162
Kishinev 80
Kissingen town 292
Kitaigorod Formation, horizon 78, 79
Kleinschierstedt I/II Mine 288
Kleybolt Peninsula 167
Klimino Formation 18
Klodowa town, well 296, 312
Kluj 80
Klyazma horizon, sulfate sequence 198, 199
Knob Lake Group 11
Kochakan Formation 67

Kodzhagul Formation 164
Kohe-Kaftarhan Range 53
Kok Tyube Dome 101
Kokhansk Beds 210
Kokiyrim Range 162
Kokomeren River 164, 165
Kokpansor sub-basin 99, 100, 162
Kokshaal River 163
Kola Peninsula 14
Kolchugino town 199
Koldin Formation 94—96
Koloshka River, Basin 125
Kolva-Vishera Region 225
Komarovichi field 111, 112
Komarovka Village 126
Komi-Permyak Arch, Uplift 122, 205, 214
Komsomol Formation 68
Kondurov Formation 210
Konga Formation 81
Koolpin Formation 12
Kopatkevichi field 105, 111—113, 117
Korab zone 323, 363
Korenev field, Formation 114, 211
Korkino No. 1-r well, field 20, 27, 33
Korotaikha Depression 204
Koryazhema well 217
Koshelevo Formation 210
Kosyu horizon 90
Kosyu-Rogovo Depression 204
Kotcho 130
Kotelnich Arch, Uplift 126, 205
Kotelnich town, district 126, 127, 197
Kotui River, Basin 16, 17, 82, 94—97
Kotuikan River, Basin 4, 13
Kovrov town 198
Kozhevnikov Dome 94
Kramatorsk Formation 209, 211, 215, 217
Krasnaya Polyana-1 borehole 197
Krasnaya Polyana Village 126, 127
Krasnokamsk town 125
Krasnopartizansk field 120
Krasnoselsk Beds 211, 213
Krasnovka Village 127
Krasnoyarsk 98
Krasnyi Beds 222—227
Krasnyi Kholm town, borehole 104, 126, 197, 198
Krekepava borehole 105
Kristallsalz (Na3γ) 255, 305—308, 311
Kristallsalz (Na4γ) 318
Krivaya Luka borehole, field 17, 28, 33
Kudeb River, Basin 125
Kudymkar District 125
Kuibyshev town, District 228
Kuleshovka area 199
Kulichkov Formation 80

Kuma-Manych Trough 205, 206
Kumsay Uplift 239
Kundyai River 94
Kungei Alatau 163
Kungurian 123, 208—211, 218, 220, 221, 227, 228, 230—235, 240, 241, 329, 330
Kuntanakha River 95
Kunya town 126
Kupferschiefer 249, 254—256, 258, 260, 265, 267—271, 275
Kurbin-Shivi Creek area 99
Kureika Formation 95, 96
Kureika River 95
Kureika-Baklanin Swell 16
Kurgantau Formation 211, 234
Kurmain Formation 210
Kurukusum Group 200
Kurunguryakh Formation 95, 196
Kutuluk borehole, field 17, 27
Kutuluk Formation 210
Kuvshinovo borehole 104
Kuwait area 47
Kuyumba borehole, field 17
Kuznetsk Basin 91, 158, 159
Kwatoboahegan Formation 149, 150, 152
Kygyltuus Formation 93—95
Kyizil 98
Kyrtyiol Anticline 128
Kyutingda River, Trough 196
Kyzylkanat sequence 162
Kzyltau Formation 211, 234

La Huerta Siltstone, Member 352
La Rioja Province 363
Laanemetsa District 105
Labrador Trough 11
Lagenanhydrit (A5) 322
Lake Brook Member 176
Lake Cohen 7
Lakefield Member 176, 178
Lalun Formation, Sandstone 5, 48—53
Langlo Embayment 155
Lapland 14
Laptevo town 197
Larapinta Group 56
Las Animas Arch 328, 329, 342
Laskowice Graben 245, 247
Latvia 90, 123
Lausitz 293, 295, 318
Lausitz Uplift 247, 249
Leba town, horizon 274, 275
Lebedyan 126
Lebedyan-Dankov horizon 108, 125
Leduk Cooking Lake 130
Ledyanopeshchera Member 210, 221, 227, 229

Lefroy Beds 9, 11
Legnica town 273, 296, 312, 319
Leicestershire 199
Leine River, Basin 290–292, 303, 304
Leine-Series 256–258, 304–314
Leine-Steinsalz (Na3) 252, 254–256, 303–311
Leipzig 266, 281, 306
Lek Formation 210, 221
Lemaisinsk Formation 210
Lemiu River 90
Lena area, Basin, District 16, 42, 82
Lena River 18, 32, 33
Lena-Kirenga watershed 41
Lena-Yenisei Basin 60, 67, 69, 70, 81, 82
Leningrad 204, 207, 208, 212, 216, 219, 220, 238
Leninogorsk field 126
Lennis Sandstone 11
Leonard Salt 130, 137
Leonard Series, Leonardian 329–332, 334–337, 341–361
Lhasa Basin 161, 196, 200
Liegendgruppe 285–288
Liepaya borehole 124
Limbla Member 6
Limbo Formation 58
Limestone Member of West Canadian Basin 130
Linienanhydrit 284
Liniensalz (Na3β) 255, 303, 305–308, 311
Lisbon Valley Anticline, area 188, 190
Listvenichnaya No. 1 borehole, field 17
Lithuania 90, 105, 123–125, 244, 274, 297
Lithuanian Depression, Beds 257, 278
Little Dal Formation, Group 12
Little Valley Limestone 182, 183
Litvintsevo Formation, sequence 18, 20, 44
Liven horizon 108, 125–127
Llandeilian 84
Llandoverian 71, 81
Llanvirnian 84
Locher Lake Fault 179
Lockhart Anticline 190
Lockport Formation 70, 71, 74
Locust Cove area, District 181–183
Lodgepole Formation 170, 171
Lodz town 276, 277, 299
Log Creek Formation 156
Loire River 246, 248, 250
Lombardia 363
London 246, 248, 250, 251, 253, 259, 279, 302, 315
London-Brabant Massif 245, 247
Long Rapids Formation 149, 150, 153
Lopushany Formation, Member, Village 102, 105, 106

Lopydin well 122
Lotsberg Formation, Salt 130–133
Louisiana 328, 333, 334
Loves Creek Member, sequence 6
Lovitos Cornwallis Resolute Bay L-41 borehole 62
Low Tatra 326
Lower Amazon Basin, Depression 194
Lower Angara area, Region 23
Lower Anhydrite 255, 273, 274, 276, 300
Lower Anhydrite Member of West Canadian Basin 130, 134, 137
Lower Anhydrite-Halite Member 211
Lower Chu Salt Dome 100
Lower Cimarron Salt 338, 339, 342–344
Lower Cornwallis Group 63
Lower Elk Point 129–132
Lower Evaporite 300
Lower Evaporite Unit 12
Lower Halite horizon 211, 232
Lower Himalayas 50
Lower Iren subhorizon 210, 227–229
Lower Lunezh Beds 228, 229
Lower Muskeg 136
Lower Pegmatite Anhydrite 255, 257
Lower Red Beds 175
Lower Red Pelite (T4a) 255, 257, 316, 319
Lower Rhine Basin, Depression 251, 252, 254, 258, 270–272, 275, 292, 293, 309, 319
Lower (Frasnian) salt sequence of Dnieper-Donets Depression 120
Lower (Upper Frasnian) salt sequence of Pripyat Depression 108, 118, 120
Lower Salt of West Canadian Basin 130, 137, 138
Lower Youngest Halite 255, 318
Lubin town 273, 296, 319
Lubsko town, barrier 272, 295, 298, 309, 311, 319
Lucas Formation 145–148
Ludlovian 71, 80, 81, 89
Ludwigshall-Immenrode town 282
Lueders Formation 330
Lugovo rhythmic unit 231
Lugovskoe Uplift 239
Luma Formation 95
Lunezh Member, Formation 210, 221, 222, 228
Lupton Beds 10
Lut Block, High, Swell 46, 47
Lut-Oman High 46, 53
Lutsk town 79
Luzhsk horizon 108
Lvov 80
Lvov Depression 79–81, 102, 105, 106

Lyons Sandstone 330, 344
Lyskovo town 126
Lyuban field 112
Lyubim town 197, 198
Lyublin town 106

Maas River 270
Mabou 180
Mabou No. 1 well 180
Macapa town 195
Maccrady Formation, Evaporite, Shale 181–183
MacDonald Platform 54
Mackenzie Basin 12, 15, 54–56, 129, 130
Mackenzie Hills 4
Mackenzie King Island 166
Mackenzie Mountains 12, 56
Mackenzie River 54
Mackenzie Trough, Arch 54–56
Macleod Lake 88
Macumber Formation 176, 179
Madagascar Island 14
Madeira River 194
Madera Formation 186
Madison Group 170, 171, 173
Madley Beds, diapir 7, 9–11
Madley sheet area 10
Magdalen Island 174
Magdalena Formation 202
Magenta Dolomite Member 353, 357
Magnesitbank 305
Magnitogorsk town 200
Maidantag Range 163
Maikor District 125
Mail-Change Limestone 59
Main Anhydrite, Member 211, 234, 255, 311, 312
Main Dolomite (Hauptdolomit Ca2) 255, 295–297, 299
Main River 246, 248, 251, 253, 259, 279, 302, 315
Makarov Bay 196
Maksakovo field 120
Maksatikha borehole 104
Makusov Formation 96
Malagash area 179
Malinovtsy Formation, Member, horizon 78–81
Mallapunyah Formation 13
Malodushin Depression 117
Maloik Formation 210
Malokinelsk Formation 210
Malyn field 115
Mamramcook Formation (Lower Red Beds) 175, 176
Manaus town 195

Mangyshlak Trough 205, 206
Manitoba 64, 66, 129, 133, 134, 138–145, 149, 153, 171
Manitoba Group 129, 139, 140
Manitoba Shelf 129
Mansfeld Syncline 318
Mansfelf Trough, Potash District 267, 280, 282, 286, 308
Mantov area 81
Manturovo Formation, sequence, town 92, 93, 95, 96, 126
Manus town 195
Manx-Furness Basin 251, 252, 254, 258, 275, 280, 301, 314, 321, 322
Marañon River, Basin 361
Marathon Uplift 328
Marchbank Syncline 178
Marfa Basin 328
Maritime Basin, Province 161, 173–181
Maritime Fold Belt 174, 175
Markha No. 2-k borehole 21
Markha-Morkoka District 67
Markovo No. 23-r borehole, field, Village 17, 27, 28, 32, 33
Marl Slate (Zechstein) 275
Marlow Formation 330
Maroon Formation 185, 186, 191–193, 331
Marpasad town 126
Marury Plateau 247
Marx-Engels I, II Mines 258
Maryland 71, 73–75
Masirah transform fault 46
Matador Arch 328, 329
Matusevich Formation 97
Matveevka Village 239
Mayan stage 18, 43, 45
Mayo River 361
Mazury Barrier, Plateau, Peninsula 298, 312
McArthur Basin, Group 13
McKean County 71
McKenzie Formation 71, 72, 75, 76
McLarty No. 1 well 84, 85
McLeary Formation 11
McNamara Group 13
McNutt potash zone 352, 355
Meadow Lake Formation, Escarpment 128, 129, 131, 133, 134
Mecklenburg 258, 293, 295, 309, 311, 318
Mecsek Basin, Mountains 203, 323, 324, 362
Meenymore Member 199
Melekes Basin, Depression 127, 198, 199, 205
Mellinjerie Limestone 84, 160
Melville Island 165, 166
Mendym Beds, Dolomite 125
Menzengraben Potash District 261
Mergelsteine (C1) 271

Meringoin Peninsula, Anticline 179
Merkbjorn Formation 13
Merkebjerg Basin 4
Merlinleigh sub-basin 88
Merrina Beds 59
Merseburg town 266, 281, 306
Mesozoic 45–47, 155
Metiger Formation 18
Mexico 332, 334, 347, 359, 360
Mexico Gulf 360
Mezen Depression 122, 123, 205, 209, 214, 215, 218, 239
Mezhsolevaya sequence of Pripyat Depression 108, 114
Mezhtsiems town 90
Michigan 69, 73–75, 202
Michigan Basin, Depression 15, 59, 69–78, 91, 145–149, 161, 196, 201
Michigan Formation 202
Michigan Lake 73–75, 77, 147, 148
Michigan-Appalachian Basin 69–78
Mid–Amazon Depression 194, 195
Mid-Basin Platform 83, 87
Mid-North Sea High 247
Mid-Tien Shan Basin 161–165, 200
Midale Member 170, 171
Midcontinent Basin 203, 327–329, 331–334, 341, 344–346, 348, 351, 357
Middle Anhydrite or Shell Lake 130, 137
Middle European Depression, Trough 245, 249, 252, 278, 301, 316, 318
Middle Halite Member 176, 178
Middle Limestone 130
Middle Moty Member 19
Middle Pharwala Member 51
Middlesbrough town 300
Midland Basin 328–330, 348, 349, 351, 354, 355, 360
Mikchanda River, Basin 93
Mike Lake, Island 152
Mikkwa Member 130, 135, 136
Mila Formation 52, 53
Mildred 130
Milianowa town, IG-1 well 273, 296, 312
Minas sub-basin 174–176, 179
Minden town 289, 304
Minhamir Formation 52
Mink Member 130, 135, 136
Minnekahta Limestone 330, 331, 344, 346
Minnelusa Formation 202, 330, 331
Minnesota 129, 134, 328, 333, 334
Minsk 204, 207, 208, 212, 216, 219, 220, 238, 246, 248, 250
Minto Inlet Formation 12
Minturn Formation 185, 186, 191, 192
Minusinsk Basin, Depression 91, 98, 158

Miocene 9
Mirny borehole, field 17
Mironovo No. 1-o borehole, field 17, 22, 33
Mirovo Village 242
Mirovo Village (OP-1) well 243
Mission Canyon Formation 170, 171
Mississippian 170, 172, 173, 176, 181, 183, 188, 191, 202
Missouri 328, 333, 334
Missourian 186, 202
Mittelberg town 292
Mittleres Werra-Steinsalz (Na1β) 254, 256, 260, 261, 264
Mittlerer Werra-Ton 255
Mitu Evaporite, Formation, Series 361, 362
Moab Valley Anticline, area 187, 188, 190
Moberly 130
Moesian Basin, Depression 203, 242–244, 323, 324
Moesian Plate 242
Moesian-Wallachian Basin 91, 158, 159
Mogilev town 105
Mogollon Rim 357, 358
Moiero River, Basin 67, 81, 82
Moiers town 270
Molas Formation 185, 186, 188, 191, 192
Moldavia 78–81, 199, 323
Moldavian Plate 78
Moldotau Range 162–164
Moncton sub-basin, Formation (= Upper Red Beds) 174–178
Moncton town 174, 177
Mongolia 99
Monino town 197
Montana 64, 66, 67, 128, 129, 131, 134, 138–144, 170–173, 327, 328, 332–334, 345–347, 349, 350
Monte Alegre Formation 194
Montenegro 323
Montgomery County 181
Moose River Basin, Formation 60, 68–70, 89–91, 148–154
Moran Formation 330
Morgan Formation 186, 191, 192
Morkvasha Village 126
Morosheshty Village 79
Morrow Series 186
Morsovo Basin 91, 102–106
Morsovo horizon, salt member 102–106
Morvan-Vosges Uplift 247
Moscow 103, 125, 126, 198, 204, 207, 208, 212, 216, 219, 220, 238
Moscow Syneclise 2, 102–106, 125–128, 197–199, 205, 209, 214, 215, 218, 221, 237, 239
Moscowian 167–169, 198

Mosolov town, horizon   103, 104
Moty Formation   18, 20
Mount Bayley Formation, sequence   167, 169, 363
Mount Cap Formation   55, 56
Mount Charlotte No. 1 borehole   7, 8, 56–58
Mount Isa area, Group   13
Mount Murrey   8
Mountrail Member   130, 139, 141
Mozyr Depression, field   112, 117
Mufulira Syncline   4, 13
Mühlhausen town   266, 281, 305, 306
Mühlhausen-Infeld Trough   245, 247
Muksha Member, horizon   78–81
Multan town   50
Muna Arch   16
Munda No. 1 well   84
Mungerebar Limestone   59
Munising Group   59
Munok Formation   18
Murbai No. 1-r borehole   17, 19, 21
Murbai-Chastin Basin   19, 21, 23
Murom town   199
Murray Island Formation   149, 150, 152
Muschelzone   282
Musgrave Block   7, 9
Muskeg Formation, sequence   130, 132–136
Muskeg-40 Marker   135, 136
Muyunkum Region   239
Myachkov horizon   198

Nabberu Basin   9
Nadbryantzevo Bed   211, 213
Nadsolevaya sequence of Pripyat Depression   108, 116
Nagaur town   50
Nakokhoz Formation, sequence   95–97
Namana No. 8-k, 2-r boreholes   17, 21
Namana horizon   18–20, 43
Namana River   21
Nambeet Formation   84
Namdyr Formation, unit   94, 95, 97
Namurian   95, 162–164, 167–169, 176, 197, 198, 201
Nansen Formation   167–169
Nappan area   179
Narimanovo horizon   232
Narovlya town   109
Narssârssuk Formation   13
Narva (Narova) horizon   102, 105, 108, 119
Naryn Depression, Trough   162, 164, 165
Naryntau Range   162
Naumburg town   266, 281, 305
Nayband Fault   47
Nazarovo deposit   98
Neale Beds   9

Nebraska   202, 327, 328, 330, 332–334, 342–347, 350, 351
Neklyudovo area   199
Nekrasov borehole   125
Nelidovo town   103, 104
Nemaha Anticline, Arch   328, 329, 342
Neman River   246, 248, 250, 251, 253, 259, 279, 302, 315
Neobolus Beds, Shale   49, 50, 125, 127
Neocomian   323
Neogene   160
Nepa Basin   16, 36, 37
Nepa No. 1 borehole   17
Nepa River   18
Nepa-Botuoba Anteclise, Uplift   16–18, 27–33, 41
Nepeitsevo town   126, 197
Neralakh River   92
Netherlands   271
Neuhof-Ellers Mine   258
Neustadt town   266, 281, 306
Neustassfurt VI/VII Mines   308
Nevada   328, 331, 357
Nevolino Member   210, 221, 228, 229
New Brunswick   173, 178
New Brunswick Platform   174, 175
New Mexico   184, 191, 202, 327, 328, 330–334, 341, 346–348, 351, 352, 354–359
New South Wales   160
New York State   69, 71, 73–78
Newfoundland   180
Newfoundland Platform   174
Nezhin Depression   205
Neyband Fault   46
Ngalia Basin   4, 13
Niagara Formation, Group   70–74
Niagaran   71–75
Nibus Basin   247
Nida River, Trough   273, 296, 319
Niederlausitz town   272
Nikitovo Formation   209, 211, 213, 214
Nikolaevka field   111, 112
Nikolaevskoe Region   239
Nikolskoe town   127
Nikulino field   112
Nima Formation   93, 95, 96
Ninmaroo Formation   59, 68
Ninnescah Shale   331, 342–344
Nippewala Group, Salt   331, 336, 342–344
Nisa-Böbr Trough   247
Nisku Formation   130
Nisporen area, District   79
Nita Formation   84
Nizhneudinsk town, borehole, field   17, 27, 28
Nizhneustie Formation   211
Nizhny Ilimsk No. 1-o borehole   17

Nizhnyaya Tunguska River, Basin, Valley 16–18, 36, 67, 82, 93, 95, 96
Nizhnyaya Tunguska-Nepa Interfluve 36
Noginsk sequence 199
Nokhtuisk Formation 18, 23
Nordhausen town 266, 281, 306
Nordvik Dome, District 94
Nordvik-Khatanga District 94
Norilsk town, District, Region 67, 81, 82, 92–97, 196
Norman Range 56
Norman Wells area 54, 55
North America 4, 11, 12, 15, 59, 64, 69, 128, 161
North Baggy Member 5
North Baikal Highland 17, 27, 36, 37
North Bobrovichi field 112, 113
North Bulgarian Uplift 243
North Carolina 182
North Dakota 64, 66, 128–131, 134, 138–144, 170–173, 327, 328, 330–334, 345–347, 350
North Danish Basin 247
North-East German Basin, Depression 249, 251, 252, 255, 256, 258, 272, 274, 278, 287, 293–295, 301, 310, 311, 314, 316, 318, 320, 322
North Greenland 62
North Harz Potash District 286, 289
North Hemeride Syncline 327
North Ireland Basin 161, 196, 199
North Italian Basin 203, 323, 324, 362, 363
North Italy 366
North Kalinovo field 112
North Kerman area 46, 47, 49
North Limestone Alps 325
North Lyzhsk Anticline 128
North Mexican Basin 203, 359–361
North Sea 244, 245, 249, 251, 252, 278, 279, 300–302, 315, 321
North Siberian Basin 2, 91–97, 161, 196
North Sudetic Trough 272, 295, 309, 311, 319
North Tien Shan 200
North-West German Basin, Depression 245, 249, 251, 252, 254, 258, 271, 278, 290, 293, 295, 301, 303, 309, 314, 316, 318–320
Northern Alberta sub-basin 128, 129, 132–134
Northern Caucasus 206
Northern Persian Gulf salt area 46
Northern Pribortovaya Depression 117
Northern Rocky Mountains Uplift 328
Northern Territory 5
Northumberland Basin 161, 196, 199

Northwest Territories 11, 12, 54, 128–130, 133–135, 143–145
Northwestern Shelf 351, 355
Norwegian-Danish Basin 251, 252, 258, 278, 301, 313
Nova Olinda Formation, town 194, 195
Nova Scotia 173, 177
Nova Scotia Platform 174
Novaya Zemlya Island 196, 197
Novgorod Region 125
Novo-Senzhara Anticline 118
Novoakmyansk Formation 255, 275, 278
Novobasovo Village 103, 104
Novomoskovsk town 197
Nugush Formation 210
Nukuty field 27
Nürnberg Uplift 247
Nyenchhen-Thangha Range 201
Nysa-Luzycka River 272
Nyuya Depression 81

Obere Letten 254
Oberer Dolomit 255, 265
Oberer Grenzanhydrit (A5r$_2$) 322
Oberer Schluffstein (T5r$_1$) 322
Oberer Werra-Anhydrit (A1r, or A1β, or A10) 254–256, 264, 265, 267, 269
Oberer Werra-Ton (T1r or T1γ) 255, 256, 264, 267, 268
Oberes Stassfurt-Steinsalz (Na2b) 291
Oberes Werra-Steinsalz (Na1γ) 254–256, 260, 263, 264, 270, 271, 275, 291
Oberlausitz 258, 309, 311
Oberste Zechstein Ton (T4r) 317
Oberste Zechsteinletten (T4r) 254–256, 314, 318
Oboshin Beds 239
Observatory Hill Beds 58, 59
Ochoa Series, Ochoan 329–332, 336, 351–353, 355, 356
Odessa town, Region 80, 81, 204, 207, 208, 212, 216, 219, 220, 238
Odra (Oder) River 246, 248, 250, 251, 253, 279, 302, 310, 315
Offensee-Almsee area 325, 326
Officer Basin 4, 7, 9–11, 15, 59
Ogr Formation 123, 124
Ohio 69, 71, 73–77
Ohre-Series 256, 257, 322
Ohre-Steinsalz (Na5) 252, 255, 322
Oka superhorizon 197, 198
Oka-Serpukhov sequence 197
Oker River 292
Oklakhoma 327, 328, 332, 334–336, 338–343, 347
Oktyabrsk field 112

Oktyabrskaya Revolyutsiya Island 89, 96, 117
Oldensaal town 270, 293
Older Halite ("Alteres" Steinsalz, Zechstein) 295–297, 299
Older Potash (Zechstein) 255, 295, 296
Oldest Halite (Zechstein) 255, 257, 272–274, 276
Olekma Formation, horizon, section 5, 18, 19, 21, 36
Olekma Nos 1-r, 3-r, 28, 40 boreholes 17, 19, 21, 23
Olenek Arch, Uplift 16, 196
Olenek River 16
Olenek-Vilyui deposit, interfluve 94–96
Olsztyńskie Wojewodztwo 297
Olympic Member 6
Oman area, Gulf 45–49, 51
Oman Block, Fault, High 46, 47
Oman Lut zone 46
Oman Zufar area, Salt Domes 46
Omulev Mountains 96
Omurtag town 243
Onkuchakh Formation 95
Onondaga Group 146
Ontario 69, 73–75, 77, 149, 153
Ontario Lake 73–75, 77
Ooraminna Anticline 6, 56, 57
Ooraminna No. 1 borehole 6
Oos Trough 245, 247
Oparino town 197
Opeche Shale, Salt 331, 344–346
Opoki Beds 211
Oquirrh Basin 328
Orange Liniensalz (Na3γ) 254, 303
Orange No. 1 well 7, 56
Orangeaugensalz 255, 308
Ordovician 8, 56, 58, 60–68, 83–88, 149, 156, 157, 167
Ore Formation of Roan Group 13
Ore Member of Solikamsk Depression 211
Oregon 328
Orekhovka Village 127
Orekhovo area, District 81, 199
Orekhovo No. 1 well 239
Orenburg Arch, District 127, 199, 205, 210, 218, 228
Orenburgian 167, 169, 199
Oriskany Formation 145
Orogrande Basin 161, 196, 202, 328, 341, 360
Osa borehole, field, Member, Village 17, 27, 28
Osh town 163
Osharovo borehole 27
Ostashkovichi field 115

Ostrovnaya Formation 18, 23
Otter Formation 173
Otto Fiord Formation 167–169
Ottoshall Beds 256
Ouachita structural belt, Mountains, Uplift 328, 329, 335
Overlying Anhydrite rhythmic unit 231, 232
Overlying Salt Member 211, 221, 224, 227, 233
Owl Canyon Formation 331, 344–346
Ozersk sequence 197

Pacific Ocean 360
Pacoota Sandstone 56, 57
Pagegyai horizon 89
Pagow IG-1 well 273, 296, 312
Pakistan 46, 49, 160
Pakistan salt basin, salt zone 46, 53
Pakryoi Formation 123, 125
Palanga town 124, 125
Paleozoic 1, 3, 5, 61, 99, 149, 160, 166, 194, 205
Palmarito Formation 202
Palo Duro Basin 328–330, 341, 342, 348, 349
Palomas 360
Pamush Formation 123, 125
Pan American State Union 1–14 well 72
Panarctic Deminex Cornwallis Central Dome K-40 borehole 61, 62
Panarctic Deminex Garnier 0–21 borehole 62
Pando Formation 200
Paprenyai horizon 89
Paprenyai-Pagegyai dolomite sequence 90
Paradox Basin, Valley 8, 161, 184–193, 202
Paradox Formation, Salt, biocherm 185–191, 202
Parda No. 1 well 68, 84–86
Parfenovo field 27
Paris 246, 248, 250, 253
Park City Formation 331
Parnaiba (Maranhão) Basin 203, 362, 363
Paslek District 274
Pasvalis town 124
Patience Lake Member 130, 135, 137, 139, 141
Patom Highland 17, 32, 36, 37, 41
Patquia Formation 363
Pavlovo Salt Dome 239
Peace River High 129
Peace River sub-basin 128
Peace River-Athabasca Arch 128, 129, 132, 133
Peace River-Athabasca Ridge 133
Pechora Basin, Depression 69, 70, 89, 90, 197, 205

Pechora Range 128
Pechora River 204, 207, 208, 212, 216, 219, 220, 225, 238
Pechora Syneclise 106, 127
Pechora-Novaya Zemlya Basin 161, 196
Pedernal positive element, Uplift 202, 328, 341
Pedra-do Fogo Formation 363
Peel River, Basin 12, 54
Peene River, Basin, Valley 310, 311
Pegmatitanhydrit (Pegmatite Anhydrite A4) 254−257, 314, 316, 317, 319
Pelcha Village 79
Peledui borehole, field 17
Pelican Island No. 1 borehole 157, 158
Pella Formation 201
Pembroke Formation 176, 179
Pendock I.D. No. 1 well 88
Penjab (Punjab) salt-bearing series 49−53
Penn No. 1, 2 boreholes 149−152
Pennine High 247
Pennington Formation 183
Pennsylvania 69, 71, 73−77
Pennsylvanian 183−186, 188, 191, 193, 194, 202, 336
Peremyshl Formation 80
Perepelitsy Member 79
Peresazhsk Formation 211
Pereslavl-Zalessky town 104
Peri-Baltic Depression 247
Perm town 125, 204, 207, 208, 211, 216, 219−221, 235, 238
Permian 2, 8−10, 69, 83, 84, 86, 87, 122, 123, 155, 157, 161, 166, 167−169, 191, 194, 197, 200, 203−212, 215, 216, 219, 220, 235, 238−253, 257, 259, 274, 279, 302, 315, 322−329, 336, 357, 359−364
Persian Gulf 5, 46, 47
Pertaoorrta Group 56, 57
Pertataka Formation 5, 6, 8, 57
Peru 361
Peru-Bolivian Basin 203, 361−363
Peruvian Basin 362
Pervomai field 115
Pervomaysk Beds 210
Petelini town 104
Peticodiac River 175, 177
Petin horizon 108
Petrikovo deposit, Depression, field 115−118
Pharwala complex of seams 51
Phosphoria Formation 331
Pichkassy town 127
Pictou Group 176, 181
Pictou town 174
Pigai Formation 79
Pigarev rhythmic unit 231, 232

Pila town 276, 277, 299
Pilbara Block 83, 87, 88
Pine Creek Geosyncline 12
Pine Point Formation 130, 133
Pine Salt 331, 350
Pinega River 199
Pinkerton Trail Formation 185−189, 191
Placer De Quadalupe 360
Plamosas Formation 360
Plasterco area 181−183
Plattendolomit (Platy Dolomite Ca3) 254−256, 303−307, 309−312
Plavsk town 104
Plyavin Formation 123, 124
Pochet borehole, field 17
Podboi Beds 211
Podbryantzevo Bed 211, 213
Podkamennaya Tunguska River 16, 96
Podlasie Barrier, Trough 245, 247, 312
Podluzhniki Beds 211
Podolia 78, 80
Podolian Saddle 78
Podolsk horizon, town 197, 198
Podonin sulfate unit, horizon 159
Podsolevaya sequence 108−110, 116
Pogozhye rhythmic unit 231, 232
Pokrovskoe Village 127
Poland 80, 106, 125, 244, 245, 272, 274, 295, 309
Polasie Trough, Barrier 273, 298
Poligus No. 1 borehole 17
Polish Trough 245
Polish Zechstein 257, 272, 295
Polish Zechstein Cycle 3 (PZ3) 311
Polish Zechstein Cycle 4 (PZ4) 316, 318
Polish-Lithuanian Lowland (Depression) 124, 249, 251, 255, 258, 274, 275, 278, 280, 297, 299, 301, 313, 319
Pollet River 175
Polotsk town 105
Poltava Depression 205
Poltvin Formation 81
Polyhalite-Sylvinite Member 211
Polyhalite zone 300
Polyhalitregion 285, 287
Pomeranian Barrier, Highland 247, 312
Pomorsko-Kujawy Anticlinorium 274, 296, 312
Pontotoc Group 335, 337
Poopo Lake 361
Popesti town 81
Poplar Member 171
Poretskoe Village 126
Porkun horizon 68
Port Hood Formation 176
Potash Member 176, 178

Potash horizon, Member of Solikamsk
  Depression 211, 221, 222, 227, 232
Poznań Graben, town 247, 276, 277, 299
Prague 246, 248, 250, 251, 253, 259, 279,
  302, 315, 324
Prairie Evaporite, Formation, Salt 130–139
Precambrian 4–14, 16, 49, 50, 53, 56, 61,
  62, 83, 87, 88, 129, 149, 157, 193, 195,
  245
Pregol Formation 255, 257, 275, 278
Prenyai town 90
Preobrazhenka 106-pr borehole, field,
  sequence, section 17, 23, 27, 28
Presayan-Yenisei Syneclise 16
Presqu'ile reef 130, 133, 134
Price Sandstone 181–183
Přídolian 71, 82, 89
Prigorodok Formation 79, 80
Priluksk area 120
Prince Alfred Bay 63
Prince Edward Island 174
Prince Patrik Island 166
Pripyat Depression 2, 102, 105–119, 127,
  128, 205
Pritok Anticline 117
Privoleno sylvinite horizon 215
Producing unit 233
Proterozoic 4–14, 55, 56
Protvin horizon 197, 198
Provadiya Village 242
Prudy Village 126
Prut River, Basin 106
Przhevalsky Range 200
Pskov Region 125, 126
Pueblo Formation 330
Pugachev town 127
Pugwash area 179
Pulaski area, Fault, Thrust 59, 181–183
Pumpenai town 124
Punkerri Beds 11
Purus Swell 194, 195
Putnam Formation 330
Pyarnu horizon 108
Pyasino Lake 93

Qeshm Island 46, 48
Quail No. 1 borehole 88, 89
Quatermaster Formation 330, 336
Quaternary 325
Quebec 149, 154
Queen Formation 330, 348, 349
Queensland 14, 155
Quill Lake Member 137, 138
Quilpie Trough 155

Rabbit Ears Anhydrite 71, 74
Rădăuti town 81
Radom-Lublin Plateau 247
Rakha Village 160
Rakhmetnura Dome 101
Rakhov 323
Rakhov Basin, zone 203, 323, 324, 362
Range and Valley Province 59
Rapid Formation, sequence 159
Raseinyai town 125
Rashkov Formation 81
Ratcliffe Member 170, 171
Ravar Evaporite, Formation 49, 52, 53
Razvedochninsk Formation 95, 96
Read Bay Formation 90
Rear Rock 130
Rechitsa field, monoclinal terrace, town 105,
  109, 112
Red Cave 330, 336, 341
Red River Formation 64–66
Red Zubers 257, 315, 319, 322
Redkino town 126
Redknife Formation 130
Redstone River Formation 12
Reggane Basin 161, 196, 201
Regina 129, 133–135, 138–142
Reichenhall Region 325
Reichenhaller Formation 326
Remte borehole 123
Rettenstein area 325, 326
Rhadames Basin 161, 196, 201
Rhenish Massif 245, 247, 264, 270
Rhine River 246, 248, 250, 251, 253, 259,
  270, 279, 302, 315, 324
Riedal Group Beds 255, 256, 303
Riga town 124, 246, 248, 250, 251, 253,
  259, 278, 279, 302, 315
Ringkøbing-Fyn High 247
Ringwood Member, Dome 6–8
Rio Blanco Basin 203, 362, 363
Rio Grande River, Basin 360, 361
Riphean 12
Riphean Evaporites 12
Riversdale Group 176, 181
Riyadh 47
Roan Group 13
Rob Roy No. 1 well 364
Robert Beds 9
Rochester Shale 71
Rock Salt 273
Rocky Mountains 128
Rodinga Sheet area 56
Rodnikovo Formation 210
Roger Mills County 338
Rogers City Formation 146–148
Roker Dolomite 254, 299, 300

Romashovka Village 79
Rome 324
Romny Anticline 118
Rondout Formation 70–72, 78
Ronnenberg Beds 256
Ronnenberg I, II 256
Ronnenberg Group 255, 256, 303
Ronnenberg Potash District 303
Ronnenberg-Hansa Potash District 280
Roscunish sequence 199
Rose Hill Formation 71, 72
Rosensalz (Na4γ) 254, 316
Rossenrei area 270
Rostock town 310
Roter Salzton (T4) 254–257, 314, 316, 317
Rotliegende 244, 245, 247–250
Rotwald Maria-Zell area 325, 326
Rowe-Mora Basin 328
Ruff Formation 71–73
Rügen Island 295, 309, 311
Rumania 78, 80, 81, 159, 242
Rundale town 124
Rusanov Formation 96
Rush Springs Sandstone 330
Rush Springs-Marlow 336
Russian Platform 127, 128, 228–230, 239
Russkaya Rechka No. 1-r, R-3, 42 boreholes 17, 21
Rustler Dome, Formation, section 190, 330, 351–353, 356, 357
Ryazan town, District 103, 104
Ryazan-Saratov Trough 102, 105, 127, 197, 198
Ryazhsk town 104
Rybinsk town 126, 197
Rydaevka Village 127

Saale River, Trough 245, 247, 281, 292
Saalfeld town 266, 281, 306
Saar-Werra Trough 245, 247, 249
Sabine Bay Formation 167, 169
Safet River 79
Safonovo Trough 122, 123, 205
Sahara No. 1 well 84, 86, 160
Sakmara Basin, Depression 205, 209, 228, 230
Sakmarian 122, 167, 169, 208–211, 215–218, 241, 329, 330, 364
Salado Formation 330, 351–356
Salaspils Formation 123, 124
Salem Limestone 201
Saliferous Marl 254, 322
Salina Group 69–72, 74, 75, 77
Salina River Formation 54–56
Salt Plain Formation 331, 342, 343, 346
Salt Range 5, 49, 50
Salt Valley Anticline 187, 190
Saltom Cycle, Dolomite, Siltstone 254, 258, 275, 301
Saltville Basin, District 161, 181–184
Saltville Thrust, Fault 181–183
Salzbrockenton (T5) 322
Salzton 305
Samagaltay Lake 99
Samara River, Basin 239
Samarskaya Luka 126
San Andres Limestone 329–332, 349
San Angelo Sandstone 330
San Juan Basin, Province 161, 184, 186, 196, 202, 328, 363
Sal Luis Basin, Uplift 184, 185, 328
San Rafael Swell 184
Sandfaserlage 305
Sandpiper No. 1 borehole 157, 158
Sandria Formation 186
Sandwith Anhydrite, Cycle, Dolomite 254, 258, 301, 314
Sangerhausen Anhydrite, Trough, facies, town 266, 267, 281, 286, 306
Santarem town 195
Sarateny-Vek Village 79
Saratov District, town 126, 204, 207–209, 212, 216, 219, 220, 236, 238
Saratov-Volgograd area 220
Sarep HQ Brion Island No. 1 well 180
Sargaev horizon 125
Sargino Formation, horizon 210, 211
Sarobil Formation 210
Sarstedt-Lehrte Potash District 280
Sarydzhaz Block 165
Sarysu area, District, River 100, 101, 240
Saskatchewan 64, 66, 128, 129, 131, 133–135, 138, 142–144, 170, 171, 173
Saskatchewan Group, sub-basin, Depression 128–136, 139, 140
Saskatoon 129, 133, 134, 138–145
Saskatoon Formation 130
Satanka Shale 330, 331, 334
Saudi Arabia 52
Saulkalne Village 123
Savich field 112, 113
Saxonian 249
Say Formation 52
Scandinavia 245
Scheidtal 1, 284
Schluffstein (T5), (T5r) 255–257, 318, 322
Schmale Lage 286
Schneesalz (Na4β) 254, 255, 316–318
Schönebeck town, Potash District 269, 280, 287, 308, 316, 318
Schramberg Trough 245, 247

Schwadensalz (Na3ϑ) 254, 255, 303, 307, 308, 311
Schwanebeck 316, 318
Schwanebeck Potash District, town 280, 287, 308
Scoodie Brook Formation 176
Scotland 199
Screening Anhydrite 255, 295–297
Screening Older Halite 255, 295
Scythian 326
Seaham Formation, residue 254, 313
Sebechan Formation, River 86
Sebkha Member 130, 143
Second Clastic Formation 1
Second Dolomite Formation 51, 52
Second Red Beds of West Canadian Basin 130, 133
Second Salp of Appalachian Basin 76
Sed-Yu River 127
Sedanovo field 23
Seine River 246, 248, 250, 253
Sekowice 272
Selwyn Range, Limestone 59
Selwyn Trough 54
Semiluki Dolomite, horizon 108, 125
Seregov Anticline, Salt Dome 121–123
Sergiev-Abdulin Basin, Depression 127, 198, 199, 205
Serpukhov superhorizon 197, 198
Serpukhov town 103, 104
Seryi Kamen Beds 211
Sette-Daban Range 95, 96
Seven Mile Anticline 187, 189, 190
Seven Rivers Formation 330, 348, 349
Severn Arch 149
Severnaya Dvina River 198
Severnaya Dvina-Sukhona River Interfluve 215, 217
Severnaya Zemlya 60, 92, 96, 97
Severnaya Zemlya Basin 68–70, 89
Severnyi Loktybai Region 239
Severodvinsk horizon 211
Severokamsk town 125
Seward 130
Sextant Formation 151
Shabb Formation 52
Shafer Dome 190
Shakarsen Formation 362
Shalashino Member 210, 221, 222, 227, 229
Shaler Group 12
Shangal well 217
Shannon Formation 56–58
Shapovalov field 120
Shark Bay 88
Sharlyk town 239
Shar'ya town 125, 126, 197, 198

Shatilkovo Depression 115–117
Shatura town 104
Shchigrov horizon 107, 108, 119, 121
Shebelinsk Formation 211
Shell Lake Gypsum 137
Shelonino Village 27
Shepody Bay 179
Sheshma Formation, horizon 210, 211
Shestovichi Depression, field 112, 117
Shidlovtsy Village 78
Shikhan Formation, horizon 210, 211, 214
Shiraz area 46–48
Shirhesth area 46
Shukuntai Nura Dome 101
Shushaktau Member 211, 234
Shvenchenas borehole 105
Siberian Platform 2, 13, 16–24, 27–33, 36, 41–43, 67, 81–82, 92–96, 196
Sidin Formation, unit 95, 96
Siegenian 95, 145, 151
Sieroszowice 273, 296, 312, 319
Sierra Grande Arch 328
Silesian Barrier 298, 312
Silurian 8, 60, 62, 64, 65, 67, 69–90, 93, 149, 156, 160
Silvia Lake 12
Sim-Inzer Basin, Depression 204, 209, 228
Sinbad Valley Anticline 190
Sinkiang 200
Sioux Arch 129
Siouxia Landmass 328
Skala horizon 78–81
Slave Point Formation 130
Slavyansk field, Formation 121, 209, 211, 213, 215
Sleights E-3 well 300
Sleights Siltstone 254, 322
Sloistyi Kamen Beds 211
Slok town 124
Slotsene River 124
Slupsk Basin, town 247, 276, 277, 299
Smolensk Region, town 103, 104, 126, 127
Snake River 12
Snowy Group 173
Society Cliffs Formation 13
Sofia 242
Sokolegorsk horizon 210, 211, 214
Soksk Formation 210
Soligalich town 125, 126, 197, 198
Solikamsk Basin, Depression 115, 122, 204, 209, 210, 221–225, 235
Solikamsk Cis-Urals area 2
Solikamsk Formation, horizon 210, 211, 235
Solikamsk town 224–227
Solon Formation 159
Soltanieh Dolomite, Formation, Mountains 52, 53

Solvychegodsk well 217
Solway Firth Basin 251, 252, 275, 279, 280, 301, 314, 321, 322
Sol'zavod borehole, field 17
Somerset Island 61, 90
Sömmerda town 266, 281, 306
Sondershausen town 266, 281, 282, 316
Sonkul Trough 164
Sonoran Basin, State 360
Sonoran Geosyncline 328
Sorbulak Lake 159
Sorkol Formation 241, 242
Sorochinsk area, field 230, 239
Sosnava Formation 255, 275, 278
Sosnino No. 1 borehole 17
Sosnovsk Formation 210, 239
Souris River Formation 129–131, 140
Souris Valley Formation 170, 171
South Alps 323, 363
South America 58, 161, 196, 202
South Antigonish-Harbour Village 180
South Appalachians 59
South Australia 9, 14, 59
South Dakota 64, 66, 129, 172, 173, 202, 327, 328, 330–334, 345–347, 349, 350
South Dobruja 243
South Eastern Fars (Group of Salt Plugs) 46
South Emba Uplift 205
South Harz 280, 282, 284, 286, 289
South Illinois Basin 60, 68
South Iowa Basin 161, 196, 201
South Iran 5, 46, 160
South Kaliningrad (Niven) well 299
South Limestone Alps 325
South Lyzhsk Anticline 128
South Moldavia 362
South Oman Dhofar salt area 46, 47
South Persian Gulf area 46, 47
South Peruvian Basin 161, 196, 202
South Pharwala Member 51
South Polish Land 298
South Radui field, Village 27, 28
South Urals 200
South Valava field 117
South Vishan field 111, 112
South Yenisei Basin 94
Southeastern Utah Basin 328
Southern Fiord 166
Southfield Member 176, 178
Southwest New Mexico Uplift 328
Sovetsk town 126
Soviet Union 80, 97, 206, 274
Spearfish Formation 331, 349, 350
Spessart Uplift, Swell 264, 265, 292
Spessart-Ruhla Swell 247
Spišska Nová Ves town 326

Spišsko-Cemerskoe Rudohoři 323, 326, 327
Spital-Pyron-Bosruk area 325, 326
Spitsbergen Basin 161, 196, 197, 203
Spivak Bed 211, 213
Springen I-V Mines 258
Squaw Bay Formation 146–148
Sredne-Botuoba 2-p borehole 17, 23, 28
St. Bees Evaporite, Shale 254, 258, 314, 322
St. George's sub-basin 174, 180
St. Lawrence Gulf, Platform 174, 180
St. Louis Limestone 201
Stachyunyai District, borehole 105
Stade Salt Dome 254, 271, 293
Stafford No. 1 well 155, 156
Staintondale Group 254, 257, 320
Stan Formation 67
Stanyuchai well 124
Stark Formation 12
Starobino deposit, Depression, town, field 105, 115–118
Starorechka Formation 13
Stassfurt 316, 318
Stassfurt Anhydrit 254
Stassfurt Basalanhydrit (A2) 291
Stassfurt Basin, area 269, 281, 292, 306
Stassfurt Cycle, Potash District, town 266, 280, 282, 308
Stassfurt Kalisalz 254
Stassfurt-Egeln District 308
Stassfurt-Series 252, 256, 257, 260, 264, 278, 279, 281, 283, 286–289, 291–295, 297, 301
Stassfurt-Steinsalz (Na2) 252, 254–256, 280, 282–284, 286, 287, 289–295, 297, 301
Stassfurt-Ton (T2) 255, 256, 280
Stattler Formation 130, 131, 144, 145
Ste Genevieve Limestone 183
Stearfish Bay 61
Steinsalzbank 261
Steinsalzmittel 311
Stellarton Cap 174
Stephanian 176, 194, 201
Sterlitamak Formation, horizon 210, 211
Sterlitamak town 229
Sterlitamak-Ishimbay District 210
Sterlitamak-Meleus Basin, Depression 205, 209, 228, 230
Stinkschiefer (Ca2st) 254, 255, 280–282, 286, 287, 289, 290, 293, 294
Stone Corral Formation 330, 331, 342–344
Stonewall Formation 64–67
Stonshkyai borehole 105
Stony Creek field 175, 176
Stony Mountain Formation 64–66
Stooping River Formation 149–152
Strain Shale Member 330

Strathlorne Formation 176
Strelichevo town 109
Stryp Formation 90
Stryp River 79
Stury District 105
Sturt Block 158
Subhercynian Depression 251, 252, 255, 258, 268, 269, 278, 286–288, 291, 292, 301, 308, 314, 316–318, 320
Sudetes High 247, 249
Südharzsteinsalz (Na2γ) 282–285
Sudoga town 199
Suduvsk Formation 257, 313, 319
Sujjowal Member 51
Sukhaya Tunguska River 31
Sukhona Formation 211
Sukhona River, Basin 217
Sulfate-carbonate sequence of Dniester-Prut Basin 79
Sulfate-carbonate sequence of East European Basin 210, 211
Sulin field 126
Sumner Group 330, 336
Sunbury Formation 146, 148
Sundyr town 126
Supai Basin 203, 357–359
Supai Formation 331, 357
Superior Lake 147
Surinda-Gazhenka Depression 36
Surkhab River 163
Sussex town 174, 175, 178
Sutpaitau Formation 211, 234
Sverdrup Basin 61–64, 161, 165–169, 203, 362, 363
Svinord Beds 125
Svyatogorsk Salt Bed 211, 213
Swidwin town 274, 312
Swidwin 2 well 274, 296
Swietokrzyskie Góry 273, 297, 319
Swift Current Platform 128, 129, 134
Sy-Vozha River 127
Syda-Erba deposit 98
Sydney sub-basin 174
Sylva-Iren area, District 210, 221
Sylvania Sandstone 145
Sylvite-bearing horizon of East European Basin 211
Sylvite-carnallite horizon of East European Basin 211
Syr-Darya River 100, 163, 240
Syracuse Formation 70–72, 77
Sysolsk, Arch, well 122
Syuren Formation 210
Syzran town 127
Szczecin town 276, 277, 299

Tabas town 47, 49
Table Hill Volcanics 10, 11
Tadzhikistan 362
Tafraout 58
Tagara iron ore deposit 22
Taggard Formation 183
Tagna Village 27
Taiga area, field, borehole, Uplift 17, 22, 27
Taimyr 93, 96
Taimyr fold area 16
Taishet No. 1 borehole 17
Talas Alatau Range 163
Talas Uplift 101, 162
Tallaringa Trough 58
Tallin horizon 68
Taloga Formation 330, 331, 351
Tandalgoo Red Beds 84–86, 160
Tangui No. 1-sp borehole 22
Tannu-Ola Range 97
Tanquary Formation 167, 169
Tansill Anhydrite, Formation 330, 348, 349
Tantreta Fault 179
Tapajos River 194
Tarim Basin, Depression 15, 59, 200
Tarma Group 202
Taryan River Basin 96
Tas-Khayakhtakh Range 96
Taseevo borehole, field 17
Tashkent 163
Tashly Village 239
Tastin Uplift 101
Tastub horizon, Formation 210, 211
Tata town 160
Tatar Arch 126, 127, 197–199, 205, 228, 229
Tataria 228
Tatarian 210, 211, 240, 241, 329, 330
Tathlina High 128–130, 132, 133
Tatnan Peninsula 149
Tatul Formation 123, 124
Tautschenthaler Swell 269
Tebra River 124
Tee-Khem River 99
Teesside Group 254, 257, 313
Tehran 47
Tekes Trough 200
Tekturmas Mountains 162
Telegraph Member 130, 135, 136
Teniz Basin, well 91, 158, 159, 161, 196, 200
Tennessee 181, 182
Tennycape Formation 180
Tensleep Sandstone 331
Terskey-Alatau Range 162, 163
Tertiary 46, 184, 325
Tesbulak sub-basin, Basin, Trough 99–101, 162

Tesbulak Creeks area 101
Tetcho 130
Texas 202, 327, 328, 332–336, 338–343, 346, 347, 351, 352, 354–356, 361
Texas Gulf Producing Well 1-x 189
Texas Panhandle 336
Third Bone Springs Sand 330
Three Forks Formation 130, 131
Thule Basin, Group 4, 13
Thumb Mountain Formation 62, 63
Thüringen Beds 256
Thüringer Wald, Uplift 260, 264–266, 281, 283, 307, 317
Thüringia Depression 251, 252, 255, 256, 258, 265–269, 275, 278, 280–282, 284–287, 290–292, 301, 304–308, 314, 316–318, 320
Tien Shan 163
Tigentourin sequence 201
Tigris River 47
Tiksi Formation 95
Tilston Member 170, 171
Timan area, Range, Uplift 121, 122, 198, 199, 205
Timan (Pritimansky) Trough 106, 107, 121–123, 127, 128, 205
Tindouf Basin, Depression 91, 158, 159, 161, 196, 201
Tinnovka Formation 18, 19
Tirana 324
Tisa River 324
Tishovka Village 105
Titicaca Group, Lake 361
Tochilnin Formation 67
Todd River Dolomite 56–58
Tohonadla 188
Tokma No. 1-r borehole 17
Tokmovo Arch 126, 127, 197–199, 205, 211
Tolbachan Formation, horizon 5, 18–20, 32, 35, 44
Tompok Formation 67
Ton 2 (T2) 254, 291
Tonbanksalz (Na4tm) 254, 316
Tonbrockensalz (Na4δ or Na4br) 254, 255, 316–318
Tonflockensalz (Na3i) 255, 307, 308, 311
Tonflockensalz and Schwadensalz (Na3ϑ + i) 255, 305, 306
Tonlöser 262, 263
Tonmittelsalz (Na3i) or (Na3tm) 254
Tonoloway Formation 71, 72, 77
Top Anhydrite 254, 322
Top Youngest Halite (Na4b) 255, 322
Tor Salt Bed (Salt Member) 211, 213
Torez horizon 217

Torgalyg deposit 97–99
Toropets borehole, town 103, 104, 125, 126
Torquay Formation 130, 131
Torun town 276, 277, 299
Totdzhal Formation 211, 234
Totleben town 242
Tournaisian 95, 101, 162, 176, 196, 197, 199, 200
Townsend Quartzite 9
Transitional Member 211, 233
Traverse Group 146–148
Tremadocian 84
Triassic 122, 157, 194, 323, 324, 326, 350, 361
Tribeč Mountains 326
Trold Fiord Formation 61, 167, 169
Trout River 130
Trümmercarnallitt 260, 261
Tsyganov Dome 239
Tubb Sand 330, 341, 342
Tubbergen town 271
Tukal horizon 81
Tuket Formation 210
Tula District, town 103, 104, 197
Tula-Serpukhov District 103
Tulun borehole, field 27, 28
Tulun Cis-Sayans 41
Tumblagooda Sandstone 89
Tundrin Formation, sequence 196
Tunguska No. 1-o borehole 17
Tunguska Syneclise 16, 17, 19, 25, 28, 31, 36, 42, 43, 82, 92–94, 96
Turgai Basin 91, 158, 159
Turgovishte 242
Turin borehole 67
Turin TO-2 borehole 82, 93
Turkey 364
Turkey Hill Beds 11
Turovo field 112
Turuk Trough 162, 164
Turukhan Region 81
Turukhansk town, Uplift 28, 29, 42
Tutaev town 197, 198
Tutoncha borehole 67
Tutrakan Trough 159
Tuva 97, 98, 126
Tuva Basin, deposit, Depression 91, 97–99
Tuz-Tag deposit 97, 99
Tuzkol Formation 241, 242
Twin Falls 130
Tynep Formation 95, 96
Tynys borehole, field 17
Typta field 27
Tyret borehole, field 17, 27
Tyrol 325
Tyup Basin, Formation, Trough 161, 163, 196, 200

Tyuya Member 210, 221, 227–229

Übergangssalz (Na3γ_1) 308
Übergangsschichten 255, 284, 285, 287, 288, 290, 291
Uel-Siktyakh River 96
Ufa Plateau 221, 228, 229
Ufa town 230
Ufimian 208–211, 235, 237, 238, 240, 241, 329, 330
Ukhta town, Region 127
Ukrainian Shield 78, 106, 107, 205
Ulagan Beds 231, 234
Ulanbel-Talas Uplift 241
Ulkunburul Mountains 162
Ulster-Manx-Furness Basin 247
Ulsterian 71
Uluksan Group 13
Ulutau Range 99, 100
Umwandlungssylvinit 260
Uncompahgre Trough, Uplift 184, 185, 189, 190, 193, 328
Underlying Salt 211, 221, 222, 227, 233
Unit A (Carribuddy Formation) 85–87
Unit B (Carribuddy Formation) 85–87
Unit C (Carribuddy Formation) 85, 86
Unit D (Carribuddy Formation) 85–87
Unit E (Carribuddy Formation) 85, 86
United States 8, 64, 67, 69, 128, 133, 139, 142, 170, 172, 201, 327, 328, 330, 331, 352, 357, 359–361
Unken Lofer area 325, 326
Unstrut 318
Unstrut area, Potash District, River 266, 269, 280–282, 284, 286–288, 292, 306
Unstrutbank 285–287
Unterer Anhydrit 267
Unterer Dolomit 255, 265
Unterer Werra-Anhydrit (A1u or A1a) 254–256, 260, 264, 265, 268–271, 275
Unteres Leine Steinsalz 255
Unteres Stassfurt-Steinsalz (Na2a) 290
Unteres Werra-Steinsalz (Na1a) 254–256, 260, 264, 270
Upgang Formation, E-6 well 254, 320
Upham Formation 178
Upper Amazon Basin, Depression 194, 195
Upper Anhydrite (Zechstein) 252, 255, 272–274, 276, 300, 320
Upper Anhydrite Member of West Canadian Basin 136
Upper Anhydrite-Halite Member of East European Basin 211, 254
Upper Castile 351
Upper Cimarron Salt 338, 339, 342
Upper Donau Uplift 247

Upper Elk Point 129, 130, 132
Upper Gypsum Valley Formation 12
Upper Gypsum-bearing Formation 363
Upper Halite horizon 211, 232, 233, 254, 258, 300, 320, 321
Upper Iren subhorizon 210, 227, 231
Upper Izhma Region 127
Upper Kama Basin, Depression, Trough 125–127, 197–199, 214
Upper Lena Depression, Formation 28, 29, 41
Upper Lunezh Beds 228–230
Upper Magnesian Limestone 313
Upper Marl 322
Upper Moty Member 21, 22
Upper Muskeg 136
Upper Pechora Depression 122, 204, 209, 211, 225–227
Upper Pegmatite Anhydrite 255, 318
Upper Pharwala Member 51
Upper Potash 320, 321
Upper Red Beds 175
Upper Red Pelite (T4$_b$) 255, 257, 322
Upper Salt Member 176, 178
Upper salt sequence of Dnieper Donets Depression 114, 115, 118, 120, 121
Upper salt sequence of Pripyat Depression 108
Upper Supai Unit 358, 359
Upper Transition Beds, Member of Solikamsk Depression 211, 221, 224
Upper Velyuchan field 23
Upper Youngest Halite 255, 318
Upperton Member 176, 178
Ural River 204, 207, 208
Urals 200, 206, 214, 237
Uralsk 236
Uren' town 126
Uritsk horizon 18, 19, 35, 36, 44
Uryuk-Nur Lake 19
Urzhum horizon 211
Ushbulak Formation 241
Uskalyk Formation 210
Usmyni Village 126
Usolye borehole, field, Formation, horizo, sequence, Village 5, 17, 18, 20, 25–32, 43, 44, 53
Ust-Kamovo borehole, field 17
Ust-Kut Nos 2–4, 4-r boreholes, field 17, 20, 27, 29, 127
Ust-Tagul Formation 22
Ust-Ukhta deposit, Formation 127
Ustie Member, horizon 78–81
Usvyati Village 126
Utah 184, 191, 328, 331, 357–359
Utokam Unit, deformation 81

Vace Triste Sandstone Member 353, 355
Vakhsh River 163
Val Verde Basin, Trough 328, 329
Vale Formation 330
Van Hauen Formation 167, 169
Vanavara No. 1 borehole 17
Varna Basin, Depression 159, 243
Varna town 242
Vayakh Formation 96
Velikie Luki town 125
Velikobubnovsk field 120
Velikozagorovsk field, area 120
Velmo borehole, Depression, field 17, 27
Vendian 5, 11, 48, 53
Venev town 197
Venezuela Basin 161, 196, 202
Venta River 124
Verkhnekuloi Formation 211, 217
Verkhnyakovtsy Village 78, 79
Verkhnyaya Chona borehole, field 17
Verkholensk Formation 18, 43
Verkhoyansk-Chukotka fold area, district 92
Verkhoyanye fold area, Range 16, 96
Vermilion Formation 130
Vernon Formation 70–72, 76
Vesa Formation 331
Veselyi Kut deposit 127
Vesla 62
Veslyansk well 122
Vetluga town 126
Vetrino 242
Vezha-Vozha River
Victoria Island 166
Victoria Lake 12
Victoria Peak Island 330
Videltsev field 120
Vienna 246, 248, 250, 323–326
Vikhtovsk Formation 211
Vilna town 246, 248, 250, 251, 253, 259, 278, 279, 302, 315
Vilyuchan Formation 94, 95
Vilyui River, Basin 16, 17, 67, 81, 82
Vilyui Syneclise 97
Vindhyan sequence 50
Virgilian 186, 202
Virginia 59, 73–75, 181–183
Visean 95, 101, 162, 163, 167, 176, 197, 199
Visean D level 201
Vishan area, field 105, 115
Vishera River 235
Vishi town 105
Vistula River 246, 248, 250, 251, 253, 279, 302, 315
Vitim River 14, 204
Vitim-Patom Highland 19, 23

Volga River, area 126, 204, 207, 208, 212, 216, 219, 220, 236, 239
Volga Monocline 105, 205, 211, 230–232, 235
Volga rhythmic unit 231
Volga-Ural Anteclise 123, 204, 205, 214, 215, 218, 221, 227–229, 235, 237, 239
Volga-Ural area, Region 2, 106, 125–128, 197, 199
Volgograd rhythmic unit 231
Volgograd town, District 204, 207–209, 212, 216, 219, 220, 238
Volkenrode town 282
Vologda Region 127
Volokon No. 1-r borehole 23
Volyno-Podolia 78
Voronezh Anteclise, horizon, Beds, Massif 106–109, 120, 125, 126, 205, 206
Voronezh town 204, 207, 208, 212, 216, 219, 220, 238
Voronezh-Evlanovo level 109
Voronezh-Liven 120
Voskresensk-Shikhany Depression 204, 209
Vyatka folding zone 198
Vyatka horizon 211
Vyazma District, town, area 103, 104, 125, 126
Vychegda Basin, River, Trough 121–123, 198, 199, 205
Vyshemirovo-Nadvin field, area 110

Wabamun Group 130, 131
Wabasca Member 130, 135, 136
Wakwaywkastic River 152
Waldo Pedlar Member 6
Wallachian Basin 159
Walton town 180
Wanamaker Formation 176
Wandagee No. 1 well 88, 89
Wanka Embayment 155
Wanna Beds 11
Wapsipinicon Formation 159
Warcha area 51
Warsaw 246, 248, 250, 251, 253, 259, 276, 277, 279, 299, 302, 315, 324
Washington 129, 134, 328
Waterways 130
Watt Mountain Formation 130, 133
Weber Sandstone 186, 191, 193
Wechelfolge 305
Weedina No. 1 well 160
Weimar town 266, 281, 292
Weldon Member 176, 177
Wenlockian 71, 72, 81, 89
Werfen area 325, 326
Werra-Anhydrit (A1) 254, 255, 268, 271, 275, 282

Werra Fulda Depression 251, 252, 254, 256, 258, 261, 262, 264, 265, 275, 278, 280, 290–292, 304, 319
Werra Potash District 258, 260, 261, 263, 267
Werra River Basin 249, 260–262, 289, 292, 303, 304
Werra Series 252, 256–258, 260, 264–272, 275
Werra-Steinsalz (Na1) 252, 255, 265, 267–270, 275
Wesel town 270
Weser Basin, River, Valley 245–248, 250–254, 256, 258, 268, 270, 278–280, 287, 289–292, 301–304, 308, 314–316, 318–320
West Alberta Arch 129, 132
West Australia 9, 11
West Canadian Basin 2, 91, 128–144
West Carpathians 323, 326, 327
West Europe 244, 247, 248, 250, 251, 253
West Germany 244, 324
West Neshinsk 120
West Netherland Basin 247
West Pakistan 5
West Siberian Plate 16
West Spring Creek Formation 68
West Sterlibashevo field 230
West Texas Permian Basin 327, 329, 335, 341, 342, 346, 348, 349, 351–353, 355, 356
West Valava field 111, 112
West Virginia 69, 71, 73–77, 182, 183
Western Afghanistan 160
Western Arizona Platform 328
Westgate Embayment 155
Westmorland Upland 174
Wheeler County 341
Whilton town 300
Whitby town 300, 320
White Bay sub-basin 174
White Bear Member 130, 137–139
Whitehorse Group, Sandstone 330, 331, 336, 349, 351
Whitewood Beds 64
Whitkow Salt 130, 137
Wichita Group, Mountains, Uplift 328–330, 332, 341
Wilkinson No. 1 well 58
Willara Formation, sub-basin 84, 86, 87
Willara No. 1 well 84, 85
Williams Island Formation 149, 150, 153
Williamsport Formation, Sandstone 71, 72, 76
Williston Basin, sub-basin, town, Depression 60, 64–67, 69, 70, 89, 90, 128, 139, 161, 170–173, 328–330, 335, 344–346, 349, 350

Wills Creek Formation 71, 72, 76
Wilson Cliffs No. 1 well 68, 84 86, 160
Windischgarsten area 325, 326
Windsor Group 176–181
Windsor town 174
Winnipegosis (Winnipeg) Formation 64, 65, 129, 130, 132–137
Winterburn Group 130, 131, 144
Wirrildar Beds 11
Wisconsin Arch 145
Wisla River 273, 324
Wiso Basin 15, 59
Wittlich Trough 245, 247
Wolfcamp Series, Wolfcampian 328–333, 336, 357, 358, 360, 361
Wolstenholme Formation 13
Wolsztyn Dissected Highland 247
Wolverine Member 130, 135, 136
Wonarah Beds 59
Woodbend Group 130, 131, 143
Woodside Formation 331
Woolnough Beds 10, 11
Woolnough Hills diapir 7, 9
Wrangel 92, 96, 97
Wright Hill Beds 9
Wroclaw town 276, 277, 299
Wschowa town, well 272, 296, 312
Wurmpartie 262
Wurmsalz 262, 264
Wymark 130
Wynniatt Formation 12
Wyoming 171, 202, 327–334, 344, 345, 347, 348, 350, 351
Wyoming Shelf 129

Yadrenyi Kamrn Beds 211
Yakushur-Bod'ya 126
Yakutsk town 16
Yanchikovo town 127
Yanga-Aul Village 126
Yangada Formation 181
Yarakta field, No. 11-p borehole 17, 23
Yarensk Trough, well 122, 123
Yargorin area 81
Yarichambi Formation 361
Yaringa Evaporite Member 89
Yaringa No. 1 well 87, 88, 160
Yaroslavl District 127
Yartsevo town 104, 126
Yasnaya Polyana 104
Yates Formation 330, 348, 349
Yelton Salt 330, 335, 337, 339–341
Yelverton area 167
Yenisei Range 16, 17, 19, 23, 29, 33, 36, 42, 43
Yenisei River 82, 98

Yenisei-Khatanga Trough 16
Yeso Formation 330
Ygyattan Trough 95, 97
Yilgarn Block 7, 9
Yorkshire Province 254, 275, 299, 300, 313, 320–322
Young Inlet No. D-21 borehole 62, 63
Young Range 7
Younger Halite 255, 257, 311, 312
Younger Potash 255, 312
Youngest Halite 257, 318
Yowalga 2 borehole 7, 10
Yuedeya (Macha) Formation 18
Yugoslavia 323
Yukon 129, 130
Yukta Formation 95, 96
Yurkovtsy Village 78
Yurovich field 118
Yuryuzan-Sylva Depression 204, 209, 222
Yuzhny Ust-Kut borehole, field 17, 27, 33

Zadon horizon 114, 126
Zadonsk-Elets horizon 108
Zagadochnino Formation 96
Zagornino Formation 67
Zagorsk town 199
Zagros Fault 46, 47
Zaigun Formation 52
Zaili Alatau 163
Zalenieki Village 124
Zambia 13
Zambian Copperbelt 13
Zaraisk town, district 103, 197
Zarechinsk field, No. 1-p well 111–114
Zaremba-Dembowec 274
Zarkow town 272
Zary Pericline 272
Zavadov Formation, section 80
Savolzhsk horizon 197
Zayarsk No. 1-o borehole, field, town 17, 23, 29, 33
Zbruch River 79
Zechstein 2, 244, 249, 251–254, 256–259, 265, 266, 271, 272, 279, 281, 289, 291–295, 301, 302, 304, 306, 310, 316, 322, 363

Zechstein cycle 256, 258, 260, 264, 268, 270–275, 290, 295
Zechstein 1 (Werra Series) 252, 254, 255, 258–260
Zechstein 2 (Stassfurt Series) 252, 254, 255, 260, 278, 279
Zechstein 3 (Leine Series) 252, 254, 255, 301–303, 308
Zechstein 4 (Aller Series) 252, 254, 255, 313–316, 319
Zechstein 5 (Ohre Series) 252, 254–256, 315, 318, 322
Zechstein Limestone 255, 257, 272, 273, 276
Zechsteinkalk (Ca1) 254–256, 258, 260, 264, 265, 267–271, 275
Zechsteinkonglomerat (C1) 254, 255, 258, 260, 265, 268–270, 275
Zeklin 272
Zeledeevo Formation, horizon 18–20, 40, 43, 44
Zema area, Member 134
Zhalgiryay Formation 255, 297
Zherba Formation 18
Zhezhmuryai town 90
Zhidelisai Formation 241, 242
Zhigalovo borehole, field 17, 27
Zhigulevskoe 127
Zhiguli-Pugachev Arch 126, 127, 197–199, 205, 218, 228, 229
Zhitkovich Depression 117
Zhvantsy River 79
Zima No. 1-r borehole 17
Zielona Gora Depression 245, 247
Zlatar 242
Zolnyi Ovrag 127
Zubers 255
Zubovaya Polyana 105
Zubovo Formation 92–96
Zubtsov town 103
Zuni positive element 328
Zvenigorod Formation 81
Zwischenlage 261
Zwischenmittel 285, 288

M. A. Zharkov

History of Paleozoic Salt Accumulation

Editor in Chief: **A. L. Yanshin**

Translated from the Russian by R. E. Sorkina, R. V. Fursenko, T. I. Vasilieva
1981. 35 figures. VIII, 308 pages
ISBN 3-540-10614-6

Contents: Distribution and Number of Paleozoic Evaporite Sequences and Basins. – Stratigraphic Position of Evaporites and Stages of Evaporite Accumulation. – Areal Extent and Volume of Evaporites. Epochs of Intense Evaporite Accumulation. – Paleogeography of Continents and Paleoclimatic Zonation of Evaporite Sedimentation. – Epilog: Evolution of Evaporite Sedimentation in the Paleozoic. – References. – Subject Index.

This book is a unique review of Paleozoic salt accumulation throughout the world. His original work is supported by valuable literature references till now unavailable to non-Russian speakers.

After a practical exposition of stratigraphy and geographical distribution of evaporites, the author presents, in great depth, the relation between accumulation of evaporites and paleogeography, plate-tectonics and paleoclimate.

Of topical importance is the significance of various salt deposits as raw materials for agriculture and chemical production etc.

The book is of both practical and theoretical interest to scientists and advanced students of geology and mining engineering.

Springer-Verlag
Berlin
Heidelberg
New York
Tokyo

L. J. Salop

Geological Evolution of the Earth During the Precambrian

Translated from the Russian by V. P. Grudina
1983. 78 figures. XII, 459 pages
ISBN 3-540-11709-1

Contents: General Problems in Division of the Precambrian. – The Katarchean: Rock Records. Geologic Interpretation of Rock Record. – The Paleoprotozoic (Archeoprotozoic): Rock Records. Geologic Interpretation of Rock Record. – The Mesoprotozoic: Rock Records. Geologic Interpretation of Rock Record. – The Neoprotozoic: Rock Records. Geologic Interpretation of Rock Record. – The Epiprotozoic: Rock Records. Geologic Interpretation of Rock Record. – The Eocambrian (Vendian sensu stricto): Rock Records. Geologic Interpretation of Rock Record. – Geologic Synthesis. – References. – Subject Index. – Index of Local Stratigraphic Units and of Some Intrusive Formation.

Geologic evolution from approximately $3.7 \cdot 10^9$ y up to the beginning of the Cambrian period is the subject of this study. The author, head of the Precambrian Geology Department of the All-Union Research Geology Institute in Leningrad, proposes herein a new, detailed division of the Earth's early history. His geologic conclusions are based on rock record data collected the world over, as well as on new methods for studying older formations and the refinement of existing methods, in particular the division, dating and correlation ocf "silent" metamorphic strata.
The scope of the book embraces such problems as
– the evolution of sedimentogenesis, of tectonic structures and of larger elements of the Earth's crust
– the periodicity of tectogenesis
– the relation between tectonic processes and magmatism
– the origin of the oldest astroblemes and their appearance on the Earth's surface, as well as possible causes for changes in organic evolution.
The author concludes with a consideration of a directed and irreversible geologic evolution of the planet.
With the great significance new data on the Precambrian bear for the theoretical and philosophical foundations of science, showing as they do previously unknown general regularities and unexpected evolutionary relations of different geologic phenomena and processes, this book will prove required reading for the Precambrian specialist and advanced student, and will also interest the researchers in other geologic branches.

Springer-Verlag
Berlin
Heidelberg
New York
Tokyo